CONCEPTS OF PARTICLE PHYSICS
Volume II

CONCEPTS OF PARTICLE PHYSICS

Volume II

KURT GOTTFRIED
Cornell University

VICTOR F. WEISSKOPF
Massachusetts Institute of Technology

OXFORD UNIVERSITY PRESS • NEW YORK
CLARENDON PRESS • OXFORD
1986

Oxford University Press

Oxford New York Toronto
Delhi Bombay Calcutta Madras Karachi
Petaling Jaya Singapore Hong Kong Toyko
Nairobi Dar Es Salaam Cape Town
Melbourne Auckland

and associate companies in
Beirut Berlin Ibadan Nicosia

Copyright © 1986 by Kurt Gottfried and Victor F. Weisskopf

Published by
Oxford University Press, Inc., 200 Madison Avenue,
New York, New York 10016

Oxford is a registered trademark
of Oxford University Press

*All rights reserved. No part of this publication may be
reproduced, stored in a retrieval system, or transmitted, in any
form or by any means, electronic, mechanical, photocopying,
recording, or otherwise, without the prior permission of
Oxford University Press.*

Library of Congress Cataloging-in-Publication Data
(Revised for vol. 2)

Gottfried, Kurt.
Concepts of particle physics.

Includes bibliographies and indexes.
1. Particles (Nuclear physics) I. Weisskopf, Victor
Frederick, 1908–   . II. Title.
QC793.2.G68   1984   539.7'21   83-17275
ISBN 0-19-503392-2 (Oxford University Press: v. 1)
ISBN 0-19-503393-0 (Oxford University Press: v. 2)

2 4 6 8 10 9 7 5 4 3 2 1

Printed in the United States of America
on acid-free paper

To the men and women who create the accelerators,
the detectors, and the experiments from which
the concepts of particle physics spring.

PREFACE

This second volume presents a more extensive and deeper treatment of the subjects treated in the first volume. It is not an independent book—Volume I is the first chapter of the complete work. The background required is the same as for the first volume: a knowledge of electrodynamics, relativity, and nonrelativistic elementary quantum mechanics.

As in Volume I, we do not attempt to convey a "professional" level of expertise to our readers—that is, the ability to carry through detailed calculations or to analyze experimental results in a definitive manner. We continue to emphasize what we have called the "oral tradition": modes of intuitive thought, inference by analogy, and semi-quantitative calculations, albeit at a more sophisticated level than in Volume I. When judiciously combined, these approaches yield a reasonably complete picture of the physical phenomena at issue. This volume does, however, reach a more demanding level on the formal side, for that cannot be avoided if one is to gain an understanding of such concepts as Feynman diagrams, renormalization, non-Abelian gauge theory, and symmetry breaking.

We have made a determined effort to write a book that can be profitably studied by readers who may be primarily interested in only a subset of the topics presented, or who do not have the time or inclination to digest the book cover to cover. For that reason, all chapters, with the exception of Chapter III, are written so that the material becomes progressively more difficult; furthermore, the earlier portions of each chapter do not depend on the later portions of other chapters. This preface will suggest the various paths that can be taken in reading this book. To that end we first give a bird's-eye view of its contents.

Chapter II, Quantum Electrodynamics, begins with a discussion of Dirac's theory of spin-$\frac{1}{2}$ fermions and the quantization of free fields (§A) and Feynman diagrams (§B), followed in §C by applications to the simplest examples of relativistic processes, such as Compton scattering and pair-annihilation, and finishes (§D) with a survey of radiative corrections and renormalization. Chapter II, Hadronic Spectroscopy, has two quite distinct purposes; §A describes the means by which spectroscopic information is extracted from the data, while §B is a fairly detailed examination of one particular, but important, topic in hadronic spectroscopy—the heavy quark-antiquark mesons. Chapter III, Quantum Chromodynamics, begins in §A with the group theory associated with the color variable, develops non-Abelian gauge fields in §B, then discusses the novel phenomena of asymptotic freedom and confinement in §C, and finishes in §D with the bag model of hadrons, which is a semiempirical application of QCD. Chapter V is devoted to deep inelastic scattering of leptons by hadrons. Chapter VI on the Electroweak Interaction develops, in §§A and B, a more sophisticated and detailed picture of the weak interaction and its connection to electromagnetism than we presented in Volume I, examines the symmetry breaking of the electroweak interaction in §C, and then in §D provides a sketch of the simplest attempt to unify the strong and electroweak interactions.

Virtually the entire volume uses the language of relativistic quantum theory, and for that reason readers are advised to first study Parts A, B, and C of Chapter II. That accomplished, they have a number of options, depending on their own priorities, and as indicated in the scheme (See facing page.) For example, those who are primarily interested in models of hadronic structure can proceed directly to §III.B and §IV.D, though there are a number of instances where some knowledge of the elements of QCD would be helpful. In the same vein, if the weak interaction is the first priority, one may proceed directly to §VI.A and VI.B, though the field-theoretic formulation of the electroweak connection requires an understanding of non-Abelian gauge theory, but those who follow this path need no group theory beyond a knowledge of angular momentum, as explained at the outset in §IV.B. There is one topic that is not a true prerequisite for other portions of the book: §III.A on data concerning hadronic properties. Furthermore, only a superficial acquantaince with deep inelastic scattering is assumed in §VI.B.

The book has two somewhat distinct levels of difficulty. The less demanding corresponds to the material depicted in the scheme. The more sophisticated is concerned with phenomena that only emerge when one goes beyond lowest-order perturbation theory, or where preturbation theory fails altogether. The material on radiative corrections (§II.D, §IV.C, §V.7, §VI.D) belongs to this category, as well as certain aspects of QCD and the electroweak interaction in §IV.C and §VI.C.

The Appendices contain considerations and calculations whose inclusion in the text would have made the argument too clumsy or elaborate. However, Appendices II and III are essentially parts of Chapter II. Whether their study is necessary to the understanding of that chapter depends on the knowledge, experience, taste and curiosity of the reader.

As in Volume I, nothing is said about the extremely sophisticated experimental techniques that have provided the empirical foundation upon which the concepts and theories are constructed. Students should recognize that the ingenuity and creativity required to design, construct, and operate the accelerators and detectors, and to master the forbidding tasks of data collection and evaluation, are at the very least the equal of the intellectual effort embodied in the theory. For a first orientation to these aspects of the subject we again refer to the excellent text by Perkins (1982).

As in Volume I, our objective has been pedagogy, not history. For that reason, our presentation does not follow the historic development. Furthermore, we have few references to the original literature, nor have we

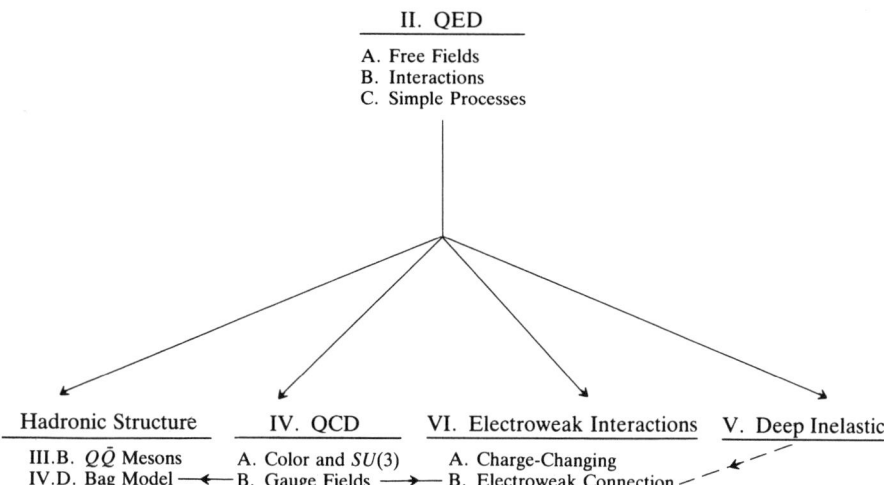

assigned credit for the theoretical and experimental work that is described, with the exception of the experimental data shown or cited. The rare references to theoretical papers are not to the original breakthrough contributions, but to pedagogic articles, to recent comparisons between theory and experiment, or to detailed calculations that elaborate on our estimates.

In the preface to Volume I we acknowledged colleagues on both sides of the Atlantic who have provided us with data, help, insights, and criticism for the totality of this work. Special thanks go to Erich Rathske, however; without his meticulous eye this book would have contained many more errors than some of our readers may find. We would have been unable to perform our task without the encouragement and generous support of CERN, and the physics departments of Cornell University and the Massachusetts Institute of Technology.

June 1986
Ithaca, New York   K.G.
Cambridge, Massachusetts   V.F.W.

NOTATION AND CITATIONS

This book is divided into Chapters, denoted by roman numerals (I, II, etc.). Chapter I constitutes Volume I. Chapters are divided into Parts, designated by A, B, etc., and Parts are subdivided into Sections and Subsections, enumerated as 1(a), 1(b), ..., 2(a), 3(b), ..., etc. The enumeration of equations begins afresh in every Part of each Chapter. A purely numerical reference to an equation, as in Eq. (41), or a Section, as in §4(a), refers to an equation or Section in the *same* Part. When an equation or Section in *another* Part (say C) of the *same* Chapter is referred to, the citation would read Eq. C(41) or §C.4(a), while if the reference is to *another* Chapter (say IV), the citations would read Eq. IV.C(41) or §IV.C.4(a).

Figures are numbered afresh in each chapter and are referred to by that number within each Chapter. A figure (e.g., No. 3) in another Chapter, say II, is denoted by Fig. II.3.

References to the Bibliography (there is one at the end of each volume) are cited using first author and year of publication, as in (Dirac, 1958). Unless there is a specific reference, experimental data are taken from the Particle Data Group, 1982; Appendix I is an abbreviated version of this data compilation.

Text in small type, which is always set between bracket symbols, [[ ]], is more advanced than the surrounding material, or only of secondary importance at that juncture, and can be skipped at a first reading.

Particle reactions are usually written in an abbreviated form, as in $\gamma A \to A e^+ e^-$, which means $\gamma + A \to A + e^+ + e^-$, where A stands for an atomic nucleus.

With but rare exceptions, we use *natural units*, wherein $\hbar = c = 1$; this system of units is explained in I.B.1(e). Other commonly used units are fm $= 10^{-13}$ cm, Å $= 10^{-8}$ cm, MeV $= 10^6$ eV, and GeV $= 10^9$ eV.

Our quantum mechanical notation is defined in §I.B.1. We designate everyday Euclidean 3-space by $\mathcal{E}_3$, while *abstract* Euclidean 3-spaces carry a superscript, such as T, which stands for weak isospin, or I, which stands for hadronic isospin. A complex N-dimensional vector space is designed by $\mathcal{C}_N$. The notation $\{\ldots\}$ refers to the set of objects . . . .

Vectors in $\mathcal{E}_3$ are in roman boldface: **E, p, ϵ**, etc.; unit vectors have a caret, as in **ê**. Minkowski 4-vectors are denoted by italic boldface, as in $\boldsymbol{x} = (t, \mathbf{r})$ or $\boldsymbol{p} = (E, \mathbf{p})$. The Lorentz-invariant scalar product of two 4-vectors is written as a dot product, as in $\boldsymbol{x} \cdot \boldsymbol{p} = Et - \mathbf{r} \cdot \mathbf{p}$. Vectors in an *abstract* $\mathcal{E}_3$ space are written as $\vec{A}, \vec{B}$, etc., and their scalar products $\vec{A} \cdot \vec{B}$, etc., have the usual meaning. Beginning with Chapter IV, we also use the notation $\vec{A}, \vec{B}$, etc., to represent color-$SU(3)$ octets.

CONTENTS

II. QUANTUM ELECTRODYNAMICS 191

A. PHOTONS AND CHARGED FERMIONS 195

1. Field operators 195
2. Photons 196
3. Fermions 198
4. Destruction and creation of particles 204
5. Positrons 208

B. THE ELECTROMAGNETIC INTERACTION; FEYNMAN GRAPHS 212

1. The interactions between field and source 212
   (a) The interaction operator 212
   (b) Helicity rules in ultrarelativistic phenomena 214
   (c) Vertices 216
2. The scattering amplitude 219
3. Compton scattering 221
   (a) Diagrams for Compton scattering 221
   (b) A spin-zero model 222
   (c) Compton amplitudes and the Klein-Nishina formula 225
   (d) Evaluation of cross sections 226
4. Feynman diagrams 228
5. The dimensionality of the coupling constant and the high-energy behavior of amplitudes 230

C. EXAMPLES OF ELECTROMAGNETIC INTERACTIONS 232

1. Scattering of two different fermions 232
2. Colliding beam processes: $e^+e^- \to \mu^+\mu^-$ and $e^+e^- \to e^+e^-$ 235
3. Data on e^+e^- collisions; tests of quantum electrodynamics 238
4. Bound states 243
5. Hydrogen, μ-mesic atoms, and muonium 244
6. Positronium 247
7. The decay of positronium: Charge conjugation 248

D. THE SELF-ENERGY OF THE ELECTRON, VACUUM POLARIZATION, AND PRECISION TESTS OF QUANTUM ELECTRODYNAMICS   252

1. Radiative corrections   253
2. The electron self-energy and mass renormalization   255
3. Vacuum polarization   258
4. The Lamb shift   266
5. The anomalous magnetic moment of the electron and muon   270
6. Other spectroscopic tests of quantum electrodynamics   272
7. Summary   273

III. HADRONIC SPECTROSCOPY   275

A. HADRONIC DIMENSIONS AND SPECTRA   278
1. The electromagnetic size of hadrons   279
2. Hadronic diffraction scattering and the size of hadrons   281
3. Excited states of hadrons   283
4. Photoproduction   288
5. Pion-nucleon scattering   291
6. Consequences of the isospin assignments   295
7. Kaon-nucleon scattering and the $S \neq 0$ excited states of the baryon   297
8. Hadron excitations observed by decay   297
9. G-parity   303

B. MESONS COMPOSED OF HEAVY QUARK-ANTIQUARK PAIRS   312
1. $e\bar{e} \to \mu\bar{\mu}$ revisited   313
2. Narrow hadron resonances in $e\bar{e}$-hadron production   317
3. The $Q\bar{Q}$ spectrum   323
4. Hadronic decays of $Q\bar{Q}$ states   333

IV. QUANTUM CHROMODYNAMICS   339

A. COLOR   342
1. The color variable   342
2. The group $SU(3)$ 347
   (a) Definition of groups. $SU(2)$ and $SU(3)$ 348

(b) An *SU(3)* primer   350
(c) Further aspects of *SU(3)*   358

B. GAUGE FIELDS   364

1. Global vs. local symmetries   364
2. The gauge field   366
3. Quanta of the gauge field and coupling to quarks   371

C. GAUGE FIELD DYNAMICS   374

1. Color analogues of the electromagnetic field strengths   374
2. Self-interactions of the gauge field   377
3. The Yang-Mills field equations   377
4. Asymptotic freedom   380
   (a) Comparison with QED   380
   (b) Antiscreening—A qualitative discussion   381
   (c) Antiscreening by an electromagnetic analogy   386
   (d) The results of an exact perturbation calculation   390
   (e) Renormalization of the coupling constant. Asymptotic freedom   394
5. Confinement   397
   (a) The QCD vacuum   397
   (b) The long-distance interaction   400
   (c) The semiempirical quark–antiquark interaction   401

D. THE BAG MODEL   404

1. The primitive bag model   404
2. The improved bag model   409
3. Quantitative test of the model   414
4. The long-range potential and the Regge slope   417

V. DEEP INELASTIC LEPTON-HADRON SCATTERING   421

1. Introduction   423
2. Kinematic variables   425
3. Deep inelastic electron and muon scattering cross sections   427
4. Deep inelastic electron and muon scattering according to the quark model   428
5. The quark momentum distributions in the nucleon   432

6. Charge-changing neutrino scattering 439
   (a) Cross sections for inelastic neutrino scattering 439
   (b) Neutrino-quark cross sections 440
   (c) Neutrino-nucleon scattering 442
   (d) Structure functions 444
7. Theoretical considerations regarding the quark distributions 449
   (a) The quark momentum distribution in the nucleon 449
   (b) The gluon field in the nucleon 452
   (c) The sea quarks 457
   (d) The structure functions 459
8. Scaling violations 460

VI. THE ELECTROWEAK INTERACTION 465

A. CHARGE-CHANGING WEAK INTERACTIONS 469

1. The charge-changing weak current 469
2. The Hamiltonian for charge-changing weak interactions 474
   (a) The Fermi interaction 474
   (b) The strength of the semileptonic interaction 475
3. Defects of the $W^{\pm}$ model of weak interactions 478

B. NEUTRAL CURRENT WEAK INTERACTIONS AND THE ELECTROWEAK CONNECTION 483

1. Conservation of the weak current: The neutral current 483
   (a) Demonstration that $J_{\pm}$ are not conserved currents 483
   (b) Weak isospin 485
   (c) Conservation of weak isospin 487
   (d) The symmetric weak interaction 488
2. The electroweak connection 491
   (a) Incorporation of electrodynamics 491
   (b) The Z^0 and the mixing angle θ_W 493
   (c) University and unification 495
3. Neutral current phenomena and the determination of the coupling constants 496
   (a) The neutral current weak interaction 496
   (b) Deep inelastic neutrino scattering 499
   (c) Parity violation in inelastic electron scattering 502

(d) The basic parameters of the electroweak theory   504
(e) Z^0-production in $e\bar{e}$-annihilation   506

C. THE HIDDEN GAUGE INVARIANCE OF THE ELECTROWEAK INTERACTION   509

1. Hidden symmetry in ferromagnetism and superconductivity   509
    (a) Ferromagnetism   509
    (b) Superconductivity   511
2. The generation of mass in the electroweak theory   516
    (a) Analogies between superconductivity and electroweak phenomena   516
    (b) The electroweak gauge group   517
    (c) The Standard Model; the masses of Z and W 519
    (d) The Higgs boson and its couplings   523

D. GRAND UNIFICATION   529

1. Basic assumptions and immediate consequences   530
2. Evaluation of the electroweak angle at low energies   535
3. Other aspects of the Grand Unified Theory   540
    (a) The gauge group   540
    (b) Symmetry breaking   541
    (c) Proton decay   542

Appendix II. Bose fields   545

1. The harmonic oscillator   545
    (a) The classical oscillator   545
    (b) Quantized oscillator   546
2. The scalar field   548
    (a) Real fields   549
    (b) Complex fields   553
3. The electromagnetic field   556

Appendix III. The Dirac field   559

1. The Lorentz transformation of spinors and the Dirac equation   559
    (a) Three-vectors as 2×2 matrices   559
    (b) Four-vectors as 2×2 matrices   560
    (c) Spinors   562
    (d) The Dirac equation   563

(e) Four-currents   566
2. The Dirac field operator   567
   (a) The field operator: Anticommutation rules   567
   (b) Energy, momentum, and charge   568
   (c) The relative parity of particle and antiparticle   570
3. Amplitudes for weak and electromagnetic scattering   571
4. Spin and statistics   573

Appendix IV. Causality and its consequences   577
1. The basic axiom of quantum field theory   577
2. The connection between spin and statistics   579
3. The need for antiparticles; Crossing symmetry   583

Appendix V. Vacuum polarization   587

Appendix VI. The magnetic susceptibility of a massless vector field   593

Appendix VII. Solutions of Dirac's equation in a spherical enclosure   597

Bibliography   600

Index   603

II

QUANTUM ELECTRODYNAMICS

II

Of all the objects known to us, only the leptons conform with Newton's conception of an elementary particle: They appear to be so "hard, impenetrable,... as never to wear or break in pieces." No experimental search—and there have been many—has ever discovered excited states of the electron or muon, and the most precise measurements have never revealed that these particles have a finite size or are composed of constituents. It goes without saying that experiment can only set limits on unobserved properties—future research may show that even the electron and muon are complex systems like the nucleon. But at this time all measurements can be understood—and understood to very high precision— by assuming that e, μ, and τ are pointlike spin-$\frac{1}{2}$ particles. While our experimental knowledge of neutrinos is nowhere as extensive as for the charged leptons, all data are consistent with the hypothesis that neutrinos are also pointlike spin-$\frac{1}{2}$ particles. Because of this pristine simplicity, it is natural to begin our more detailed study of particles and fields with the leptons.

One needs but little theory to understand how we know that leptons have no excited states. If, say, the electron had an excited state e^*, it should be formed in a collision, just as in atomic or nuclear physics. No collision involving electrons has ever been seen to lead to a final state having an electron and photon that can be ascribed to the decay $e^* \to e\gamma$. In particular, the muon is not an excited electron, for it does not undergo the decay $\mu \to e\gamma$.

Our claim that e and μ are pointlike does, however, stem from theory. For the observation of an object's structure involves not only a probe but also a detailed knowledge of that probe's interaction with the object. The most useful probe for studying the electron and muon is the electromagnetic field. Thus one studies the motion of e and μ in the nuclear Coulomb field, through laboratory magnets, and in more dramatic processes, such as $e^+e^- \to 2\gamma$ and $e^+e^- \to \mu^+\mu^-$. In such experiments one measures atomic-energy levels, precession frequencies of magnetic moments, and cross sections. One then compares these data to a theory that assumes a conventional Maxwell-Lorentz interaction between the electromagnetic field and point charges. Our assertion that e and μ appear to be structureless rests on the detailed and quantitative success of this comparison.

The phenomena we have just mentioned occur in a regime where classical theory is inapplicable; relativistic and quantum effects are both of paramount importance. The theory that handles electromagnetic phenomena of point charges in the quantum domain is called *quantum electrodynamics* (QED).

The remarkable agreement between all predictions of quantum electrodynamics and experiment, to be summarized in §C, has had an enormous intellectual impact on the evolution of elementary particle physics. When applied to processes in which the energies are well under 100 GeV, the theory works to a precision comparable to Newtonian mechanics in the realm of planetary motion. For processes involving energies of order 100 GeV and higher, the electroweak connection, discussed in §I.E.10, must be taken into account. More of this is discussed in Chap. VI. With but few exceptions, the present chapter confines itself to phenomena where weak interaction effects can be neglected.

The great success of QED has been bought at a price, however. As is shown in §D, the evaluation of the interaction between the electromagnetic field and its sources gives rise to infinite terms. These infinities stem from the interaction of point charges with field components of exceedingly high frequencies corresponding to energies far above 100 GeV. The incorporation of the electroweak connection does not, however, remove these infinities, because the electroweak theory is also based on pointlike sources. A careful examination of the infinite expressions reveals that they can be separated unambiguously into divergent portions that can be attributed to the mass and charge of the sources, and a finite remainder. The so-called renormalization procedure replaces the infinite quantities by the observed values of the mass and charge, which then appear as free parameters in finite expressions that describe all other observable quantities, such as cross sections and energy levels.

Because of the impressive predictive power of renormalized quantum electrodynamics, physicists have sought descriptions of other interactions between particles and fields molded in the same image. Indeed, our understanding of the weak and strong interactions has been revolutionized by a quantum field theory—the so-called Yang-Mills theory—that is a natural generalization of quantum electrodynamics. The Yang-Mills theory is developed in Chap. IV, where it is applied to the strong interaction; its application to the electroweak interaction is described in Chap. VI.

A. PHOTONS AND FERMIONS

Quantum electrodynamics has two essential ingredients beyond Maxwell's classical theory of electromagnetic fields and their interaction with charged particles. The first is a quantum version of Maxwell's equations. The second is the Dirac equation, that describes the quantum mechanics of free spin-$\frac{1}{2}$ particles in accordance with the requirements of special relativity.

1. Field operators[1]

The quantum theory of the electromagnetic field is constructed from Maxwell's equations by a procedure that is essentially the same as the one that leads to ordinary quantum mechanics. In the latter case, the coordinates and momenta that appear in Hamilton's classical equations—the $q_i(t)$ and $p_i(t)$—are replaced by operators that satisfy simple commutation rules. In the electromagnetic case, the analogous recipe leads to a reinterpretation of the vector potential $\mathbf{A}(\mathbf{r}t)$ as a set of operators. By this one means that to each spatial point $\mathbf{r}$ there are associated three operators $\tilde{A}_x(\mathbf{r}t)$, $\tilde{A}_y(\mathbf{r}t)$, $\tilde{A}_z(\mathbf{r}t)$. Such entities are called *field operators*. Operators for the electric and magnetic fields are obtained from $\mathbf{A}$ in the conventional manner

$$\tilde{\mathbf{E}}(\mathbf{r}t) = -\partial \tilde{\mathbf{A}}(\mathbf{r}t)/\partial t, \tag{1}$$

$$\tilde{\mathbf{B}}(\mathbf{r}t) = \nabla \times \tilde{\mathbf{A}}(\mathbf{r}t). \tag{2}$$

The equations of motion for $\tilde{\mathbf{E}}$ and $\tilde{\mathbf{B}}$ are Maxwell's equations, just as the equations of motion of ordinary quantum mechanics retain the formal structure of Hamilton's equations. In both cases, however, the time derivative [as, for example, in Eq. (1) above] means $d\tilde{O}/dt = i[\tilde{H}, \tilde{O}]$ for any operator $\tilde{O}$ where $\tilde{H}$ is the Hamiltonian operator of the system in question. In the case of quantum electrodynamics, the Hamiltonian is the

---

[1] In Part A of this Chapter an operator is designated by a symbol with a tilde, e.g., the operator for the electric charge is $\tilde{Q}$. In the remainder of this book we do not distinguish operators by any special notation, for it should then be evident from the context whether an object is an operator or a number.

same function of the fields as in classical electrodynamics, i.e., the total energy:[2]

$$\tilde{H}_\gamma = \frac{1}{2}\int (\tilde{E}^2 + \tilde{B}^2)\, d^3r. \tag{3}$$

This is the Hamiltonian operator of the free field, i.e., of the field in the absence of any charged objects.

The detailed construction of the operators $\tilde{\mathbf{E}}, \tilde{\mathbf{B}}$, etc., and the derivation of many of the equations that follow, will be found in Appendix II.

2. Photons

The most compelling feature of the quantization recipe is that the photon emerges as a natural and inevitable consequence of the theory. We shall now sketch how this comes about.

As we know from classical electrodynamics, an arbitrary electromagnetic disturbance in a region totally devoid of charges (vacuum) can be viewed as a superposition of plane waves. A *plane wave* is characterized by a unit vector $\hat{\mathbf{n}}$ in the direction of propagation, a circular frequency ω, and a mode of polarization. For brevity's sake one introduces the propagation vector $\mathbf{k} = \hat{\mathbf{n}}\omega$; as $c = 1$ in our system of units, the wavelength is then $2\pi/k$. There are two independent modes of polarization for each $\mathbf{k}$. We use left- and right-circularly polarized waves as our independent modes, and designate them by $h = +1$ and $h = -1$, respectively, for we shall soon learn that in the quantum theory h becomes the photon's helicity.[3]

The point in breaking the field up into plane waves is that each set of numbers $(\mathbf{k}, h)$ specifies an independent degree of freedom of the electromagnetic field *in vacuo*. That is to say, if at some instant we have an electromagnetic disturbance in empty space that is a superposition of waves specified by $(\mathbf{k}_1, h_1), (\mathbf{k}_2, h_2), \ldots$, and having amplitudes $a_1, a_2, \ldots$, respectively, then this field is given by precisely the same superposition for all future times. If, in particular, the original field did not contain a wave $(\mathbf{k}, h)$, it will never do so.

We now return to the quantum theory, and ask for the eigenstates of the Hamiltonian (3). Since these are the stationary states, they are intimately related to the "stationary" Fourier series that describes the free electromagnetic field. First of all, we ask what follows from the dynamical independence of the modes with different values of $\mathbf{k}$ and h. The wave function of a system that has two independent noninteracting parts described by the

---

[2] Note that we use rationalized units, so that there are no factors of 4π in Maxwell's equations, nor in the energy or momentum densities. On the other hand, the Coulomb energy is $e_1 e_2/4\pi r$ in these units.

[3] We may use the letter h for helicity since there will be no confusion with Planck's constant, which is unity with our choice of units.

dynamical variables q_1 and q_2 can be written as a product function[4] $\psi_1(q_1)\psi_2(q_2)$. In our case, we have an infinite set of independent modes, the set of all the **k**'s, and $h = \pm 1$. Each separate mode has its own set of states, which we write as $|n_h(\mathbf{k})\rangle$ where, for the moment, $n = 0, 1, 2, \ldots$, is just a label to distinguish different eigenstates of the mode $\mathbf{k}, h$ [as did the suffix on, for example, $\psi_2(q_2)$]. A state vector of the whole system—of the electromagnetic field—is then a product like $\psi_1\psi_2$, but it has an infinite number of factors:

$$|\rangle = \prod_\mathbf{k} \prod_h |n_h(\mathbf{k})\rangle. \tag{4}$$

The energy of this system is the sum of the energies of each independent part. Since a plane wave is dynamically equivalent to a harmonic oscillator, the energy eigenvalue of a given mode with frequency ω_k is $n_k(\mathbf{k})\omega_k$ apart from the zero point energy $\frac{1}{2}\omega_k$, which we omit because the zero of energy is arbitrary (see Appendix II). Therefore Eq. (4) is an eigenstate of Eq. (3) with eigenvalue

$$E = \sum_\mathbf{k} \sum_h n_h(\mathbf{k})\omega_k. \tag{5}$$

Furthermore, Eq. (4) is also an eigenstate of the total momentum $\int (\mathbf{E} \times \mathbf{B})\, d^3r$ of the electromagnetic field, with eigenvalue

$$\mathbf{P} = \sum_\mathbf{k} \sum_h n_h(\mathbf{k})\mathbf{k}. \tag{6}$$

Consider two especially simple states. First, the one where all $n_h(\mathbf{k})$ vanish. This state has neither momentum nor energy, and for that reason we call it the vacuum state $|\Omega\rangle$. Secondly, consider the state where $n_k(\mathbf{k}) = 1$, and *all* other n's vanish. Call this state $|\mathbf{k}h\rangle$. From Eqs. (5) and (6) we see that $|\mathbf{k}h\rangle$ has energy ω and momentum $\mathbf{k}$. Since $\omega = k$, we see that for the state $|\mathbf{k}h\rangle$ the energy and momentum eigenvalues satisfy the relation $E = P$ characteristic of zero mass. Thus $|\mathbf{k}h\rangle$ has all the attributes of a single photon, and for that reason one calls $|\mathbf{k}h\rangle$ a one-photon state.

The physical significance of the numbers $n_k(\mathbf{k})$ is now clear: They specify how many photons are "in" the arbitrary stationary state, and what their momenta and polarizations are. For that reason they are called *occupation numbers*.

Finally, we turn to the label h. Let $\tilde{\mathbf{M}}$ be the angular momentum of the electromagnetic field, $\tilde{\mathbf{M}} = \int \mathbf{r} \times (\tilde{\mathbf{E}} \times \tilde{\mathbf{B}})\, d^3r$, and form its projection $\hat{\mathbf{k}} \cdot \tilde{\mathbf{M}}$ along the direction of propagation of the one-photon state $|\mathbf{k}h\rangle$. It turns

---

[4] If these "parts" are indistinguishable particles, the wave function must be symmetrized or antisymmetrized. It is one of the prime virtues of quantum field theory that this requirement is automatically satisfied. In the case of photons, see the discussion in Appendix II, especially Eqs. (34) and (58).

out[5] that $|\mathbf{k}h\rangle$ is also an eigenstate of this operator, with eigenvalue h:

$$\hat{\mathbf{k}} \cdot \tilde{\mathbf{M}} |\mathbf{k}h\rangle = h |\mathbf{k}h\rangle, \tag{7}$$

and for that reason h is indeed the photon's helicity: $h = \pm 1$ corresponds to an angular momentum ± 1 along the direction of propagation. There is no one-photon state with helicity eigenvalue zero because electromagnetic waves only have transverse fields.

3. Fermions

In this and in the following sections we deal with spin-$\frac{1}{2}$ particles such as leptons, quarks, and their antiparticles. We refer to them as fermions. Frequently the electron is used as a typical example of a fermion.

The Schrödinger equation for a free particle,

$$\left(i\frac{\partial}{\partial t} + \frac{\nabla^2}{2m}\right)\Psi = 0, \tag{8}$$

does not satisfy the principle of relativity. Any equation that satisfies that principle must, in essence, be symmetric in the time and space coordinates, for a Lorentz transformation replaces t and $\mathbf{r}$ by linear combinations of t and $\mathbf{r}$. As Eq. (8) is of first order in t, and second order in $\mathbf{r}$, it stands no chance of conforming with the principle of relativity.

There is another (and equivalent) way of expressing the trouble with the Schrödinger equation. The general solution of (8) is a superposition of plane waves $\exp[i(\mathbf{p}\cdot\mathbf{r} - Kt)]$; according to (8) the momentum $\mathbf{p}$ and *kinetic* energy K are related by $K = p^2/2m$, whereas the correct relativistic relationship between the *total* energy E and momentum $\mathbf{p}$ is

$$E = \sqrt{p^2 + m^2}. \tag{9}$$

It is not difficult to write a differential equation that results in the energy-momentum relation (9). We first square Eq. (9), $E^2 - p^2 - m^2 = 0$, and then make the familiar substitutions

$$E \to i\frac{\partial}{\partial t} \qquad \mathbf{p} \to -i\nabla. \tag{10}$$

This gives us the so-called Klein–Gordon equation

$$\left(\frac{\partial^2}{\partial t^2} - \nabla^2 + m^2\right)\phi = 0. \tag{11}$$

---

[5] See Jackson (1975) p. 333; Gottfried (1966), pp. 412–414.

3. FERMIONS

In the case of a plane wave $\phi = \exp[i(\mathbf{p}\cdot\mathbf{r} - Et)]$, Eq. (11) requires $E^2 = p^2 + m^2$, and therefore $E = \pm\sqrt{p^2 + m^2}$. Thus the Klein–Gordon equation leads not only to Eq. (9), but also to the "negative-energy" solutions with $E = -\sqrt{p^2 + m^2}$. As we shall see, these unexpected solutions are of great physical importance.

We now come to a crucial turn in the argument—a turn that is far from obvious. We ask: Does the wave function of a relativistic particle satisfy more fundamental equations than the Klein–Gordon equation? The motivation for this question is provided by classical vacuum electrodynamics, where $\mathbf{E}$ and $\mathbf{B}$ satisfy the wave equation

$$\left(\frac{\partial^2}{\partial t^2} - \nabla^2\right)\mathbf{E} = 0, \qquad \left(\frac{\partial^2}{\partial t^2} - \nabla^2\right)\mathbf{B} = 0, \tag{12}$$

but where these are *not* the fundamental equations of the theory. The reason for this is that Eqs. (12) fails to relate $\mathbf{E}$ to $\mathbf{B}$, and therefore do not specify the relative orientation of $\mathbf{E}$ and $\mathbf{B}$. For that one needs Maxwell's equations, which are of first order:

$$\frac{\partial \mathbf{E}}{\partial t} = \nabla \times \mathbf{B}, \qquad \nabla \cdot \mathbf{E} = 0, \tag{13}$$

$$\frac{\partial \mathbf{B}}{\partial t} = -\nabla \times \mathbf{E}, \qquad \nabla \cdot \mathbf{B} = 0 \tag{14}$$

These equations imply Eqs. (12), but not vice versa. In particular, Maxwell's equations require $\mathbf{E}$ and $\mathbf{B}$ to be perpendicular to each other, and to the direction of propagation. They therefore specify the polarization state of the electromagnetic field. The wave equation (12), on the other hand, does not even require the fields to be vectors. What Eq. (12) does specify is that the fields are superpositions of plane waves $\exp[i(\mathbf{k}\cdot\mathbf{r} - \omega t)]$ whose momenta $\mathbf{k}$ and energy ω are related by $\omega = |\mathbf{k}|$, and it therefore implies that photons have zero mass.

In the light of these observations, we now rephrase our question: Are there first-order equations analogous to Maxwell's that specify the spin state of a spin-$\frac{1}{2}$ particle, and that also lead to the Klein–Gordon equation? An affirmative answer to this was supplied by Dirac, but the argument that we use to arrive at his equation is rather different from the one that he used originally. To exploit the analogy to Maxwell's theory to the utmost, we begin by considering a spin-$\frac{1}{2}$ fermion of zero mass. This leads us to the equations of motion for massless fermions.[6] Once these are understood, the generalization to nonzero mass is relatively easy.

---

[6] The upper limits on the masses of ν_e and ν_μ are so small that they can be described by those equations under (almost) all circumstances. Furthermore, we are often concerned with ultrarelativistic phenomena, where particle masses can be ignored.

For a mass-zero particle, the helicity h is a Lorentz invariant quantum number, as one can see from the following argument. Recall that $h = \hat{\mathbf{p}} \cdot \mathbf{s}$, where $\hat{\mathbf{p}}$ is a unit vector along the momentum, and $\mathbf{s}$ is the spin. The latter has the same transformation properties as an orbital angular momentum. Now if we set $\hat{\mathbf{p}}$ along the z-axis, say, then $h = s_z$, which transforms like $(xp_y - yp_x)$, and does not change under a Lorentz transformation along z. Thus the only factor in h affected by a Lorentz transformation along the direction of motion is $\hat{\mathbf{p}}$: If the transformation fails to "overtake" the particle, $\hat{\mathbf{p}}$ is unchanged, whereas if it does "overtake," $\hat{\mathbf{p}} \to -\hat{\mathbf{p}}$. But a zero mass particle can never be overtaken, and therefore its helicity h is invariant.[7] The most familiar example of this is offered by the photon, as already pointed out on page 130.

We now know the essential ingredients that must enter the basic equation for a massless spin-$\tfrac{1}{2}$ particle. We write this equation as

$$\mathscr{D}\chi = 0, \tag{15}$$

where

$$\chi = \begin{pmatrix} \chi_1 \\ \chi_2 \end{pmatrix} \tag{16}$$

is a two-component wave function (a spinor), because the spin is $\tfrac{1}{2}$. Concerning the differential operator $\mathscr{D}$ we know that:
1. It must be of first order in space and time;
2. $\mathscr{D}\chi = 0$ must guarantee that the helicity is $h = \pm\tfrac{1}{2}$;
3. $\mathscr{D}\chi = 0$ must, on iteration, lead to the wave equation

$$\left(\frac{\partial^2}{\partial t^2} - \nabla^2\right)\chi = 0, \tag{17}$$

which guarantees the mass-zero energy-momentum relation.

In view of (16), $\mathscr{D}$ is a 2×2 matrix. The most general such matrix is of the form $\mathscr{D} = A + \sum B_i \sigma_i$, where the σ_i are the 3 Pauli matrices, and A and B_i are first-order differentials. Since $\mathscr{D}$ must be invariant under rotations, the structure is actually $\mathscr{D} = A + \mathbf{B} \cdot \boldsymbol{\sigma}$, where A is an invariant and $\mathbf{B}$ is a vector under rotations. The first-order differential operators $\partial/\partial t$ and $\boldsymbol{\nabla}$ that must occur in $\mathscr{D}$ are therefore proportional to $\mathbf{B}$ and A, respectively. Thus (17) must have the form

$$\left(\frac{\partial}{\partial t} + \lambda\boldsymbol{\sigma} \cdot \boldsymbol{\nabla}\right)\chi = 0. \tag{18}$$

The unknown constant λ is determined by requiring that the wave equation (17) follows from (18). For this purpose, we multiply (18) from the left by

---

[7] We have been careful to call this an argument, not a proof. We only proved that h is invariant under Lorentz transformations along $\hat{\mathbf{p}}$, while it is actually invariant under any (proper) transformation when $m = 0$.

$[(\partial/\partial t) - \lambda \boldsymbol{\sigma} \cdot \boldsymbol{\nabla}]$, and use the identity $(\boldsymbol{\sigma} \cdot \mathbf{a})(\boldsymbol{\sigma} \cdot \mathbf{a}) = a^2$. This yields $[(\partial^2/\partial t^2) - \lambda^2 \nabla^2]\chi = 0$, so that $\lambda = \pm 1$. There are, therefore, two equations of the form (18):

$$\left(\frac{\partial}{\partial t} \pm \boldsymbol{\sigma} \cdot \boldsymbol{\nabla}\right)\chi^{(\pm)} = 0. \tag{19}$$

These are often referred to as the Weyl equations.

The two solutions $\chi^{(\pm)}$ have helicities $h = \pm\frac{1}{2}$. To demonstrate this, consider a plane wave

$$\chi^{(\pm)}(\mathbf{r}, t) = \eta^{(\pm)}(\mathbf{p}) \, e^{i(\mathbf{p} \cdot \mathbf{r} - Et)}, \tag{20}$$

where $\eta^{(\pm)}$ are constant spinors, while $\mathbf{p}$ and E are the momentum and energy. According to (19)

$$(E \mp \boldsymbol{\sigma} \cdot \mathbf{p})\eta^{(\pm)}(p) = 0. \tag{21}$$

Since the spin $\mathbf{s} = \boldsymbol{\sigma}/2$, and $E = |\mathbf{p}|$ when $m = 0$, (21) can be written as $\mathbf{s} \cdot \hat{\mathbf{p}} \eta^{(\pm)} = \pm\frac{1}{2}\eta^{(\pm)}$; but $\mathbf{s} \cdot \hat{\mathbf{p}}$ is the helicity operator, so $h = \pm\frac{1}{2}$, as claimed. (Here we have taken E as positive throughout; later (see p. 207) we shall also have to consider solutions with $E < 0$.)

The two first-order equations (19) are therefore the analogues[8] of Maxwell's equations for spin $\frac{1}{2}$: They describe free (massless) neutrinos, and also spin-$\frac{1}{2}$ particles with mass in the limit of ultrarelativistic motion. The helicity $-\frac{1}{2}$ object $\chi^{(-)}$ is called the left-handed fermion state, the other, $\chi^{(+)}$, the right-handed state. These states play an essential role in the theory of weak interactions, where they are usually denoted by χ_L and χ_R.

We must still learn how to extend the theory to the case of a massive spin-$\frac{1}{2}$ particle. Here we have several guideposts: (i) Any new terms must have the same dimension as those already in (19), i.e., that of (time)$^{-1}$, or mass when $\hbar = c = 1$, and therefore they must be proportional to m; (ii) the new equations must reduce to (19) when $m \to 0$; (iii) the new equations must imply the Klein–Gordon equation; and (iv) the new equations must, in contrast to (19), couple $\chi^{(+)}$ and $\chi^{(-)}$. This last requirement can be seen as follows. When $m \neq 0$, one can go to a frame that overtakes the particle, and there h will be seen with the opposite value. In other words, *whereas the helicity is an invariant when $m = 0$, it is not an invariant when $m \neq 0$*, and there is no way of writing the equations so that $\chi^{(+)}$ and $\chi^{(-)}$ are uncoupled.

---

[8] This analogy goes even further than is evident from the conventional formulation. If one defines the two vector fields $\mathbf{F}^{(\pm)} \equiv \mathbf{E} \pm i\mathbf{B}$, one sees from (1) and (2) that $\partial_t \mathbf{F}^{(\pm)} \pm \boldsymbol{\nabla} \times \mathbf{F}^{(\pm)} = 0$, where $\partial_t = \partial/\partial t$. These are the true analogues of (19). [One readily verifies that $\mathbf{F}^{(\pm)}$ is (left/right) circularly polarized, i.e., $\mathbf{F}^{(\pm)}$ has helicity ± 1.] If one writes the curl in terms of the $j = 1$ angular momentum matrices $\mathbf{L}$, one can cast Maxwell's equations into the Weyl form (19): $(\partial_t \pm \mathbf{L} \cdot \boldsymbol{\nabla})\mathbf{F}^{(\pm)} = 0$, where $\mathbf{F}^{(\pm)}$ are to be considered as three-dimensional column vectors acted on by the 3×3 matrices $\mathbf{L} \cdot \boldsymbol{\nabla}$ and $\mathscr{I}\partial_t$, $\mathscr{I}$ being the 3×3 unit matrix.

From these four observations one can derive the $m \neq 0$ generalization of (19):

$$i\left[\frac{\partial}{\partial t} + \boldsymbol{\sigma} \cdot \boldsymbol{\nabla}\right]\chi^{(+)} = m\chi^{(-)}, \tag{22}$$

$$i\left[\frac{\partial}{\partial t} - \boldsymbol{\sigma} \cdot \boldsymbol{\nabla}\right]\chi^{(-)} = m\chi^{(+)}. \tag{23}$$

These are the *Dirac equations* for a massive particle of spin $\frac{1}{2}$. Equations (22) and (23) imply the Klein–Gordon equation. For $\chi^{(+)}$ this is verified by applying $(\partial_t - \boldsymbol{\sigma} \cdot \boldsymbol{\nabla})$ to (22), and by using (23) as well as $(\boldsymbol{\sigma} \cdot \boldsymbol{\nabla})^2 = \nabla^2$. The same holds for $\chi^{(-)}$, so

$$\left(\frac{\partial^2}{\partial t^2} - \nabla^2 + m^2\right)\chi^{(\pm)} = 0. \tag{24}$$

One often calls $\chi^{(-)}$ and $\chi^{(+)}$ the left- and right-handed portions of the Dirac wave functions, even though these are not eigenstates of helicity when $m \neq 0$. The name is still meaningful because the ultrarelativistic solutions of (22) and (23) have the property that the Weyl functions $\chi^{(\pm)}$ are almost exact eigenstates with $h = \pm\frac{1}{2}$. As we shall see, the weak interactions of massive fermions are most conveniently described by splitting their wave functions into left- and right-handed components.

Dirac's equations require the existence of both $\chi^{(+)}$ and $\chi^{(-)}$, unless $m = 0$. We therefore have the surprising result that a massive relativistic spin-$\frac{1}{2}$ particle requires two 2-component functions for its description, or four components in all. For the sake of brevity it is often convenient to combine these into a single 4-component object

$$\psi \equiv \begin{pmatrix} \chi^{(+)} \\ \chi^{(-)} \end{pmatrix} \equiv \begin{pmatrix} \psi_1 \\ \psi_2 \\ \psi_3 \\ \psi_4 \end{pmatrix}. \tag{25}$$

A similar compact notation is often used in electrodynamics, where $\mathbf{E}$ and $\mathbf{B}$ are combined into the field tensor $F_{\mu\nu}$.

One of the salient features of the Dirac equation is that it implies the existence of a conserved current. Such a current exists for the Schrödinger equation. In that case, we recall that the continuity equation,

$$\frac{\partial}{\partial t}(\Psi^*\Psi) + \boldsymbol{\nabla} \cdot \left(\frac{1}{2mi}\Psi^*\boldsymbol{\nabla}\Psi + \text{c.c.}\right) = 0, \tag{26}$$

follows from (8), and therefore $e\Psi^*\Psi$ and $(e/2mi)\Psi^*\boldsymbol{\nabla}\Psi + \text{c.c.}$ are the charge and current densities of the particle described by Ψ. In the case of the Dirac equation, a derivation similar to the one that leads from (8) to

3. FERMIONS

(26) shows that there also exist a charge density $e\rho$, and a current density $e\mathbf{j}$, that satisfy the continuity equation

$$\frac{\partial \rho}{\partial t} + \nabla \cdot \mathbf{j} = 0. \tag{27}$$

The charge density has the simple form

$$\rho = \sum_{\alpha=1}^{2} [\chi_\alpha^{(+)*}\chi_\alpha^{(+)} + \chi_\alpha^{(-)*}\chi_\alpha^{(-)}], \tag{28}$$

where the index α labels the two components of the spinors. The expression for $\mathbf{j}$ is

$$\mathbf{j} = \sum_{\alpha=1}^{2} [\chi_\alpha^{(+)*}\boldsymbol{\sigma}_{\alpha\beta}\chi_\beta^{(+)} - \chi_\alpha^{(-)*}\boldsymbol{\sigma}_{\alpha\beta}\chi_\beta^{(-)}], \tag{29}$$

where $\boldsymbol{\sigma}$ are again the Pauli matrices. If we use the compact notation (25), we see that Dirac's ρ is actually the sum of four Schrödinger-like terms:[9]

$$\rho = \sum_{\alpha=1}^{4} \psi_\alpha^* \psi_\alpha. \tag{30}$$

[In many instances it is useful to have a more compact and elegant notation for the quantities that appear in Dirac's theory.[10] First we want to write the two equations (22) and (23) as a single differential operator acting on the Dirac spinor (25). For this purpose we introduce a set of 4×4 matrices, which we express in the form

$$\begin{pmatrix} a & b \\ c & d \end{pmatrix},$$

where a, b, c, d are 2×2 matrices:

$$\gamma_0 = \begin{pmatrix} 0 & 1 \\ 1 & 0 \end{pmatrix}; \quad \gamma_i = \begin{pmatrix} 0 & -\sigma_i \\ \sigma_i & 0 \end{pmatrix}. \tag{31}$$

Using them we can write (22) and (23) as[11]

$$\left(i\gamma_\mu \frac{\partial}{\partial x_\mu} - m\right)\psi = 0, \tag{32}$$

---

[9] Here and in the following equations α labels the four components of ψ.

[10] We use a representation of the Dirac spinors and matrices that leads to a simple ultrarelativistic limit (or equivalently, to a simple $m = 0$ limit). The literature [compare, for example, Bjorken and Drell (1964)] uses a "standard" representation that makes the nonrelativistic limit simple. In the latter, one uses the two combinations $2^{-1/2}(\chi^{(+)} \pm \chi^{(-)})$ as the components of the Dirac 4-spinor, because $\chi^{(+)} = \chi^{(-)}$ in the nonrelativistic limit. The current $\mathbf{j}$ in the "standard" notation mixes the upper and lower components of the 4-spinor, in contrast to Eq. (29).

[11] From Eq. (32) one might suspect that the matrices γ_μ form a 4-vector. This is true in the same sense as the statement that $\boldsymbol{\sigma}$ is a 3-vector under spatial rotations. These matters are taken up in Appendix III.1(e).

where

$$\gamma_\mu \frac{\partial}{\partial x_\mu} \equiv \gamma_0 \frac{\partial}{\partial t} + \sum_{i=1}^{3} \gamma_i \frac{\partial}{\partial x_i}.$$

Equation (32) is to be understood as a 4×4 matrix acting on the four-spinor ψ: In terms of the components of γ_μ and ψ, $(\gamma_\mu \psi)_\alpha = \sum_{\beta=1}^{4} (\gamma_\mu)_{\alpha\beta} \psi_\beta$. The charge and current densities also take on an elegant form in this notation. For this purpose we define the adjoint Dirac spinor $\bar{\psi}$ with components

$$\bar{\psi}_\alpha \equiv \sum_{\beta=1}^{4} \psi_\beta^*(\gamma_0)_{\beta\alpha}. \tag{33}$$

Then we recall that $(\rho, \mathbf{j})$ form the components of a 4-vector j_μ, and that in covariant form the continuity equation (27) is $\partial j_\mu / \partial x_\mu = 0$. From (28) and (29) one finds that

$$j_\mu = \bar{\psi} \gamma_\mu \psi. \tag{34}$$

Here we have introduced the standard compact notation for expressions bilinear in the Dirac spinors, namely, if M is a 4×4 matrix, then

$$\bar{\psi} M \psi \equiv \sum_{\alpha,\beta=1}^{4} \bar{\psi}_\alpha M_{\alpha\beta} \psi_\beta. \tag{35}$$

Further details concerning the Dirac equation can be found in Appendix III.]].

4. Destruction and creation of particles

After the Dirac equation was discovered, it was taken for granted that it described a single electron, just as the Schrödinger equation (8) does. But it soon became clear that the solutions with $E = -\sqrt{p^2 + m^2}$ could not be ignored, nor could they be interpreted as one-electron states. The fundamental insight that these peculiar solutions must imply the existence of the positron, a spin-$\frac{1}{2}$ particle having precisely the same mass as the electron, but the opposite charge and magnetic moment, was confirmed in 1933.

The easiest way to understand the significance of the "negative energy" solutions is to return to electrodynamics. The vector potential of a plane wave, $\mathbf{A} \propto \exp[i(\mathbf{k} \cdot \mathbf{r} - \omega t)]$, is a solution of the wave equation if $\omega^2 = |\mathbf{k}|^2$, or

$$\omega = \pm |\mathbf{k}|. \tag{36}$$

The usual interpretation of $\hbar \omega$ as the energy also creates the puzzle of "negative energies." What are we to make of this?

4. DESTRUCTION AND CREATION OF PARTICLES

To begin with, let us write out the vector potential for a plane wave in the classical theory. The fields **E** and **B** are real, and therefore **A** is also. Thus a plane wave must have the form

$$\mathbf{A}(h\mathbf{k}) = \frac{a_h(\mathbf{k})\boldsymbol{\varepsilon}_h}{\sqrt{2V\omega}} e^{i(\mathbf{k}\cdot\mathbf{r}-\omega t)} + \frac{a_h^*(\mathbf{k})\boldsymbol{\varepsilon}_h^*}{\sqrt{2V\omega}} e^{-i(\mathbf{k}\cdot\mathbf{r}-\omega t)}, \qquad (37)$$

where $\boldsymbol{\varepsilon}_h$ is a (complex) unit vector that specifies the polarization $h = \pm 1$ of the wave, and $a_h(\mathbf{k})$ is a complex number that determines its amplitude and phase. Here V is the volume of a large cube on whose surface the plane waves are required to satisfy periodic boundary conditions. The length $V^{1/3}$ is taken to be very large in comparison to all dimensions of interest, and V must therefore disappear from all observable quantities, such as lifetimes, cross sections, etc. The factor $\omega^{-1/2}$ occurs because it is convenient to normalize the a's so that the energy of the wave (37) is ω when $|a| = 1$. Both roots (36) appear in (37). The first term is proportional to $\exp(-i\omega t)$, and the second to $\exp(+i\omega t)$; the latter must be the complex conjugate of the former if **A** is to be real.

We now turn to the quantum theory. As we know, photons can be emitted (i.e., created) and absorbed (i.e., destroyed). The quantization procedure sketched in §2, and described in detail in Appendix II, leads directly to operators that create and destroy photons. Remarkably enough, the field operator $\tilde{\mathbf{A}}$ introduced in §1 is obtained from the classical expression (37) by simply replacing the amplitude $a_h(\mathbf{k})$ by the *destruction operator* $\tilde{a}_h(\mathbf{k})$, and $a_h^*(\mathbf{k})$ by the *creation operator* $\tilde{a}_h^\dagger(\mathbf{k})$.

The action of these operators on the one-photon states $|h\mathbf{k}\rangle$ is especially simple. When acting on $|h\mathbf{k}\rangle$, $\tilde{a}_h(\mathbf{k})$ produces the vacuum state $|\Omega\rangle$, viz.

$$\tilde{a}_h(\mathbf{k})|h\mathbf{k}\rangle = |\Omega\rangle. \qquad (38)$$

Thus $\tilde{a}_h(\mathbf{k})$ destroys a photon of helicity h and momentum $\mathbf{k}$. As $\tilde{a}_h(\mathbf{k})$ is, in general, a complex number in the classical theory, its quantum mechanical equivalent is not a Hermitian operator. Indeed, the adjoint operator $\tilde{a}_h^\dagger(\mathbf{k})$ has the opposite role to $\tilde{a}_h(\mathbf{k})$: when acting on the vacuum it produces the one-photon state:

$$\tilde{a}_h^\dagger(\mathbf{k})|\Omega\rangle = |h\mathbf{k}\rangle. \qquad (39)$$

Naturally the transitions (38) and (39) cannot occur in isolation; light can only be emitted or absorbed if charges are involved. For that reason the operators $\tilde{a}$ and $\tilde{a}^\dagger$ appear in the Hamiltonian that describes the interaction between the electromagnetic field and matter [see Eq. B(1)].

The action of the creation and destruction operators on multiphoton states is the obvious generalization of what has already been said: $\hat{a}_h^\dagger(\mathbf{k})$ adds a photon of momentum $\mathbf{k}$ and helicity h to the state, while $\tilde{a}_h(\mathbf{k})$ removes such a photon (provided it is present in the first place, of course).

Thus $\tilde{a}_h^\dagger(\mathbf{k})$ increases the energy of an arbitrary state by ω, and $\tilde{a}_h(\mathbf{k})$ decreases it by ω. This statement provides the correct interpretation of the "negative energies": After quantization the frequencies that arise from the classical field equation become energy *differences*, not energies, because the classical field becomes an operator that can both excite and de-excite the state of the field. The "negative" energies are simply energies of de-excitation, and therefore negative; there is nothing mysterious or magical about them.[12]

The plane wave field (37) can now be reexpressed in operator form:

$$\tilde{\mathbf{A}}(h\mathbf{k}) = \frac{\tilde{a}_h(\mathbf{k})}{\sqrt{2V\omega}}\boldsymbol{\varepsilon}_h e^{i(\mathbf{k}\cdot\mathbf{r}-\omega t)} + \frac{\tilde{a}_h^\dagger(\mathbf{k})}{\sqrt{2V\omega}}\boldsymbol{\varepsilon}_h^* e^{-i(\mathbf{k}\cdot\mathbf{r}-\omega t)}. \quad (40)$$

Only $\tilde{a}$ and $\hat{a}^\dagger$ are operators here; all other quantities are numbers. The vector nature of $\tilde{\mathbf{A}}$ resides in $\boldsymbol{\varepsilon}_h$. The complete field operator $\tilde{\mathbf{A}}(\mathbf{r}, t)$ is obtained by summing over all the photon eigenvalues:

$$\tilde{\mathbf{A}}(\mathbf{r}, t) = \sum_{h\mathbf{k}} \tilde{\mathbf{A}}(h\mathbf{k}). \quad (41)$$

One should not harbor the misconception that a sum of photon creation operators creates a multiphoton state. Any linear combination $\sum c_h(\mathbf{k})\tilde{a}^\dagger(h\mathbf{k})$ creates a coherent superposition of one-photon states of definite helicity and momentum. For example, the operator that describes dipole emission is a sum over all creation operators having a definite magnitude $|\mathbf{k}|$, with an angular distribution characteristic of dipole radiation. The probability of finding anything but one photon in this state vanishes.

We are now adequately prepared to come to grips with the interpretation of the Dirac equation and its "negative energy" solutions. The parallel to electromagnetism is rather close. In analogy to (40), let us write the part of ψ that corresponds to particles with definite momentum $\mathbf{p}$ and helicity $h = \pm\frac{1}{2}$:

$$\psi(h\mathbf{p}) = \frac{1}{\sqrt{2VE}}[b_h(\mathbf{p})u_h(\mathbf{p})e^{i(\mathbf{p}\cdot\mathbf{r}-Et)} + d_h^*(\mathbf{p})v_h(\mathbf{p})e^{-i(\mathbf{p}\cdot\mathbf{r}-Et)}]. \quad (42)$$

Here $E = \sqrt{p^2 + m^2}$, while u_h and v_h are 4-component spinor amplitudes, like (25), which do not depend on $\mathbf{r}$ and t. Their actual form is unimportant

---

[12] The following may be of help to some readers. One can represent the electromagnetic field as a system of uncoupled oscillators, one for each plane wave $(h, \mathbf{k})$. When the oscillators are quantized, each amplitude becomes an operator that increases or decreases the energy $n\omega$ of the oscillator by one unit ω. The operator $\tilde{a}_h(\mathbf{k})$ in (40) is the part of the amplitude that decreases the energy, whereas $\tilde{a}_h^\dagger(\mathbf{k})$ increases the energy. For further details, see Appendix II.

4. DESTRUCTION AND CREATION OF PARTICLES

here;[13] it suffices to understand that they are the analogue of the photon polarization vectors $\boldsymbol{\varepsilon}$.

Note one essential difference between (42) and (40): Because the Dirac equation is intrinsically complex, $\psi(h\mathbf{p})$ need not be real, as $\mathbf{A}(h\mathbf{k})$ was. Therefore $d_h^*(\mathbf{p})$ need not be the complex conjugate of $b_h(\mathbf{p})$. This will have the important consequence that when d^* and b are replaced by operators, they are not each others Hermitian conjugates.

We have just learned that $\mathbf{A}(h\mathbf{k})$ is not a wave function of a photon of momentum $\mathbf{k}$ and helicity h, but an operator $\tilde{\mathbf{A}}$ that creates and destroys photons. Similarly, Dirac's $\psi(h\mathbf{p})$ cannot be regarded as a Schrödinger-like wave function of a particle with momentum $\mathbf{p}$ and helicity h; it, too, is an operator $\tilde{\psi}(h\mathbf{p})$ that can either create or destroy particles of *positive* energy.

Using this insight, we now rewrite the wave function $\psi(h\mathbf{p})$ of a particle with helicity h and momentum $\mathbf{p}$ as an operator:

$$\tilde{\psi}(h\mathbf{p}) = \frac{1}{\sqrt{2VE}}[\tilde{b}_h(\mathbf{p})u_h(\mathbf{p})e^{i(\mathbf{p}\cdot\mathbf{r}-Et)} + \tilde{d}_h^\dagger(\mathbf{p})v_h(\mathbf{p})e^{-i(\mathbf{p}\cdot\mathbf{r}-Et)}]. \quad (43)$$

Here only $\tilde{b}_h(\mathbf{p})$ and $\tilde{d}_h^\dagger(\mathbf{p})$ are operators; they are analogous to the operators $\tilde{a}_h(\mathbf{k})$ and $\tilde{a}_h^\dagger(\mathbf{k})$ in (40). When a particle is destroyed, the energy of the system is decreased. The "negative energy" solutions describe this de-excitation of the system. They are only paradoxical if one labors under the misconception that Dirac's ψ is a Schrödinger-like wave function of a single particle.[14] As in electrodynamics (cf. (41), the complete Dirac field operator is the sum of the Fourier components, viz.,

$$\tilde{\psi}(\mathbf{r}, t) = \sum_{ph} \tilde{\psi}(h\mathbf{p}). \quad (44)$$

---

[13] To obtain u_h one sets $\psi = u_h(\mathbf{p}) \exp[i(\mathbf{p}\cdot\mathbf{r} - Et)]$, $E > 0$, in (32), which then leads to an algebraic equation for u_h. This equation (see Eq. 54) has two solutions, corresponding to the two possible helicities. The spinor v_h is the coefficient of the negative frequency solution of (32), i.e., $\psi = v_h(\mathbf{p}) \exp[-i(\mathbf{p}\cdot\mathbf{r} - Et)]$. Our choice of norm is $u^*u = v^*v = 2E$. In contrast to $\hat{\mathbf{e}}$, u and v are not dimensionless, because $(\dot{\mathbf{A}})^2$ is an energy density, whereas $\psi^*\psi$ is a probability density. Consequently, $\mathbf{A}$ has the dimension of mass, while ψ has the dimension of $(\text{mass})^{3/2}$. The norm of (42) is chosen so that $\int |\psi(\mathbf{r}, t)|^2 d^3r = 1$ when $b = 1$ and $d = 0$, or vice versa. For further details, see Appendix III.

[14] What is the Schrödinger-like wave function in quantum field theory? Consider a one-particle state $|A\rangle$ that is a linear combination of states of given $(h\mathbf{p})$, with (complex) amplitudes $f_h(\mathbf{p})$. The state $|A\rangle$ can be expressed in the form

$$|A\rangle = \sum_{ph} f_h(\mathbf{p})\tilde{b}_h^\dagger(\mathbf{p})|\Omega\rangle,$$

and $f_h(\mathbf{p})$ is the wave function of $|A\rangle$ in momentum space.

To summarize this situation, we now write

$$\tilde{\mathbf{A}}(h\mathbf{k}) = \tilde{\mathbf{A}}_{\text{destroy}} + \tilde{\mathbf{A}}_{\text{create}}, \tag{45}$$

$$\tilde{\psi}(h\mathbf{p}) = \tilde{\psi}_{\text{destroy}} + \tilde{\phi}_{\text{create}}. \tag{46}$$

Here the operators $\tilde{\mathbf{A}}_{\text{destroy}}$ and $\tilde{\psi}_{\text{destroy}}$ have the time dependence in the form $\exp(-i\omega t)$, whereas $\tilde{\mathbf{A}}_{\text{create}}$ and $\tilde{\phi}_{\text{create}}$ have it in the form $\exp(+i\omega t)$, with $\omega \geq 0$. Note that $\tilde{\mathbf{A}}_{\text{create}} = (\tilde{\mathbf{A}}_{\text{destroy}})^\dagger$, so that $\tilde{\mathbf{A}}^\dagger = \tilde{\mathbf{A}}$. (Since the classical field $\mathbf{A}$ is real, $\tilde{\mathbf{A}}$ is a Hermitian operator.) However, $\tilde{\phi}_{\text{create}} \neq (\tilde{\psi}_{\text{destroy}})^\dagger$, because $\tilde{\psi}$ is not a Hermitian operator. This is why we use the symbol $\tilde{\phi}$ instead of $\tilde{\psi}$ in the second part of $\tilde{\psi}(h\mathbf{p})$. Let us construct the Hermitian conjugate of this operator:

$$\tilde{\psi}^\dagger(h\mathbf{p}) = \tilde{\psi}^\dagger_{\text{destroy}} + \tilde{\phi}^\dagger_{\text{create}}. \tag{47}$$

The Hermitian conjugate of a destruction operator is the creation operator of the same particle, and vice versa; hence

$$\tilde{\psi}^\dagger(h\mathbf{p}) = \tilde{\psi}_{\text{create}} + \tilde{\phi}_{\text{destroy}}. \tag{48}$$

But keep in mind that whatever $\tilde{\phi}_{\text{create}}$ creates is *not* destroyed by $\tilde{\psi}_{\text{destroy}}$, only by $\tilde{\phi}_{\text{destroy}}$. We will now see what it is.

5. Positrons

Despite the close similarity between $\tilde{\psi}$ and $\tilde{\mathbf{A}}$, there is a profound difference between the particles they describe: Electrons are electrically charged, whereas photons are neutral. Because of this, it is useful to classify operators by the effect they have on the charge of a system. This classification is also of importance in the theory of weak interactions (see Chap. VI).

Let Q be the charge of an arbitrary state $|Q\rangle$, in units of e. When acting on $|Q\rangle$, $\tilde{\mathbf{A}}$ does not change Q, because adding or subtracting a photon does not alter Q. For that reason, we call $\tilde{\mathbf{A}}$ a $\Delta Q = 0$ operator. We also know another important $\Delta Q = 0$ operator: the charge operator $\tilde{Q}$, whose eigenvalues are the numbers Q:

$$\tilde{Q}|Q\rangle = Q|Q\rangle. \tag{49}$$

By its very definition, $\tilde{Q}$ is a $\Delta Q = 0$ operator. The charge operator is an integral over space of the charge density operator $\tilde{\rho}$:

$$\tilde{Q} = \int \tilde{\rho}(\mathbf{r}, t)\, d^3 r. \tag{50}$$

Therefore $\tilde{\rho}$ is also a $\Delta Q = 0$ operator.

We shall now use the $\Delta Q = 0$ property of $\tilde{\rho}$ to infer that the Dirac field $\tilde{\psi}$ cannot describe electrons alone—that the theory must also contain particles of the opposite charge.

We have the expression (30) for $\tilde{\rho}$, with ψ_α^* replaced by $\tilde{\psi}_\alpha^\dagger$, the Hermitian adjoint of $\tilde{\psi}_\alpha$, for we have now learned that ψ_α is an operator. Thus

$$\tilde{\rho}(\mathbf{r}, t) = \sum_{\alpha=1}^{4} \tilde{\psi}_\alpha^\dagger(\mathbf{r}, t) \tilde{\psi}_\alpha(\mathbf{r}, t) = \tilde{\psi}^\dagger \tilde{\psi}, \tag{51}$$

and this must be a $\Delta Q = 0$ operator.

It is obvious that $\tilde{\psi}$ cannot be a $\Delta Q = 0$ operator, for it creates and destroys charged particles. Hence the only way that we can make $\tilde{\rho}$ a $\Delta Q = 0$ operator is to have $\tilde{\psi}^\dagger$ undo any change in Q produced by $\tilde{\psi}$. This is not farfetched. It occurs frequently in quantum mechanics, and the best-known example is afforded by the angular momentum $\mathbf{J}$. Let $J_- = J_x - iJ_y$, $J_-^\dagger = J_x + iJ_y$. Acting on an eigenstate $|m\rangle$ of J_z with eigenvalue m, J_- lowers m by one unit [see Eq. I.B(19)]; we may call it a $\Delta m = -1$ operator. $J_-^\dagger$ does the opposite, it is an $\Delta m = 1$ operator, and therefore $J_-^\dagger J_-$ and $J_- J_-^\dagger$ are both $\Delta m = 0$ operators. (Indeed, $2J_z = J_-^\dagger J_- - J_- J_-^\dagger$.) This is an example of a general theorem: If $\theta^\dagger$ is a non-Hermitian operator that creates something, its Hermitian adjoint destroys the same thing, and therefore $\theta^\dagger \theta$ and $\theta \theta^\dagger$ are operators that do not change the quantity in question. In the case of J_- and $J_-^\dagger$, "something" is the z-component of angular momentum; in the electromagnetic case, θ is $\tilde{a}_h(k)$, and "something" is a photon.

Bearing this lesson in mind, we now face the problem of making $\tilde{\rho}$, as expressed by Eq. (51), into a $\Delta Q = 0$ operator. Clearly, the "something" in the preceding theorem is charge. Since $\tilde{\psi}$ destroys electrons, it is a $\Delta Q = 1$ operator; our theorem then guarantees that $\tilde{\psi}^\dagger$ is a $\Delta Q = -1$ operator, and $\tilde{\rho}$ a $\Delta Q = 0$ operator, as desired. But now recall Eq. (46): $\tilde{\psi} = \tilde{\psi}_{\text{destroy}} + \tilde{\phi}_{\text{create}}$. By the argument we have just given, $\tilde{\phi}_{\text{create}}$ and $\tilde{\psi}_{\text{destroy}}$ must *both* be $\Delta Q = 1$ operators. This is only possible if $\tilde{\phi}_{\text{create}}$ creates an object having the *opposite* charge to whatever $\tilde{\psi}_{\text{destroy}}$ destroys. By convention, we define $\tilde{\psi}_{\text{destroy}}$ as the electron destruction operator; then $\tilde{\phi}_{\text{create}}$ is a positron creation operator. Let us now look at Eq. (47). It is clear that this operator creates an electron and destroys a positron. It, therefore, is a $\Delta Q = -1$ operator, and the $\tilde{\psi}$'s operate on electrons, the $\tilde{\phi}$'s on positrons.

The foregoing conclusions are stated most succinctly in terms of the operators $\tilde{b}$ and $\tilde{d}$: $\tilde{b}_h(\mathbf{p})$ destroys an electron, $\tilde{b}_h^\dagger(\mathbf{p})$ creates an electron, $\tilde{d}_h(\mathbf{p})$ destroys a positron, and $\tilde{d}_h^\dagger(\mathbf{p})$ creates a positron. In each case $\mathbf{p}$ and h are the momentum and helicity of the particle destroyed or created.

We now have a consistent formulation of the relativistic quantum theory of charged spin-$\frac{1}{2}$ particles. As our argument shows, such a theory must contain particles and antiparticles of opposite charge.[15] That the particles and antiparticles must have precisely the same mass is demonstrated in Appendix III.

⟦There is an important property of the Dirac theory that remains to be established: That fermions and antifermions have opposite intrinsic parities. As we know, the quark model relies on this fact to account for the parity of mesons. Our consideration of the behavior of fermions under reflection begins with the Dirac equation in the form given in Eqs. (22) and (23). If we reflect the coordinates, $\nabla \to -\nabla$; therefore, those equations show that space reflection amounts to

$$\chi^{(+)}(\mathbf{r}, t) \leftrightarrow \chi^{(-)}(-\mathbf{r}, t). \tag{52}$$

Since γ_0 interchanges the upper and lower 2-spinors when acting on a Dirac 4-spinor [see Eq. (31)], the Dirac field operator (44) undergoes the following transformation under reflection

$$\psi(\mathbf{r}, t) \to \gamma_0 \psi(-\mathbf{r}, t). \tag{53}$$

This determines the behavior of the 4-spinors u and v under reflection. As $u(\mathbf{p})$ and $v(\mathbf{p})$ are the amplitudes of the plane waves $\exp[\pm i(\mathbf{p} \cdot \mathbf{r} - Et)]$, respectively, they satisfy the algebraic equations

$$(\gamma_0 E - \boldsymbol{\gamma} \cdot \mathbf{p} - m)u(\mathbf{p}) = 0, \tag{54}$$

$$(-\gamma_0 E + \boldsymbol{\gamma} \cdot \mathbf{p} - m)v(\mathbf{p}) = 0. \tag{55}$$

If we multiply (54) by γ_0, and use $\gamma_0 \gamma_i = -\gamma_i \gamma_0$, we find

$$(\gamma_0 E + \boldsymbol{\gamma} \cdot \mathbf{p} - m)\gamma_0 u(\mathbf{p}) = 0. \tag{56}$$

Hence $\gamma_0 u(\mathbf{p}) = Nu(-\mathbf{p})$. As N cannot depend on $\mathbf{p}$, we can determine it at $p = 0$, where (54) reduces to $(\gamma_0 - 1)u = 0$; hence $N = 1$. The same argument applied to (55) gives $(\gamma_0 + 1)v(0) = 0$. In short, the 4-spinors transforms as follows under reflection:

$$\begin{aligned} u(\mathbf{p}) &\to u(-\mathbf{p}), \\ v(\mathbf{p}) &\to -v(-\mathbf{p}). \end{aligned} \tag{57}$$

If the Dirac field operator is to undergo the transformation (53), the difference of signs in (57) must be compensated by a corresponding sign

---

[15] One may ask why the $\Delta Q = 0$ operators $\tilde{\psi}^\dagger_{\text{destroy}} \tilde{\psi}_{\text{destroy}} = \tilde{\psi}_{\text{create}} \tilde{\psi}_{\text{destroy}}$ or $\tilde{\phi}^\dagger_{\text{destroy}} \tilde{\phi}_{\text{destroy}}$ will not do as charge density operators, for if they were acceptable, we would have no need for antiparticles. These possibilities are eliminated by the causality principle: The requirement that two observables that cannot be connected by a light signal can be measured simultaneously to arbitrary accuracy. As shown in Appendix IV, this principle leads to the conclusion that all observables involving fermions must be built from operators having the structure (43), with coefficients u_h and v_h having precisely the same normalization.

difference when the destruction and creation operators undergo space reflection:[16]

$$b_h^\dagger(\mathbf{p}) \to b_{-h}^\dagger(-\mathbf{p})$$
$$d_h^\dagger(\mathbf{p}) \to -d_{-h}^\dagger(-\mathbf{p}). \tag{58}$$

Here the helicity and momentum change signs because they are, respectively, a pseudoscalar and a polar vector. We can now see what (57) implies for the parity of an e^+e^- ^1S state. Such a state is a spherically symmetric superposition of states of zero total momentum, and zero net spin along any direction, and therefore has the form

$$|^1S\rangle = \int d^3p\, \phi(|\mathbf{p}|) \sum_{h=\pm 1/2} d^\dagger(\mathbf{p})b^\dagger(-\mathbf{p})|\Omega\rangle, \tag{59}$$

where ϕ is the wave function in momentum space. In view of (58), this state has *odd* parity. Quite generally, a fermion-antifermion state of orbital angular momentum L has the parity $(-1)^{L+1}$. This is to be contrasted with a boson-antiboson state, which has the parity $(-1)^L$.]

---

[16] Further remarks about these transformations and related matters are found in Appendix III.2.

B. THE ELECTROMAGNETIC INTERACTION; FEYNMAN GRAPHS

We have now described the theory of free photons, and of fermions. Observable phenomena are the consequence of interactions, and a dynamical theory must also describe them. Therefore we must incorporate the interactions between charged fermions and photons. It is this complete theory that is called *quantum electrodynamics*.

1. The interactions between field and source

(a) The interaction operator

In classical electrodynamics the energy of interaction between a current density $\mathbf{j}(\mathbf{r}t)$ and the electromagnetic field is given by[1,2]

$$H_{\text{int}} = e \int \mathbf{A}(\mathbf{r}, t) \cdot \mathbf{j}(\mathbf{r}, t) \, d^3r, \qquad (1)$$

where $\mathbf{A}$ is the vector potential. In quantum electrodynamics $\mathbf{A}$ and $\mathbf{j}$ become the operators discussed in the preceding Section. An explicit factor of the electronic charge appears in (1), so that $\mathbf{j}$ is the flux of charge in units of e. Let us then look in greater detail at the form of these operators.

The operator $\mathbf{A}$ in (1) is a sum of creation and destruction operators [cf. Eqs. A(40) and (41)]. We introduce the convenient abbreviation

$$\mathbf{A} = \sum_k (\mathbf{A}_k + \mathbf{A}_k^\dagger), \qquad (2)$$

where $\mathbf{A}_k$ and $\mathbf{A}_k^\dagger$ are the first and second terms of Eq. A(40), respectively. Note that we now use the notation k as a shorthand for the momentum $\mathbf{k}$ and helicity h.

The current $\mathbf{j}$ in (1) is the current of electrons and positrons. It is constructed from the Dirac operator ψ and its adjoint $\psi^\dagger$, These operators are sums of the $\psi(h\mathbf{p})$ given by A(43). We also write them in a shorthand

---

[1] See Jackson (1975) §12.8, where the Lagrangian interaction density L_{int} is given in covariant form. To obtain (1), one must use the Coulomb gauge and $H_{\text{int}} = -\int L_{\text{int}} d^3r$.
[2] Henceforth, we usually do not distinguish operators by a tilde.

1. THE INTERACTIONS BETWEEN FIELD AND SOURCE

analogous to (2):

$$\psi = \sum_p (B_p + D_p^\dagger),$$
$$\psi^\dagger = \sum_p (B_p^\dagger + D_p), \tag{3}$$

with

$$B_p = \frac{1}{\sqrt{2VE}} b_h(\mathbf{p}) u_h e^{i(\mathbf{p}\cdot\mathbf{r} - Et)},$$
$$D_p = \frac{1}{\sqrt{2VE}} d_h(\mathbf{p}) v_h^* e^{i(\mathbf{p}\cdot\mathbf{r} - Et)}; \tag{4}$$

their Hermitian conjugate partners are $B_p^\dagger$ and $D_p^\dagger$. Again, the index p includes the helicity h, while $E = +\sqrt{p^2 + m^2}$. The operators B_p and D_p are exactly what we called ψ_{destroy} and ϕ_{destroy} before, while $B_p^\dagger$ and $D_p^\dagger$ are ψ_{create} and ϕ_{create}: B_p destroys an electron of momentum $\mathbf{p}$, and D_p destroys a positron of momentum $\mathbf{p}$, while $B_p^\dagger$ and $D_p^\dagger$ are the corresponding creation operators. The current densities are essentially of the form $\psi^\dagger \psi$, as we see from A(34); thus

$$\mathbf{j} \doteq \sum_{pp'} (B_p^\dagger + D_p)(B_{p'} + D_{p'}^\dagger). \tag{5}$$

Here we use the notation $\doteq$ to indicate that $\mathbf{j}$ has the form of the right-hand side, but that certain details [the matrices $\boldsymbol{\sigma}$ in A(29), for example] that need not concern us here have been omitted. For example, in (5) the left-hand side is a vector, while the vectorial index is not shown on the right-hand side.

The current–field interaction operator is therefore

$$H_{\text{int}} \doteq e \int d^3r \sum_{kpp'} (A_k + A_k^\dagger)(B_p^\dagger B_{p'} + D_p D_{p'}^\dagger + D_p B_{p'} + B_p^\dagger D_{p'}^\dagger). \tag{6}$$

When the integral over $\mathbf{r}$ is carried out, the plane waves give Dirac δ-functions that guarantee momentum conservation. Thus H_{int} has the simple form

$$H_{\text{int}} \doteq e \sum_{kpp'}{}' [a(k) + a^\dagger(k)][b^\dagger(p)b(p') + d(p)d^\dagger(p')$$
$$+ d(p)b(p') + b^\dagger(p)d^\dagger(p')], \tag{7}$$

where the prime means that the sum over the momenta (k, p, p') is restricted by momentum conservation. For example, in the term $a(k)b^\dagger(p)b(p')$, an electron of momentum $\mathbf{p}'$ and a photon of momentum $\mathbf{k}$

are destroyed, while an electron of momentum $\mathbf{p} = \mathbf{p}' + \mathbf{k}$ is created:

$$\sum{}' a(k)b^\dagger(p)b(p') = \sum a(k)b^\dagger(p)b(p')\,\delta^3(\mathbf{p}-\mathbf{p}'-\mathbf{k}). \qquad (8)$$

Let us examine the processes generated by the various terms of H_{int}. The operator $b^\dagger(p)b(p')$ destroys an electron of momentum $\mathbf{p}'$ and creates an electron of momentum $\mathbf{p}$; in short, it causes an electron to scatter. The deflection of a charge is accompanied by the emission or absorption of light. Here these processes are represented by the two terms $a^\dagger b^\dagger b$ and $ab^\dagger b$, respectively, for they add or subtract a photon. The operators $a^\dagger dd^\dagger$ and $add^\dagger$ have precisely the same role for positrons: they describe the deflection of a positron due to emission or absorption of a photon. The last two terms of Eq. (7) describe creation and destruction of an electron-positron pair; $a^\dagger db$ causes pair-destruction accompanied by photon creation, while $ab^\dagger d^\dagger$ creates a pair and destroys a photon. This discussion, and especially Eq. (6), reveals that the theory has a remarkable feature: *complete symmetry between electrons and positrons*. In the language of Chap. I.C.5, quantum electrodynamics is invariant under charge conjugation.

(b) Helicity rules in ultrarelativistic phenomena

In general, the interactions between electrons, positrons, and photons have a complicated dependence on the various helicities. But in particle physics we are usually interested in ultrarelativistic (UR) motions, and in this kinematical regime striking simplifications arise. These may be summarized as follows: *In the ultrarelativistic (UR) domain, the helicity of any e or ē is conserved (in a scattering process), and when an eē pair is created or destroyed, the two particles always have opposite helicities*. This theorem is of importance in electromagnetic and weak interaction phenomena. It follows from two properties of the Dirac theory: in the UR domain, the $\chi^{(\pm)}$ portions of the Dirac field [cf. Eq. A(25)] destroy electrons with $h = \pm\tfrac{1}{2}$ and create positrons with $h = \mp\tfrac{1}{2}$ [recall the discussion following Eq. A(21)]; and secondly, the expression A(29) for the current $\mathbf{j}$ does not mix $\chi^{(+)}$ and $\chi^{(-)}$.

[To prove the statement in italics in the preceding paragraph one must first construct the UR limits of the spinors u and v that appear in the expression A(43) for the Dirac field operator. Quite generally,

$$u = \begin{pmatrix} \eta^{(+)} \\ \eta^{(-)} \end{pmatrix}, \quad v = \begin{pmatrix} \bar{\eta}^{(+)} \\ \bar{\eta}^{(-)} \end{pmatrix}.$$

Here $\eta^{(\pm)}$ is the coefficient of a plane wave solution $\chi^{(\pm)}$ of the Dirac equations, Eqs. A(22) and (23), for 'positive frequencies' (time-dependence e^{-iEt}), and $\bar\eta^{(\pm)}$ the coefficient of a plane wave with time-dependence e^{iEt}. But in the UR case, the Dirac equations reduce to the Weyl equations, and $\eta^{(\pm)}$ are the constant spinors defined by Eq. A(21). The Weyl equation $(E - \boldsymbol{\sigma}\cdot\mathbf{p})\eta^{(+)} = 0$ only has a solution for $h = \tfrac{1}{2}$, and the other Weyl equation,

1. THE INTERACTIONS BETWEEN FIELD AND SOURCE

$(E + \boldsymbol{\sigma} \cdot \mathbf{p})\eta^{(-)} = 0$, only for $h = -\frac{1}{2}$, where $E = |\mathbf{p}|$ throughout. The UR limits of the u-spinors are therefore

$$u(h = \tfrac{1}{2}) = \begin{pmatrix} \xi_{1/2} \\ 0 \end{pmatrix}, \qquad u(h = -\tfrac{1}{2}) = \begin{pmatrix} 0 \\ \xi_{-1/2} \end{pmatrix},$$

where $\xi_{\pm 1/2}$ is a Pauli spinor satisfying $\boldsymbol{\sigma} \cdot \hat{\mathbf{p}} \xi_{\pm 1/2} = \pm \xi_{\pm 1/2}$, and 0 stands for the null spinor. If $\mathbf{p}$ is along z,

$$\xi_{1/2} = \begin{pmatrix} 1 \\ 0 \end{pmatrix}, \qquad \xi_{-1/2} = \begin{pmatrix} 0 \\ 1 \end{pmatrix}.$$

For the negative energy solutions, $E \to -E$ in A(21), which reverses the helicity. Hence[3]

$$v(h = \tfrac{1}{2}) = \begin{pmatrix} 0 \\ \xi_{-1/2} \end{pmatrix}; \qquad v(h = -\tfrac{1}{2}) = \begin{pmatrix} \xi_{1/2} \\ 0 \end{pmatrix}.$$

Then the solutions of (22) and (23) are

$$\left.\begin{aligned} \chi^{(+)} &= Nb\xi_{1/2}e^{i\alpha} \\ \chi^{(-)} &= Nd^\dagger \xi_{-1/2}e^{-i\alpha} \end{aligned}\right\} h = \tfrac{1}{2}$$

$$\left.\begin{aligned} \chi^{(+)} &= Nd^\dagger \xi_{1/2}e^{-i\alpha} \\ \chi^{(-)} &= Nb\xi_{-1/2}e^{i\alpha} \end{aligned}\right\} h = -\tfrac{1}{2}$$

where $\alpha = (\mathbf{p} \cdot \mathbf{r} - Et)$ and $N = (2EV)^{-1/2}$.]

We have now established our claim that in the UR domain the field operator $\chi^{(\pm)}$ only destroys e with $h = \pm\frac{1}{2}$, and only creates $\bar{e}$ with $h = \mp\frac{1}{2}$. The preceding sentence applies to $\chi^{(\pm)\dagger}$, if the words "destroys" and "creates" are interchanged. Consequently, we have the following *helicity selection rules in the UR regime*:[4]

$$\chi^{(\pm)\dagger}\chi^{(\pm)} \begin{cases} \text{creates or destroys a pair with } h_e = \pm\tfrac{1}{2},\ h_{\bar{e}} = \mp\tfrac{1}{2} \\ \text{scatters } e \text{ with } h_e = \pm\tfrac{1}{2} \\ \text{scatters } \bar{e} \text{ with } h_{\bar{e}} = \mp\tfrac{1}{2} \end{cases} \text{ and conserves } h \qquad (9)$$

As there are no terms of the type $\chi^{(+)\dagger}\chi^{(-)}$ in the current operator, Eq. (9) takes care of all possible situations in the UR limit.

---

[3] We have purposely passed over a subtlety that is of no consequence for us here: $v_h(\mathbf{p})$ is the coefficient of $\exp[-i(\mathbf{p} \cdot \mathbf{r} - Et)]$ in the solution of the Dirac equation, and $\mathbf{p}$ must therefore be rotated by π about an axis normal to itself. As shown in Appendix III.1(d), this means that $\eta_\pm$ should be replaced by $i\sigma_y \eta_\pm$ in v.

[4] As we show in Appendix III [see its Eq. (28)] these statements have corrections of order m/E. For example, there is an amplitude of relative order m/E for creating a pair with equal helicity, a fact that we already used in discussing $\pi \to ev$ decay (see Vol. I, p. 131).

We also note that because of reflection invariance, the electromagnetic current contains the (+)(+) and (−)(−) terms symmetrically [cf. Eq. A(29)]. Therefore, transitions between states where *all* helicities are reversed have precisely the same amplitudes. As we know, this is not true of the weak interaction. As we shall learn in §VI.A.1, the charge-changing weak interaction only contains terms of the (−)(−) type, and as we see from Eq. (9), this accounts for the helicity rule enunciated on p. 130 of Vol. I.

(c) Vertices

The various terms in the interaction H_{int} can be depicted diagrammatically as vertices. These vertices form the basic building blocks of *Feynman diagrams* (or Feynman graphs). Consider first $a^\dagger(k)b^\dagger(p)b(p')$. When acting on a state that contains an electron of momentum $\mathbf{p}'$, it replaces that state with one containing an electron of momentum $\mathbf{p}$ and an additional photon of momentum $\mathbf{k}$. We depict this process by

$$\tag{10}$$

Here straight lines represent electrons, wavy lines photons. The state before the operator acts is represented by the lines below the vertex, and the state subsequent to the action of the operator by the lines above the vertex. One may, if one finds that helpful, think of the diagram as a picture of the process in time, with time running from the bottom to the top of the page.

In the term $a(k)b^\dagger(p)b(p')$ the photon is destroyed, so we represent this term by

$$\tag{10'}$$

which shows the electron absorbing the photon.

We now turn to terms in H_{int} involving positrons. These could be drawn as in (10) or (10'), with some other (e.g., dashed) lines to differentiate them from electrons. Consider, for example, the terms $a^\dagger(k)d^\dagger(p')d(p)$ and

1. THE INTERACTIONS BETWEEN FIELD AND SOURCE

$a(k)d^\dagger(p')d(p)$

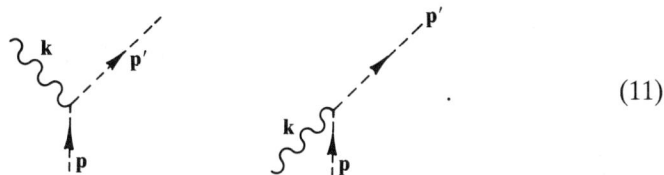
(11)

However, there is a far more economical way to do this: One replaces any positron line $- \to -$ by an electron line in the opposite direction. Thus (11) is replaced by

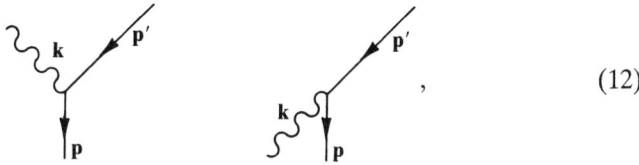
(12)

i.e., a positron transition from the initial to the final state is represented by an electron line running from the final to the initial state, or going "backward in time."

We shall not try to offer a complete explanation of how the mathematical formalism leads to this as the most economical rule for handling positrons, and confine ourselves to some brief observations. H_{int} is invariant under the substitution

$$b(p)e^{-iEt} \leftrightarrow d^\dagger(p)e^{iEt}, \qquad b^\dagger(p)e^{iEt} \leftrightarrow d(p)e^{-iEt},$$

where $E > 0$ throughout. Thus positron creation carries a time-dependence that is obtained from electron destruction by setting $t \to -t$. Hence the creation of a positron into the final state is on a symmetric footing with electron destruction in the initial state. The depiction of positrons as "backward running electrons" captures this symmetry of H_{int} in an elegant manner. One should not, however, attach any physical meaning to the "backward running" lines; they are merely a convenient symbolism. Positrons are not electrons running backward in time.

Having established how electrons and positrons are to be depicted, we now draw the diagrams for pair creation and annihilation:

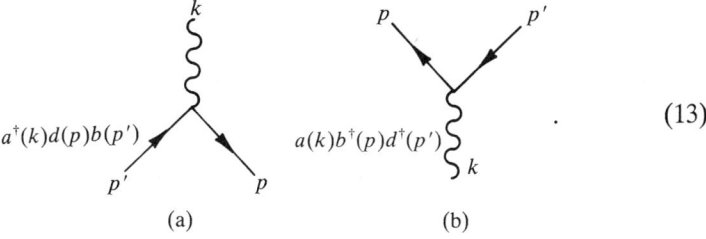
(13)

On examining the vertices (10)–(13), we observe that all these terms in H_{int} have one fermion line entering and one fermion line leaving the vertex, and one photon line attached to the vertex. If the fermion line leaving the vertex is in the final state, it represents an electron; if in the initial state, a positron. Conversely, if the fermion line entering the vertex is in the initial state, it represents an electron; if in the final state, a positron. The photon line need not carry an arrow because there is no distinction between photon and antiphoton.

The attentive reader may have noted that we have left out the diagrams corresponding to terms $a^\dagger d^\dagger b^\dagger$ and abd in (7). They are depicted by the two diagrams:

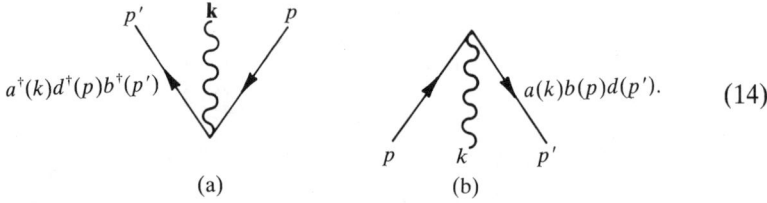

$$a^\dagger(k)d^\dagger(p)b^\dagger(p') \qquad\qquad a(k)b(p)d(p'). \qquad (14)$$

(a) \qquad\qquad (b)

Such operators allow one to create a pair plus a photon out of the vacuum, or conversely, to return to the vacuum from a state containing a pair plus a photon. These seemingly bizarre transitions also contribute to the processes that we shall soon consider. In fact, they are no more bizarre than any of the others.

The operator H_{int} is invariant under spatial translation, and for that reason it only can cause transitions that conserve the total momentum. Therefore, all parts of H_{int}, as described by diagrams (10) to (14), must conserve total momentum if the electrons and positrons are free. This was shown explicitly in the discussion following Eq. (6). (By "free" or "isolated" particles we mean particles that are *not* moving under the influence of externally applied fields, e.g., the Coulomb field of a nucleus, or the magnetic field in a synchrotron.) The energy, however, is not conserved. This is seen immediately for the "bizarre" processes (14), which do, however, conserve momentum. We quickly verify this in one other case, (13). In the frame where $\mathbf{p} + \mathbf{p}' = 0$, momentum conservation requires $\mathbf{k} = 0$. Since the photon has zero mass, the photon energy is also zero. But the energy of the pair is at least $2m$, where m is the electron mass, and so energy cannot be conserved. In a similar way, one can show that it is impossible for a free electron to absorb or emit a photon.

This illustrates a general result: in quantum electrodynamics there are no scattering processes of first order in H_{int} when the charged particles in the initial and/or final states are free, i.e., are in eigenstates of momentum. There is, however, a very important class of processes where the charged particles are not free. The most familiar example is that of the electron moving in the Coulomb field of a nucleus. Under this circumstance, the electron's states fall into two categories: bound states having a discrete energy spectrum, and scattering states having a continuous spectrum. None of these are momentum eigenstates: they are superpositions of states with various electron momenta. There can then be transitions from one state $|i\rangle$ to another state $|f\rangle$ accompanied by the emission or absorption of a single photon. If $|i\rangle$ and $|f\rangle$ both belong to the discrete spectrum, this gives the emission and absorption lines of atomic spectroscopy. The photoelectric effect is the case where $|i\rangle$ is a bound state and $|f\rangle$ a continuum state. The case where $|i\rangle$ and $|f\rangle$ are both continuum states is also a common occurrence: when an electron is deflected by a nucleus it loses energy by radiation, as in an X-ray machine. This process therefore has the German name Bremsstrahlung, which means braking radiation. Electrons bent into circular orbits by synchrotron magnets also lose energy by photon emission.

How is it that momentum and energy conservation prevent the emission or absorption of single photons by free electrons, but allow these processes for electrons that are not free? Of course, total momentum is conserved in all processes, but the momentum balance includes the other body (or bodies) with which the electron interacts. In a Bremsstrahlung process, for example, the nucleus with which the electron interacts recoils and thereby supplies the energy and momentum that would be missing if a free electron were to "try" to radiate the photon.

2. The scattering amplitude

As we have seen, to first order in H_{int} there are no processes between free particles in quantum electrodynamics. So we must go to higher orders. The general expression for the transition probability per unit time (i.e., the transition rate) from an initial state $|i\rangle$ to a final state $|f\rangle$ is given by[5]

$$\Gamma_{fi} = 2\pi\, \delta(E_f - E_i)\, |\langle f|\, T\, |i\rangle|^2, \qquad (15)$$

where $\langle f|\, T\, |i\rangle$ is called the scattering amplitude. The evaluation of T will occupy us for some time. The cross section σ_{fi} for the process $|i\rangle \to |f\rangle$ is, by definition,

$$\sigma_{fi} = \frac{\Gamma_{fi}}{\text{Incident flux}}. \qquad (16)$$

Incident flux is the number of beam particles crossing a unit area per unit time.

It is important to realize that $\langle f|\, T\, |i\rangle$ is Lorentz invariant. This is seen as follows: Since $\int dE\, \delta(E) = 1$, $\delta(E_f - E_i)$ is an inverse energy, and transforms as a reciprocal time, because E and t have the same behavior under Lorentz transformations. But Γ_{fi} is also a reciprocal time, and so the factor $\langle f|\, T\, |i\rangle$ in Eq. (15) must be an invariant; the same is true of σ_{fi}.

---

[5] See, for example, Gottfried (1966), §56, and Messiah (1966), Chap. XIX.

Standard perturbation theory[6] shows that

$$\langle f|T|i\rangle = \langle f|H_{\text{int}}|i\rangle + \sum_n \frac{\langle f|H_{\text{int}}|n\rangle\langle n|H_{\text{int}}|i\rangle}{E - E_n}$$
$$+ \sum_{nm} \frac{\langle f|H_{\text{int}}|n\rangle\langle n|H_{\text{int}}|m\rangle\langle m|H_{\text{int}}|i\rangle}{(E - E_n)(E - E_m)} + \cdots. \quad (17)$$

Here $|n\rangle$, $|m\rangle$, $\cdots$ are eigenstates of the unperturbed Hamiltonian, with eigenvalues E_n, E_m, ..., and $E = E_i = E_f$. In our case $|n\rangle$, etc., are states having some number of free photons, electrons, and positrons, and

$$E_n = \sum_i |\mathbf{k}_i| + \sum_j \sqrt{\mathbf{p}_j^2 + m^2}, \quad (18)$$

where $\mathbf{k}_i$ and $\mathbf{p}_j$ are the momenta of photons and electrons and/or positrons in $|n\rangle$.

On the face of it, the expansion (17) converges rapidly, because each operator H_{int} carries a factor of e, the electronic charge [cf. Eq. (1)]. In our units

$$\frac{e^2}{4\pi} = \alpha = \frac{1}{137.036}, \quad (19)$$

or $e = 0.303$. While this does not look so small, it turns out that all quantities of physical interest involve power series in α, not e.

The different terms in Eq. (17) decompose the transition $i \to f$ into sequences of steps; for example, the second term corresponds to $i \to n \to f$, the third term to $i \to m \to n \to f$, etc. We have already seen that the first term of Eq. (17) vanishes in quantum electrodynamics if $|i\rangle$ and $|f\rangle$ contain free particles, and we therefore focus on the second-order term

$$\langle f|T_2|i\rangle = \sum_n \frac{\langle f|H_{\text{int}}|n\rangle\langle n|H_{\text{int}}|i\rangle}{E - E_n}. \quad (20)$$

As we shall now see, many familiar processes are described by this formula.

Before turning to these applications, we comment on the interpretation of the so-called "intermediate states" $|n\rangle$ in Eq. (17). For this purpose it is best to think of the collision $|i\rangle \to |f\rangle$ as a process in time: for $t \ll 0$, the system is in the state $|i\rangle$; for $t \gg 0$, Eq. (17) gives the probability amplitude for finding it in $|f\rangle$. During the actual collision ($t \approx 0$) the interaction H_{int} perturbs the initial state $|i\rangle$. This change of $|i\rangle$ can be described by a linear superposition of a complete set $|n\rangle$. As the intuitive picture correctly indicates, these intermediate states $|n\rangle$ only have a transitory presence in the state vector, and for that reason they are often called *virtual states*. A

---

[6] See Gottfried (1966), p. 438.

time τ_n characterizes the duration of this transitory presence; because of the uncertainty principle, it is related to the energy difference between $|n\rangle$ and $|i\rangle$ by $\tau_n \sim (E - E_n)^{-1}$, where $E = E_i(= E_f)$. Observe that $1/\tau_n$ is just the energy denominator in Eq. (17). Therefore, the intermediate states closest in energy to the initial state have the longest transitory presence, and tend to dominate the sum of Eq. (17).

3. Compton scattering

(a) Diagrams for Compton scattering

In classical physics the scattering of light by a charge—*Thomson scattering*—is described as follows. An electromagnetic wave impinges on a stationary charge; the **E**-field of the wave exerts a force on the charge; the charge moves in reaction to that force; as a result of the charge's acceleration, it radiates. This radiation is what we call *scattered light*. The source-field coupling acts twice: first, when the incident field moves the charge, and a second time when the accelerated charge radiates. As a consequence, the amplitude of the scattered light is proportional to the square of the charge, and the energy in the scattered wave is proportional to its fourth power.

There is a rather close parallel between the classical and quantum theories of photon-electron scattering (*Compton scattering*). We consider an initial state $|i\rangle \equiv |\mathbf{k}, \mathbf{p}\rangle$, having a photon and electron of momenta $\mathbf{k}$ and $\mathbf{p}$, respectively, and ask for the probability of finding a state $|f\rangle \equiv |\mathbf{k}', \mathbf{p}'\rangle$, having a photon and electron with the indicated momenta long after the collision. The scattering amplitude $\langle f| T_2 |i\rangle$ for this process is given by Eq. (20), where H_{int} also acts twice.

We must now enumerate the "intermediate" states $|n\rangle$ that contribute to the sum in Eq. (20). This can be done by merely "fitting" together some of the diagrams already introduced. Our initial state is given by one photon and one electron

 (21)

The action of H_{int} on this state is then found by either applying the operator H_{int} to $|i\rangle$, or equivalently, by allowing (21) to flow into the vertices of the preceding section in all possible ways:

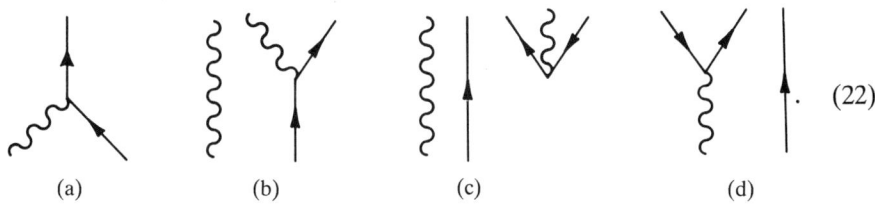 (22)

(a)   (b)   (c)   (d)

Hence the intermediate states $|n\rangle$ are just those shown as flowing upward in (22). Equation (20) instructs us to also apply H_{int} to $|n\rangle$, and since we are concerned with elastic scattering the final state again looks like (21). But now recall that the action of H_{int} on any state is one of the processes depicted in diagrams (10)–(14), and *no* others. Hence for each of the intermediate states in (22), there is a unique way of reaching the desired final state:

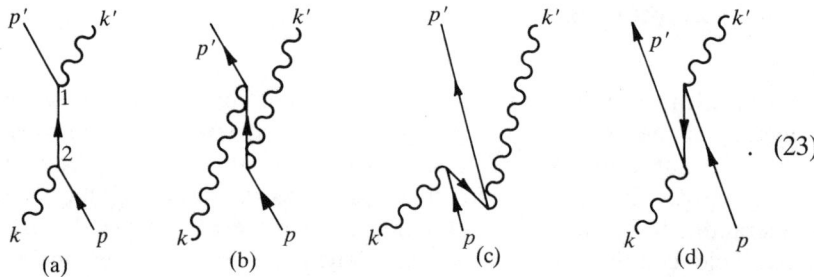

(23)

Diagrams (23a, b, c, d) are obtained from (22a, b, c, d) by joining a term in H_{int} depicted in (10), (10'), (14b), and (13a), respectively. In constructing the diagrams in (23) all terms in H_{int} were used at least once, *including* the "bizarre" terms (14).

Diagram (23a) has precisely the structure that one would expect from Thomson scattering: First the electron absorbs the photon, and subsequently radiates another photon. The other parts in (23) do not have so simple a description. In (23b) the electron first changes its momentum by emitting the final photon k', and changes it again by absorbing the initial photon k. In (23c) the situation is rather odd: The vacuum produces an electron-positron pair, together with the final photon [process (14a)], the electron having the final momentum p'; then the positron annihilates the initial electron and simultaneously absorbs the initial photon k [process (14b)]. In (23d) the initial photon k produces a pair wherein the electron has the final momentum p' [process (13b)], then the positron annihilates with the initial electron with the emission of the final photon k' [process (13a)]. The four diagrams in (23) contain all possible ways of combining the "first-order" diagrams when the initial and final states contain one electron and one photon.

It should be understood that to each diagram in (23) there corresponds a contribution to the sum (20). These terms must be added to form the full scattering amplitude T_2, and it is T_2 itself that is squared to form Γ_{fi}. Therefore, the transition probability is not the sum of probabilities for the various processes in (23).

(b) A spin-zero model

The complete evaluation of the mathematical expressions that correspond to the diagrams in (23) is rather complicated because of the electron and photon helicities. For the sake of clarity, we therefore consider a rather simple analogue of Compton scattering that has some of the most essential features of an electrodynamic process, but avoids the intricacies due to spin. Once this is understood, we shall return to photon-electron scattering.

We consider a fictitious world consisting of three kinds of particles, *all* of which are spin zero bosons: An uncharged particle α, and another charged particle β and its antiparticle $\bar{\beta}$. The analogy we have in mind is $\alpha \leftrightarrow$ photon, $\beta \leftrightarrow$ electron. As shown in Appendix II, Eq. (28), the field operator that creates α's is

$$A(\mathbf{r}t) = \sum_k \frac{1}{\sqrt{2\omega_k V}} [a_k e^{i(\mathbf{k}\cdot\mathbf{r}-\omega_k t)} + a_k^\dagger e^{-i(\mathbf{k}\cdot\mathbf{r}-\omega_k t)}]. \tag{24}$$

Like the vector potential, it is Hermitian, but there are no polarization vectors because there is no spin. The analogue of the Dirac field is the charged scalar field of Appendix II, Eq. (38):

$$\Phi(\mathbf{r}t) = \sum_p \frac{1}{\sqrt{2E_p V}} [b_p e^{i(\mathbf{p}\cdot\mathbf{r}-E_p t)} + d_p^\dagger e^{-i(\mathbf{p}\cdot\mathbf{r}-E_p t)}]. \tag{25}$$

Like the Dirac field, it is not Hermitian, but it contains no spinors u and v. The masses of α and β are μ and m, respectively, so that $\omega_k = \sqrt{k^2 + \mu^2}$, $E_p = \sqrt{p^2 + m^2}$. Of course, when we identify α with the photon, we will put $\mu = 0$.

The analogue of the electromagnetic interaction, Eq. (1), is

$$H_1 = g \int \Phi^\dagger(\mathbf{r}t)\Phi(\mathbf{r}t)A(\mathbf{r}t)\, d^3r, \tag{26}$$

because **j** is of the form $\Phi^\dagger \Phi$. We shall call this the *scalar model*.

There is one difference between Eq. (26) and the electromagnetic interactions that appears to be quite innocent at first sight: While e is dimensionless, g is not. That this is so follows from Eqs. (24) and (25), according to which Φ and A have the dimension of a mass when $\hbar = c = 1$, because the volume V has the dimension m^{-3}. Since the element of volume has the same dimension, the integral in (26) is dimensionless, and g must have the dimension of mass because H_1 is an energy. As we will see in §5, the dimension of the coupling constant is very significant in quantum field theory.

We can now turn to the exercise of interest, the calculation of the scattering amplitude of the analogue process to Compton scattering, $\alpha\beta \to \alpha\beta$. As before, the leading contribution to the amplitude occurs in second order, and has four terms that can be represented by the four diagrams of (23). Because of the simplicity of this scalar model, their evaluation is straightforward. The contribution of diagram (23a) is

$$T_a^0 = \frac{\langle \mathbf{k}', \mathbf{p}' | H_1 | \mathbf{k} + \mathbf{p} \rangle \langle \mathbf{k} + \mathbf{p} | H_1 | \mathbf{k}, \mathbf{p} \rangle}{\omega_k + E_p - E_{k+p}}. \qquad (27)$$

Here $|\mathbf{k}, \mathbf{p}\rangle$ is the state of a particle α with momentum $\mathbf{k}$ and a particle β with momentum $\mathbf{p}$, whereas $|\mathbf{p} + \mathbf{k}\rangle$ denotes the intermediate state of only one particle, β, with momentum $\mathbf{p} + \mathbf{k}$. The substitution of (24) and (25) into (26) shows that the operator H_1 consists of a sum of products of three destruction or creation operators. A characteristic term of H_1 is

$$\frac{V}{\sqrt{2\omega_k V}\sqrt{2E_{p'} V}\sqrt{2E_{p} V}} a_k b_{p'}^\dagger b_p, \qquad (28)$$

where V in the numerator comes from the spatial integration of the exponentials, which also guarantees that $\mathbf{p}' = \mathbf{p} + \mathbf{k}$ (momentum conservation). The matrix element $\langle \mathbf{k}', \mathbf{p}' | a_k b_{p'}^\dagger b_p | \mathbf{k} + \mathbf{p} \rangle$ is unity. It then follows that the amplitude (27) becomes

$$T_a^0 = g^2 \left\{ \frac{V^2}{\sqrt{2VE_p 2V\omega_k (2VE_{k+p})^2 2VE_{p'} 2V\omega_{k'}}} \right\} \frac{1}{E_p + \omega_k - E_{k+p}}. \qquad (29)$$

The amplitude for graph (23c), which was described on p. 222, is

$$T_c^0 = \frac{\langle \Omega | H_1 | \mathbf{k}', \mathbf{p}', -\mathbf{k}' - \mathbf{p}' \rangle \langle \mathbf{k}, \mathbf{p}, -\mathbf{k} - \mathbf{p} | H_1 | \Omega \rangle}{\omega_k + E_p - [\omega_{k'} + E_p + \omega_{k'} + E_{p'} + E_{k+p}]}, \qquad (30)$$

where the momenta $\mathbf{k}'$, $\mathbf{p}'$, $-(\mathbf{k}' + \mathbf{p}')$, in the intermediate state, are those of α, β and $\bar{\beta}$, respectively, while the expression in [ ] is the energy of the intermediate state. The evaluation of the matrix elements is entirely similar, and yields

$$T_c^0 = g^2 \{ \quad \} \frac{1}{-\omega_{k'} - E_{p'} - E_{k+p}}, \qquad (31)$$

where $\{ \ \}$ is the expression in braces in (29), because $\mathbf{k} + \mathbf{p} = \mathbf{k}' + \mathbf{p}'$.

Now we combine (29) and (31) and use energy conservation:

$$T_{a+c}^0 = g^2 \frac{1}{V} \frac{1}{\sqrt{2E_p 2\omega_k 2E_{p'} 2\omega_{k'}}} \frac{1}{(E_p + \omega_k)^2 - E_{k+p}^2}. \qquad (32)$$

The total energy and momentum of the collision are $E_p + \omega_k$ and $\mathbf{k} + \mathbf{p}$, and constitute the 4-vector P. Furthermore, $P^2 = (E_p + \omega_k)^2 - (\mathbf{k} + \mathbf{p})^2$, while $(\mathbf{k} + \mathbf{p})^2 + m^2 = E_{k+p}^2$. Thus (32) becomes

$$T_{a+c}^0 = g^2 \frac{1}{V} \frac{1}{\sqrt{2E_p 2\omega_k 2E_{p'} 2\omega_{k'}}} \frac{1}{P^2 - m^2}. \tag{33}$$

This is an important result, for it typifies what always occurs when one combines a judiciously chosen set of terms in the perturbation expansion (20): the complicated energy denominators, such as those of (29) and (30), combine to a Lorentz-invariant expression that can be written down by inspection, because it is

$$\frac{1}{(\text{4-momentum of internal line})^2 - (\text{mass of internal line})^2}. \tag{34}$$

Here "internal line" means one that does not represent particles in the initial or final states. The "internal" 4-momentum is found by 4-momentum conservation at either of its terminating vertices; because of overall conservation, both vertices give the same internal 4-momentum.

The other diagrams in (23), namely (b) and (d), combine to the following expression

$$T_{b+d}^0 = g^2 \frac{1}{V} \frac{1}{\sqrt{2E_p 2\omega_k 2E_{p'} 2\omega_{k'}}} \frac{1}{K^2 - m^2}, \tag{35}$$

where $K = (E_p - \omega_{k'}, \mathbf{p} - \mathbf{k'})$ is the 4-momentum of the corresponding internal line ($K = p - k'$).

(c) Compton amplitudes and the Klein–Nishina formula

Now we can return to Compton scattering itself, i.e., to the evaluation of the terms in the perturbation expansion depicted by (23), but using the electromagnetic interaction (1). We will designate the amplitude by the superscript $e\gamma$. One finds that

$$T_{a+c}^{e\gamma} = (e/g)^2 T_{a+c}^0 \Xi_{a+c} \tag{36}$$

$$T_{b+d}^{e\gamma} = (e/g)^2 T_{b+d}^0 \Xi_{b+d} \tag{37}$$

This shows that the basic structure of the amplitudes in QED and our scalar model are closely related. The factor $(e/g)^2$ is trivial; it merely says that now the coupling constant is e, not g. The Ξ's contain all the helicity information:

$$\Xi_{a+c} = \bar{u}_f(\varepsilon_f^* \cdot \gamma)(\slashed{P} + m)(\varepsilon_i \cdot \gamma) u_i \tag{38}$$

$$\Xi_{c+d} = \bar{u}_f(\varepsilon_i^* \cdot \gamma)(\slashed{K} + m)(\varepsilon_f \cdot \gamma) u_i. \tag{39}$$

Here u_f and u_i are the Dirac spinors of the scattered and incident electrons, defined as in Eq. A(42); ε_f and ε_i are the polarization vectors of the corresponding photons; the components of γ are the Dirac matrices defined in Eq. A(31), $\not{P}$ is defined as $P_0\gamma_0 - \mathbf{P}\cdot\boldsymbol{\gamma}$ and so is $\not{K}$; $\bar{u} = u^*\gamma_0$, as in Eq. A(33); and the notation $\bar{u}\ldots u$ has the same meaning as in Eq. A(35). We do not provide a derivation of Eqs. (38) and (39), and refer the reader to Bjorken (1964), §7.7, and Berestetskii (1971), §86.

As e is dimensionless, whereas g has the dimension of energy, the Ξ's must have the dimension (energy)^2. This is indeed so, because our u's are normalized to $\bar{u}u = 2m$ [see Appendix III, Eq. (44)], and there is an explicit factor of energy in Eqs. (38) and (39).

Our next task is the evaluation of the Compton cross sections by use of Eqs. (15) and (16). To evaluate Γ_{fi}, we must take the absolute square of the sum, $|T^{e\gamma}_{a+c} + T^{e\gamma}_{b+d}|^2$, and divide by the incident flux. This procedure is described at the very end of this section. In most experiments (though certainly not all), one has an unpolarized photon beam, an electron target that has random spin orientation, and measures no helicities of particles in the final state. In this case, the measured cross section is found from the one where all helicities are specified by an average over initial helicities and a sum over final helicities. The algebraic technique needed for this averaging is quite intricate unless the momenta are large, and will not be presented here (see Bjorken, and Berestetskii, both loc. cit.).

The final result is the Klein–Nishina formula for the Compton cross section. After intergrating over all angles, one finds that the total cross section is[7]

$$\sigma_{e\gamma} = \frac{\pi\alpha^2}{km_e}\left[\ln\left(\frac{2k}{m_e}\right) + \frac{1}{2}\right], \tag{40}$$

when the photon energy is "high" (i.e., $k \gg m_e$).

(d) Evaluation of cross sections

[Once the amplitude T is known, one must still evaluate the cross section. To illustrate how this is done, we consider the simple problem of elastic $\alpha\beta$ scattering with our pedagogic scalar model. Let $(\boldsymbol{k}, \boldsymbol{p})$ and $(\boldsymbol{k}', \boldsymbol{p}')$ be the initial and final 4-momenta, with $k^2 = \mu^2 = k'^2$, $p^2 = m^2 = p'^2$. The initial 4-momenta are fixed, but the final ones are spread over an interval when we are interested in the scattering into a certain solid angle element, or the total cross section. We work in the c.o.m., and evaluate the total transition rate, which is a sum of transition rates into the final states of interest. It is not necessary to sum over $\boldsymbol{p}'$ and $\boldsymbol{k}'$ separately, since $\boldsymbol{k}' = -\boldsymbol{p}'$. According to Eq. (15),

$$\Gamma = 2\pi\sum_{k'}\langle|T|^2\rangle\,\delta(E_0 - E' - \omega'),$$

---

[7] This is only the high-energy limit in lowest order perturbation theory. At extremely high energies, when $\alpha\ln(k/m_e)$ is of order unity, higher order terms dominate, and the cross section has the behavior $\sim[\ln(k/m_e)]^2$.

3. COMPTON SCATTERING

where $E' = \sqrt{k'^2 + m^2}$, $\omega' = \sqrt{k'^2 + \mu^2}$, and E_0 is the total c.o.m. energy. If the particles have spins, which are undetected in the final state, and randomly oriented in the initial state, one must sum over the former and average over the latter; this is indicated by the notation $\langle \ldots \rangle$. The sum is converted to an integral by means of $\Sigma_{k'} = V(2\pi)^{-3} \int k'^2 \, dk' \, d\Omega$, where $d\Omega$ is the element of solid angle. The energy δ-function is then integrated over by use of $k' \, dk' = E' \, dE' = [E'\omega'/(E' + \omega')] d(E' + \omega')$. This results in[8]

$$\Gamma = \frac{V}{(2\pi)^2} \frac{k'}{E_0} E'\omega' \int d\Omega \, \langle |T|^2 \rangle \tag{41}$$

The last step is to evaluate the incident flux F [cf. (16)]. The two particles α and β are spread over the volume V. If v_1 and v_2 are the initial velocities, $F = (v_1 - v_2)/V$, or

$$F = \frac{1}{V}\left(\frac{p}{E} - \frac{k}{\omega}\right) = \frac{k}{V}\left(\frac{1}{E} + \frac{1}{\omega}\right),$$

when we are in the c.o.m. frame, where $\mathbf{p} = -\mathbf{k}$. In evaluating (16) we also use $E = E'$, $\omega = \omega'$, $k = k'$, and get the formula for the cross section

$$\frac{d\sigma}{d\Omega} = \frac{V^2}{(2\pi)^2}\left(\frac{E\omega}{E_0}\right)^2 \langle |T|^2 \rangle . \tag{42}$$

The cross section for elastic $\alpha\beta$ scattering is obtained by substituting (33) and (35) into (42):

$$\frac{d\sigma_0}{d\Omega} = \left(\frac{g^2}{4\pi}\right)^2 \frac{1}{4P^2} \left| \frac{1}{P^2 - m^2} + \frac{1}{K^2 - m^2} \right|^2. \tag{43}$$

We want to understand the dependence of this cross section on the scattering angle and the incident energy when the latter is high, i.e., $k^2 \gg m^2$. To have as close an analogy to Compton scattering as possible, we set the mass μ of the uncharged particle α to zero. In the c.o.m., $P = (2k, 0)$ and $K^2 = p^2 + k'^2 - 2p \cdot k' = m^2 - 2(p_0 k - \mathbf{p} \cdot \mathbf{k}')$, where $\mathbf{p} \cdot \mathbf{k}' = -k^2 \cos\theta$, θ being the c.o.m. scattering angle (see Fig. 1, p. 234). When $k \gg m$, $p_0 = \sqrt{k^2 + m^2} \simeq k + (m^2/2k)$, and therefore $K^2 - m^2 \simeq -m^2 - 2k^2(1 + \cos\theta)$, which is $-m^2$ for backward scattering ($\theta = 180°$). The first denominator in (43) is $k^2 - m^2 \simeq k^2$ for all angles. When $k^2 \gg m^2$ the dominant contribution to the cross section therefore comes from the second term in the backward scattering region. This statement also holds for high-energy Compton scattering itself: the amplitude $T^{e\gamma}_{b+d}$ dominates the cross section, which is sharply peaked towards $\theta = 180°$.

Taking advantage of what we have learned, we can approximate (43) by

$$\frac{d\sigma_0}{d\Omega} = \left(\frac{g^2}{4\pi}\right)^2 \frac{1}{64k^6}\left(\frac{1}{1 + \cos\theta + (m^2/2k^2)}\right)^2,$$

[8] Upon setting $E_0 = M$, this formula gives the total rate (or *width*) for decay of a particle of mass M into the objects α and β.

which yields the high-energy total cross section

$$\sigma_0 \simeq \left(\frac{g^2}{4\pi}\right)^2 \frac{\pi}{32m^2k^4}, \qquad (k^2 \gg m^2). \tag{44}$$

This expression has the dimension of an area because the coupling constant g in the scalar model has the dimension of a mass. One should note that the cross section for scattering of a scalar field quantum by a spin-zero falls off much more rapidly at high energy than does the Klein–Nishina formula, Eq. (40), which describes the scattering of a vector field quantum by a spin-$\frac{1}{2}$ particle. As we shall see in §5, the dimensionality of the coupling constant has a crucial impact on the high-energy behavior of amplitudes.]

4. Feynman diagrams

Let us take stock of what we have learned in the last section: Two diagrams, (23a) and (23c), which are related to each other by a displacement of the vertices, have amplitudes that combine to form a simple Lorentz invariant expression, Eq. (33). This is a general result, and allows one to define a far more compact and efficient bookkeeping method for the perturbation expansion.

The Feynman technique differs from the one we have just described in two intimately related ways: The Feynman diagrams have a different meaning from the "old-fashioned" diagrams drawn thus far, and the mathematical expression that corresponds to a Feynman diagram is constructed from a simpler rule.

In order to combine "old-fashioned" diagrams into Feynman diagrams we introduce the notion of "topologically distinct" diagrams.

A single Feynman diagram with n vertices corresponds to several (sometimes $n!$) "old-fashioned" graphs; this shortens calculations greatly. We must specify a rule that tells us which "old-fashioned" graphs belong to one Feynman graph. For this purpose, imagine that all the lines in a diagram [say (23a)] are rubber bands connected at the vertices, and that the external ends of the incoming and outgoing lines are nailed down. Two diagrams that can be transformed into each other by moving the vertices, while leaving the nails untouched, are said to be topologically equivalent; two diagrams that cannot be related in this manner are *topologically distinct*.

In the Feynman expansion, one only considers topologically distinct graphs. Thus the Feynman graphs for Compton scattering are

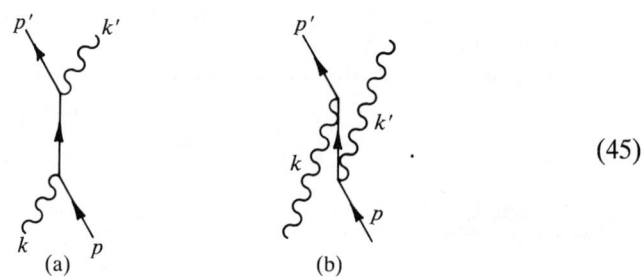

(45)

Feynman graph (45a) contains "old-fashioned" graphs (23a) and (23c), while (45b) contains (23b) and (23d). They differ by the fact that in (45a) the incoming photon and incoming fermion meet at the same vertex, whereas in (45b) the outgoing photon and incoming fermion meet at the same vertex.

In the "old-fashioned" expansion the contribution of a diagram to the amplitude has an energy denominator to which all particles in the intermediate state contribute. We found out that the contribution of Feynman graphs (45a) and (45b) are given by (32) and (33), respectively, and noted the remarkable fact that the energy denominators in the "old-fashioned" graphs combined to give a factor that can be completely expressed in terms of the 4-momentum and mass of the internal line [cf. Eq. (34)].

With this experience behind us, it should be easy to accept *the Feynman rules* for constructing the *lowest order* approximation to any amplitude:

(i) Draw the photon and fermion lines for the initial and final states $|i\rangle$ and $|f\rangle$ at the bottom and top of the paper.

(ii) Connect $|i\rangle$ to $|f\rangle$ with the minimum number of vertices in all topologically distinct ways; consider only connected diagrams that cannot be separated into two (or more) parts without cutting one (or more) lines.

(iii) Using 4-momentum conservation at each vertex, assign a 4-momentum to each internal line.[9]

(iv) The contribution $T(G)$ to the amplitude due to a graph G with n vertices is given by

$$T(G) = e^n \Phi(G) \prod_{j=0}^{n-1} \frac{1}{Q_j^2 - m_j^2}, \qquad (46)$$

where j labels the internal lines, and Q_j and m_j are the 4-momentum and mass of the jth internal line. In the case of $n = 2$ (second-order processes) there is only one internal line in each diagram.

(v) The factor $\Phi(G)$ contains all information about the helicities of the colliding photons and leptons, as exemplified by the explicit expressions (38) and (39). Since T has the dimension of energy, and e is dimensionless, Φ has the dimension[10] (energy)$^{2n-1}$.

(vi) The full amplitude is

$$T = \sum_G T(G), \qquad (47)$$

a sum over all topologically distinct graphs.

---

[9] Connected diagrams having the minimal number of vertices have no closed loops, and as a consequence momentum conservation determines the momenta of all internal lines uniquely (note the analogy with electrical circuit theory).

[10] In a fictitious world having only spin zero particles, coupled as in Eq. (26), the Feynman rules would be given by (i) through (vi), where Φ has dimension (energy)$^{n-1}$. In QED Φ has an angular dependence arising from the spins of the photons and fermions, whereas in the scalar model Φ depends only on the particle energies, and *not* on the angles between their momenta.

In our development of the theory, the factor $(Q_j^2 - m_j^2)^{-1}$ associated with each internal line arose from the energy denominators of conventional perturbation theory. As discussed in §3, the size of these denominators is given by the difference in energy between the initial (or final) state and the transitory intermediate states. The expressions $(Q_j^2 - m_j^2)^{-1}$ that replace the energy denominators in the Feynman method give an equivalent, though slightly different, characterization of the transitory nature of the intermediate states. If a particle of mass m_j is in a stationary (i.e., nontransitory) eigenstate of momentum, the time and space components of its 4-momentum Q_j satisfy $(Q_{j0})^2 - (\mathbf{Q}_j)^2 \equiv Q_j^2 = m_j^2$. The expression $Q_j^2 - m_j^2$ therefore gives a Lorentz-invariant characterization of the difference between the transitory state and a stationary free-particle state. This is often stated by saying that the latter state has a momentum 4-vector lying "on the mass shell," i.e., $Q_j^2 = m_j^2$, whereas an internal line representing the transitory intermediate state has a momentum "off the mass shell," i.e., $Q_j^2 \neq m_j^2$. As in old-fashioned perturbation theory, the probability that a certain internal line will be a pathway from the initial to the final state increases as its distance from the mass shell decreases. Roughly speaking, therefore, the dominant diagrams are those whose internal lines are closest to their mass shell.

It is also common to refer to the functions $(Q^2 - m^2)^{-1}$ in Eq. (46) as *propagators*. This name is appropriate because the 4-dimensional Fourier transform

$$\Delta(x - x') = \int e^{iQ \cdot (x-x')} \frac{1}{Q^2 - m^2} \frac{d^4 x'}{(2\pi)^4} \tag{48}$$

describes how a particle of mass m that is created at the space-time point x' propagates to the point x. To verify this we apply the Klein–Gordon operator $(\partial^2/\partial t^2 - \nabla^2 + m^2)$ to Δ; the factor $e^{iQ \cdot x}$ in Eq. (48) then yields $m^2 - Q^2$, canceling the denominator, and leaving the Fourier representation of the 4-dimensional δ-function. In short

$$\left(\frac{\partial^2}{\partial t^2} - \nabla^2 + m^2\right) \Delta(x - x') = -\delta^4(x - x'). \tag{49}$$

Thus $\Delta(x - x')$ is the solution of the Klein–Gordon equation at the point x when a unit point source exists momentarily at x'.

5. The dimensionality of the coupling constant and the high-energy behavior of amplitudes

[In §3 we already remarked that the coupling constant g of the scalar model, whose interaction is given by Eq. (26), has the dimension of energy, in contrast to the dimensionless coupling constant e of QED. One can glimpse the significance of this by comparing the total Compton cross section σ_{ey} with that of the scalar model, σ_0. The former has the high-energy behavior $k^{-1} \ln k$, the latter k^{-4} [see Eqs. (40) and (44)]. Both have the dimension (energy)$^{-2}$, since they are areas. In the scalar case the large negative power of k compensates the factor (energy)^4 of g^4.

To analyze this further, we define the *dimension* of any quantity Q, dim Q, by
$$Q \propto (\text{energy})^{\dim Q} \tag{50}$$

Thus dim $T = 1$, as we see from (17). Let p be the dimension of the coupling constant; $p = 0$ for QED, and $p = 1$ for the scalar model of §3. For the purpose of this discussion we also contemplate theories with other values of p.

As we improve the expression for T by using the perturbation expansion (17), we introduce higher powers of g. Since dim $T = 1$, for every term in the perturbation series, there must be factors of energies and/or masses to compensate for this. At high energies, masses tend to be irrelevant, so the dimensionality p of the coupling determines the energy dependence. If, as in the scalar model, $p > 0$, higher orders will involve ever higher negative powers of energy, and therefore be negligible in the high-energy limit. Because the lowest order term in the cross section is $0(g^4)$, it itself involves a negative power of energy, and in such a theory all scattering cross sections are small in the high-energy limit, no matter what value g has.

If p is negative, the opposite happens: the powers of g imply *positive* powers of energy. Higher order corrections dominate at high energies, and cross sections grow without bound. This, once again, is true for any value of g. But cross sections cannot grow forever, because they are constrained by probability conservation. In short, theories with coupling constants that have a negative dimension are of questionable validity. Indeed, the Fermi theory of weak interactions [see §VI.A.3] has this character: its coupling constant has $p = -2$. The electroweak gauge theory, to be discussed in Chap. VI, is designed to cure this disease.

Quantum electrodynamics, with $p = 0$, is on the border between theories that have vanishing and growing cross sections at high energies.

There is an intimate relation between the high-energy behavior of scattering amplitudes, and the nature of self-energy effects. We discuss those effects in some detail in §D, so confine ourselves to a few remarks here. The self-energy is due to the interaction of a charge with its own field; the magnitude of the self-energy depends on the geometric size a that one ascribes to the source, and, in general, it diverges as $a \to 0$. The self-energy is therefore a short distance, or equivalently, a high momentum phenomenon. For that reason, a theory with scattering amplitudes that grow with energy has self-energies that diverge violently as $a \to 0$. Once again, QED is on the borderline: the electromagnetic self-energy behaves similarly to $\ln a$ as $a \to 0$, and this is sufficiently benign to permit a redefinition of the theory, called *renormalization*, that removes all divergences from every observable quantity. Theories with $p < 0$ have self-energies that cannot be removed by renormalization, and one does not know how to construct meaningful predictions from such theories.[11]]

---

[11] Theories with $p > 0$, like the scalar model, have finite self-energies, but they are only of academic interest in a world composed of quarks and leptons. The electromagnetic interaction of charged spin zero particles is actually more singular as $a \to 0$ than conventional QED with spin-$\frac{1}{2}$ sources, but the former is still renormalizable.

C. EXAMPLES OF ELECTROMAGNETIC PROCESSES

1. *The scattering of two different fermions*

The Feynman rules make it possible to calculate a large number of processes involving photons, electrons, muons, and their antiparticles. In this and in the following sections we select a few that either illustrate important interaction mechanisms, or provide good tests of quantum electrodynamics. As our first process, we choose the relativistic version of *Rutherford scattering*: the scattering of two different charged fermions. Examples are electron-muon, electron-quark, and muon-quark scattering; the latter underlie the scattering of charged leptons by hadrons (Chap. V).

It is important to realize that different particles of opposite charge cannot annihilate one another, or equivalently, one particle cannot emit a photon and change into a different one of the same charge. In particular, the decay $\mu \to e\gamma$ does not seem to exist (the present experimental limit is a branching ratio $<10^{-10}$). Hence there is no vertex of the type

$$\tag{1}$$

and the Feynman rules for the scattering of two different charged fermions leads to just one diagram:

$$\tag{2}$$

The initial and final momenta of the first particle are **k** and **k**′, of the second **p** and **p**′. Their masses and charges are (e_1, m_1) and (e_2, m_2), respectively. Energy-momentum conservation requires that

$$p + k = p' + k'. \tag{3}$$

1. THE SCATTERING OF TWO DIFFERENT FERMIONS

The 4-momentum transfer q is given by

$$q = k - k' = p' - p. \tag{4}$$

According to B(46) the amplitude for this process is

$$T = e_1 e_2 \Phi_1 \frac{1}{Q^2 - m^2},$$

where e_1 and e_2 are the charges of the two fermions. Q is the 4-momentum of the internal line in diagram (2), which represents a photon, so that $m = 0$. We have $Q^2 = (k - k')^2 \equiv q^2$, and therefore

$$T = e_1 e_2 \frac{\Phi_1}{q^2}. \tag{5}$$

The factor Φ_1 carries the dependence on the fermion helicities; it contains the expression $(\bar{u}_{p'} \gamma^\mu u_p)(\bar{u}_{k'} \gamma_\mu u_k)$, where the u's are the Dirac spinors for the indicated particles.

The cross section is found from B(16), and by averaging over the incident spin states and summing over the outgoing ones. Specifically, we are interested in the cross section $d\sigma/dt$ for those collisions in which the square of the 4-momentum transfer $q^2 \equiv t$ is between t and $t + dt$.

The differential cross section $d\sigma/dt$ is Lorentz invariant because both σ and t are invariants. For that reason it is convenient to express $d\sigma/dt$ in a manifestly invariant manner. We are especially interested in collisions at high energy, i.e., when the total c.o.m. energy $\sqrt{(k + p)^2}$ is large compared to $(m_1 + m_2)$. Under this circumstance $d\sigma/dt$ is given by[1]

$$\frac{d\sigma}{dt} = \frac{1}{8\pi} \frac{e_1^2 e_2^2}{t^2} \left(1 + \frac{(p \cdot k')^2}{(p \cdot k)^2}\right) \tag{6}$$

The factor $(e_1 e_2/t)^2$ is evident from squaring the amplitude (5). The second factor comes from $|\Phi_1|^2$ after averaging over the spins. These formulas are used in Chap. V to discuss the deep inelastic scattering of leptons by hadrons.

In Chap. V we also need the cross section for various helicities. As we have seen in §B.1(b), the helicity of an ultrarelativistic fermion is conserved in all QED processes. Furthermore, thanks to reflection invariance QED amplitudes do not change if all helicities are reversed. There are therefore just two distinct contributions to Eq. (6):

$$\frac{d\sigma}{dt} = \frac{1}{8\pi} \frac{e_1^2 e_2^2}{t^2} \cdot \begin{cases} 1 & (h_1 = h_2) \\ \frac{(p \cdot k')^2}{(p \cdot k)^2} & (h_1 = -h_2). \end{cases} \tag{7}$$

---

[1] The derivation is in Appendix III.3.

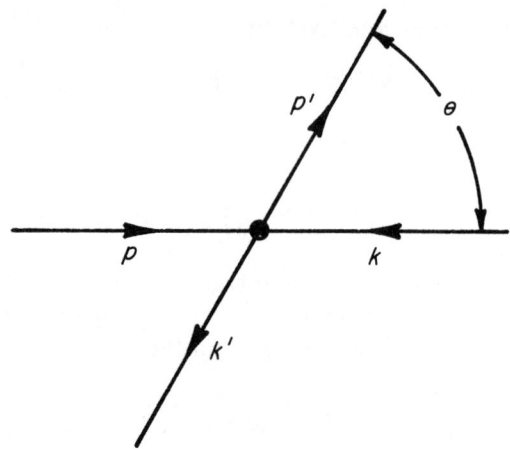

FIG. 1. Elastic scattering in the c.o.m. frame. Here **p** and **k** are the incident, **p'** and **k'** the outgoing 3-momenta and θ is the scattering angle.

Let us evaluate Eq. (6) in the c.o.m. frame. In this frame the incident and the target particle have equal and opposite momenta and, in the high-energy limit, equal energies:

$$\mathbf{p} = -\mathbf{k}, \qquad p_0 = k_0 = |\mathbf{p}| = \tfrac{1}{2}W, \tag{8a}$$

where W is the total energy of the two incident particles. The same is true after scattering:

$$\mathbf{p'} = -\mathbf{k'}, \qquad p_0' = k_0' = |\mathbf{p'}| = \tfrac{1}{2}W. \tag{8b}$$

It is easy to see that $p \cdot k = 2p_0^2$ and $p \cdot k' = 2p_0^2 \cos^2 \tfrac{1}{2}\theta$, where θ is the scattering angle between **p** and **p'** (see Fig. 1). We then get from (6)

$$\frac{d\sigma}{dt} = \frac{1}{8\pi} \frac{e_1^2 e_2^2}{t^2} [1 + \cos^4 \tfrac{1}{2}\theta]. \tag{9}$$

In the c.o.m. system, the 4-momentum transfer q has a zero time component. An exercise in geometry shows that

$$t = |q|^2 = -|\mathbf{q}|^2 = -4p_0^2 \sin^2 \tfrac{1}{2}\theta. \tag{10}$$

It is useful to have an expression for the scattering cross section per solid angle element $d\Omega = \sin\theta\, d\theta\, d\phi$ instead of per dt. Since t depends only on θ and not on ϕ, a transition from $d\Omega$ to dt implies an integration over ϕ, i.e., a factor of 2π, so that $d/d\Omega = (p_0^2/\pi)d/dt$. Then we express the charges in terms of the electron's charge, $e_i = z_i e$, and use $e^2/4\pi = \alpha$.

Equation (9) then yields the cross section for elastic collisions of two distinguishable fermions in terms of the c.o.m. scattering angle θ and the total c.o.m. energy W:

$$\frac{d\sigma}{d\Omega} = \frac{z_1^2 z_2^2 \alpha^2}{2W^2 \sin^4 \tfrac{1}{2}\theta}[1 + \cos^4 \tfrac{1}{2}\theta], \qquad W \gg m_1, m_2. \tag{11}$$

This is the generalization to extreme relativistic energies, of the nonrelativistic Rutherford scattering cross section[2]

$$\frac{d\sigma}{d\Omega} = \frac{z_1^2 z_2^2 \alpha^2}{16W^2 \sin^4 \tfrac{1}{2}\theta}, \qquad W \ll m_1, m_2. \tag{11a}$$

The physical picture that the Feynman diagram evokes is that the internal photon line in (2) represents the electromagnetic field responsible for the interaction between the two charges. In the nonrelativistic theory, the collision mechanism is the instantaneous Coulomb interaction, whereas in reality the interaction is a propagating electromagnetic disturbance. The Lorentz-invariant factor $1/q^2$ in (5) is the Fourier transform of this electromagnetic interaction, as was already demonstrated in the discussion pertaining to Eq. B(48). Other relativistic effects are contained in Φ. They describe a variety of interactions involving the fermion spins, such as spin-orbit coupling. The expressions in the brackets of (6) and (9) come from these spin effects.

2. Colliding beam processes: $e^+e^- \to \mu^+\mu^-$ and $e^+e^- \to e^+e^-$

We now discuss processes arising from the collisions of electrons and positrons, and start with the production of muon pairs. Such processes are studied in e^+e^- storage rings. Because the vertex (1) does not exist, the amplitude is represented by only one electromagnetic[3] Feynman diagram:

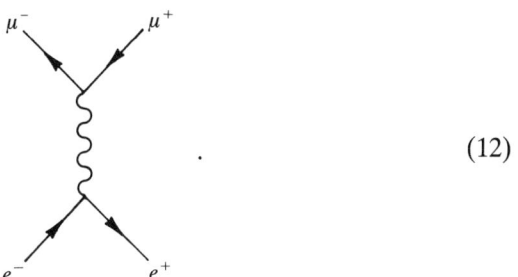

(12)

---

[2] Note that expressions (9) and (11) are valid only if the two colliding particles are different. In case of identical particles ($e - e$ or $\mu - \mu$ scattering) the Pauli principle requires antisymmetry of the wave function. This introduces interference terms between the scattered waves that change the results.

[3] As we know from §I.E.10(d), the weak interaction can also mediate this process.

Intuitively, we can think of (12) as an annihilation of the e^+e^- pair, followed by a materialization of the "virtual" photon. The wavy line does, in fact, represent an electromagnetic field that is momentarily present. It is not equivalent to a real photon, however. This intermediate photon line carries a 4-momentum P that is equal to the total 4-momentum of e^+ and e^-, whereas a real photon would have $P^2 = 0$. The internal line represents an electromagnetic field that is not a light wave. This is also a familiar situation in classical electromagnetic theory where the fields in the neighborhood of sources are not light waves. The transitory electromagnetic field never propagates to large distances from the lepton sources. According to B(46), the amplitude is

$$T(e^+e^- \to \mu^+\mu^-) = (e^2/P^2)\Phi_a, \tag{13}$$

where P is the total 4-momentum of the initial state and Φ_a is the spin factor. As T is Lorentz invariant, we can evaluate it in any frame. Choose the e^+e^- center of mass; this is actually the laboratory frame in a colliding-beam experiment. Then the photon's 4-momentum is $P = (W, \mathbf{P} = 0)$, where W is the total incoming energy—the sum of the e^+ and e^- energies in that frame. Since $P^2 = W^2$ in the c.o.m. frame,

$$T(e^+e^- \to \mu^+\mu^-) = \frac{e^2}{W^2}\Phi_a. \tag{14}$$

A few words about the form of Φ_a are in order. In the c.o.m. frame the total momentum is zero. The intermediate state is therefore a single virtual photon of momentum zero and mass W. Thus the intermediate state is analogous to a massive particle of spin 1 at rest. This "particle" then transforms into the final lepton pair, whose total angular momentum must also be one. The angular distribution of a $J = 1$ state is given by an amplitude that is a linear combination of $\cos\theta$, $\sin\theta$, and a constant where θ is the angle between the direction of the emerging muon pair and the incident electron pair (in the c.o.m. frame the members of either pair move in opposite directions). No higher power of $\sin\theta$ or $\cos\theta$ can appear in a $J = 1$ state. Hence

$$\Phi_a = A + B\cos\theta + C\sin\theta, \tag{15}$$

where the coefficients depend on the helicities of the incoming and outgoing leptons. The spin-averaged cross section for the $e^+e^- \to \mu^+\mu^-$ reaction is[4]

$$\frac{d\sigma}{d\Omega} = \frac{\alpha^2}{2W^2}\left[\frac{1 + \cos^2\theta}{2}\right]. \tag{16}$$

---

[4] A derivation of this angular distribution is given in Appendix III.3.

2. COLLIDING BEAM PROCESSES

The relatively weak angular variation only arises from the square of Eq. (15), since the 4-momentum of the intermediate line in graph (12) does *not* depend on θ. However, the scattering cross section (11) has a strong angular variation, due to the factor t^{-2} in Eq. (7) or (9). The 4-momentum q of the intermediate photon line in the corresponding graph (2) depends strongly on θ as is seen in Eq. (10). The two amplitudes (13) and (5), and the corresponding cross sections (16) and (11), are prototypes of the interaction mechanisms that occur repeatedly throughout particle physics. The first of the two represents an annihilation-recreation process, whereas the second is an exchange of a field quantum.

Another purely leptonic process that can be studied in an e^+e^- colliding ring is elastic scattering. According to the rules, there are now two Feynman graphs:

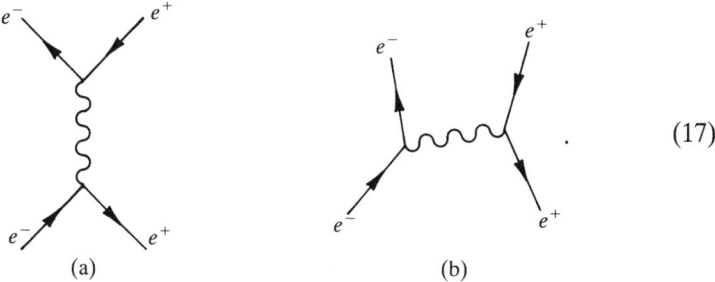

(17)

Graph (b) cannot occur in $e^+e^- \to \mu^+\mu^-$ because the vertex (1) does not exist. Let p and k be the four momenta of e^+ and e^- before scattering, and p', k' afterwards. $P = p + k$ is again the total e^+e^- 4-momentum, and the graph (17a) leads to the amplitude (13), while the graph (17b) is similar to (2). The internal photon line carries the 4-momentum q given by (4) and therefore leads to an amplitude similar to (5), having the form $(e^2/q^2)\Phi_b$. The e^+e^- elastic amplitude then consists of two terms

$$T(e^+e^- \to e^+e^-) = e^2\left(\frac{\Phi_a}{P^2} + \frac{\Phi_b}{q^2}\right), \tag{18}$$

where Φ_a and Φ_b are the spin factors[5] of the graphs (17a) and (17b). The factor Φ_a is the same as in (13) or (15), and the factor Φ_b is the same as Φ_1 in (5), as long as we restrict ourselves to high energies where the mass difference between e^+e^- and $\mu^+\mu^-$ is irrelevant.

Once more we evaluate (18) in the c.o.m. frame, which for e^+e^- collisions is the lab system. In that frame the relations (8a), (8b), and (10) are valid; therefore, $P^2 \simeq W^2$, and $q^2 \simeq -W^2 \sin^2\frac{1}{2}\theta$, or

$$T(e^+e^- \to e^+e^-) = \frac{e^2}{W^2}\left[\Phi_a - \frac{\Phi_b}{\sin^2\frac{1}{2}\theta}\right], \tag{19}$$

---

[5] The factor Φ_b contains the expression $(\bar{u}_p\gamma^\mu u_p)(\bar{v}_k\gamma_\mu v_{k'})$, where u, v are the Dirac spinors defined in Eq. A(42).

where, again, W is the total energy of the initial pair, and θ is the scattering angle in the c.o.m. frame. The two terms in (19) have completely different angular dependences: Φ_a depends weakly on θ, as indicated by (15), whereas the second term is proportional to $\sin^{-2}\frac{1}{2}\theta$.

The complete formula for the spin-averaged elastic e^+e^- cross section per unit solid angle at high energies is

$$\frac{d\sigma}{d\Omega} = \frac{\alpha^2}{2W^2}\left[\frac{1+\cos^4\frac{1}{2}\theta}{\sin^4\frac{1}{2}\theta} + \frac{1+\cos^2\theta}{2} - \frac{2\cos^4\frac{1}{2}\theta}{\sin^2\frac{1}{2}\theta}\right] \quad (20)$$

The first term is the same as (11), the second is the same as (16), while the last term describes the interference between Rutherford scattering and annihilation.

3. Data on e^+e^- collisions; tests of quantum electrodynamics

With the advent of colliding-beam facilities, it has become possible to test the validity of quantum electrodynamics in processes that involve the interactions of leptons at very short distances, i.e., large momentum transfers. No deviations from theory have been discovered up to the energies currently available at the level of precision attained thus far. As we show, it is possible to translate this agreement into a limit on the possible size of the electron and muon.

The processes we consider are those of the preceding section: $e^+e^- \to l^+l^-$. For these reactions we have already discussed the scattering amplitudes in detail. The data measured at PETRA are shown in Figs. 2 to 4.

Before discussing the significance of these results, we briefly comment on the qualitative features of the data. Observe that in e^+e^- elastic scattering there is a sharp forward peak [$\cos\theta \to 1$ in Fig. 2(a)]. This is just the familiar small-angle divergence of Rutherford scattering. As we saw in the preceding section, $e^+e^- \to \mu^+\mu^-$ only occurs with relative angular momentum $J = 1$, and it therefore has a very undramatic angular distribution, as is evident in Fig. 3. The energy dependence of the total $e^+e^- \to \mu^+\mu^-$ and $\tau^+\tau^-$ cross sections is shown in Fig. 4. All these data are in excellent agreement with the predictions of pure QED, except the angular distribution of $e^+e^- \to \mu^+\mu^-$, where the expected interference between γ and Z^0 exchange must be taken into account [see Eq. I.E(74b) and §VI.B.3].

One would like to translate this agreement between theory and experiment into a statement about lepton structure—or the lack thereof. What would the cross section be if, say, e or μ had an extended charge distribution having a radius of order a? To gain an insight into this we first ask a slightly simpler, and closely related question: How is the Rutherford nonrelativistic scattering law modified if the charges are not pointlike?

This question was already answered in the first subatomic collision experiment—the scattering of α particles by metallic foils—and yielded an

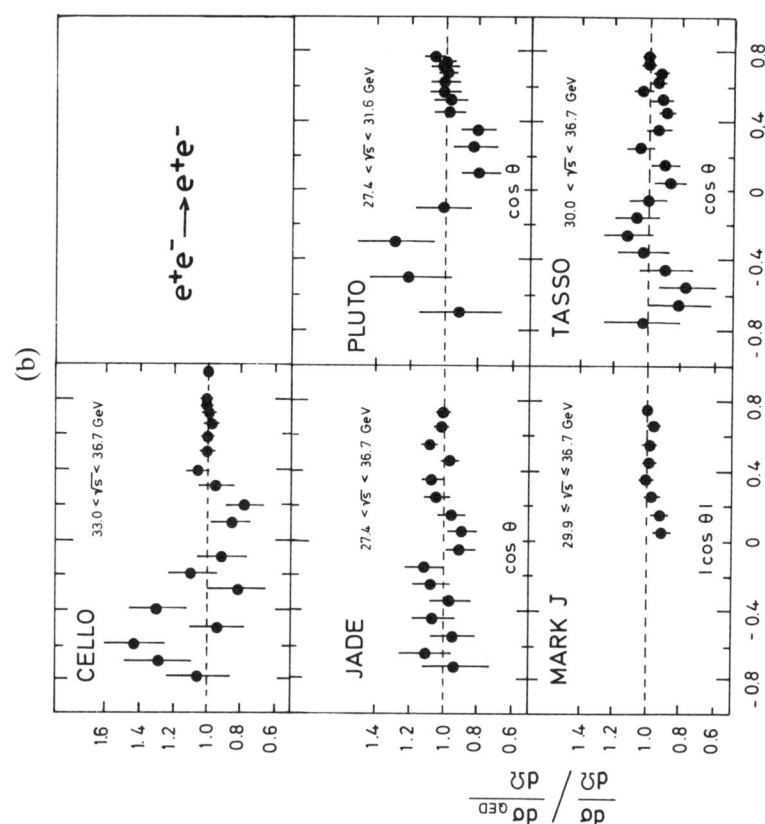
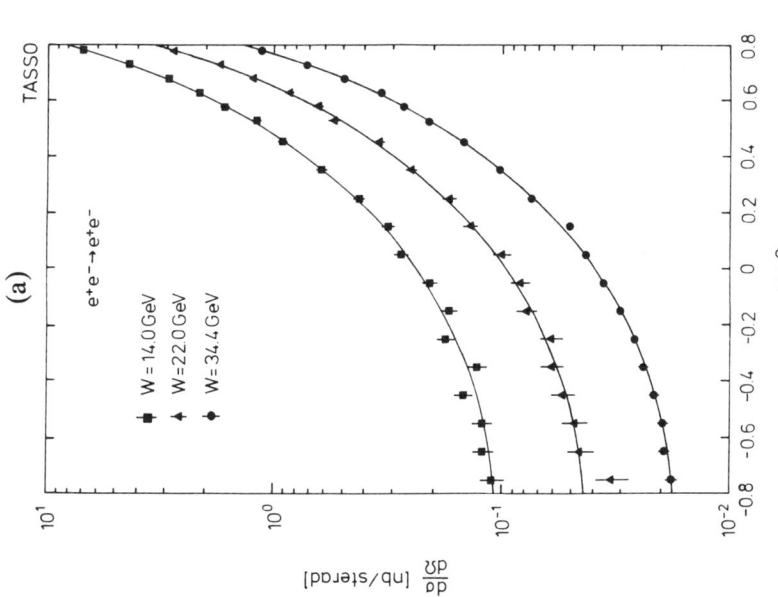

FIG. 2. Angular distribution of e^+e^- elastic scattering in the c.o.m. frame. Part (a) shows the experimental cross section; nb is the abbreviation for nanobarn (nanobarns = 10^{-9} barns = 10^{-33} cm^2), and W is the total energy (TASSO collaboration, 1982). Part (b) shows the ratio of experiment to theory at different angles and energies. The total energy is $\sqrt{s}$, and the code names refer to different experimental groups (Dittmann, 1981).

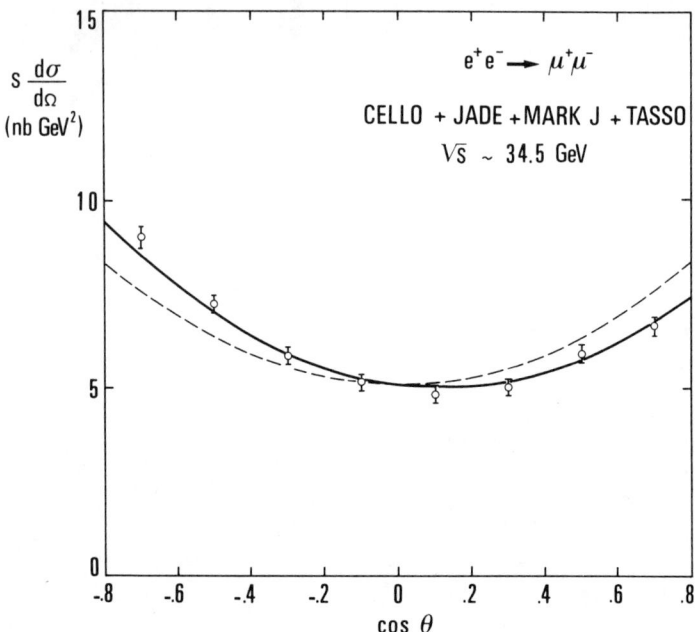

FIG. 3. Angular distribution of the process $e^+e^- \to \mu^+\mu^-$ in the c.o.m. frame, as measured at DESY by different groups (Davier 1982). The forward-backward asymmetry is due to the interference between the electromagnetic and weak interaction amplitudes (see §VI.B.3). The dashed curve is the prediction of pure QED, and the solid curve that of the electroweak theory with the mixing angle θ_W given by $\sin^2 \theta_W = 0.217$.

estimate of the nuclear radius. An extended charge distribution removes the singularity from the Coulomb field, and therefore leads to weaker forces at the smallest impact parameters. At large angles there will therefore be less scattering from an extended charge distribution than from a point charge. All that remains then is to translate these qualitative considerations into precise statements.

To this end, consider the scattering of a point charge e_1 by an extended charge distribution ρ_2. The electrostatic interaction energy between the charges is

$$U(r) = \frac{e_1}{4\pi} \int \frac{\rho_2(r') \, d^3r'}{|\mathbf{r} - \mathbf{r}'|} \tag{21}$$

when e_1 is at $\mathbf{r}$. If $(e_1 e_2 / 4\pi v) \ll 1$, where v is the incident velocity, the lowest order approximation to the scattering amplitude T suffices. This is given by Born's formula, which states that T is proportional to the Fourier transform of U,

$$T \propto \int e^{i\mathbf{q}\cdot\mathbf{r}} U(r) \, d^3r, \tag{22}$$

with $\mathbf{q}$ being the momentum change suffered by the projectile.

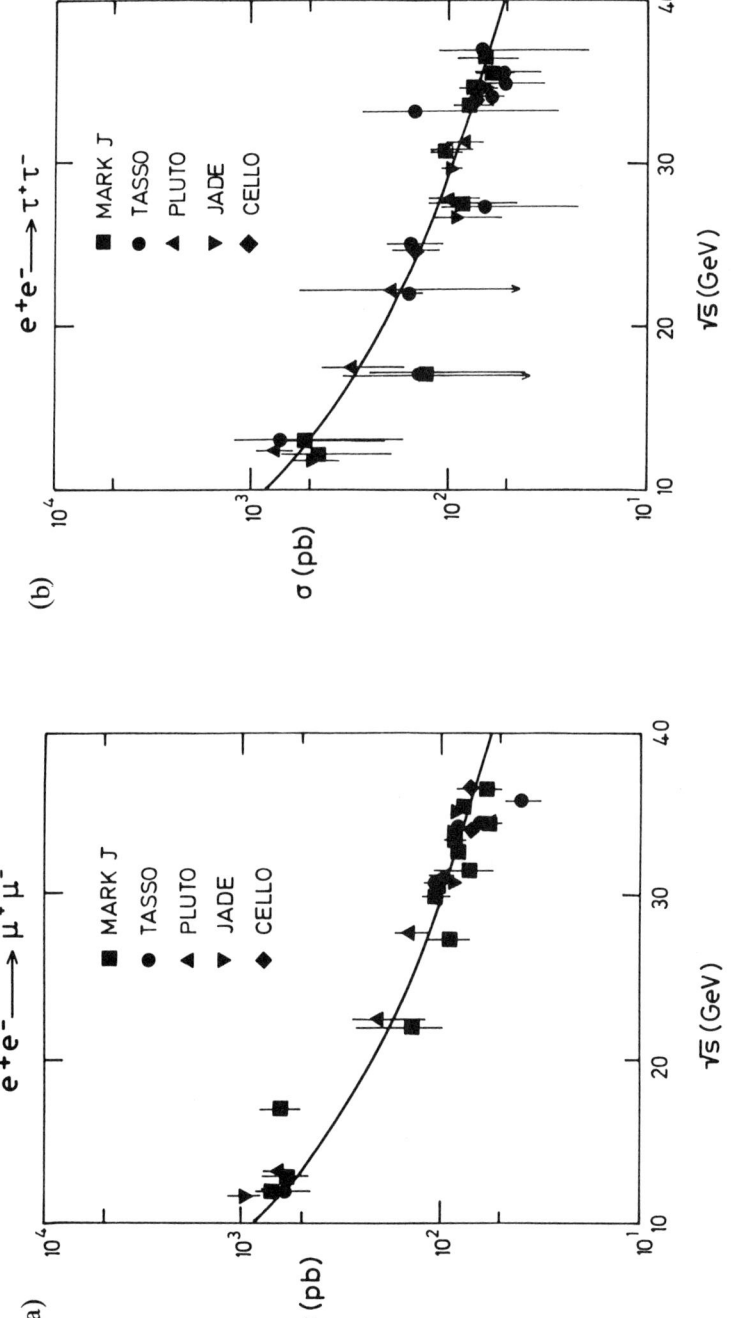

FIG. 4. The total cross sections for $e^+e^- \to \mu^+\mu^-$ and $e^+e^- \to \tau^+\tau^-$, measured by different groups at DESY (Mess, 1982). Here pb = 10^{-12} barns = 10^{-36} cm^2.

In evaluating (22), it is convenient to introduce the Fourier transform of the charge distribution:

$$e_2 F(\mathbf{q}) = \int e^{i\mathbf{q}\cdot\mathbf{r}} \rho_2(r)\, d^3r. \tag{23}$$

Here e_2 is the total charge of the target, and therefore

$$F(0) = 1. \tag{24}$$

The function $F(\mathbf{q})$ is called the target's form factor. It depends only on $|\mathbf{q}|$ if ρ_2 is spherically symmetric. We shall encounter such form factors again when we discuss the structure of hadrons in §III.A.1. When (21) is substituted into (22), and (23) is used, one readily finds that

$$T \propto \frac{e_1 e_2}{q^2} F(q^2). \tag{25}$$

As $q^2 = 4p^2 \sin^2\tfrac{1}{2}\theta$, where p is the incident momentum and θ the scattering angle, we see that what multiplies the form factor in (25) is just the amplitude for scattering by a point charge. The cross section is therefore

$$\frac{d\sigma}{d\Omega} = \frac{d\sigma_R}{d\Omega} |F(q^2)|^2, \tag{26}$$

where $d\sigma_R/d\Omega$ is the differential cross section for Rutherford scattering, Eq. (11a).

Hence the departure of the form factor from unity is a measure of the target's extension. Indeed, returning to (23) we see that $F(q^2) = 1$ for all q if the charge distribution is pointlike, $\rho_2 = e_2 \delta^3(\mathbf{r})$. If the target is small compared to the length q^{-1}, the departure of the form factor from unity is approximated by[6]

$$F(q^2) = 1 - \tfrac{1}{6} q^2 a^2, \tag{27}$$

---

[6] For small q, we can expand the phase factor in (23) in a Taylor series: $\exp(i\mathbf{q}\cdot\mathbf{r}) \simeq 1 + i\mathbf{q}\cdot\mathbf{r} - \tfrac{1}{2}(\mathbf{q}\cdot\mathbf{r})^2 + \cdots$. When substituted into (23), the first term gives the total charge e_2. The second term is zero if the charge distribution is spherically symmetric, and the third term gives

$$\tfrac{1}{2}\int (\mathbf{q}\cdot\mathbf{r})^2 \rho_2(r)\, d^3r = (2\pi q^2/3) \int \rho_2(r) r^4\, dr.$$

On the other hand, the mean-square charge radius is

$$\langle r^2 \rangle \equiv a^2 = (1/e_2) \int \rho_2(r) r^2\, d^3r = (4\pi/e_2) \int \rho_2(r) r^4\, dr,$$

which then gives (27). As one sees, the next term in the expansion is of order $q^4 \langle r^4 \rangle$, and the approximation (27) is only valid if (28) is satisfied.

which is valid for $q^2a^2 \ll 1$, or

$$4a^2p^2 \sin^2 \tfrac{1}{2}\theta \ll 1, \qquad (28)$$

where a is the rms radius of the change distribution. Observe that the scattering by an extended charge is indeed smaller than from a point charge, as explained in the opening paragraphs of this Section.

If the incident momentum p is large enough to violate (28), the departure from Rutherford scattering determines not only the rms radius but the actual shape of the charge distribution. This fact is used to great advantage to measure the detailed charge distribution of the proton and neutron, and of heavier nuclei, by means of high-energy electron scattering.

So far we have only discussed nonrelativistic scattering of a point charge e_1 by an extended distribution $\rho_2(r)$. The treatment can be generalized to the case where the projectile and target are both extended objects, and the momenta are relativistic. The scattering amplitude is then found to have form factors for the projectile as well as the target, and these are functions of q^2. In the c.o.m. system, the magnitudes of the 3- and 4-momentum transfers are the same since the energies of the scattering particles do not change. The relativistic form factors have very similar features to the nonrelativistic one; in particular, $F(q^2)$ decreases as $|q^2|$ increases from zero.

If the electron were an extended object, the preceding discussion leads to the conclusion that it would have a form factor $F_e(q^2)$; in e^+e^- scattering the scattering amplitude would therefore contain the extra factor $[F_e(q^2)]^2$. Consequently, the pointlike formula (20) is to be multiplied by $|F(q^2)|^4$. Given that the data and point-formula agree, one can then use our form (27) for $F_e(q^2)$ to set an upper bound on the electron's radius. A similar analysis can be applied to the processes $e^+e^- \to \mu^+\mu^-$ and $\tau^+\tau^-$. The bounds on the lepton radii that emerge from the data in Figs. 2–4 are [Wu (1984), p. 236]

$$a_e \sim a_\mu \lesssim 2 \times 10^{-16} \text{ cm} \qquad (29)$$

$$a_\tau \lesssim 1 \times 10^{-15} \text{ cm}. \qquad (30)$$

4. Bound states

Having discussed the scattering of leptons, we now turn to their bound states. The bound-state wave functions in a static potential are determined by the Schrödinger equation when the nonrelativistic approximation suffices, and by the Dirac equation when relativistic accuracy is needed.

There is a logical connection between the phenomena of scattering and binding. The same basic mechanism is responsible for both. One can think of a bound state as an everlasting sequence of deflections that prevent the bound particles from escaping to infinity. Consequently, perturbation theory to any finite order cannot describe a bound state.[7] Nevertheless, the language of perturbation theory—Feynman diagrams—can still be used to describe bound states, provided one does not truncate the expansion.

We illustrate this with the hydrogen atom. First, consider electron scattering by a proton. Because of the huge mass ratio, we treat the proton as an infinitely heavy object moored to the origin. In this approximation the proton cannot propagate, and so we do not represent it by a line in the Feynman graph. All that the proton does is to supply a Coulomb attraction for the electron. The Feynman diagram for ep scattering is therefore

$$\tag{31}$$

where the cross represents the immovable proton.

One may depict the bound state as a succession of scattering processes:

$$\tag{32}$$

The infinite series (32) is a somewhat peculiar way of writing the Schrödinger or Dirac equation for an electron in a Coulomb field. Nevertheless, it emphasizes that the simple scattering diagram (31) is the fundamental building block of the bound state.

5. Hydrogen, μ-mesic atoms, and muonium

We now classify the bound states of leptons that are of importance to the study of fundamental interactions. For this purpose, we confine ourselves to one- and two-body systems. These are simple enough so that one can compute the consequences of the theory with complete confidence. Any discrepancies that arise are then unambiguous. They cannot be ascribed to doubts about the precision of the solutions to the equations of motion.

---

[7] The perturbation expansion for the amplitude is a power series in α/v, where v is the characteristic velocity. In a hydrogenic bound state, $v \sim \alpha$, so all terms in (32) are of the same magnitude.

Ideally one would want to test quantum electrodynamics with systems that only contain leptons, as this would avoid any possible ambiguity that might arise from hadronic structure. Remarkably enough, Nature and the ingenuity of experimenters have cooperated to provide us with two such atoms: positronium and muonium. Positronium is a bound electron-positron system, while muonium is a bound μ^+e^- system.[8] Because these exotic atoms are very difficult to study, only a handful of their levels have been measured with sufficient precision to provide tests of quantum electrodynamics. For that reason, most of the precision tests employ systems of one e^- or one μ^- bound to an atomic nucleus.

The electronic atoms are hydrogen itself, as well as deuterium and singly ionized helium. In these systems the nuclei are so small in comparison to the Bohr radius that their structure cannot be a source of any appreciable uncertainty. Muonic atoms (often called μ-mesic atoms) have a quite different spatial structure. Because of the large muon mass, the low-lying levels have radii about 200 ($\simeq m_\mu/m_e$) times smaller than the smallest electronic orbits. Thus a muon in a low-lying state is far inside the electron shell of the atom, and the electrons have hardly any influence on the muon. By the same token, however, the size of the nucleus is no longer very small compared to the muon orbits. For high Z the muon moves literally within the nucleus when it is in the lowest orbits. Fortunately, elastic electron-nucleus scattering gives an independent measurement of the nuclear charge distribution, and this then allows a very reliable interpretation of μ-mesic spectra, especially for the larger muonic orbits.

Let us now study the spectrum of these systems. If we neglect relativistic and spin effects, and assume that all particles, including the nuclei, have no spatial extension, the spectrum is given by the Bohr formula,

$$E_n = -\frac{m'\alpha^2 Z^2}{2n^2}, \tag{33}$$

where m' is the reduced mass, $n = 1, 2, \ldots$ is the principal quantum number, and Z is the nuclear charge.

Corrections to Eq. (33) stem from a large variety of effects that fall into two broad categories. The first are those that do not involve quantum field theory as such: finite nuclear size, relativistic kinematics, and magnetic interactions involving the spins of the particles. These are especially important when Z is large. For the sake of brevity, we call these the *obvious corrections*. The second category arises when one treats the electromagnetic field as a quantum mechanical system and takes the possibility of pair

---

[8] Unfortunately, this conventional nomenclature is very misleading, for the definition of "positronium" would lead one to think that "muonium" refers to the $\mu^+\mu^-$ system. While $\mu^+\mu^-$ and $e^+\mu^-$ are leptonic systems that have hydrogenic spectra, it is not yet possible to study them in the laboratory because neither of their constituents occur abundantly in nature. In positronium and muonium, on the other hand, the e^- is abundantly available, and therefore these systems can be produced by stopping e^+ and μ^+ beams in ordinary matter.

creation into account. These latter effects, which are called *radiative corrections*, are so novel and important that we devote all of §D to them. Here we confine ourselves to a cursory overview of the "obvious corrections." As spin is important here, we must understand the multiplet structure, which is not displayed in (33).

We first divide the systems that we have mentioned into two categories: (I) unequal-mass systems; and (II) equal-mass systems. In category I, the heavier particle (nucleus in the case of electronic and μ-mesic atoms, μ^+ in the case of muonium) is, to a good approximation, situated at the system's center of mass. We will call it the central particle. About this there circulates a spin-$\frac{1}{2}$ particle. The only equal-mass system that will be considered is positronium. Here the positron and electron are on a completely equal footing. It is therefore not suprising that the angular momentum coupling scheme is not the same in the two categories, and for that reason we postpone the description of category II to the next Section.

Consider, then, the unequal mass systems (category I), and let m and M be the masses of the light and heavy particles, respectively. We proceed in three steps: (A) we first treat the heavier central particle as the fixed-point source of a Coulomb field and ignore the effects of both spins; (B) we take the effects of the lighter particle's spin into account; and (C) we incorporate the spin and magnetic moment of the central particle.

In step (A) all angular momenta, i.e., spin, orbital (l), and total $(j = l \pm \frac{1}{2})$, and their z-projections, are constants of the motion. So is the parity, which is given by $(-1)^l$. The spectrum has the familiar Bohr degeneracy: the nth level contains $l = 0, 1, \ldots, n - 1$, and therefore $j = \frac{1}{2}, \frac{3}{2}, \ldots, n - \frac{1}{2}$. Thus the $n = 2$ level contains the states $2s_{1/2}$, $2p_{1/2}$, $2p_{3/2}$, where s and p stand for $l = 0, 1$, respectively, and the subscript is j.

Now we take step (B). The corrections to (33) due to relativistic kinematics and spin effects conspire to split the degeneracy in such a way that the corrected energy eigenvalues depend only on n and j, but not on l. Thus the $2s_{1/2}$ and $2p_{1/2}$ levels remain degenerate, and slightly below $2p_{3/2}$ by an amount[9] $\frac{1}{32}\alpha^4 Z^4 m$. This is called the *fine structure* splitting. The hydrogen spectrum reveals that the $2p_{1/2}$ level is about 1000 MHz below $2s_{1/2}$, in contradiction with the Dirac theory of the "obvious" corrections.[10] As we shall see, this discrepancy is removed by the "radiative corrections," which account completely for this "Lamb shift" (see §D.4).

Finally, we take step (C), and consider the effects of the spin $\mathbf{J}_c$ of the central particle. The total angular momentum is $\mathbf{J} = \mathbf{j} + \mathbf{J}_c$. Each of the

---

[9] The contribution of the spin effect to this energy difference is best understood by going to the rest frame of the lighter particle, where the moving central particle produces a magnetic field $\mathbf{B}$. There is therefore a contribution to the energy $-\boldsymbol{\mu} \cdot \mathbf{B}$, where $\boldsymbol{\mu} = e\boldsymbol{\sigma}/2m$ is the spin magnetic moment of a Dirac particle. In order of magnitude $|\mathbf{B}| \sim Ze(v/a^2)$, where $a \sim (Z\alpha m)^{-1}$ and $v \sim Z\alpha$ are the radius and velocity of the orbit. Thus $\boldsymbol{\mu} \cdot \mathbf{B}$ is of order $\alpha^4 Z^4 m$.

[10] In studies of fine (and hyperfine) splittings, microwave techniques are widely used, and energy differences are therefore given in frequencies. The conversion to eV is 1 MHz = 0.4136×10^{-8} eV.

levels for which the angular momentum of the lighter particle was j therefore splits into states with $J = |j - J_c|, \ldots, j + J_c$. The splitting of this degeneracy is due to the interaction between the magnetic moments of the two particles. As magnetic moments are of order $e\hbar/Mc$, the magnetic moment of the heavy central particle is far smaller than that of the orbiting particle, and the splitting due to the interaction of the two magnetic moments is smaller than the fine structure splitting by the ratio m/M of the two masses, i.e., of order $Z^4\alpha^4 m(m/M)$. For that reason one calls the gaps between levels of the same J the *hyperfine* splitting. The hyperfine splitting between the $J = 0$ and $J = 1$ ground-state $1s$ levels in hydrogen is the most precisely known number in physics. The formula for this splitting is $\Delta E = 2\alpha^4 m g_p(m/M_p)$, where $g_p \equiv 2M_p\mu_p/e$ is the g-factor of the proton, and μ_p is its magnetic moment. The transition between these two levels is the 21-cm line of hydrogen so essential to radio astronomy. The ground-state hyperfine splitting of muonium has also been measured to high precision.

All the formulas of this Section are only correct to the indicated order in α. Higher order terms, due to radiative corrections, will be discussed in §D.

6. Positronium

We now turn to positronium. On the one hand, it provides precision tests of quantum electrodynamics; on the other, we have already learned in Vol. I that all the known mesons appear to be quark-antiquark bound states. One would therefore expect a certain correspondence between the excitation spectrum of positronium and the meson spectrum, and this is already evident in Fig. I.19. In Chap. III, we shall learn that this correspondence is well-established.

Because the two constituents of positronium have the same mass, we describe the system in the c.o.m. frame, where the positions and momenta of the particles are $(\frac{1}{2}\mathbf{r}, -\frac{1}{2}\mathbf{r})$ and $(\mathbf{p}, -\mathbf{p})$, respectively. The orbital angular momentum due to their relative motion is then $\mathbf{L} = \mathbf{r} \times \mathbf{p}$. In contrast to the hydrogen atom, the magnetic moments of both particles are of equal magnitude, so the two spins must be treated on an equal footing. Therefore, we introduce the total spin $\mathbf{S} = \mathbf{s}_{e^-} + \mathbf{s}_{e^+}$, whose eigenvalues S are 0 or 1. The total angular momentum $\mathbf{J}$ of positronium is then $\mathbf{J} = \mathbf{L} + \mathbf{S}$. States are designated by the spectroscopic symbol $n^{2S+1}L_J$, where for L one writes $S, P, D, \ldots$, when $L = 0, 1, 2, \ldots$. For example, 2^3P_2 stands for $n = 2$, $S = 1, L = 1, J = 2$.

If one ignores everything but the Coulomb interaction, the levels are given by the Bohr formula (33) with $m' = \frac{1}{2}m_e$. As in other atoms, the spin-dependent interactions break the degeneracies in (33). The $n = 1$ ground state splits into 1^3S_1 and 1^1S_0. This is called the hyperfine structure in analogy with nuclear atoms, but it is of the same order of magnitude as the "fine" structure splittings between the n^3P_2, n^3P_1, and n^3P_0 states, namely, $\sim \alpha^4 m$.

Positronium differs from all the other atoms that we have considered in that the electron and positron can annihilate into photons. This has two consequences: (1) there is an effective interaction between e^+ and e^- due to virtual annihilation; and (2) even the ground state of positronium is unstable.

The annihilation interaction is represented by Feynman diagram (17a). The virtual annihilation can only occur in states having the quantum numbers of the photon, i.e., $J = 1^-$, where the superscript $(-)$ specifies the parity. At first sight it would appear that this selection rule picks out positronium states with odd orbital angular momentum, because such wave functions are odd under spatial reflection. But now we must remember a remarkable fact already established in §A.5 [see Eq. A(52) et seq.]: The intrinsic parity η_f of a fermion is opposite to that of its antiparticle $\bar{\eta}_f$! That is to say, the parity of a positronium state is $\eta_{e^+}\eta_{e^-}(-1)^L$, where $(-1)^L$ is the parity of the spatial portion of the wave function. Thus the parity of a positronium state is $\Pi = (-1)^{L+1}$. In particular, $^3S_1$ has odd parity, and therefore transforms similarly to a one-photon state under rotations and reflections!

Thus the one-photon annihilation force acts only[11] on $^3S_1$ states, and produces a shift of these states that is of the same order as the spin splittings we have just discussed.

7. The decay of positronium: Charge conjugation

Finally we turn to the decay of positronium into photons. As we know, 4-momentum conservation does not allow the process $e^+e^- \to \gamma$. One might therefore expect that all positronium levels undergo the decay $e^+e^- \to 2\gamma$, since two photons moving in opposite directions can have total momentum zero. There is another fundamental conservation law at play here, however, that forbids the 2γ-decay of certain positronium levels.

The symmetry behind this conservation law is charge conjugation (recall §I.C.5). Within the limited context of QED, charge conjugation is the operation that replaces all charged particles by their antiparticles, and changes the sign of the electric and magnetic fields:

$$e^- \leftrightarrow e^+, \quad \mathbf{A}(\mathbf{r}, t) \to -\mathbf{A}(\mathbf{r}, t). \tag{34}$$

The symmetry of Dirac's theory under the interchange of particle and antiparticle was pointed out already. That this interchange must be accompanied by $\mathbf{A} \to -\mathbf{A}$ is also clear, for charge conjugation changes the

---

[11] The $^3D_1$ states also have the same quantum numbers as the photon. But wave functions with $L \neq 0$ vanish at $r = 0$, and are therefore insensitive to the annihilation interaction.

7. THE DECAY OF POSITRONIUM: CHARGE CONJUGATION

sign of the charge of the sources, and therefore of the fields resulting from them.

We introduce an operator $\tilde{C}$ that generates the transformation (34). That is, if $\tilde{C}$ acts on a state, it produces a new state where all leptons are replaced by their antiparticles, and each photon operator $\tilde{A}$ is replaced by its negative. Obviously, the only systems that can be eigenstates of $\tilde{C}$ are those that are "truly neutral," that is, that do not change when every particle is transformed into its antiparticle. Positronium is such a system, and so is an assembly of photons, since photons are their own antiparticles. Clearly, a one-electron state is not an eigenstate of $\tilde{C}$, nor is muonium (μ^+e^-), for under the operator $\tilde{C}$ it turns into antimuonium (μ^-e^+). To each eigenstate $\tilde{C}$ we can assign an eigenvalue C, which can only be $+1$ or -1 because $\tilde{C}^2 = 1$. Since the equations of quantum electrodynamics are invariant under charge conjugation, $\tilde{C}$ must be a constant of motion.

We first show that for a state containing n photons, $C = (-1)^n$. Consider first a one-photon state $|1\gamma\rangle$, created from the vacuum by applying the vector potential operator $|1\gamma\rangle \propto \tilde{A}(h\mathbf{k})|\Omega\rangle$. When we apply $\tilde{C}$ to $|1\gamma\rangle$, $\tilde{A} \rightarrow -\tilde{A}$, and therefore[12] $\tilde{C}|1\gamma\rangle = -|1\gamma\rangle$. A two-photon state $|2\gamma\rangle$ is made by applying two creation operators to the vacuum: $|2\gamma\rangle \propto \tilde{A}(h_1\mathbf{k}_1)\tilde{A}(h_2\mathbf{k}_2)|\Omega\rangle$. Applying $\tilde{C}$ once more, we see that it leaves this state alone. Quite generally

$$\tilde{C}|n\gamma\rangle = (-)^n |n\gamma\rangle \qquad (35)$$

as claimed.[13]

We now determine C for the positronium levels. In positronium, charge conjugation is just the interchange $e^+ \leftrightarrow e^-$. Let $\psi(m_+, m_-, \mathbf{r})$ be a positronium wave function, $\mathbf{r}$ the $e^+ - e^-$ separation, and $m_\pm$ the z-components of the $e^\pm$ spins. Then

$$\tilde{C}\psi(m_+, m_-, \mathbf{r}) = \psi(m_-, m_+, -\mathbf{r}) \qquad (36)$$

because $e^+ \leftrightarrow e^-$ means $\mathbf{r} \rightarrow -\mathbf{r}$ and an interchange of spins. We claim that

$$\psi(m_-, m_+, -\mathbf{r}) = -(-1)^{L+S+1}\psi(m_+, m_-, \mathbf{r}), \qquad (37)$$

and therefore

$$C = (-1)^{L+S}. \qquad (38)$$

Equation (37) is derived as follows: $\mathbf{r} \rightarrow -\mathbf{r}$ is just space reflection, and therefore carries the ordinary parity factor $(-1)^L$; the spin singlet ($S = 0$) is antisymmetric under spin exchange, whereas the triplet ($S = 1$) is sym-

---

[12] Here we adopt the natural convention that the vacuum $|\Omega\rangle$ has eigenvalue $C = 1$.

[13] Charge conjugation, like space reflection, therefore has a multiplicative composition law. Thus, if two systems $|a\rangle$ and $|b\rangle$ have C-eigenvalues C_a and C_b, the composite (product) state $|a\rangle \cdot |b\rangle$ has eigenvalue $C_a C_b$.

metric, and therefore $m_+ \leftrightarrow m_-$ carries the factor $+1$ if $S = 1$, -1 if $S = 0$, or $(-)^{S+1}$ in general; beyond that the interchange $e^+ \leftrightarrow e^-$ introduces another factor of -1 because e^+ and e^- are fermions, and each other's antiparticles.[14]

We now apply this knowledge to the decay of positronium into photons. A two-photon state has the eigenvalue $C = +1$, and therefore the only states that can decay into two photons must have $L + S$ even. In particular, 1^3S_1 cannot undergo 2γ-decay; it must undergo 3γ-decay. On the other hand, the lower ground-state $C = 1$ level 1^1S_0 can undergo 2γ-decay.

Annihilation is only an important process for the $n = 1$ levels. Levels with $n > 1$ cascade to lower levels by emitting a single photon with a decay rate of order α, in contrast to the 2γ and 3γ annihilation rates, which are of order α^2 or α^3 for $L = 0$ states, respectively, and negligible for all $L \neq 0$ states for reasons already stated.

We shall now make a rough estimate the rate of decay of the $^1S_0$ positronium ground state into two photons. For this purpose, we first consider the annihilation of a free electron-positron pair into two photons. The term *free* means that they are assumed to move as if there is no Coulomb attraction between them. The members of the pair have 4-momenta p_1 and p_2, and the photons have the 4-momenta k_1 and k_2. Consider the Feynman diagram

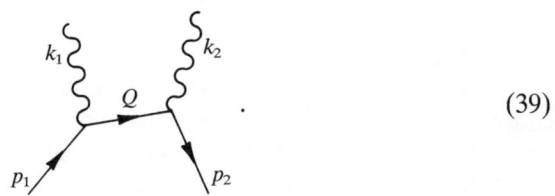

(39)

In positronium we only need to consider nonrelativistic motions, so the spatial part of the leptons' 4-momentum are negligible compared to their energies; furthermore, $k_{10} \simeq k_{20} \simeq m$. Hence, $Q^2 = m^2 - 2p_1 \cdot k_1 \simeq m^2 - 2mk_{10} \simeq -m^2$, which shows that the internal line is far off its mass shell. The transitions at each vertex therefore involve momentum transfers that are large compared to the initial momenta of the pair ($\sim \alpha m$). As a consequence the motion of the pair and the influence of the binding field can be ignored during the annihilation process.

The annihilation rate Γ_{ann} of an e^+e^- pair at rest in a volume V can be estimated by a dimensional argument. From diagram (39) we know that the amplitude is of order α, so $\Gamma_{\text{ann}} \sim \alpha^2$. This rate must also be proportional to the probability for finding the e^+ and e^- in close proximity, which is $1/V$.

---

[14] This last factor of -1 is not the relative intrinsic parity of e^+ and e^-. An e^+e^- state can be written as $d_p^\dagger b_{p'}^\dagger |\Omega\rangle$, where $d_p^\dagger$ and $b_{p'}^\dagger$ are the e^+ and e^- creation operators defined in Eq. A(43). Applying $\hat{C}$ yields $b_p^\dagger d_{p'}^\dagger |\Omega\rangle$. As shown in Appendix IV.2, Eq. (18), the causality principle requires $b_p^\dagger d_{p'}^\dagger = -d_p^\dagger b_{p'}^\dagger$; this is the source of the minus sign.

Finally Γ_{ann} is a rate, and must have the dimension of (time)$^{-1}$. The only other quantity involved in the problem having a dimension is m, and we therefore conclude that $\Gamma_{ann} \sim \alpha^2/m^2 V$.

In positronium the pair is in a volume V of order $a_0^3 = (1/\alpha m)^3$, and therefore the two-photon annihilation rate $\Gamma_{2\gamma}$ of positronium is of order $\alpha^5 m \sim 10^{10}$ sec^{-1}. A detailed calculation gives the result

$$\Gamma_{2\gamma} = \tfrac{1}{2}\alpha^5 m = 0.802 \times 10^{10} \text{ sec}^{-1}. \tag{40}$$

The rate for the three-photon decay of the $^3S_1$ state is far smaller, as there is one more factor α, since one more vertex is necessary to produce the additional photon, and there is a further reduction because there is less phase space available. Refined calculations, including radiative corrections not considered in (40), give the results[15]

$$\Gamma(1^1S_0 \to 2\gamma) = 0.798 \times 10^{10} \text{ sec}^{-1}$$
$$\Gamma(1^3S_1 \to 3\gamma) = 0.70386(16) \times 10^7 \text{ sec}^{-1}. \tag{41}$$

The observed values are[16]

$$\Gamma(^1S_0) = 0.799(11) \times 10^{10} \text{ sec}^{-1}$$
$$\Gamma(^3S_1) = 7.051(5) \times 10^6 \text{ sec}^{-1}. \tag{42}$$

If the small discrepancy between Eqs. (41) and (42) for the ^3S state withstands the test of time, it would pose a serious problem for quantum electrodynamics.

---

[15] 1^3S_1: W. E. Caswell, G. P. Lepage, and J. Sapirstein, *Phys. Rev. Lett.* **38**, 488 (1977). G. Adkins, *Ann. Phys.* **146**, 78 (1983).
1^1S_0: I. Harris and L. M. Brown, *Phys. Rev.* **105**, 1656 (1957). All these calculations include radiative corrections.

[16] 1^3S_1: D. W. Gidley, A. Rich, E. Sweetman, and D. West, *Phys. Rev. Lett.* **49**, 525 (1982).
1^1S_0: E. D. Theriot, R. H. Beers, V. W. Hughes, and K. O. H. Ziock, *Phys. Rev. A2*, 707 (1970).

D. THE SELF-ENERGY OF THE ELECTRON, VACUUM POLARIZATION, AND PRECISION TESTS OF QUANTUM ELECTRODYNAMICS

In the preceding Part we discussed a number of processes in lowest order perturbation theory. Some of these are familiar phenomena in classical physics; for example, the emission, absorption and scattering of light, or the scattering of charged particles. But we also encountered processes that have no counterpart in classical physics, such as pair creation and annihilation. As we had mentioned, perturbation theory provides an expansion for scattering amplitudes and energy eigenvalues in powers of $\alpha = \frac{1}{137}$. While this is a small number, it is by no means negligible, and if one supposes that quantum electrodynamics is a fundamental theory, one must address the question of whether it correctly describes phenomena to an accuracy that is much better than one part per 137.

Quantum electrodynamics passes these tests with flying colors, but only after some new and rather daring methods are introduced to circumvent fundamental difficulties connected with the electromagnetic self-energy of the electron and muon, and with the existence of virtual pairs in the vacuum. In these more refined calculations we again encounter new phenomena that have no counterpart in classical electrodynamics; for example, a change in the Coulomb field of a point charge at small distances, the scattering of light by light, and the scattering of light by a Coulomb field. Experiments have established the existence of these novel phenomena, and shown that they are in detailed quantitative agreement with the theory.

We are now interested in high precision. It is clear from a glance at the experimental curves in the preceding Part that conventional scattering experiments are too crude for this purpose. For that reason the relevant experimental techniques now include precision spectroscopy. Here ingenious measurements of incredible refinement have been compared with very sophisticated and difficult calculations. Despite the vast difference in precision of spectroscopic and collision measurements, it should be recognized that these two types of experiments complement each other. As we saw in §C.3, collision experiments are very sensitive to rather large deviations from quantum electrodynamics at very short distances (large momentum transfers). On the other hand, low-lying levels of bound systems involving electrons (or muons) have dimensions that are of order 10^{-8} cm

1. RADIATIVE CORRECTIONS

(or $\sim 10^{-10}$ cm), and the spectroscopic tests are therefore sensitive to minuscule departures from the theory over rather long distances.

The complementary character of the two types of experiments is most clearly seen in the case of the charge distribution of a fermion (recall §C.3). As we will learn, quantum electrodynamics predicts that what we have loosely called a point charge in the vacuum is actually surrounded by an induced charge distribution $\rho_{in}(r)$ that stems from transient e^+e^- pairs produced by the electromagnetic field of the point charge. This phenomenon is called *vacuum polarization*. The distribution $\rho_{in}(r)$ extends out to distances of order the electron Compton wavelength m^{-1} ($\sim 10^{-11}$ cm), and is a small and diffuse effect except at distances vastly shorter than those currently attainable by experiment ($\sim 10^{-16}$ cm). Measurements at distances large compared to m^{-1} yield the "actual" total charge Q that we normally ascribe to the particle. The total induced charge *outside* any radius $r_0 < m^{-1}$ has the opposite sign to Q, and a magnitude of order $-Q/137$. In order for the total charge to be Q, the charge *inside* r_0 must be somewhat larger than Q.

High-energy experiments are not yet accurate enough to observe the slight increase of the charge at the center mentioned in the preceding paragraph. Atomic experiments are able to do this and to study the small correction term ρ_{in}. The high-energy experiments are akin to a microscope with a very high resolution of $\sim 10^{-16}$ cm, but with poor "illumination"; they cannot see the small correction ρ_{in}. What they have taught us is that the dominant charge in the center looks like a δ-function when viewed with a resolution down to $\sim 10^{-16}$ cm. The atomic experiments are akin to a microscope with a low resolution, but excellent "illumination"; even though the effects of ρ_{in} are so small, they can determine them to less than 1%. Because of their low resolving power, the atomic experiments are insensitive to a breakdown of the theory at small values of r_0 that the cruder high-energy collision experiments would detect with ease if the breakdown occurs at a distance r_0 comparable to the inverse of the momentum transfers in the collisions.

1. Radiative corrections

When we treated various scattering processes in §C, we always confined ourselves to the Feynman graphs with the least number of vertices possible. Consider e^-e^- scattering, where we previously considered the graph

 (1)

This represents the first term in an infinite series of terms whose sum is presumably the exact scattering amplitude.

In discussing bound states in §C.4, we already saw that it is essential to retain the simplest iterations of (1), such as

$$\tag{2}$$

This does not represent a new dynamical mechanism; it is simply a repetition of the basic scattering act. But to the same order in e as (2), there are several other graphs that bear no simple relation to (1):

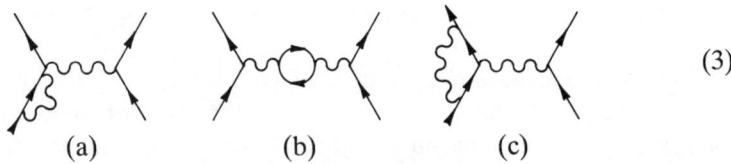

$$\tag{3}$$

(a)   (b)   (c)

The graphs of (3) depict various *radiative corrections* to the basic process (1). In (3a), an electron emits and reabsorbs a photon. For reasons that we explain below, this is called an electron self-energy graph. In (3b), the exchanged virtual photon materializes transiently into an e^+e^- pair, which then reannihilates into a photon. This is called a vacuum polarization graph. Finally in (3c) we have a correction to the basic electron-photon vertex. All higher order corrections can be built by nesting or combining the three basic graphs:

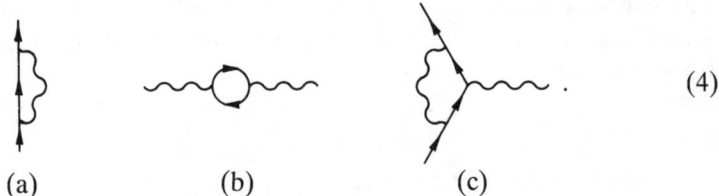

$$\tag{4}$$

(a)   (b)   (c)

As an example of a more complex radiative correction to (1), we have

$$\tag{5}$$

2. ELECTRON SELF-ENERGY AND MASS RENORMALIZATION

These corrections are always of higher order in e than the simplest amplitude describing the process.

Unfortunately, a straightforward application of perturbation theory using the electromagnetic interaction of Eq. B(1) leads to unacceptable results: the radiative corrections turn out to be infinite, because they contain integrals that diverge. It was eventually recognized that these divergences stem from two relatively simple physical effects. First, the theory implies that the "mechanical" mass m_0 of the electron, which is put into the original Hamiltonian, must be augmented by an additional mass Δm due to the effects of the electromagnetic field. Second, the theory implies that the vacuum is akin to a polarizable dielectric medium. Because of this vacuum polarization, the force observed between two distant and equal test charges Q_0 "inserted into the vacuum" is not $Q_0^2/4\pi r^2$, but $Q^2/4\pi r^2$, where the observed charge Q is not equal to Q_0. The expressions for both Δm and Q/Q_0 contain divergent integrals. It turns out that *all* the divergences that appear in higher approximations can be ascribed to the mass and charge corrections: $m_0 \to m = m_0 + \Delta m$ and $Q_0 \to Q$, where Δm and $Q_0 - Q$ are infinite. By judiciously removing the input mass m_0 and input charge Q_0 in favor of the actually observed quantities m and Q, *all* the infinities can be removed. This procedure is called mass and charge renormalization. It results in well-defined expressions for all scattering amplitudes and energy eigenvalues, and leads to predictions in superb agreement with experiment.

In the following sections we describe the physical origins of mass and charge renormalization, sketch how the finite results for radiative corrections are found, and how well they compare with experiment.

2. The electron self-energy and mass renormalization

The self-energy of a charge is already a deep problem in classical electrodynamics.[1] There are two closely related questions at issue, and both have their counterparts in the quantum theory. First, when we measure the mass of a charged particle, do we not automatically include the energy carried by its field as well as its "mechanical" mass? Second, when a charge radiates, should we not take the interaction of this radiation with the emitting charge into account in computing the evolution of the system?

Clearly, the answer to both questions is yes! The real problem occurs when one tries to evaluate the effects implied by this answer. This is most easily seen by considering the total energy m of a stationary charged particle having a mechanical mass m_0 and an electric field $\mathbf{E}$:

$$m = m_0 + \frac{1}{2}\int |\mathbf{E}(r)|^2 \, d^3r. \qquad (6)$$

Naturally, $\mathbf{E}$ depends on the particle's charge distribution ρ. Assume ρ to

---

[1] See Jackson (1975), Chap. 17.

be spherically symmetric, and to vanish outside a sphere of radius a. Then $|\mathbf{E}| = e/4\pi r^2$ for $r > a$, and therefore

$$m = m_0 + \frac{e^2}{8\pi a} + \frac{1}{2}\int_{(r<a)} |E(r)|^2 \, d^3r, \tag{7}$$

where the second term is the "outside" field energy, while the last integral is the energy due to the electric field "inside" the particle. As the particle radius a tends to zero, both terms diverge like $1/a$. Since all terms in (7) are positive, the smallest radius r_0 that we can possibly ascribe to the particle is that for which the whole mass m is due to the particle's own electric field. To find r_0, we set $m_0 = 0$ in (7), and find

$$r_0 \simeq \frac{e^2}{8\pi m},$$

where the symbol $\simeq$ indicates that the "inside" term depends on the details of the charge distribution. The length r_0 is called the *classical electromagnetic radius* of the particle. For the electron, $r_0 = 1.4 \times 10^{-13}$ cm.

The existence of the radius r_0 raises a serious paradox for the classical theory. On the one hand, the particle must have at least the size $\sim r_0$, or its electromagnetic energy will exceed its total energy. On the other hand, ascribing a radius to a particle requires one to specify its internal structure, and that leads to unresolved difficulties [Jackson (1975), §17.4].

On the experimental side, we know (recall §C.3) from the colliding beam experiments that the electron radius is smaller than 10^{-16} cm, and may even be zero. It is therefore clear that the classical radius cannot have any physical significance.

Quantum electrodynamics does not resolve the fundamental problem of the electron's radius, but it demonstrates that that problem is irrelevant to all measurements at distances attainable for the forseeable future. As a consequence the quantum theory of the electron's self-energy represents a large step forward in comparison to the classical theory.

The Feynman graph that describes the influence on an electron due to its own field is shown in (4a). That is to say, if we have a process wherein an unadorned electron (or positron) line appears, and we replace that line by (4a), we have the lowest order correction due to that electron's (or positron's) self-field.

In §B.4, we learned that the Feynman graphs are a shorthand for mathematical expressions. This continues to be true of radiative corrections, and the rules continue to be the same except for one simple but important difference. In all graphs we encountered previously, the 4-momenta carried by the internal lines are found by energy-momentum conservation at the vertices from the 4-momenta of the external lines. Examples are the graphs [Eq. B(23)] for the Compton effect. In graphs with loops, 4-momentum conservation at the vertices does not suffice to determine the 4-momenta of the internal lines. In particular, in all the graphs of (3) there is an internal

2. ELECTRON SELF-ENERGY AND MASS RENORMALIZATION

line momentum that remains undetermined. For example, in (3a) one photon may have any momentum, since the electron in the loop can carry the difference; in (3b), the momenta of the electron and positron may have any value, only their sum is fixed. When diagrams have closed loops, the rules of §B.4 are augmented by a further simple statement: *If the 4-momentum k of an internal line is not fixed by conservation, the expression (B.46) is to be integrated over that momentum.* This integration is just the sum over intermediate states familiar from the perturbation theory formula [B(17)].

When graphs of the type (4a) are taken into account, it is found that the mass of a charged lepton (electron or muon) in free space is changed from m_0 to m. This shift can also be written as an integral like the one in (6):

$$m = m_0 + \tfrac{1}{2}e^2 \int f(r)\, d^3r. \tag{8}$$

The integrand $f(r)$ turns out to be the square of the electric field only for distances $r > m^{-1}$. For $r < m^{-1}$ the quantum mechanical expression $f(r)$ has a different behavior from the classical $|\mathbf{E}(r)|^2$; instead of increasing like $1/r^4$ as $r \to 0$, it behaves like $1/r^3$.

This reduction of the electromagnetic energy density relative to the classical value is due to the existence of antiparticles. Even in empty space the electromagnetic interaction can produce transitory e^+e^- pairs, depicted by the graph

which is a combination of graphs B(14a and b). These pairs produce charge fluctuations. When an electron is present, the Pauli principle reduces the probability that the e^- member of such a transitory pair will be in the neighborhood of the aforementioned electron. Hence a dearth of electric charge exists in the immediate vicinity of the electron under investigation, thereby reducing the electric field and hence the energy density $f(r)$. As a consequence, $f(r) \sim 1/r^3$ as $r \to 0$, and the integral in (8) has the form $\int dr/r$ in the small r region. Thus it still diverges at the lower limit, but only logarithmically, whereas the classical self-energy diverged linearly. In order to get the observed finite electron (or muon) mass we may either postulate that the two terms in (8) add up to the observed mass, or we may give the electron a small but finite radius a. Taking the latter approach, and carrying out a more careful evaluation of (8), one finds that as $a \to 0$

$$m = m_0\left(1 + \frac{3\alpha}{2\pi}\ln\frac{1}{am_0} + 0(\alpha)\right), \tag{9}$$

where the last term stems mainly from large values of r and is finite as $a \to 0$.

From a fundamental point of view we have made no progress because we have been forced to introduce a particle radius once again. But the appearance of the logarithm in (9) makes an enormous difference when compared with the classical theory. To see this, we evaluate the mass from (9) when $a = 10^{-16}$ cm, the present limit on the electron radius. We find $m \simeq 1.04 m_0$, a very small correction, whereas the classical theory gave a $\sim 100\%$ correction for the much larger radius of 10^{-13} cm. Indeed, in order to get a correction of $\sim 20\%$ one would have to choose[2] $a \sim 10^{-35}$ cm!

Thus, while it is still doubtful that quantum electrodynamics is a completely self-consistent theory, in contrast to classical electrodynamics the fundamental paradoxes occur at distances that are incredibly short compared to all dimensions accessible to experiment in the foreseeable future. Indeed, there are many that entertain the hope that when the relationship between electrodynamics and Nature's other interactions (the weak, strong, and gravitational) are understood, these divergence problems at incredibly small distances will disappear.

The renormalization procedure removes the blemish of the remaining divergence in Δm. It does so by identifying the expression for the total mass (9) with the *experimentally observed mass* of the electron, even though these expressions contain an integral that diverges logarithmically. As we shall see in the next Section, a similar replacement of the theoretical charge by the actually observed charge will also be required. *The success of the renormalization procedure resides in the fact that this replacement of theoretical masses and charges by their observed counterparts yields finite theoretical expressions for all the other observable quantities, i.e., scattering amplitudes, energy eigenvalues, and lifetimes.*

3. Vacuum polarization

We now come to radiative corrections due to fermion closed loops, i.e., to graphs of type (3b). These loops, we recall, describe the transient presence of e^+e^- pairs created by photons out of the vacuum, and for that reason this effect is called *vacuum polarization*. Insertion of fermion loops again leads to infinite results if one integrates over the loop momenta—the momenta of the transient fermions. Once more it is possible to remove these infinities by requiring that the charge of the electron has its observed value. From both the experimental and conceptual point of view, the most important consequence of this procedure is a correction to the interaction energy between two charges separated by distances smaller than the electron's

---

[2] Thus a is smaller than the Planck length $\sqrt{\hbar G/c^3}$ (see Vol. I, p. 167). At such distances quantum fluctuations of the gravitational field are expected to become significant, but that is a phenomenon beyond our present understanding.

3. VACUUM POLARIZATION

Compton wavelength—in other words, a modification of Coulomb's $1/r$ law when $r \lesssim m_e^{-1} \sim 10^{-11}$ cm. This alteration of Coulomb's law has been verified experimentally to a precision of better than 1%.

Vacuum polarization appears in its simplest form if one studies the interaction energy $\phi(r)$ of two point charges $Q^{(1)}$ and $Q^{(2)}$ that are fixed in space at a separation r. This "fixing" will actually occur if we suppose that the charges are borne by very massive particles. In this case, the Feynman diagram that describes the correction to Coulomb's law should be drawn as

$$\text{\small x}\!\!\sim\!\!\sim\!\!\bigcirc\!\!\sim\!\!\sim\!\!\text{\small x} \tag{10}$$

in analogy to (3b). The crosses at both ends symbolize the charged particles and indicate that they cannot change their states, since we have given them an infinite mass. The evaluation of diagram (10), i.e., of $\phi(r)$, and the charge renormalization that is involved, are quite complicated. Appendix V sketches the details of the calculation. In this section we only describe the physical picture underlying the calculation, and discuss the final results.

Vacuum polarization can be described by considering the vacuum to be a dielectric medium. The polarization is due to the fluctuations of the fermion fields, which create virtual particle-antiparticle pairs. The distance between the members of a pair never exceeds a distance of order the Compton wavelength m^{-1}. The reason for this finite size lies in the uncertainty principle: The creation of a virtual e^+e^- pair requires a surplus energy of at least $2m_e$; thus energy conservation must be violated by $\Delta E \gtrsim 2m_e$. This violation can persist for a time $\sim(\Delta E)^{-1}$. But nothing can move faster than light, so the virtual pair must remain within a distance $d \lesssim (2m_e)^{-1}$. This argument also demonstrates that vacuum polarization is dominated by the lightest charged particles in Nature: heavy pairs require a larger ΔE, and thus their effect is concentrated at correspondingly smaller distances. Since the electron is some 200 times lighter than any other charged particle, we need only consider e^+e^- pairs. For that reason we now drop the subscript e from m_e—we always[3] deal with e^+e^- pairs. From these arguments we conclude that the fluctuations with a wavelength $\lambda \gtrsim m^{-1}$ do not occur, since they cannot be located within $\sim m^{-1}$. We shall make use of this fact later on.

What happens when one places a charge Q_0 into a dielectric medium at a point P? The electric field induces dipoles in the medium oriented in such a way that the ends with the opposite charge to Q_0 point toward P. As a consequence, the effective charge Q_{eff} *is smaller* than Q_0, $Q_{\text{eff}} = Q_0/\varepsilon$, where $\varepsilon > 1$ is the dielectric coefficient[4] of the medium. Here we assume that ε has the same value throughout the medium.

---

[3] This is not true of the very precise higher order calculation cited in later portions of this chapter, where the tiny effects of muon and hadron pairs must be included.

[4] ε is usually called the dielectric "constant," even though it is usually a function of several variables. That is why we prefer the terminology "coefficient."

The reduction of the charge occurs as follows. Assume that the charge Q_0 fills a small sphere of radius a from which the dielectric medium is excluded. Then an induced charge of opposite sign appears on the surface of that sphere, canceling part of the charge of Q_0. At the outer surface of the dielectric medium an equal charge of opposite sign appears. As long as the charge is measured within the medium (e.g., by a Gaussian surface) it will appear to be reduced. This familiar example should warn us that we must exercise care in defining "charge" once a medium is involved—there is a distinction between the "bare" charge Q_0 that describes a totally isolated system and a "dressed" charge that takes into account screening by dipoles in the medium.

In contrast to an ordinary dielectric, the vacuum is always present, its outer surface being at infinity, and the whole notion of "placing" a charge "in" the vacuum appears to be sophistry. Nevertheless, the theory necessarily starts out with a naïve picture where there are isolated charges, and then, via perturbation theory, treats the interaction between the charges of interest and the vacuum pairs. Hence we must distinguish between the "bare" charges Q_0 that our massive particles would have if there were no polarizable e^+e^- vacuum pairs, and the "dressed" or "effective" charges that they acquire when they polarize the vacuum pairs.

In order to understand the "dressing" process in more detail we must realize that the dielectric coefficient ε of the vacuum is not constant but depends on the distance r from P. This is so because at r only those virtual pairs whose members have wavelengths $\lambda \lesssim r$ contribute to the dielectric effect. For distances $r < m^{-1}$, ε therefore decreases as P is approached. Indeed, when r becomes equal to the assumed small radius[5] a of the charge, ($a \ll m^{-1}$), ε becomes unity, since an assumed radius a of a fermion is equivalent to a cutoff of all fields with wavelengths $\lambda < a$. For $r > m^{-1}$, ε remains constant at a value ε_v since, as we argued before, only electron fields with $\lambda \lesssim m^{-1}$ contribute to the virtual pairs (see Fig. 5).

A variation in space of the dielectric coefficient produces an induced charge density

$$\rho_{\text{in}} = -\mathbf{E} \cdot \boldsymbol{\nabla}\varepsilon, \tag{11}$$

where $\mathbf{E}$ is the electric field. Thus a positive charge at P induces a negative charge density around it, since ε increases from unity to its full value ε_v at $r > m^{-1}$. Hence, $\rho_{\text{in}}(r)$ does not extend beyond m^{-1}; it drops exponentially to zero[6] for $r > m^{-1}$.

Let us introduce an effective charge $Q_{\text{eff}}(r)$ contained in a sphere of

---

[5] We keep a finite but small in our qualitative discussions, because the polarization effects diverge logarithmically for $a \to 0$, as we see later on.

[6] It is instructive to compare the ε in QED, which increases between $r = a$ and $r \sim m^{-1}$, with the example of a dielectric medium where ε jump jumps from unity to a constant value at a. One may say that the induced opposite charge at $r = a$ when $\varepsilon =$ const. is smeared out between $r = a$ and $r \sim m^{-1}$ in the case of QED.

3. VACUUM POLARIZATION

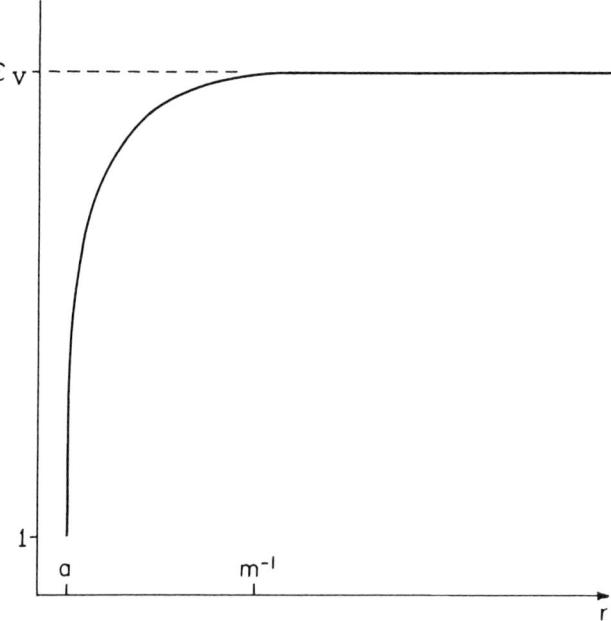

FIG. 5. A qualitative sketch of the dielectric coefficient ε as function of the distance r from a bare charge Q_0 of radius a.

radius r:

$$Q_{\text{eff}}(r) = Q_0 + \int_a^r \rho_{\text{in}}(r) \, d^3r, \tag{12}$$

where Q_0 is the bare charge placed at $r = 0$. This dressed charge, $Q_{\text{eff}}(r)$, is less than Q_0 because ρ_{in} has the opposite sign to Q_0. Since ρ_{in} vanishes for $r \gg m^{-1}$, $Q_{\text{eff}}(r)$ reaches a fixed value independent of r when $r \gg m^{-1}$, namely

$$Q = Q_0 + \int_a^\infty \rho_{\text{in}}(r) \, d^3r. \tag{13}$$

This asymptotic value Q is what, *by definition*, we call the *observed* charge, since we usually measure the effect of a charge at distances that are large compared to m^{-1}. At distances smaller than m^{-1}, the effective charge would be larger than Q. This indeed is the observed effect of vacuum polarization. The resulting overall charge density around a point charge is then given by

$$\rho(r) = Q_0 \delta^*(r) + \rho_{\text{in}}(r), \tag{14}$$

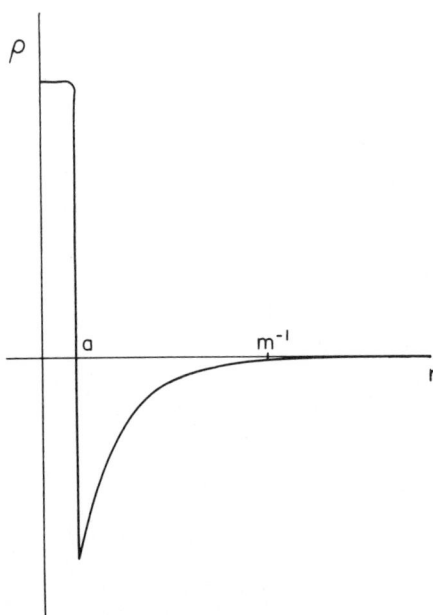

FIG. 6. A qualitative sketch of the charge density ρ near a positive charge Q_0 which is uniformly distributed within a sphere of radius a. The integral over ρ from a to infinity is $Q - Q_0$ where Q is the charge measured at large distances.

where $\delta^*(r)$ is a "delta-function" of width a (i.e., $\delta^* = 3/4\pi a^3$, if $r < a$; $\delta^* = 0$, if $r > a$). The function $\rho(r)$ is sketched in Fig. 6.

Our description of the vacuum polarization led us to the conclusion that the dielectric coefficient of the vacuum ε_v is very large and constant for $r \gg m^{-1}$. As we see below, this value diverges as $\log a^{-1}$ when the electron radius a goes to zero. The same holds for the ratio of the bare charge Q_0 to the observed one Q; it is equal to ε_v at $r \gg m^{-1}$.

Instead of assigning a large coefficient ε_v to most of the vacuum, it is simpler to describe the situation without regard to a dielectric coefficient, by using only the induced charge density. Consider an electrically charged particle of dimension $a \ll m^{-1}$. Its charge measured at distances $r \gg m^{-1}$ is Q. Let us assume Q is positive. Due to the vacuum fluctuations of virtual electron pairs, the particle is surrounded by a negative charge density ρ_{in} for $r < m^{-1}$, which increases with decreasing r. The region $r \leq a$, however, contains a positive charge $Q_0 \gg Q$ such that Eq. (13) is fulfilled. The resulting effective charge density is given by (12), and sketched in Fig. 6.

3. VACUUM POLARIZATION

[We now turn to a more quantitative description of the effects of vacuum polarization. Consider two fixed point sources, whose observed charges are $Q^{(1)}$ and $Q^{(2)}$, in the sense of (13). We call $\phi(r)$ their interaction energy when they are separated by r, and write ϕ as follows:

$$\phi(r) = \frac{Q^{(1)}Q^{(2)}}{4\pi r}[1 + C(r)]. \tag{15}$$

$C(r)$ tends to zero rapidly when $r \gg m^{-1}$, as one expected from the preceding discussion. For $r \ll m^{-1}$, on the other hand, $C(r)$ grows logarithmically. If one restricts oneself to the leading order [order e^2, as is clear from diagram (10)], these limiting forms of $C(r)$ are quite simple:

$$C(r) \sim \frac{\alpha}{4\sqrt{\pi}} \frac{e^{-2mr}}{(mr)^{3/2}} \qquad (r \gg m^{-1}), \tag{16}$$

$$C(r) \sim \frac{2\alpha}{3\pi}\left[\ln\left(\frac{1}{mr}\right) - \gamma - \frac{5}{6}\right] \quad (r \ll m^{-1}), \tag{17}$$

where $\gamma = 0.577$ is Euler's constant [see Lifshitz and Pitaevski (1974), §111].

The mathematical expression for $C(r)$ when r is of order m^{-1} is quite complicated, and for that reason we only show a plot of $C(r)$ in Fig. 7. As we see from the limiting expressions for $C(r)$ and Fig. 7, the correction to the Coulomb energy is very small (of order $\frac{1}{137}$) unless $r \to 0$, for there $C(r)$ diverges logarithmically. It must be emphasized, however, that $C(r)$ becomes large compared to $\frac{1}{137}$ only if r is small enough to make $\ln(1/mr)$ much larger than unity. Even for the Planck length, $r = 2 \times 10^{-33}$ cm, $C(r)$ is only 0.08.

The logarithmic singularity of $C(r)$ at $r = 0$ indicates that Q_0 would be logarithmically infinite if the artificially introduced radius a goes to zero. It follows, therefore, from (12) that

$$\int_a^r \rho_{\text{in}}(r)\, d^3r \sim \ln\frac{a}{r} \quad \text{for } a < r \ll m^{-1}. \tag{18}$$

Thus $\rho_{\text{in}}(r)$ must have a r^{-3} dependence for $r \ll m^{-1}$.

Both the bare charge and the induced charge density ρ_{in} at $r = 0$ diverge in our theory, but the effective charge $Q_{\text{eff}}(r)$ at any distance $r > 0$ is finite because the infinite bare charge is canceled by the infinite opposite charge of density ρ_{in} contained inside r. The deviation of $Q_{\text{eff}}(r)$ from the observed charge Q is small and of order αQ as long as $\log(1/mr)$ is of the order unity. As with the electron self-energy, the renormalization procedure has succeeded in reducing the radiative effects to small corrections for all distances of physical interest for the foreseeable future.]

How can one observe the effects of vacuum polarization? In ordinary atoms the effects are very small, since they are concentrated in a region of linear dimensions m_e^{-1}, which is much smaller than the dimensions of the atomic K-shell ($137/m_e Z$), if Z is small. Vacuum polarization does produce

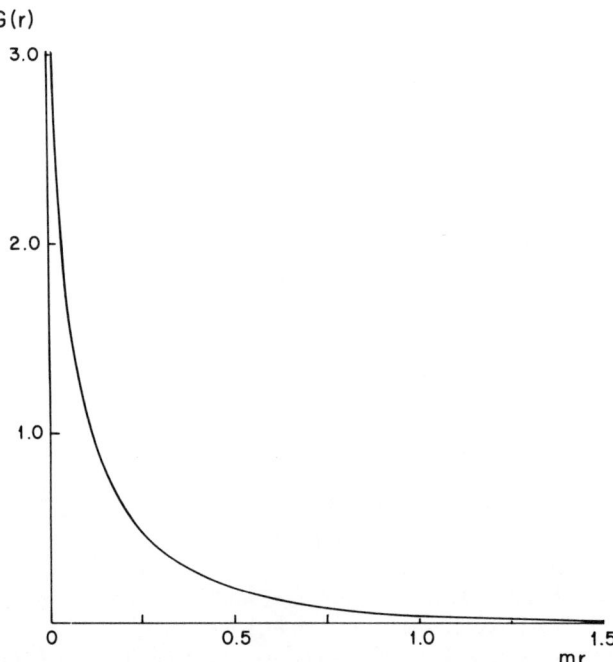

FIG. 7. A plot of the function $G(r) \equiv (3\pi/2\alpha)C(r) = 646 C(r)$.

a split between the $s_{1/2}$ and $p_{1/2}$ levels in hydrogen, since only in s-states do the electrons come near enough to the nucleus to feel the departure from the Coulomb field. But the effect is small and overshadowed by another that is discussed in the next section.

Vacuum polarization plays a much more important role in μ-mesic atoms. The characteristic Bohr radius of the μ^- is $(Z\alpha m_\mu)^{-1}$; since $m_\mu = 207 m_e \sim 2m_e/\alpha$, the μ^- Bohr radius is $\sim (2Zm_e)^{-1}$, which is smaller than the electron's Compton wavelength, especially when Z is large. Hence, μ-mesic orbits, even when $l \neq 0$, can lie in a region where the vacuum polarization is significant. The study of μ-mesic atoms has therefore provided a detailed check of the theory of vacuum polarization. For example, consider a transition between two $n = 2$ levels of a μ^- bound to a helium nucleus. The energy difference is measured to be[7]

$$\Delta E^{ex}(2s_{1/2} - 2p_{3/2}) = 1\,527.4(0.9)10^{-3}\,\text{eV}, \tag{19}$$

---

[7] A. Bertin et al., *Phys. Lett.* **55B** (1975) 411.

TABLE 1
Muonic X-ray transitions and vacuum polarization[a]

	Pb($5g_{9/2} \to 4f_{7/2}$)	Ba($4f_{7/2} \to 3d_{5/2}$)
ΔE_0	429,334.1	431,590.2(1.0)
δE_{vp}	2,078.7(2.2)	2,325.7(1.4)
δE_{other}	−70.4(7.0)	−3.6(2.6)
E^{th}	431,342(7)	433,912(3)
E^{ex}	431,360(11)	433,926(8)
$E^{th} - E^{ex}$	−17 ± 13	−14 ± 8

[a] All energies are in eV.

while the calculated value is[8]

$$\Delta E^{th}(2s_{1/2} - 2p_{3/2}) = 1\,525(9)10^{-3} \text{ eV}. \quad (20)$$

The large error in the calculation is due to uncertainties in the charge distribution of the He^4 nucleus. Furthermore, a number of effects besides vacuum polarization are responsible for the level splitting. Nevertheless, an analysis of the comparison of (19) and (20) shows that the vacuum polarization contribution is checked to about 1%.

One can reduce the uncertainties due to nuclear structure by studying muon transitions between states of high n and l, because such states have very little overlap with the nucleus, yet they range over the region of vacuum polarization ($r < m_e^{-1}$) if one chooses a heavy nucleus that produces a great deal of vacuum polarization as well as small μ-orbits. For that reason the muonic X-ray transitions $5g \to 4f$ in Pb and $4f \to 3d$ in Ba have been studied. The experimental results[9] are in Table 1. There, ΔE_0 is the energy difference computed from the Dirac equation for a muon in the nuclear Coulomb field, δE_{vp} is the correction due to vacuum polarization, and δE_{other} are other smaller corrections (e.g. nuclear recoil, influence of atomic electrons, etc.). E^{th} is the final theoretical result for the X-ray energy, and E^{ex} the observed value. As one sees, vacuum polarization accounts for about $\frac{1}{2}$% of the X-ray energy, and these measurements therefore verify the theory of this effect to about 5×10^{-3}.

Aside from having implications for spectroscopy, vacuum polarization also leads to scattering processes that have no classical counterpart. Maxwell's vacuum equations are strictly linear in the fields, and as a consequence a superposition of any number of electromagnetic waves is a solution of the classical equations of motion. Hence, in empty space there is no scattering of light by light in classical electrodynamics. Not so in quantum electrodynamics. The following diagram provides for photon–

[8] R. Barbieri, *Phys. Lett.* **56B** (1975) 266.
[9] L. Tauscher et al., *Z. Phys.* **285A** (1978) 139.

photon scattering:

$$\text{(diagram)} \quad (21)$$

This process is a consequence of vacuum polarization: One photon scatters from the transient charge fluctuations induced by the other. Unfortunately, this is a difficult process to observe, though there is good reason to hope that it will become possible as lasers become more powerful.

Scattering of light by an electrostatic field also does not occur in Maxwell's classical theory, but it too is possible in quantum electrodynamics because of vacuum polarization. Indeed, this process is closely related to photon–photon scattering in that two of the photons in diagram (21) are replaced by the Coulomb field of a large-Z nucleus.[10]

$$\text{(diagram)} \quad (22)$$

This process has been observed by using high-energy γ-rays; the data agree with the cross section computed from diagram (22).

4. The Lamb shift

If one computes the hydrogen spectrum by solving the Dirac equation for the motion of an electron in a Coulomb field, one finds a "fine structure" splitting of the multiplets that are degenerate in the nonrelativistic theory (see §C.5). But this splitting does not remove all degeneracies. For

---

[10] One might suppose that the simpler graph

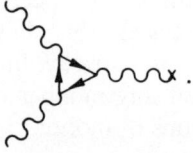

would dominate over (22), as it is of lower order in α. It turns out, however, that the sum of all such 3-vertex graphs vanishes because of charge conjugation invariance. One photon ($C = -1$) cannot turn into two because a 2-photon state has the quantum number $C = (-1)^2 = 1$ (recall §C.7).

4. THE LAMB SHIFT

example, in the $n = 2$ multiplet the $s_{1/2}$ and $p_{1/2}$ levels have the same energy to all orders in α. Actually the $2s_{1/2}$ level lies ~ 1000 MHz above the $2p_{1/2}$ level. As we see later in this Section, the $2s_{1/2} - 2p_{1/2}$ splitting (the Lamb shift) is due to radiative corrections.

The major part of the Lamb shift can be explained by a relatively simple physical picture because it is due to the response of the electron to the zero-point fluctuations of the radiation field. These fluctuations contribute to the self-energy of the electron, and also provide a correction to the electron-photon vertex:

 (23)

This is to be viewed as a correction to the amplitude for ordinary Coulomb scattering. As in the phenomena discussed previously, divergent integrals arise, but as we are now concerned with the energy levels of a bound electron, the renormalization procedure is somewhat simpler. The binding energy is, by definition, the *difference* in energy of a bound and a free electron. In calculating that difference the infinities associated with the electron's charge and mass can be arranged to drop out.

As we have said, the Lamb shift is largely due to the electron's response to the vacuum fluctuations of the electric field. The electric field $\mathbf{E}$ is not zero in the vacuum: the expectation value of $\mathbf{E}$ vanishes, but not its fluctuations. We can estimate the strength E_{fl} of these fluctuations as follows. Imagine that the field is enclosed in a very big cavity of volume V. There are then $aV\omega^2 d\omega$ modes in the interval, $d\omega$, where a is some numerical constant. Each mode acts like a quantized oscillator, and has an energy $E_0 = \frac{1}{2}\omega$ when unexcited (no photon present). To compute E_{fl}, recall that the electromagnetic energy density is $\frac{1}{2}(E^2 + B^2)$, and that $|\mathbf{E}| = |\mathbf{B}|$ for propagation in the vacuum. Hence $E_{\text{fl}}^2 V = E_0 = \frac{1}{2}\omega$ for one eigenfrequency, and for all frequencies in the interval $d\omega$, $E_{\text{fl}}^2 \sim \omega^3 d\omega$. This fluctuating field forces the electron to oscillate. For a nonrelativistic electron, the amplitude Δx of these oscillations is estimated from $m\ddot{x} = eE_{\text{fl}}$, since $\ddot{x} \sim \omega^2 \Delta x$. Thus

$$(\Delta x)^2 \sim e^2 E_{\text{fl}}^2 / m^2 \omega^4 \sim \frac{e^2}{m^2} \frac{d\omega}{\omega}. \tag{24}$$

This expression is only valid for low frequencies, $\omega \lesssim m$. When ω becomes larger than m, the existence of e^+e^- pairs, and other relativistic effects, reduce the previous expression by a factor $(m/\omega)^2$:

$$(\Delta x)^2 \sim e^2 E_{\text{fl}}^2 / \omega^6 \propto \frac{e^2}{\omega^3} d\omega, \quad (\omega > m). \tag{25}$$

We can now estimate the energy shift of a bound electron due to vacuum fluctuations. Consider an electron in a region where there is a potential $V(\mathbf{r})$ (e.g., a nuclear Coulomb field), and let $\psi_n(\mathbf{r})$ be one of its wave functions. When the vacuum fluctuations are ignored, the mean potential energy is

$$\langle V \rangle = \int V(\mathbf{r}) |\psi_n(\mathbf{r})|^2 \, dr^3. \tag{26}$$

The vibrations of the electron caused by E_fl change $\langle V \rangle$ by a small amount. To calculate this change we replace $V(\mathbf{r})$ in (26) by $V(\mathbf{r} + \Delta \mathbf{x})$, where $\Delta \mathbf{x}$ is the amplitude of the vibration, and expand:[11]

$$V(\mathbf{r} + \Delta \mathbf{x}) \simeq V(r) + \boldsymbol{\nabla} V \cdot \Delta \mathbf{x} + \tfrac{1}{6}(\Delta x)^2 \nabla^2 V.$$

The second term has no expectation value because $\langle \Delta \mathbf{x} \rangle = 0$. Hence there is an addition to the potential energy (26), which produces an energy shift δE_n:

$$\delta E_n = \tfrac{1}{6}(\Delta x)^2 \int \nabla^2 V |\psi_n(r)|^2 \, dr^3. \tag{27}$$

According to Gauss' Law, the Laplacian of the nuclear Coulomb potential is equal to the charge density of the nucleus. Since the latter lies within a radius much smaller than the extension of the wave function, the integration in (27) is reduced to an integration over the nuclear charge density, and gives Ze. Therefore

$$\delta E_n = \tfrac{1}{6} Z e^2 |\psi_n(0)|^2 (\Delta x)^2. \tag{28}$$

From this expression we see that only atomic s-states ($l = 0$) will have this positive energy shift, since only in these states is the probability $|\psi_n(0)|^2$ of being at the nucleus nonzero. For hydrogenlike atoms (that is, where the electron or muon sees a point charge Ze), the Schrödinger wave functions have the property

$$|\psi_n(0)|^2 = Z^3 / \pi n^3 a_0^3 \qquad (l = 0 \text{ only}), \tag{29}$$

where $a_0 = (m\alpha)^{-1}$ is the Bohr radius.

Finally, we determine $(\Delta x)^2$ by integrating the expressions (24) and (25) over frequency. At first sight one might integrate (24) from $\omega = 0$ to $\omega = m$, and (25) from $\omega = m$ to infinity. But as we are dealing with a bound state, the electron cannot vibrate freely for $\omega \lesssim \omega_0$, where ω_0 is a characteristic excitation energy. Hence the integral over (24) begins at

---

[11] The amplitudes do not depend on their direction, and that leads to the simple form of the third term.

$\omega \simeq \omega_0$, and gives a contribution of order $(e^2/m^2) \ln(m/\omega_0)$ to $(\Delta x)^2$. The integration over (25) converges and yields Ce^2/m^2, where C is some numerical constant. Thus the total value of $(\Delta x)^2$ is

$$(\Delta x)^2 \sim \frac{e^2}{m^2}\left(\ln\frac{m}{\omega_0} + C\right). \tag{30}$$

Inserting this and (29) into (28) we obtain

$$\delta E_n = A\frac{Z^4 \alpha^5 m}{n^3}\left(\ln\frac{m}{\omega_0} + C\right), \tag{31}$$

where A is another numerical constant. For $n = 2$ this energy difference is the $2s_{1/2} - 2p_{1/2}$ splitting—the Lamb shift. By simply setting $A \simeq C \simeq 1$, and $m/\omega_0 \sim (137)^2$, we find $\delta E_2 \simeq 10^3$ MHz for the $n = 2$ levels of hydrogen ($Z = 1$).

There are other ingredients in the theoretical calculations of the Lamb shift apart from the one we have estimated, but they are all considerably smaller, and only add an increment to the constant C. These contributions can be divided into three parts. The first comes from that part of the difference between the electron self-energy in the free and bound states which is not due to the electron vibrations, which we have just estimated. The second comes from the change of the Coulomb field due to vacuum polarization; it is negative because the effective nuclear charge is larger within a radius m^{-1}; however, it contributes only about 2% of the total effect. The third contribution is due to the radiative correction to the electron's magnetic moment, which is described in the next section. When all these effects are evaluated, one finds that the level shift is given by (31), with $A = 4/3\pi$ for all n, whereas C varies slowly with n; for $n = 2$, $C = -2.87$. The size of these separate contributions to the Lamb shift is shown in Table 2.

The two most recent measurements give the following results:[12,13]

$$\Delta\nu(2s_{1/2} - 2p_{1/2}) = \begin{array}{l} 1057.845(9) \text{ MHz} \\ 1057.862(20) \text{ MHz} \end{array}$$

The most recent theoretical calculation gives[14]

$$\Delta\nu(2s_{1/2} - 2p_{1/2}) = 1057.860(9) \text{ MHz}$$

This theoretical value differs a little from the sum of the tabulated

---

[12] S. R. Lundeen and F. M. Pipkin, *Phys. Rev. Lett.* **46**, 232 (1981).
[13] D. A. Andrews and G. Newton, *Phys. Rev. Lett.* **37**, 1254 (1976); G. Newton, D. A. Andrews, and P. J. Unsworth, *Phil. Trans. Roy. Soc.* **290**, 373 (1979).
[14] J. Sapirstein, *Phys. Rev. Lett.* **47**, 1723 (1981).

TABLE 2
Lowest order contributions to the Lamb shift in hydrogen (in MHz)

	$2p_{1/2}$	$2s_{1/2}$
Electron self-energy = vacuum fluctuations	4.07	1015.52
Vacuum polarization	0	−27.13
Anomalous magnetic moment	−16.95	50.86

contributions because it also contains the contribution of higher approximations in α. The agreement with experiment is most impressive.

5. The anomalous magnetic moment of the electron and muon

Another consequence of the radiative modification of the interaction of a lepton and an external field, as depicted by diagram (23), is the anomalous magnetic moment. In this context, the external field is a uniform magnetic field $\mathbf{B}_0$. It is customary to write the magnetic moment of a particle as

$$\mathbf{\mu} = \frac{e}{2m} g \mathbf{s}, \tag{32}$$

where $\mathbf{s}$ is the spin angular momentum, and g is the gyromagnetic ratio—the so-called g-factor. For a lepton, $\mathbf{s} = \frac{1}{2}\mathbf{\sigma}$. One can find g by computing the energy $-\mathbf{\mu} \cdot \mathbf{B}_0$ of a lepton in the magnetic field. If one retains only the lowest order diagram [C(31)], one finds that $\mathbf{\mu} \cdot \mathbf{B}_0 = \pm eB_0/2m$, or $g = 2$. This is called the Dirac value of g because it emerges directly from the Dirac equation itself. This value of $|\mu|$ agrees well with all low-precision spectroscopic measurements, and for that reason any departure from $e/2m$ is called an *anomolous magnetic moment*.

If one calculates the lowest radiative correction (23) to the spin-field coupling, one finds that the lepton's g-factor is slightly altered from its Dirac value:

$$g = 2\left(1 + \frac{\alpha}{2\pi}\right). \tag{33}$$

Precision studies of the fine structure and Zeeman effect in hydrogenic atoms have shown that the gyromagnetic ratio is, to a good approximation, given by this formula.

Today, extremely accurate measurements of $g - 2$ are available for both the electron and the muon. It is conventional to define $a \equiv \frac{1}{2}(g - 2)$. The

5. MAGNETIC MOMENT OF THE ELECTRON AND MUON

experimental results for this quantity are[15]

$$a_{e^-} = 1{,}159{,}652{,}220(40) \times 10^{-12},$$

$$a_{e^+} = 1{,}159{,}652{,}222(50) \times 10^{-12},$$

$$a_{\mu^-} = 11{,}659{,}370(120) \times 10^{-10},$$

$$a_{\mu^+} = 11{,}659{,}110(110) \times 10^{-10}.$$

Note that the *departure* of g from 2 is known to a part per 10^7 for the electron, and ~ 2 parts per 10^5 for the muon!

As indicated in Appendix IV, the most basic principles of quantum field theory alone require a particle and its antiparticle to have equal g factors (i.e., equal but opposite magnetic moments). That the anomalies for leptons and antileptons agree to such high precision therefore provides a highly significant check of those principles.

To compare these remarkable measurements with theory, one must go far beyond the simplest graph (23), and the corresponding expression (33). All terms to order α^3 in $g - 2$ must be kept, i.e., graphs such as

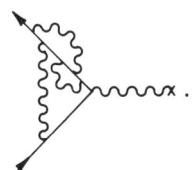

The electron and muon $g - 2$ values differ once one goes beyond the order (23). Larger virtual momenta are involved in the muon case, and therefore muon vacuum polarization becomes important in a_μ. These very lengthy and intricate calculations give the following results[16]

$$a_e = 1{,}159{,}652{,}379(261) \times 10^{-12}$$

$$a_\mu = 11{,}659{,}202(20) \times 10^{-10}$$

Errors in the theoretical values are due to experimental uncertainties in the fine structure constant, uncertainties as to the numerical importance of α^4 radiative corrections, and uncertainties in hadronic contributions to vacuum

---

[15] e: P. B. Schwinberg, R. S. Van Dyck, and H. G. Dehmelt, *Phys. Rev. Lett.* **47**, 1679 (1981).
   μ: J. Bailey et al., *Phys. Lett.* **68B**, 191 (1977). F. J. M. Farley and E. Picasso, *Ann. Rev. Nuc. Sc.* **29**, 243 (1979).

[16] e: M. J. Levine and J. Wright, *Phys. Rev.* **D8**, 3171 (1973). P. Cvitanović and T. Kinoshita, ibid., **D10**, 4007 (1974). R. Carroll, ibid., **D12**, 2344 (1975).
   μ: T. Kinoshita, B. Nižić, and Y. Okamoto, *Phys. Rev. Lett.* **52**, 717 (1984).

polarization, as in

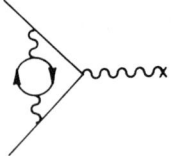

where the loop is due to quarks.

The remarkable agreement between theory and experiment[17] must be taken as strong evidence that the leptons conform with our present notion of "elementarity," which implies no extension in space and no internal structure: It is only for such an elementary Dirac particle that $g = 2$ apart from radiative corrections, and it is only for such a particle that one can compute the radiative corrections precisely. To underscore this point, recall that spin-$\frac{1}{2}$ hadrons (in particular, the proton and neutron) have g-factors that are very far from their Dirac values (2 and 0, respectively). This is no longer the mystery it once was, for we now have a vast body of data that clearly says that hadrons are objects of finite extension with internal structure. Thus, if leptons do have finite extension or internal structure, it must be restricted to very small dimensions. Today the limit is less than 10^{-16} cm, as we have seen.

6. Other spectroscopic tests of quantum electrodynamics

In addition to the tests of radiactive correction calculations that we have now discussed in some detail, there exists a larger body of spectroscopic data that bears on the validity of quantum electrodynamics. We have no intention of providing an exhaustive summary of this information, [see Kinoshita (1984)], and will only quote several results that serve to provide a more rounded picture of the present state of our knowledge.

As we learned in §C.5, the ground state of the hydrogen atom actually consists of four states, a triplet ($J = 1$) where the electron and proton spins are parallel, and a singlet ($J = 0$) where these spins are antiparallel. The splitting Δv between these levels is called the hyperfine structure (hfs). It has been measured to the incredible precision of 1 part in 10^{12} by the hydrogen maser:[18]

$$\Delta v_{\text{ex}} = 1{,}420.405\,751\,706\,7(10) \text{ MHz.}$$

This is so precise that one cannot foresee when theory will achieve comparable accuracy, because that would require a knowledge of the

---

[17] Note that the quoted results are for the anomaly; consequently, the agreement between theory and experiment for the whole magnetic moment holds to a part per 10^{10} for the electron, and 10^8 for the muon.

[18] L. Essen et al., *Nature* **229**, 110 (1971).

proton's electromagnetic properties that we are unlikely to attain. The most sophisticated calculation to date, which already includes a variety of radiative corrections, agrees with experiment but still falls seven orders of magnitude short of the experimental precision!

The spectra of positronium and muonium (μ^+e^-) have the virtue that they are not marred by uncertainties concerning hadronic structure. While these measurements cannot match that of the hydrogen hfs, they nevertheless provide tests of the theory of comparable significance. The ground state hfs of muonium gives the best test of theory at present:[19]

$$\Delta\nu_{ex} = 4{,}463{,}302.88(16) \text{ kHz,}$$

$$\Delta\nu_{th} = 4{,}463{,}304.7(2.0) \text{ kHz.}$$

Here only the first three significant figures are unaffected by radiative corrections, and this therefore provides a very severe test of quantum electrodynamics.

As a final example, consider a 2-electron system, the He atom, and focus on the fine structure splittings in the 2^3P multiplet. Let E_J be the energies of the three ($J = 0, 1, 2$) levels in the multiplet, and define $\nu_{JJ'} = (E_J - E_{J'})/h$. Then:[20]

$$\nu_{01}^{ex} = 29{,}616.864(36) \text{ MHz}$$

$$\nu_{01}^{th} = 29{,}616.834(110) \text{ MHz}$$

$$\nu_{12}^{ex} = 2{,}291.196(5) \text{ MHz}$$

$$\nu_{12}^{th} = 2{,}291.28\,(21) \text{ MHz}$$

7. Summary

We have now quoted a most impressive body of data that confirms the predictions of quantum electrodynamics: high-energy cross sections, magnetic moments, and atomic energy differences. From this we conclude that the photon, electron, and muon are structureless particles to the precision attained thus far, in the sense that they obey Maxwell's and Dirac's equations. Furthermore, the successful verification of the radiative corrections must mean that the basic concepts of relativistic quantum field theory have captured essential elements of physical reality.

---

[19] Experiment: F. G. Mariam et al., *Phys. Rev. Lett.* **49**, 993 (1982). Theory: G. P. Lepage, *Phys. Rev.* **A16**, 865 (1977); G. T. Bodwin, D. R. Yennie, and M. Gregorio, *Phys. Rev. Lett.* **48**, 799 (1982); J. Sapirstein, ibid. **51**, 985 (1983).
[20] Experiment: A. Kponou et al., *Phys. Rev. Lett.* **26**, 1613 (1971). Theory: J. Daley et al., ibid., **29**, 12 (1972); also M. L. Lewis, *Proc. 4th Int. Conf. Atomic Physics*, Heidelberg (Plenum Press, New York, 1974).

III

HADRONIC SPECTROSCOPY

This chapter is devoted to static properties of hadrons, and to their low-lying excited states. In Part A we summarize the data that determine the dimensions of hadrons, and describe techniques that are used to deduce the hadronic level schemes shown in Figs. I.6 and 8. This analysis of the spectrum does not assume any model of hadronic substructure.

As we learned in §I.E.6, the quark model gives a comprehensive rendition of the hadronic spectrum. But one must ask whether this success is only descriptive, or whether it is quantitatively accurate. This question is best answered by examining the ψ- and Υ-family of mesons. These are composed of heavy quark-antiquark pairs, and their approximately non-relativistic motions are therefore more amenable to a detailed theoretical treatment [recall §I.E.6(e)]. Part B is devoted to this topic, and as we shall see, the quark model does describe the rich ψ- and Υ-spectra in a quantitatively satisfactory manner. Indeed, a large portion of the model's consequences was known before the data were available. This circumstance led to a widespread recognition that the quark model provides a dynamical description of hadronic structure, and is not just an economical algorithm for constructing the correct quantum numbers of hadronic states. Furthermore, it is also clear that hadrons containing light quarks are relativistic systems, and therefore not as accessible to theoretical analysis. The most incisive confrontations between theory and experiment are therefore to be expected in the spectroscopy of heavy quark-antiquark systems; it is there that one can hope eventually to see tests of quantum chromodynamics (QCD) analogous to those that simple atoms provide for QED. The theory of hadrons containing light quarks is based on models that lean quite heavily on QCD, and this topic is therefore left to Chap. IV.

A. HADRONIC DIMENSIONS AND SPECTRA

The experiments that reveal the structure and the excitation spectra of hadrons are similar in character to those that revealed the structure of atoms. It was the scattering of charged particles that led to the planetary model of the atom; it is the scattering of high-energy electrons and neutrinos that gives us the most unambiguous knowledge of the spatial extent of hadrons, and of the existence of small charged entities (the quarks) within the hadrons. These experiments are discussed in Chap. V. The spectra of atoms were most clearly revealed by the study of the radiation emitted from excited states, and by the Franck–Hertz experiment. In analogy, we use the emissions from excited hadrons, and scattering, to examine the hadronic excitation spectrum.

To our best knowledge the proton is the only completely stable hadron. All other baryons decay into the proton by some combination of strong, electromagnetic, or weak decays, while all mesons eventually decay into photons and/or leptons. Decays due to the weak and electromagnetic interactions are analogous to the decay of atoms by radiation. They are very slow compared to the time scale of hadronic dynamics, a time defined by the reciprocal of the typical level spacing. One can easily imagine a world in which those interactions are "switched off". Decay due to the strong interaction—such as the decay of an excited baryon state by meson emission—presents greater conceptual difficulties. One cannot switch off the strong interactions without dissolving the hadrons; there is no clear analogy in the atomic realm.[1] But there are analogies in the nuclear realm: There are excited states of nuclei that can emit a nucleon or an α-particle. The strong interaction leads to quasi-stationary hadronic states with a reasonably large lifetime that is long compared to the hadronic time scale. These states are sufficiently excited to allow the emission of groups of quarks that bond into hadrons, as we learned in §I.E.7. The time required for this to occur is usually quite comparable to the characteristic time of the internal hadronic motions. Therefore, the widths of hadronic levels are often appreciable, and sometimes not much smaller than the level spacing. For example, the width of the first Δ-level is 120 MeV, whereas its distance above the ground state is 300 MeV.

[1] Atoms are kept together by the Coulomb force. If one ignores relativity, it is possible to separate the Coulomb force from the radiation field, and one can imagine a world with Coulomb forces but without radiation.

1. The electromagnetic size of hadrons

In §II.C.3 we learned that one determines the charge distribution of an object T by scattering a point charge off T, and measuring the deviations from the Rutherford formula. As electrons are known to have a spatial extent smaller than $\sim 10^{16}$ cm, they are the ideal projectiles for this purpose. If we ignore spin for the moment, we recall [cf. Eq. II.C(26)] that the differential cross section can be written as

$$\frac{d\sigma}{d\Omega} = \frac{d\sigma_{\text{pt}}}{d\Omega}|F(q)|^2, \qquad (1)$$

where $d\sigma_{\text{pt}}/d\Omega$ is the cross section when T is a point charge, in which case the form factor F is identically one. If T has an extended charge distribution $\rho(r)$, the dependence of F on q, the momentum transfer suffered by the projectile, is given by

$$F(q) = \int e^{i\mathbf{q}\cdot\mathbf{r}} \rho(r) \, d^3r. \qquad (2)$$

When T has spin, the differential cross-section formula becomes somewhat more complicated than (1). This is hardly astonishing. Take the important case when T is a proton, which carries not only a charge, but also a magnetic moment. One must expect the deflection suffered by a passing electron to depend on how both the charge and the magnetization are distributed. For that reason the electron-proton scattering formula involves two form factors G_E and G_M, the first for the charge density as given by (2), and the second given by an analogous expression involving the density of magnetization. However, (1) is correct when T has spin zero; for example, for a pion.

In Fig. 1 we show these form factors G_E and G_M/μ_p for the proton, where $\mu_p = 2.79$ is the proton's magnetic moment in units of the nuclear Bohr magneton $e/2m_p$. If the proton had a point charge and magnetic moment, one would have $G_E = G_M = 1$, but as one sees, enormous deviations from this are observed. By using Eq. II.C(27), one extracts the rms radii for the proton's charge and magnetization distributions from the G's, and finds

$$\sqrt{\langle r^2 \rangle}_p = (0.82 \pm 0.02) \text{ fm} \qquad (3)$$

for both distributions, where fm $\equiv 10^{-13}$ cm. The neutron's magnetization distribution has an rms radius that also agrees with (3) to within errors.[2]

The instability of the π's and K's makes it impossible to scatter electrons from them. But the converse is possible: One can scatter a pion or kaon beam off bound electrons in matter. These experiments have been per-

---

[2] The neutron's form factors are found from electron-deuteron scattering.

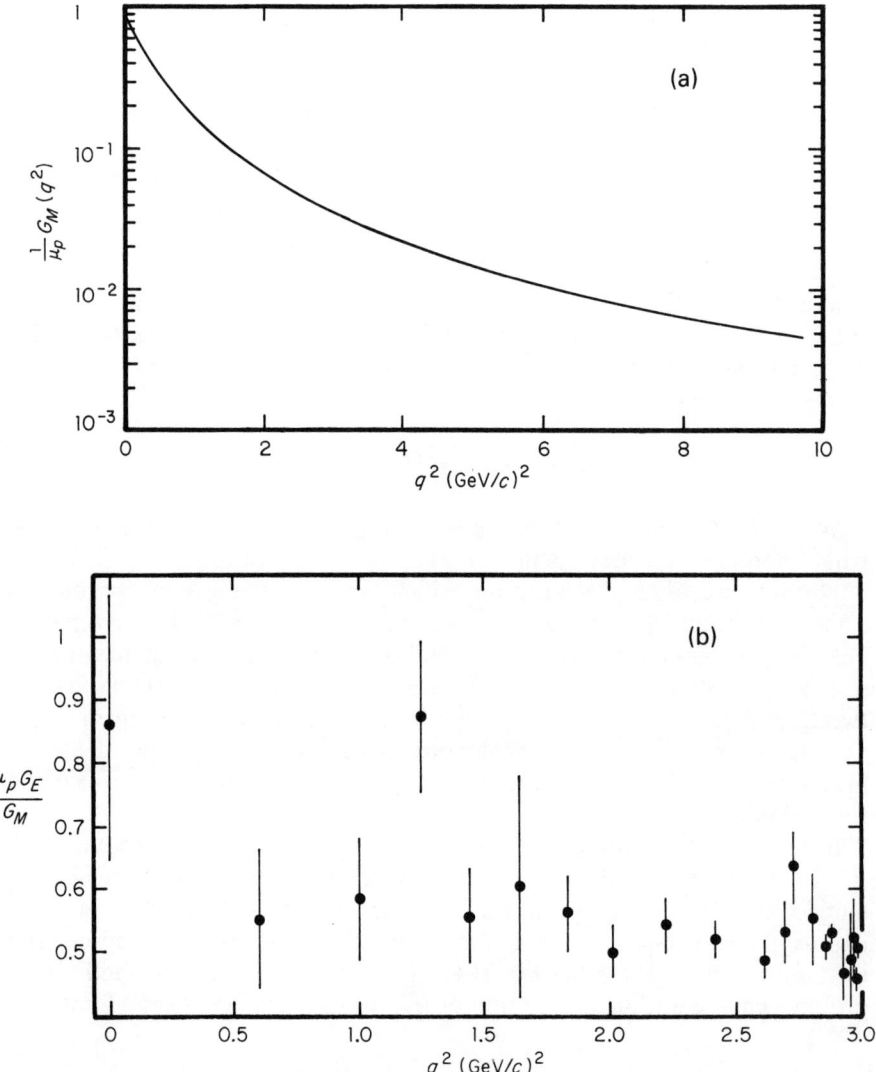

FIG. 1 (a) The form factor of the proton's distribution of magnetization divided by the proton's magnetic moment, $G_M(q^2)/\mu_p$. The form factor G_M is defined to have the value μ_p at $q^2 = 0$. The curve shown is $[1 + (q^2/0.71)]^{-2}$, where q^2 is in units of $(\text{GeV}/c)^2$. It gives an excellent fit to the data [see, for example, Weber (1967)] for the momentum transfers shown here, though there are deviations at larger values of q^2.

(b) The measured ratios of the form factors for the charge and magnetization distributions, $\mu_p G_E(q^2)/G_M(q^2)$, as a function of q^2. This compilation is taken from Höhler et al. (1976).

formed [Dally et al. (1980, 1982)], with the results

$$\sqrt{\langle r^2 \rangle_\pi} = (0.64 \pm 0.04) \text{ fm},$$
$$\sqrt{\langle r^2 \rangle_{K^\pm}} = (0.53 \pm 0.05) \text{ fm}.$$
(4)

It is therefore an established fact that hadrons have an extended charge distribution.

2. Hadronic diffraction scattering and the size of hadrons

When light impinges on an opaque sphere, a shadow is cast beyond the sphere. But this shadow does not extend to infinity; at large distances it collapses and evolves into the diffraction pattern shown in Fig. 2(a). By studying this pattern, i.e., the intensity of scattered light as a function of angle, one can infer the radius of the sphere.

A similar diffraction pattern is formed whenever a plane wave impinges upon an object that is partially or completely absorbing, and large compared to the wavelength. The wave may be electromagnetic, as in the preceding example, or it may be a quantum-mechanical wave function. The shape and opacity of the absorbing object can, in all cases, be deduced from the diffraction pattern by techniques familiar from wave optics.

In scattering between hadrons, the conditions for the formation of a diffraction pattern are met when the energy rises above ~ 10 GeV. The wavelength is then short compared to the size of hadrons, and the almost certain outcome of the collision is the production of many particles (mainly pions). Therefore the hadron acts as an absorbing object to the incoming wave. To see the consequences of the latter more clearly, consider a proton beam impinging on a proton. The state that evolves out of this initial configuration is described by many wave functions: ψ_{pp}, which describes pp elastic scattering, and the wave functions $\psi_{n\pi}$ for configurations containing n pions produced in the collision. (We ignore kaons, etc., as they do not affect the argument.) The production amplitudes $\psi_{n\pi}$ are only nonzero in a cylinder on the downstream side of the target. In this same cylinder the probability of finding no pions, and a proton of the incident energy, is virtually zero, so ψ_{pp} vanishes in this region. Therefore, the target casts a shadow, which results in a proton diffraction pattern at large distances where the elastically scattered protons are detected.

In Fig. 2(b) we show the diffraction pattern observed in high-energy elastic pp scattering. There is a secondary maximum, but it is far less pronounced than in the case of an opaque sphere [Fig. 2(a)], showing that hadrons have a rather diffuse edge, as is also known from the detailed analysis of the ep scattering data of Fig. 1. The pp data can be analyzed by techniques that are essentially optical; the hadronic opacity is found to be

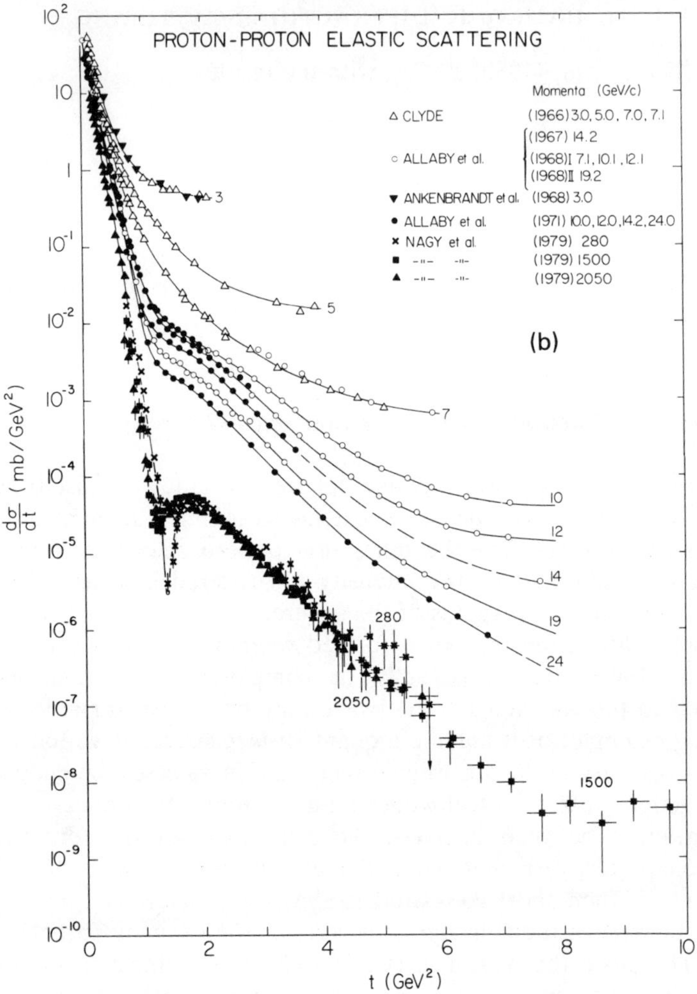

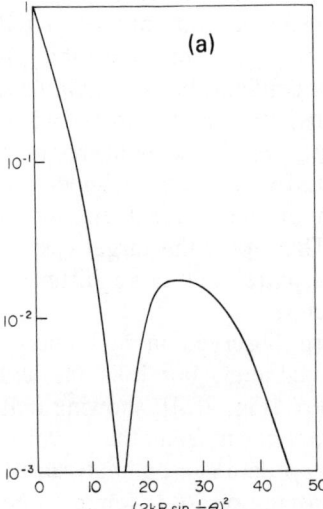

FIG. 2. (a) The diffraction pattern of an opaque sphere of radius R for light of wavelength $2\pi/k$. The vertical scale gives the scattered intensity per unit solid angle normalized to one in the forward direction.

(b) The diffraction pattern as observed in pp elastic scattering. The angular variable is $t = 4p^2 \sin^2 \frac{1}{2}\theta$, the square of the momentum transfer, with p and θ the momentum and scattering angle in the c.o.m. frame. This compilation includes data from fixed target experiments with laboratory momenta p_{lab} from 3 to 24 GeV, as indicated, and from the CERN-ISR pp colliding beam facility, in which case the equivalent p_{lab} ranges from 280 to 2050 GeV, as shown. The diffraction minimum only emerges at high energies because the absorptive region has a diffuse edge, as shown in Fig. 3. For the same reason, the secondary maximum is far smaller than in (a). This compilation is from Nagy et al. (1979), where references to the original data can be found.

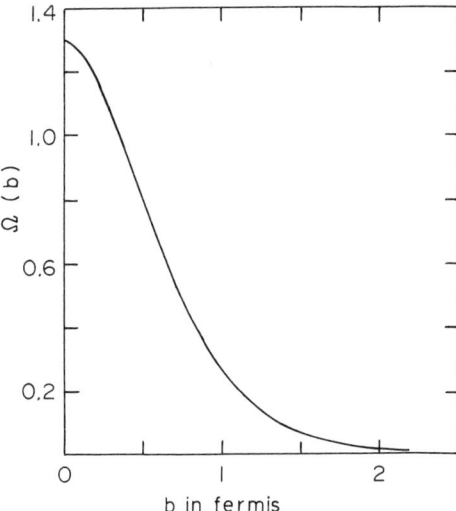

FIG. 3. Radial distribution of opacity, $\Omega(b)$, inferred from pp scattering data of Fig. 2(b). The definition of $\Omega(b)$ is that a ray at impact parameter b will be decreased in amplitude by the factor $\exp[-\Omega(b)]$. The rms value of b characterizes a two-dimensional projection of the hadronic density, whereas the rms charge radii pertain to a three-dimensional distribution. This graph is taken from Amaldi (1973).

roughly Gaussian in shape (see Fig. 3) with an rms radius of 0.65 fm. Similar radii are found from the analysis of πp and Kp elastic scattering.

3. Excited states of hadrons

Having established that hadrons are spatially extended objects, we now turn to evidence concerning their excitation spectra.

There are two ways of observing the existence of excited hadron states: (a) by "formation," and (b) by "decay." We illustrate this with atomic excitations.

(a) An atom in the state A is exposed to photons of variable frequencies ω; in the immediate neighborhood of a certain frequency ω_R, a strong reaction with the atom takes place (resonance). This is interpreted as absorption of the photon by the atom, and the formation of an excited state B, according to the scheme

$$\gamma + A \to B \dashrightarrow \text{decay}.$$

The broken arrow indicates that B is not the end product of the process; it will decay back to A, or in other ways. The excited state is found in the *formation reaction* (solid arrow) by determining the frequency at which resonance occurs:

$$\omega_R \cong E_B - E_A, \tag{5}$$

where E_B and E_A are the energies of these states. This relation is not exact since the finite lifetime τ_B of B prevents its energy from being sharply defined. The spread of energy has a width $\Gamma_B = 1/\tau_B$.

(b) An excited state B of the atom is created in some way or other. Then the decay of B is observed by measuring the energy ω_R of the light emitted. This energy will be $\omega_R \cong E_B - E_C$, where C is a state of the atom lower than B. This process is described by

$$\text{Production} \dashrightarrow B \to C + \gamma.$$

The excited state B is determined by observing the decay process (solid arrow).

We now describe the two analogous ways of observing excited states of hadrons.

(a) A nucleon in the state A is exposed to a beam of particles a (a can be any particle, including photons). In the immediate neighborhood of a certain definite momentum p_a of a we find a large cross section for interaction with the target. We conclude that an excited state B is produced by the coalescence of a and A, which decays after ward in one or more ways:

$$a + A \to B \dashrightarrow \text{decay}. \tag{6}$$

Let us look at the energy relations. We ascribe an "invariant mass" M_{aA} to the initial particle pair, which is defined as the total energy E_i of the pair in the center-of-mass frame. The term "invariant mass" is quite natural, since the mass is the energy in a frame where the momentum vanishes. In such a frame $\mathbf{p}_a = -\mathbf{p}_A$, and

$$E_i \equiv M_{aA} = \sqrt{p_a^2 + m_a^2} + \sqrt{p_a^2 + M_A^2}. \tag{7}$$

The large cross section (resonance) occurs when E_i is close to the energy E_B of B. In the c.o.m. frame E_B is equal to the mass M_B of B, since a has coalesced with A, and therefore B is at rest. Hence, the resonance condition becomes

$$M_{aA} \approx M_B. \tag{8}$$

This energy relation is only approximately true, since B has a finite lifetime, which gives rise to a width, as already mentioned.

(b) The excited state B of a hadron is created in some way. Then the decay of B into $C_1, C_2, C_3, \ldots$ is observed, and the energy and momentum of the products measured:

$$\text{Production} \dashrightarrow B \to C_1 + C_2 + \cdots. \tag{9}$$

Here we again define the invariant mass $M_{C_1 C_2 \ldots}$ of all the decay products as their total energy in the rest frame of B. This need not be the c.o.m. frame of the reaction, since B may have been produced together with other particles.

As an example, consider B decaying into two particles with masses m_1 and m_2 and momenta $\mathbf{p}_1$ and $\mathbf{p}_2 = -\mathbf{p}_1$ measured in the rest frame of B. Their invariant mass is

$$M_{C_1 C_2} = \sqrt{p_1^2 + m_1^2} + \sqrt{p_1^2 + m_2^2}, \tag{10}$$

which can be computed from the observed decay momenta. Energy conservation then allows one to find M_B:

$$M_B \approx M_{C_1 C_2}. \tag{11}$$

Again this is only an approximate equality because of the finite lifetime of B. Note that the decay $B \to C_1 + C_2$ implies that in a collision of C_1 and C_2, B appears as a formation resonance. However, such collisions cannot be performed in the numerous cases where C_1 and C_2 are short-lived excited states (e.g., in the reaction $pK^+ \to K^{*+}p$, see Fig. 11 below).

There are reactions in which the decay process is the inverse of the formation process. The simplest example is resonance fluorescence—the elastic scattering of a photon by a system at resonance. A photon γ is absorbed by an object in the state A, and excites it into the state B; then B emits γ and returns to A: $\gamma A \to B \to \gamma A$. More generally, the role of the photon is played by any incident particle a. The corresponding Feynman diagram is

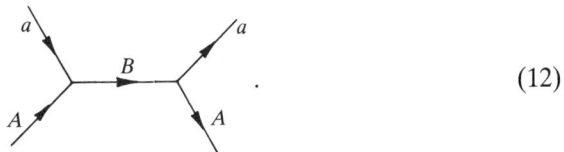

$$\tag{12}$$

A comment is in order concerning this diagram, and others to follow.[3] In Chaps. I and II the lines in Feynman diagrams always represented pointlike particles: quarks, leptons, γ, $W^\pm$, and Z^0. When one is analyzing spectroscopic phenomena, as we are here, one can describe collision and decay processes as if the various hadrons were point particles, and represent them as lines in Feynman diagrams. In contrast to the Feynman diagrams of QED

---

[3] Another comment: In this Chapter we often draw diagrams as if time flows from left to right, not bottom to top. As already mentioned in §I.C.4, these are equivalent conventions.

and QCD, the vertices do not involve fundamental constants such as e, but structure-dependent transition amplitudes for $B \to Aa$, etc. These can, at least in principle, be computed from the underlying quark structure, in the manner indicated in Fig. I.24 for the example of the nuclear force.

The scattering amplitude corresponding to diagram (12) follows from a generalization of the perturbation theory described in §II.B. We recall that in a process involving the initial state $|i\rangle$ and the intermediate state $|n\rangle$, the scattering amplitude is proportional to $(E_i - E_n)^{-1}$. In our example,

$$E_i = M_{aA}, \quad E_n = M_B. \tag{13}$$

The energy eigenvalue of the intermediate state B requires further discussion since it decays and is not stationary. If the state $|B\rangle$ is prepared at $t = 0$ (e.g., by bombardment with a nearly monochromatic wave packet), it evolves into $|B(t)\rangle$, and the probability $P(t)$ for still finding the latter in $|B\rangle$ subsequently is given by

$$P(t) = |\langle B | B(t)\rangle|^2 = e^{-t/\tau}, \tag{14}$$

where τ is called the *lifetime* of $|B\rangle$. The Schrödinger equation tells us that the time-dependence of any state has the form e^{-iWt}; furthermore, $P(t) = |e^{-iWt}|^2$. Therefore, (14) says that for an unstable state the eigenvalue W is not real,[4] but has an imaginary part $-i/2\tau$. The real part of W is, as usual, given by the energy of the state, which is M_B in our case. Thus $W = M_B - i/2\tau$. It is now convenient to define the quantity $\Gamma_B = 1/\tau$, which is called the *width* of $|B\rangle$, for reasons that will soon be clear. From (14) we also have

$$-\frac{1}{P}\frac{dP}{dt} = \Gamma_B, \tag{15}$$

and therefore Γ_B is the *decay rate* (recall that $\hbar = 1$).

In terms of Γ_B the eigenvalue W of $|B\rangle$ is

$$W = M_B - \tfrac{1}{2}i\Gamma_B. \tag{16}$$

This is also the energy E_n of the intermediate state in (12). Thus, the energy

---

[4] As the Hamiltonian of any system is Hermitian, one may wonder at the appearance of a complex eigenvalue. To understand this, consider an atom in an excited state B, and the radiation field which, at $t = 0$, contains no photons. The complete wave function describing both the atom and photons has a time-independent norm, and therefore a real energy eigenvalue. But if we confine our attention to the atom, it *seems* that probability is lost: the probability for finding B decreases exponentially. For that reason the subsystem B has a complex eigenvalue, with the imaginary part accounting for this "loss." Overall probability is strictly conserved, of course; the "loss" is exactly compensated by the gain in probability for finding $A + \gamma$.

denominator that occurs in the scattering amplitude corresponding to graph (12) is

$$[E_i - W]^{-1} = [M_{aA} - M_B + \tfrac{1}{2}i\Gamma_B]^{-1}. \tag{17}$$

The elastic scattering cross section σ_{aa} of particle a by A is proportional to the absolute square of this amplitude:

$$\sigma_{aa} \propto \frac{1}{(M_{aA} - M_B)^2 + \tfrac{1}{4}\Gamma_B^2}. \tag{18}$$

Equation (18) confirms our expectation: σ_{aa} as function of the incident energy $E_i = M_{aA}$ shows a resonance of width Γ_B centered at the resonant energy M_B. In atomic physics ($a = \gamma$, A = atom), Γ_B is minute compared to level spacings, and the resonance is exceedingly sharp. But the resonance formula (18) is also applicable in situations where Γ_B is appreciable.

Let us now look at the decay side of the process (12). Clearly, the energy of the emitted particle must be the same as that of the absorbed one. In addition to the energy dependence, the angular distribution of the scattered particle is also of great interest. The angular momentum J_B of B can actually be determined from the angular distribution at resonance, for it merely involves the square of the angular momentum wave functions of the scattered particles. To illustrate this point, consider a ground state with $J_A = 0$, and a particle a of spin zero. In this special case, the angular distribution of resonance scattering is just $|P_L(\cos\theta)|^2$, with $L = J_B$, where P_L is the Legendre polynomial of order L, and θ the scattering angle. When the spins of a and A are not zero, the angular distribution is not very different; P_L is just replaced by a somewhat more complicated polynomial of $\cos\theta$ and $\sin\theta$.

We must generalize the process described by diagram (12) to cases in which the decay of the resonant state B is not the inverse of its formation process. We simplify the situation by assuming that the decay of B leads only to two particles c and C. The Feynman diagram is

$$\tag{19}$$

Examples of such processes are $\gamma p \to B \to \pi^+ n$, $\pi^- p \to B \to n\pi^0$, and $\pi^- p \to B \to \gamma n$. In these examples, B is an excited baryon. As (19) has the structure of the resonance fluorescence graph (12), the denominator that governs the energy dependence of the amplitude has the same structure as in Eq. (17), and the cross section will be proportional to $|E_i - W_B|^{-2} = [(M_{aA} - M_B)^2 + \tfrac{1}{4}\Gamma_B^2]^{-1}$, where E_i is the energy of the initial state as given by (7), and W_B is the complex energy (16) of the intermediate state B.

There must also be factors associated with the vertices in (19) that distinguish the different reactions that can occur in the formation and decay of B. The complete expression for the cross section of process (19) is given by the *Breit–Wigner formula*,[5]

$$\sigma(aA \to cC) = \frac{2J_B + 1}{(2s_1 + 1)(2s_2 + 1)} \frac{\pi}{k^2} \frac{\Gamma(B \to aA)\Gamma(B \to cC)}{|M_{aA} - M_B + \tfrac{1}{2}i\Gamma_B|^2}, \quad (20)$$

where M_{aA} is given by (7), k is the relative momentum of the colliding particles, and s_1 and s_2 are their spins. The factors $\Gamma(B \to f)$ are the so-called *partial widths* of B for decay into f, where f can be (a, A) or (c, C). The ratio $\Gamma(B \to f)/\Gamma_B$ is the probability that B will decay into the final state f, and is called the *branching ratio*, or *branching fraction*. The partial widths $\Gamma(B \to f)$ sum to give the total width:

$$\Gamma_B = \sum_f \Gamma(B \to f). \quad (21)$$

Equation (20) also holds for the elastic process (12), and determines the proportionality factor in (18):

$$\sigma_{aA} = \frac{2J_B + 1}{(2s_1 + 1)(2s_2 + 1)} \frac{\pi}{k^2} \frac{[\Gamma(B \to aA)]^2}{|M_{aA} - M_B + \tfrac{1}{2}i\Gamma_B|^2}. \quad (20')$$

It should be realized that (20) and (18) are valid only if B is a well-isolated intermediate state of the combined system $a + A$. Other states of this system must be energetically separated from B by more than Γ_B. Highly excited hadronic energy levels have large widths because of the increasing number of decay channels. Consequently, the Breit–Wigner formula (20) is only applicable to relatively low-lying states.

4. Photoproduction

We now consider the formation process (6) in photon-proton collisions ($a = \gamma, A = p$). For excited states that emit mesons, the branching ratio for photon decay is generally of order $\tfrac{1}{137}$. It therefore follows that resonant production of pions is far easier to observe accurately than resonance

---

[5] See Blatt and Weisskopf (1952), §VIII.10. In general

$$k = \sqrt{[M_{aA}^2 - (m_a + M_A)^2][M_{aA}^2 - (m_a - M_A)^2]}/2M_{aA}.$$

This reduces to a's laboratory momentum if M_A is large compared to the incident energy, and A is at rest. At very high energy, where M_{aA} is large compared to m_a and M_A, $k \to \tfrac{1}{2}M_{aA}$. For photons $(2s + 1)$ is replaced by 2, because there are only two helicities.

fluorescence. As we see from the resonance formula (20), the widths and locations of resonances do not depend on the particular manner in which they are produced, nor on the decay mode detected; the denominator of (20) is the same in all processes that involve the resonance, and gives a universal function of energy that determines its basic parameters Γ_B and M_B. Thus the reaction $\gamma p \to \pi N$ serves as well for determining these quantities as the far more elusive process $\gamma p \to \gamma p$.

Figure 4 shows the total γp cross section, $\sigma_{\gamma p}$. For total c.o.m. energies $M_{\gamma p}$ below ~1.8 GeV, the process that dominates $\sigma_{\gamma p}$ is $\gamma p \to \pi N$, as detailed measurements of $\gamma p \to \pi^0 p$ and $\gamma p \to \pi^+ n$ have shown. Three resonances are in clear evidence: the first at $M_{\gamma p} \simeq 1232$ MeV, the others at $M_{\gamma p} \simeq 1500$ and 1700 MeV. The first is actually the lowest excited state of the nucleon. As we recall from §I.E.2, this state is called $\Delta(1232)$, and has the quantum numbers $I = \frac{3}{2}, J^P = \frac{3}{2}^+$. The second and third bumps correspond to the $I = \frac{1}{2}$ states $N(1520, J^P = \frac{3}{2}^-)$ and $N(1680, J^P = \frac{5}{2}^+)$. As the total cross section gives no angular information, the J^P assignments quoted here must be established by other data.

The levels just discussed conform with the expectation that all hadronic states formed in γp collisions, no matter how complex, must have $I = \frac{1}{2}$ or $\frac{3}{2}$. This follows from the selection rule $\Delta I = 0$ or 1 for one-photon transitions [Eq. I.E(9b)], and the fact that the proton has $I = \frac{1}{2}$.

Let us briefly look at the magnitude of $\sigma_{\gamma p}$. At the first Δ resonance it is about 500 μb, where b stands for *barn*:

$$b = 1 \text{ barn} = 10^{-24} \text{ cm}^2 = 100 \text{ fm}^2;$$

and $\mu b = 10^{-6}$ b (microbarn). Thus, at this first resonance $\sigma_{\gamma p} \sim 5 \times 10^{-2}$ fm^2, while a typical nonresonant value is smaller by a factor of about 4. In §1 we showed that hadrons have dimensions of order 1 fm. The small value of $\sigma_{\gamma p}$ is due to the weakness of the electromagnetic interaction: hadrons are highly transparent to photons. This is merely another expression of our earlier statement that the branching fraction for the decay $\Delta \to \gamma p$ is very small.

Baryon resonances can also be excited by inelastic electron scattering. The Feynman diagram for this process is

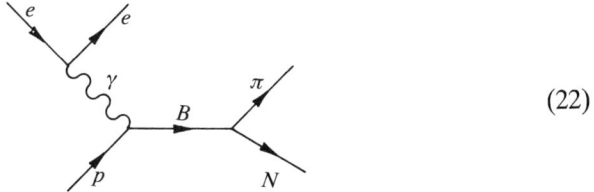

(22)

Obviously this is very closely related to the photoprocess ($a = \gamma$ in (19)). The only difference between photo- and electroproduction is that in the latter case the electron suffers a large 4-momentum transfer, and so the

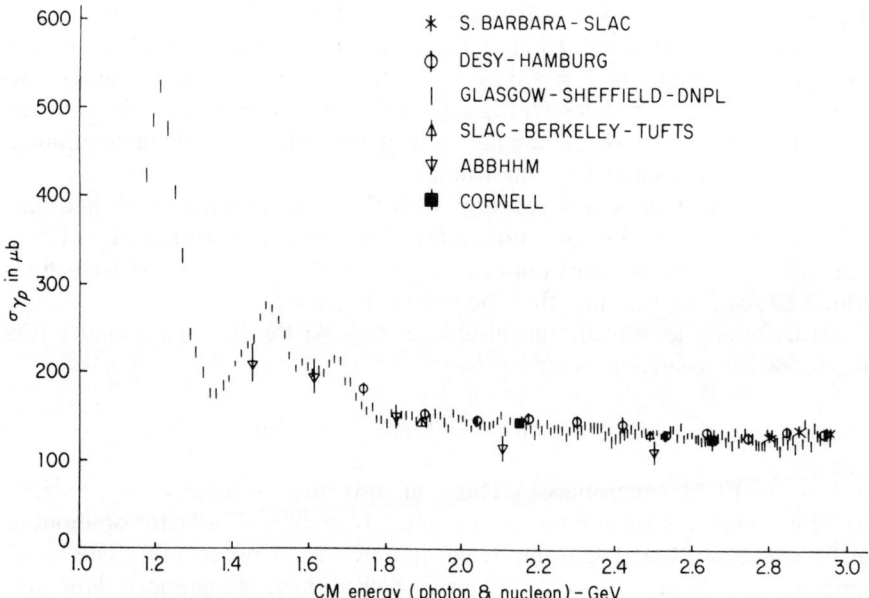

FIG. 4. The total γp cross section in microbarns (μb) as a function of the total c.o.m. energy $M_{\gamma p}$. This compilation is provided by courtesy of G. M. Lewis. For a fit to three Breit-Wigner resonances, see Fig. 8 in Armstrong et al. (1972).

4-momentum q of the virtual photon in (22) does not satisfy $q^2 = 0$, as it does for a real photon. We expect the baryon resonances seen in photoproduction to also show up in electroproduction. That this is so can be seen in Fig. I.5.

5. *Pion-nucleon scattering*

Now we discuss processes of type (19), initiated in hadron-hadron collisions. We first consider pion-nucleon scattering:

$$\pi^+ p \to \pi^+ p, \tag{23}$$

and

$$\pi^- p \to \pi^- p \tag{24}$$

$$\to \pi^0 n. \tag{25}$$

The Feynman graphs for these reactions are all subsumed in (19).

One glance at the total $\pi^+ p$ cross section (Fig. 5) reveals that we are dealing with a far stronger interaction than in photoproduction. At the first maximum the $\pi^+ p$ cross section is $\sim 10^3$ larger than in $\gamma p \to \pi^+ n$. At higher energies, where there is no obvious resonance structure, there is still a difference by a factor of order 10^3, and $\sigma_{\text{tot}}(\pi^+ p) \simeq 30$ mb $\simeq 3$ fm^2. If we naively set this[6] equal to $2\pi R^2$, the total cross section for a black disk of radius R, we obtain $R \simeq 0.7$ fm, in qualitative agreement with the radius of the proton as measured electromagnetically. Clearly these processes are due to an interaction that is highly effective within distance of order 10^{-13} cm.

As the processes we are now concerned with are the result of strong interactions, isospin is conserved. Thus (23) is a pure $I = \frac{3}{2}$ channel ($Q = 2$, and therefore $I_3 = \frac{3}{2}$), whereas $\pi^- p$ is a combination of $I = \frac{1}{2}$ and $I = \frac{3}{2}$ ($Q = 0$, and therefore $I_3 = -\frac{1}{2}$, which may be $I = \frac{1}{2}$ or $I = \frac{3}{2}$).

Let us then examine the total cross section in the pure $I = \frac{3}{2}$ $\pi^+ p$ channel (Fig. 5). The most striking feature is the enormous resonance at $M_{\pi p} \sim 1230$ MeV. Both the location and the shape of this resonance concur with the first resonance in photoproduction shown in Fig. 4. As we recall from the discussion pertaining to Eq. (20), a resonantly excited state must display this kind of universal dependence on the mass M_{aA}, and this therefore strengthens our belief that we are indeed dealing with a well-defined excited state. That it is an $I = \frac{3}{2}$ resonance is then a foregone conclusion if isospin conservation is accepted. Confirming evidence comes from observing the same resonance in the other channels that form the $I = \frac{3}{2}$ quartet, i.e., $\pi^+ n$ or $\pi^0 p$, $\pi^- p$ or $\pi^0 n$, and $\pi^- n$. Figure 6 shows that this resonance appears in the $\pi^- p$ channel. An even more incisive test of the I-spin assignment is discussed in §6.

---

[6] One should not use the cross section at resonance to estimate an interaction radius, for at a resonance σ only depends on the relative momentum, the spins of the particles, and the branching fractions, as one sees from Eq. (20).

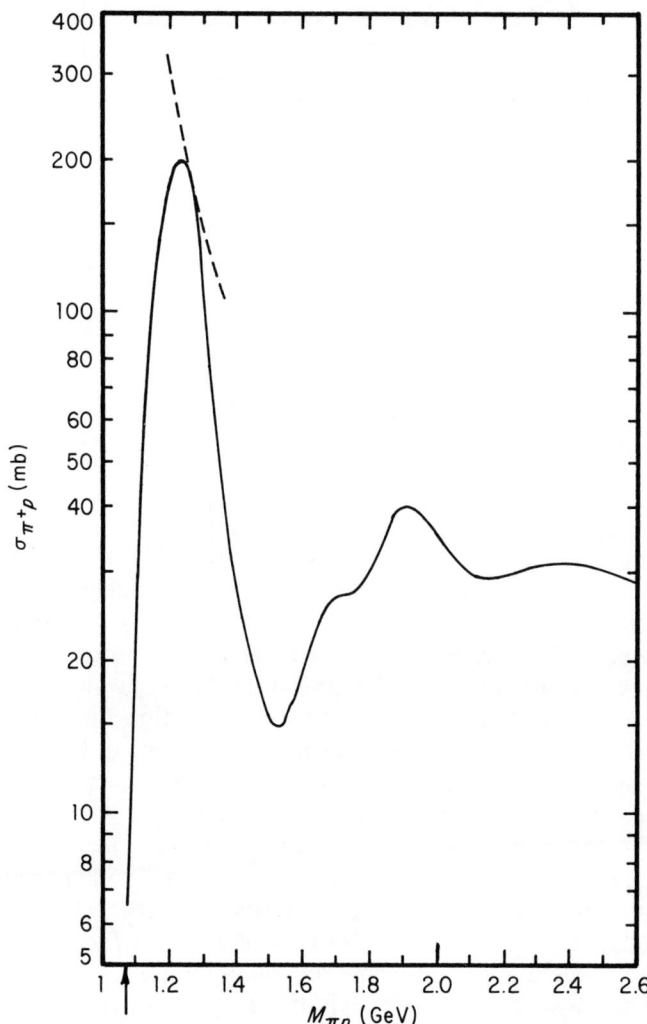

FIG. 5. The total $\pi^+ p$ cross section as a function of the c.o.m. energy. The curves shown here and in Fig. 6 are drawn through the experimental data compiled in Flaminio et al. (1979a). The arrow at $M_{\pi p} = 1.08$ is the threshold at $m_\pi + m_p$. The first maximum is due to $\Delta(1232)$. The dashed curve is the bound imposed by probability conservation on elastic scattering in the $J = \frac{3}{2}$ state [see discussion related to Eq. (26)]. The structure in the vicinity of 1.9 GeV is due to several overlapping Δ resonances, with the dominant contribution coming from the $J = \frac{7}{2}^+$ level $\Delta(1950)$.

Observe that the maximum at $M \approx 1500$ in photoproduction (Fig. 4) has no counterpart in $\pi^+ p$ elastic scattering. This establishes that if there is a well-defined level at $M \approx 1500$, it must have $I = \frac{1}{2}$, because $\pi^+ p$ has $I = \frac{3}{2}$.

From the shape of the angular distribution at resonance, as shown in Fig. 7, one can determine the orbital πp angular momentum, as well as the total angular momentum J. As we see, at $M = 1.24$ GeV the angular distribu-

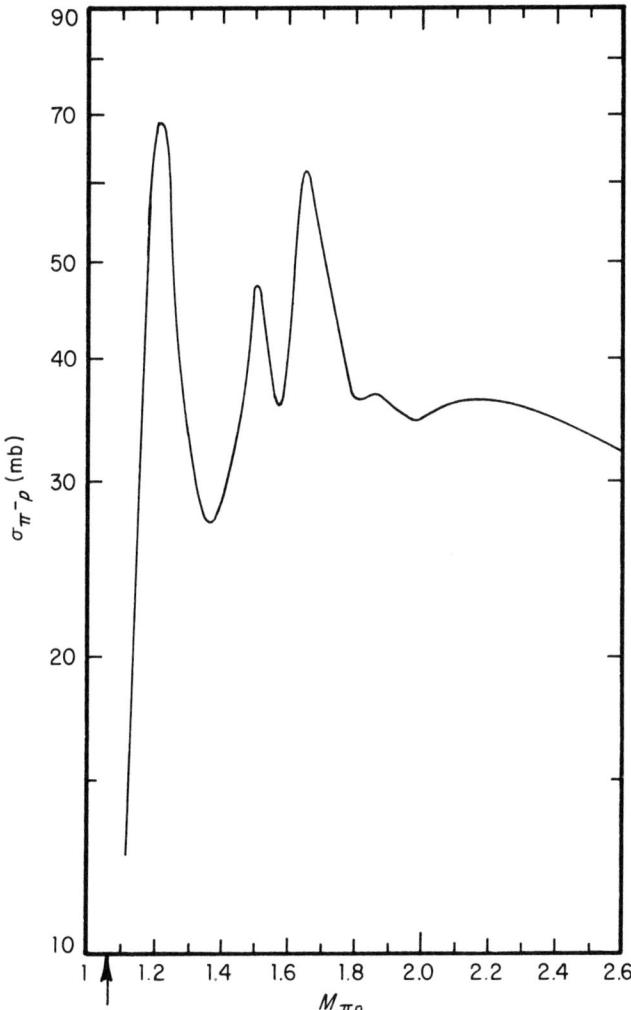

FIG. 6. The total π^-p cross section [Flaminio et al. (1979a)]. The first peak is $\Delta(1232)$, while the structures at higher masses are due to several overlapping N- and Δ-resonances, since π^-p feeds both $I = \frac{1}{2}$ and $I = \frac{3}{2}$ states (see Appendix I.7 and 8).

tion, which is quadratic in the angular momentum wave function, is fitted well by $1 + 3\cos^2\theta$. As $\cos\theta$ is just an $l = 1$ eigenfunction, we see that we are dealing with a p-wave. A more complete analysis shows that it is a p-state with $J = \frac{3}{2}$ and even parity,[7] i.e., $J = \frac{3}{2}^+$.

[7] The reason for the even parity assignment is that the pion has odd intrinsic parity, as explained in §I.E.3(b); this then combines with the odd parity of the p-wave to give an overall even parity. If one wishes, one can turn the argument around to demonstrate that the pion has odd intrinsic parity. This intrinsic parity assignment is confirmed in photoproduction, for $\gamma p \to \Delta$ is a magnetic (and not an electric) dipole transition.

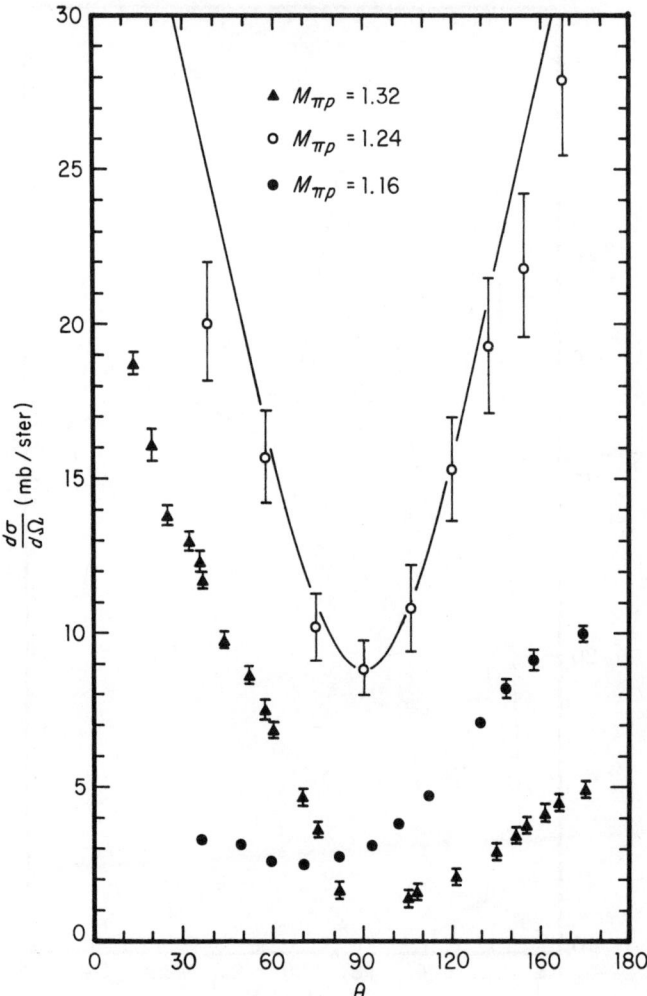

FIG. 7. The angular distribution of elastic $\pi^+ p$ scattering in the c.o.m. frame at, and on either side of, the $\Delta(1232, \frac{3}{2}^+)$ resonance. The data are taken from the compilation of Giacomelli et al. (1969), and the c.o.m. energy is as indicated. The solid curve is proportional to $1 + 3\cos^2\theta$, and is fitted to the data at $M_{\pi p} = 1.24$ GeV.

Having established J_Δ, we can now compare the total cross section $\sigma(\pi^+ p)$ with the theoretical expectation. First, note that the threshold for inelastic scattering, $\pi^+ p \to \pi^+ p \pi^0$, is at $M = 1213$ MeV. Thus inelastic scattering is barely possible throughout the whole Δ-resonance, and we can safely treat the process as if it were exclusively elastic. Hence expression (21) for the width of Δ contains only one term: $\Gamma_\Delta = \Gamma(\Delta \to p\pi)$. According to (20'), the elastic cross section becomes ($J_B = \frac{3}{2}$,

$s_1 = \frac{1}{2}, s_2 = 0$):

$$\sigma(\pi p \to \pi p) = \frac{2\pi}{k^2} \frac{\Gamma_\Delta^2}{(M_{\pi p} - M_\Delta)^2 + \frac{1}{4}\Gamma_\Delta^2}. \tag{26}$$

This assumes its maximum value[8] of $8\pi/k^2$ at resonance, $M_{\pi p} = M_\Delta$. Figure 5 shows that this bound is reached by the data, and that Δ is therefore an elastic resonance.

In principle this type of analysis is necessary to establish the existence of an excited level. In practice the analysis is frequently very difficult because other levels may lie nearby, and also because several decay modes may compete with single pion emission. For example, the second $I = \frac{3}{2}$ maximum visible in Fig. 5 at $M \simeq 1900$ MeV is really a superposition of three levels having the following masses, spin-parities, and one-pion decay branching ratios: $\Delta(1905, \frac{5}{2}^+)$, ~10%; $\Delta(1910, \frac{1}{2}^+)$, 20–25%; and $\Delta(1950, \frac{7}{2}^+)$, ~40%. Obviously, a very elaborate set of experiments is required to extract this information.[9] A situation of similar complexity occurs in the fourth bump observed in electroproduction (Fig. I.5).

Let us now look at $\pi^- p$ elastic scattering, which involves both $I = \frac{1}{2}$ and $I = \frac{3}{2}$ states. The total cross section is shown in Fig. 6. We see $\Delta(1232)$ again, but there are now two bumps at $M \simeq 1530$ and $M \simeq 1700$ that have no counterpart in $\pi^+ p$ scattering. These must therefore be $I = \frac{1}{2}$ levels if they are resonances (recall the discussion of γp at $M \simeq 1500$ on p. 289). Once more, each of these maxima is due to more than one level. The first is due to the levels $N(1520, \frac{3}{2}^-)$ and $N(1535, \frac{1}{2}^-)$; their $N\pi$ branching fractions are ~55% and ~40%, respectively. The second bump also comprises two levels, $N(1675, \frac{5}{2}^-)$ and $N(1680, \frac{5}{2}^+)$, with branching fractions of ~35% and ~60%, respectively.

It goes without saying that as M_B increases, the level density also increases, the branching fractions into simple decay modes decrease, and the confidence with which one can determine excited baryon states becomes correspondingly poorer. At this time (1984), the most highly excited states of $I = \frac{1}{2}$ and $I = \frac{3}{2}$ seen with good confidence in πp scattering are $N(2600, \frac{11}{2}^-)$ and $\Delta(2420, \frac{11}{2}^+)$.

6. Consequences of the isospin assignments

Until now we have assigned I to levels seen in πN scattering by merely asking whether they occur in the pure $I = \frac{3}{2}$ $p\pi^+$ configuration, or only in the mixed $I = (\frac{1}{2}, \frac{3}{2})$ configuration $p\pi^-$. But there is more to isospin. The

---

[8] There are purely geometrical reasons why $(4\pi/k^2)(2J + 1)/[(2s_1 + 1)(2s_2 + 1)]$ is the maximum possible cross section of any reaction in which two particles of spin s_1 and s_2 and total angular momentum J collide. If it were larger, more particles would be scattered than came in. In an exclusively elastic process, this maximum is reached at resonance.

[9] The bump is dominated by $\Delta(1950, \frac{7}{2}^+)$ because the branching fraction appears quadratically in the elastic πp cross section—cf. Eq. (20'), and because of its high spin.

strong interaction Hamiltonian is invariant in the isospin space $\mathscr{E}_3^I$, and therefore commutes with all the components of the isospin $\vec{I}$. Once this symmetry is assumed, relations between the various cross sections involving a level having a definite I follow from the standard laws of angular momentum addition, since isospin obeys the same rules as angular momentum (see §I.B.3 and §I.D.2).

Consider the set of reactions (23)–(25). Let us call $|I\,I_3\rangle$ the state with the indicated quantum numbers of the total isospin of the pion-nucleon system. Remember further that (π^-, π^0, π^+) are the $I_3 = (-1, 0, 1)$ members of an $I = 1$ triplet, and that (p, n) are the $I_3 = (\tfrac{1}{2}, -\tfrac{1}{2})$ members of an $I = \tfrac{1}{2}$ doublet. Using the vector addition (Clebsch–Gordan) coefficients for angular momentum $\tfrac{1}{2}$ and 1, one finds

$$|\pi^-p\rangle = \sqrt{\tfrac{1}{3}}|\tfrac{3}{2}, -\tfrac{1}{2}\rangle - \sqrt{\tfrac{2}{3}}|\tfrac{1}{2}, -\tfrac{1}{2}\rangle,$$
$$|\pi^0 n\rangle = \sqrt{\tfrac{2}{3}}|\tfrac{3}{2}, -\tfrac{1}{2}\rangle + \sqrt{\tfrac{1}{3}}|\tfrac{1}{2}, -\tfrac{1}{2}\rangle. \tag{27}$$

Let T be the operator whose matrix elements $\langle\alpha|T|\beta\rangle$ give the scattering amplitude from α to β; that is, $|\langle\pi^-p|T|\pi^0 n\rangle|^2$ determines the differential cross section for $\pi^-p \to \pi^0 n$ at some definite energy and through a given angle of scattering. It follows from isospin symmetry that T cannot connect states of different I and I_3, and that T has the *same* matrix element T_I for scattering of any and all members of a definite I-multiplet.[10] The only nonzero matrix elements of T are therefore

$$\langle I\,I_3|T|I'I_3'\rangle = T_I\,\delta_{I_3 I_3'}\,\delta_{II'}. \tag{28}$$

Thus

$$\langle\pi^-p|T|\pi^-p\rangle = \tfrac{1}{3}T_{3/2} + \tfrac{2}{3}T_{1/2}$$
$$\langle\pi^-p|T|\pi^0 n\rangle = \sqrt{2}/3\,(T_{3/2} - T_{1/2}) \tag{29}$$

and, of course,

$$\langle\pi^+p|T|\pi^+p\rangle = T_{3/2}. \tag{30}$$

If at $M = 1232$ MeV the $I = \tfrac{3}{2}$ level is indeed resonant, while $T_{1/2}$ is not, we can ignore $T_{1/2}$ in these expressions, and obtain the result that at the resonance the differential cross sections are in the ratios[11] of the squares of T:

$$d\sigma(\pi^-p \to \pi^-p):d\sigma(\pi^-p \to \pi^0 n):d\sigma(\pi^+p \to \pi^+p) = 1:2:9.$$

---

[10] Note the complete analogy with angular momentum: If scattering is due to a spherically symmetric interaction, collisions cannot change the angular momentum, and the scattering amplitude depends only on the total angular momentum, not its orientation.

[11] Note that this implies that $\sigma_{\text{tot}}(\pi^+p):\sigma_{\text{tot}}(\pi^-p) = 3:1$, because $\sigma_{\text{tot}}(\pi^-p)$ includes π^-p elastic scattering as well as $\pi^-p \to \pi^0 n$.

Similarly, at an $I = \frac{1}{2}$ resonance

$$d\sigma(\pi^-p \to \pi^-p) : d\sigma(\pi^-p \to \pi^0 n) = 2:1.$$

We see that isospin assignments, and the concepts that underlie them, provide information about the ratios of scattering cross sections at resonance. These have been verified by many experiments.

7. Kaon-nucleon scattering and $S \neq 0$ excited states of the baryon

We shall only concern ourselves with the qualitative features of the $K^\pm p$ cross sections, as shown in Fig. 8. There is structure only at lower energy; at higher energy the total cross section is $\sim 2 \text{ fm}^2$, and therefore only slightly smaller than the corresponding $\pi^\pm p$ cross sections. Clearly the KN and πN interactions do not differ markedly in strength.

The most striking feature of K^+p and K^-p interactions is that the former has a totally featureless cross section, while the latter displays a very rich structure. This indicates that there are no excited levels of the baryon with strangeness $S = +1$, whereas there are a number of levels with $S = -1$. Detailed studies of the angular distributions of decay modes and of their energy dependence, along the lines discussed above, fully confirm this. Indeed, one calls $B = 1, S = 1$ systems *exotic* because they appear to have no well-defined levels. As we know from §I.E.6(a), the hypothesis that baryons are 3-quark systems leads to the conclusion that there are no baryons with $B = 1$ and $S = 1$.

The $S = -1$ resonances found in K^-p collisions are denoted by $\Lambda(M, J^P)$ and $\Sigma(M, J^P)$, depending on whether they have $I = 0$ or 1 (recall Table I.E.1). The manner in which M, J^P, and I are found has already been explained in connection with the πN system. Some of the identified resonances are shown in Fig. I.6, and tabulated in Appendix I.9.

8. Hadron excitations observed by decay

We now come to the second technique for establishing the existence of excited states, the observation of the decay of a state: production $\to B \to$ decay. In the case of atoms, the excitation of B may be due to heat, spark discharge, collisions, or exposure to suitable radiation. The decay is most frequently observed by the emitted radiation (recall p. 285). In the case of hadrons, an excited state can also be produced in many ways. We group these production mechanisms into two categories: peripheral collisions and central collisions. They are depicted in Fig. 9. In peripheral collisions, two hadrons a and A pass by each other at a distance that is comparable to or larger than the characteristic size found from diffraction scattering. Then one (or both) may get excited, and leave the encounter in an excited state

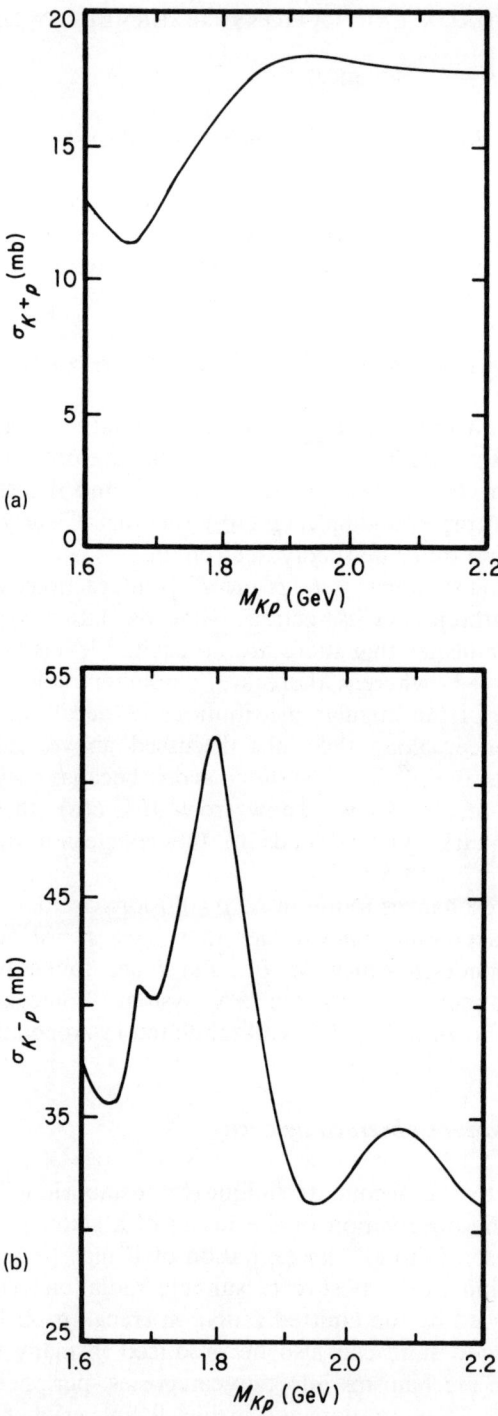

FIG. 8. Kaon-proton total cross sections: (a) shows the $S = 1$ channel, K^+p, which has no resonances, while (b) is the $S = -1$ channel K^-p, which has $I = 0(\Lambda)$ and $I = 1(\Sigma)$ resonances. [This data compilation is due to Flaminio et al. (1979b).]

8. HADRON EXCITATIONS OBSERVED BY DECAY

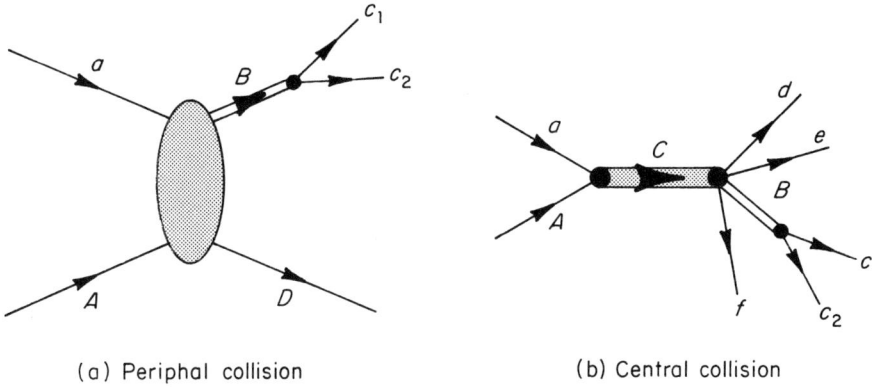

(a) Periphal collision   (b) Central collision

FIG. 9. Mechanisms responsible for peripheral and central collisions. The blob in (a) represents a rather long-range interaction, which is usually mediated by exchange of a light meson, as in the example of the long-range nuclear force (recall Fig. I.24). The compound system C in (b) need not be a well-defined level as in the resonance reactions; it may be a superposition of many levels.

B. In a central collision, the two particles a and A produce a compound system C, which disintegrates into a number of different particles, of which one is the excited hadron, B. The compound system C will usually not be a hadron in a well-defined level.

There are two important differences between the excited hadrons made by what we have called "formation," and the production by peripheral or central collisions. First, in the formation process $a + A = B$ the particle in the excited state of interest is an amalgamation of the two incident particles, and B is at rest in the center-of-mass frame. In central or peripheral production of B, there is always at least one particle present in addition to B, and therefore, B is not at rest in the c.o.m. frame. Second, in the formation process, B always is a state of an excited baryon, since the target A is a baryon, unless the incident particles are antiprotons. In central and peripheral production the excited hadron can also be a meson. In the peripheral case, the incident particle can be a meson and may emerge in an excited state; in the central case, one of the decay products of C may be an excited meson.

As mentioned before, the excited state B is observed by determining the invariant mass M_i of the decay products $C_1, C_2, \ldots$, as given by Eq. (10) in the case of two-body decay. According to (11), M_i will lie in the neighborhood of M_B. The probability of finding M_i around M_B is also given by a Breit–Wigner distribution:

$$P(M_i) \propto \frac{1}{(M_i - M_B)^2 + \tfrac{1}{4}\Gamma_B^2}. \tag{31}$$

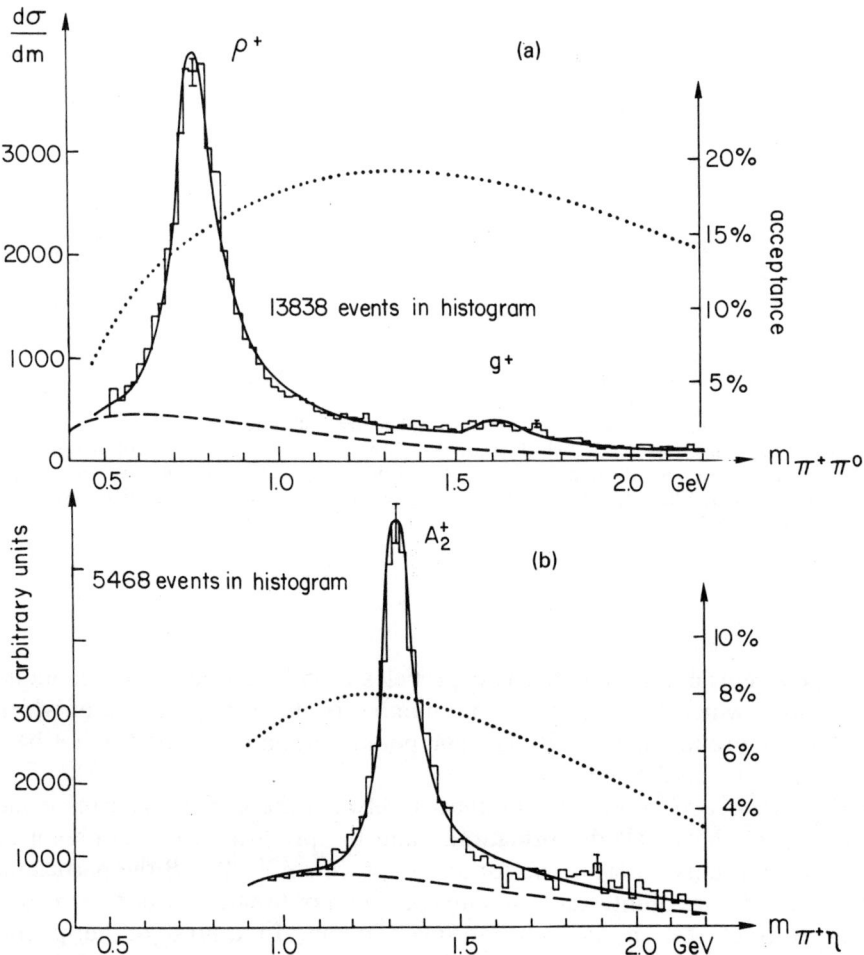

FIG. 10. Resonances produced by 50 GeV pions on a hydrogen target at CERN [Delfosse et al. (1981)]. This experiment detects π^0 and η, and measures their momenta, by observing the 2γ decay mode. The number of events as a function of $\pi^+\pi^0$ invariant mass in the final state $\pi^+\pi^0 p$ is shown in (a); this displays the $I = 1$ levels ρ^+ and $g^+(J^P = 3^-)$. The $\pi^+\eta$ invariant mass distribution for the final state $\pi^+\eta p$ is plotted in (b), and shows $A_2^+(1318)$, a $J^P = 2^+$, $I = 1$ level of width 100 MeV. The solid curves are fits to the data using Breit-Wigner formulas, and a nonresonant background, shown as a dashed line. The dotted curve shows the acceptance efficiency of the detector for the final state in question.

Typical examples of peripheral production are the reactions

$$\pi p \to \pi^* N, \quad \pi^* \to \pi\pi, \qquad (32)$$

$$Kp \to K^* N, \quad K^* \to K\pi, \qquad (33)$$

where π^* and K^* are generic symbols for excited mesons. To establish their

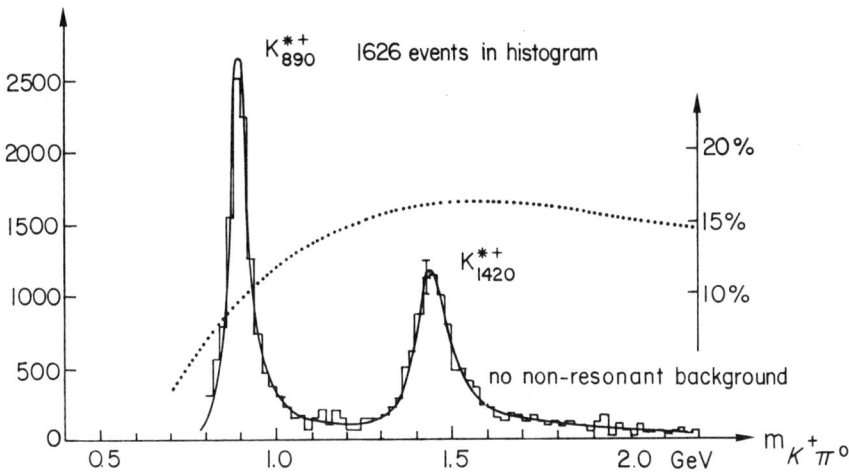

FIG. 11. Resonances produced at CERN by 50 GeV kaons in the reaction $K^+p \to K^+\pi^0 p$ [Delfosse et al. (1981)]. The $K^+\pi^0$ invariant mass distribution shows two levels: $K^{*+}(892)$ and $K^{*+}(1430)$, the latter having $J^P = 2^+$. The dotted curve has the same meaning as in Fig. 10.

existence, one must analyze the $\pi\pi$ and $K\pi$ invariant mass distributions in the final states $\pi\pi N$ and $K\pi N$, which, via (31), then determine the masses and widths of the meson levels. Their isospin is found by, for example, establishing whether the resonance exists in both the $\pi^\pm \pi^0 (I=1)$ and $\pi^+\pi^- (I=0$ or $1)$ combinations. The spin-parity of the states requires an analysis of the angular distribution of, for example, the $K\pi$ decay in the rest frame of K^*, because if K^* has $J \neq 0$, it will, in general, be produced with a nonrandom spin orientation (a phenomenon explained in Fig. I.35).

Figure 10(a) shows the $\pi^+\pi^0$ mass distribution in the reaction (32), and displays a large resonance at $M_{\pi\pi} \simeq 770$ MeV, the ρ^+ vector meson ($J^P = 1^-$), as well as a very weak resonance $g^+ (J^P = 3^-)$ at $M_{\pi\pi} \simeq 1700$. An analysis of the $\pi\pi$ angular distribution in the ρ-resonance region establishes the quoted spin-parity. Figure 10(b) shows a more complicated decay process, in which π^* in (32) decays into $\pi^+\eta$, where η is a member of the same nonet as π [see Fig. I.11(a)], with a mass of 549 MeV; η is quite stable ($\Gamma_\eta = 0.9$ keV), and is observed via its 2γ decay mode. The $\pi^+\eta$ mass plot reveals the meson $A_2^+(1318)$, and further analysis establishes its spin-parity as 2^+. Results from the K^+p reaction (33) are shown in Fig. 11; the $K^+\pi^0$ mass plot reveals two excited kaons: $K^*(892)$ and $K^*(1430)$, with $J^P = 1^-$ and 2^+, respectively.

A $\pi^+\pi^-$ mass distribution, observed in $\pi^-p \to \pi^-\pi^+ n$, is shown in Fig. 12. It demonstrates that ρ^+ and g^+ have $I_3 = 0$ partners, ρ_0 and g^0, with the same mass and width as observed in the $\pi^+\pi^0$ system. But Fig. 12 also shows a $\pi^+\pi^-$ resonance that is never observed in any $Q = \pm 1$ multipion system; this is therefore an $I = 0$ object, called $f(1270)$, and it has $J^P = 2^+$.

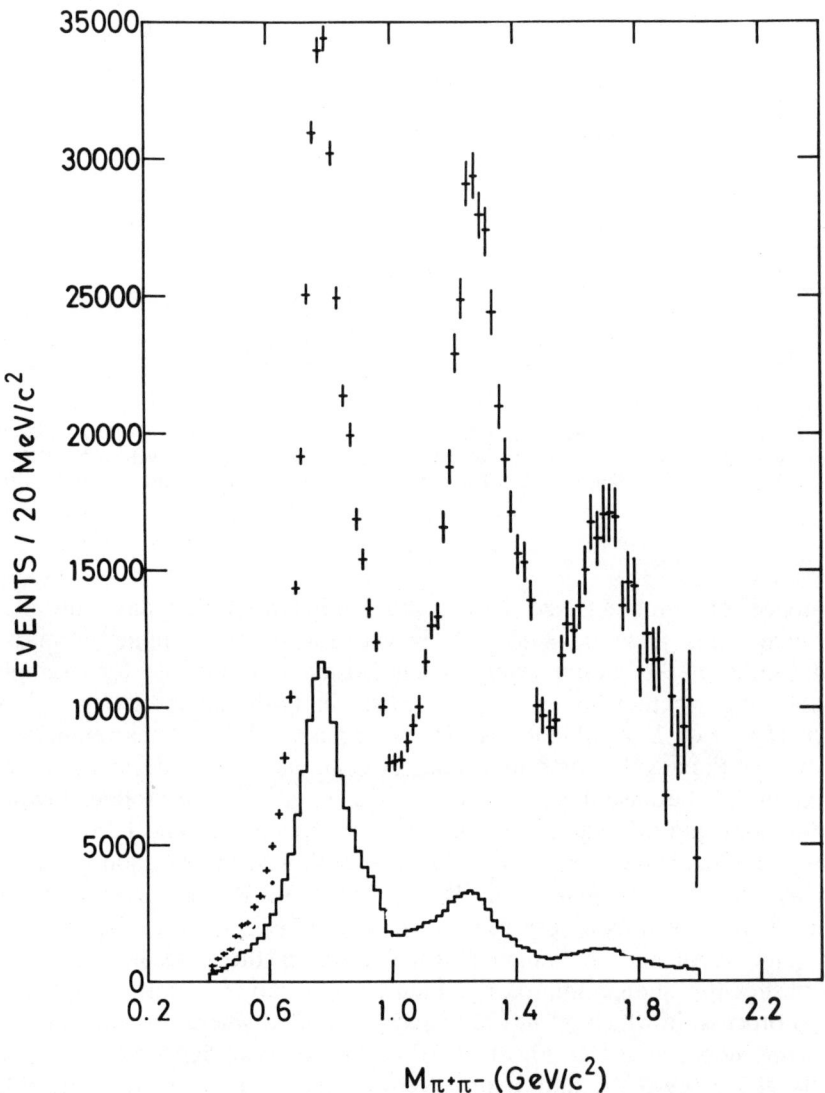

FIG. 12. Resonances that decay into $\pi^+\pi^-$ observed at CERN in the reaction $\pi^- p \to \pi^-\pi^+ n$ [Greyer et al. (1974)]. The solid histogram is the observed spectrum, which is degraded from what is actually produced (indicated by crosses) due to the acceptance of the detector and pion decay. In order of increasing mass, the resonances are $\rho(J^P = 1^-, I = 1)$, $f(J^P = 2^+, I = 0)$, and $g(J^P = 3^-, I = 1)$.

Baryon resonances can also be produced in peripheral collisions. We quote two examples:[12]

$$K^-p \to \pi^- \Sigma^+(1385), \tag{34}$$

$$\Xi^-p \to \Xi^0(1530)X. \tag{35}$$

In reaction (34), $\Sigma^+(1385)$ is found as a peak in the $\Lambda\pi^+$ mass distribution, whereas in (35), $\Xi^0(1530)$ is observed in the $\Xi^-\pi^+$ mass distribution, both of which are shown in Fig. 13. $\Sigma^+(1385)$ and $\Xi^0(1530)$ are both members of the $J = \frac{3}{2}$ decuplet discussed in §I.E.6(d). Their angular momentum is found by analyzing the $\Lambda\pi^+$ and $\Xi^-\pi^+$ angular distributions.

As emphasized previously, a bump in a cross section or mass distribution is only established definitively as a resonance if it has a mass and width that do not depend on the manner in which it is formed. Furthermore, it must have a definite spin-parity. We illustrate this with the vector mesons ϕ and ρ^0, by examining their photoproduction in peripheral collisions: $\gamma N \to V^0 N$, where V^0 is ϕ or ρ^0, which are observed via their $K\bar{K}$ or $\pi\pi$ decay, respectively. The $M_{K\bar{K}}$ distribution measured in $\gamma A \to \phi A$, where A is a complex nucleus, is shown in Fig. 14(a). A ϕ peak at the same mass and of the same width also appears in the quite different process $pp \to pp\phi$, as shown in Fig. 14(b).

As a detailed example of how the spin of a resonance can be determined, Fig. 15 shows the angular distributions of the $\pi^+\pi^-$ decay of ρ^0 produced by polarized photons in the reaction $\gamma p \to \rho^0 p$.

The neutral vector mesons also appear as resonances in e^+e^- annihilation, a subject that is discussed in detail in Part B of this Chapter. Here we confine ourselves to a few remarks. As we learned in §I.E.8(c), $e^+e^- \to H$, where H is any hadronic system, proceeds through the steps $e^+e^- \to \gamma$, followed by $\gamma \to H$. Consequently, H must have $J^P = 1^-$, which means that resonances in $e^+e^- \to H$ can only reveal neutral vector mesons. Furthermore, the mass spectrum of M_H is simply given by the variation of the cross section $\sigma(e^+e^- \to H)$ as the e^+e^- energy is changed. In the case of ϕ, one has the reaction chain $e^+e^- \to \gamma \to \phi \to K\bar{K}$, and one observes [see Fig. 14(c)] an $M_{K\bar{K}}$ spectrum identical to those shown in Fig. 14(a) and (b), which arise from photoproduction and purely hadronic collisions.

9. G-parity

The strong interaction is invariant under charge conjugation, and also conserves isospin. By combining these two conservation laws one can define

---

[12] It is hardly obvious that these are peripheral collisions. To see that they can be, consider a crossed counterpart of (34), $K^-\pi^+ \to \bar{p}\Sigma^+(1385)$. By the argument of Fig. I.24 concerning the nucleon-nucleon interaction, we can see that (34) can be mediated by the exchange of K^{*0}, i.e., the lightest meson that can decay strongly into $K\pi$. As with the π-exchange contribution to the NN force, the exchange of such a relatively light object gives an interaction of moderately long range. That this peripheral mechanism is actually operative in reaction (34) can only be established by detailed analysis of the angular distribution of $\Sigma^+(1385)$.

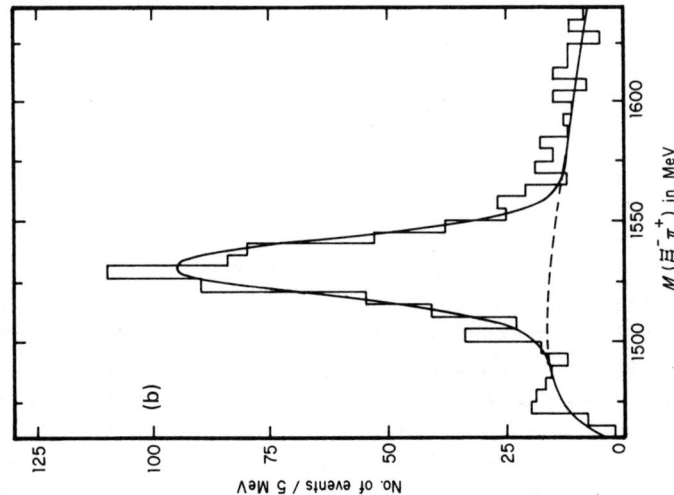
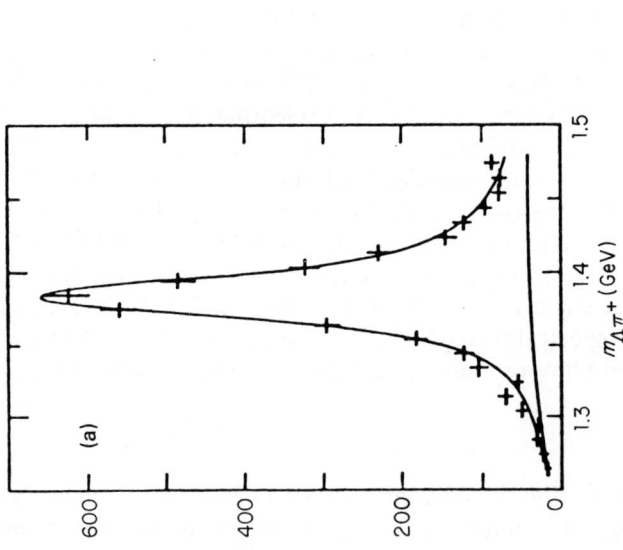

FIG. 13. (a) The resonance $\Sigma^+(1385)$ as observed in the reaction $K^-p \to \pi^-\Lambda\pi^+$ at an incident K^--momentum of 4.2 GeV/c [Holmgren et al. (1977)].
(b) The resonance $\Xi^0(1530)$ as observed in $\Xi^-N \to \Xi^-\pi^+X$ [Biagi et al. (1981)]. This reaction is produced by a charged hyperon beam, which is made at the CERN 400-GeV proton synchrotron. The hyperon in the final state is observed through the decay chain $\Xi^0(1530) \to \Xi^-\pi^+, \Xi^- \to \Lambda\pi^-, \Lambda \to p\pi$.

9. G-PARITY

a certain reflection operation in the isospin space, to which one associates a multiplicative quantum number called G-parity. A large number of mesons turn out to have a definite G-parity, and their hadronic decays are then restricted by G-parity selection rules. These selection rules are nontrivial; they cannot be derived from isospin conservation or charge conjugation separately.

The motivation for introducing G-parity is readily understood if one examines the pion isotriplet (π^+, π^0, π^-). The charge conjugation operator C interchanges π^+ and π^-, but leaves π^0 unchanged. Hence C maps the pion triplet into itself, but in so doing it reflects the sign of the I_3 eigenvalue. In other words, C and $\vec{I}$ do not commute, and cannot be specified simultaneously. Our goal is to find an operator that somehow involves C, but which commutes with $\vec{I}$, so that its eigenvalue G can be specified together with I^2 and I_3.

To see how this is to be done, we must know the commutation rules for C and $\vec{I}$. To this end we define[13]

$$C|\pi^\pm\rangle = |\pi^\mp\rangle. \tag{36}$$

Since $\pi^0 \to 2\gamma$, and γ is odd under C, we know that π^0 is even:

$$C|\pi^0\rangle = |\pi^0\rangle. \tag{36'}$$

Consequently, $CI_3|\pi^\pm\rangle = \pm|\pi^\mp\rangle$, but $I_3 C|\pi^\pm\rangle = \mp|\pi^\mp\rangle$, or

$$CI_3 = -I_3 C. \tag{37}$$

From Eq. I.B(19) we have $I_\pm|\pi^0\rangle = \sqrt{2}|\pi^\pm\rangle$, so that $I_\pm C|\pi^0\rangle = CI_\mp|\pi^0\rangle$, or $I_\pm C = CI_\mp$, which imply

$$CI_1 = I_1 C, \tag{38}$$

$$CI_2 = -I_2 C. \tag{39}$$

These commutation rules (37)–(39) can be derived from any multiplet.

We can find the operator $\mathscr{G}$ that incorporates C, and which commutes with $\vec{I}$, by taking advantage of the symmetry under all rotations in the isospin space $\mathscr{E}_3^I$. Note, in particular, that a rotation of 180° about the 1-axis, produced by the operator $\exp(i\pi I_1)$, reflects the 2- and 3-components of any isovector. By combining this rotation with C we leave (38) alone, but reverse the signs in (37) and (39). With this in mind, we define[14]

$$\mathscr{G} = e^{i\pi I_1}C. \tag{40}$$

---

[13] One is free to introduce arbitrary phase factors $\eta_\pm$ into $C|\pi^\pm\rangle = \eta_\pm|\pi^\mp\rangle$. These will not affect the final results, and are therefore set to unity.

[14] A phase convention different from (36) is frequently used, which leads to $\mathscr{G} = C \exp(i\pi I_2)$. The G-parity assignments and selection rules are, however, independent of the phase convention.

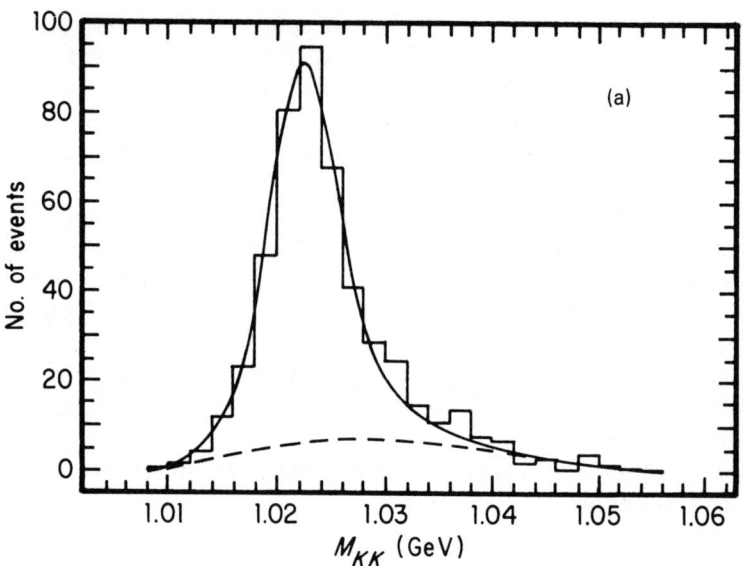

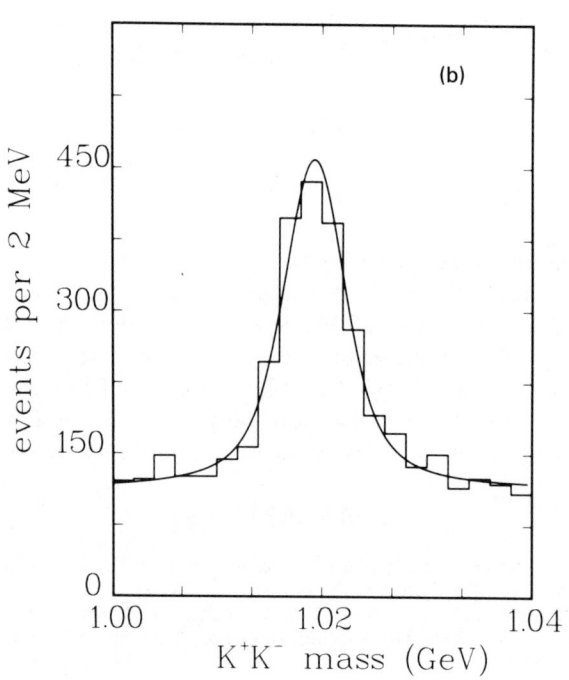

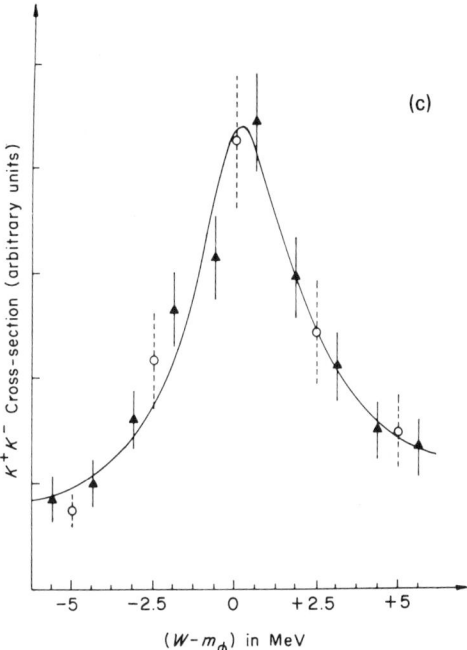

FIG. 14. (a) Photoproduction of K^+K^- pairs in copper, showing the vector meson ϕ at $M_{K\bar{K}} = 1020$ MeV, as observed at the Cornell Electron Synchrotron [McClellan et al. (1971)].
(b) Production of ϕ in the reaction $pp \to ppK^+K^-$ at an incident momentum of 11.75 GeV/c, as observed at Argonne National Laboratory [Arenton et al. (1982)].
(c) Production of ϕ in the reaction $e\bar{e} \to K^+K^-$, as observed at the Orsay Storage Ring [LeFrancois (1971)]. W is the total $e\bar{e}$ energy.

Then

$$[\mathcal{G}, \vec{I}] = 0, \tag{41}$$

and therefore $\mathcal{G}$ can be specified simultaneously with I^2 and I_3. Since $[C, I_1] = 0$, $\mathcal{G}^2 = \exp 2\pi i I_1 = 1$ for integer isospin which, as we see in the next paragraph, is the only case of interest. Consequently, the eigenvalues of $\mathcal{G}$ are $G = \pm 1$. These are called the G-parity.

A hadron can only have a definite G-parity if it has no attribute beyond isospin, such as baryon number or strangeness, which is reflected by C. For example, for the isoscalar baryon Λ^0, $\mathcal{G}|\Lambda^0\rangle = |\bar{\Lambda}^0\rangle$, and therefore Λ^0 cannot be an eigenstate of $\mathcal{G}$. As a consequence, *only mesons that carry no net flavor beyond isospin can have a definite G-parity*. All such mesons have $I = 0$ or 1.

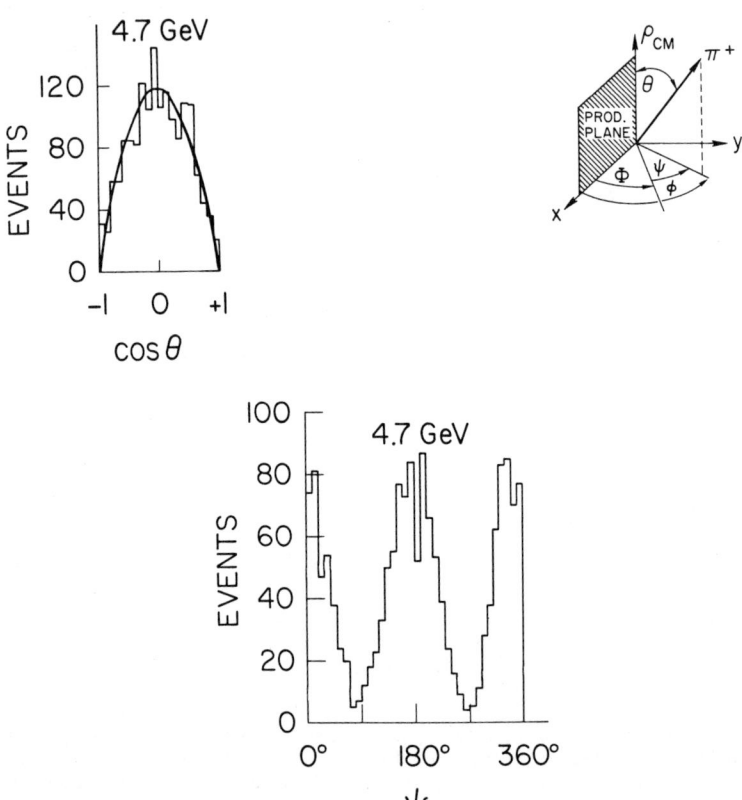

FIG. 15. Determination of the spin of ρ^0 in the reaction $\gamma p \to p\pi^+\pi^-$ [Ballam et al. (1970)]. In this experiment the 4.7-GeV photon beam provided by the Stanford Linear Accelerator is linearly polarized. The coordinate frame shown is the ρ^0 rest frame, and the xz-plane is the plane in which this quasi two-body (i.e., $\gamma p \to \rho^0 p$) collision occurs. In this frame the ρ^0 lab momentum is along z, while the photon polarization is in the xy-plane, and makes an angle Φ, as shown. The histograms show the polar and azimuthal distributions of the π^+ momentum, which are as expected for a two-body decay of a $J = 1$ particle.

Since the strong interaction commutes with $\vec{I}$ and C, it does so with $\mathcal{G}$, and therefore *G-parity is conserved in strong interaction processes.* On the other hand, the electromagnetic interaction does not conserve $\vec{I}$, while the weak interaction violates C as well, and therefore *G-parity is not conserved in electromagnetic or weak processes.*

Consider a state $|a_1 a_2 \cdots\rangle$ of noninteracting hadrons $a_1, a_2, \ldots$. These could be the products of a hadronic decay, particles about to undergo a collision, etc. As they are noninteracting, their state vector is a product: $|a_1\rangle |a_2\rangle \cdots$. If G_i is the G-parity of a_i, it then follows from the definition of $\mathcal{G}$ that $|a_1 a_2 \cdots\rangle$ has the G-parity $\prod_i G_i$. In short, G-parity is a multiplicative quantum number, as is parity. If an object A, with G-parity G_A, decays hadronically into $|a_1 a_2 \cdots\rangle$, the conservation law is

$$G_A = \prod_i G_i. \tag{42}$$

Let us first determine the G-parity of the pions. Because of the commutation rule (41), all members of any isomultiplet have the same G-parity, so we only need to find G for π^0. From (36') we know that $C|\pi^0\rangle = |\pi^0\rangle$. Since $|\pi^0\rangle$ is the $I_3 = 0$ member of an isotriplet, it transforms under rotations in $\mathscr{E}_3^I$ in the same way as the 3-component of a vector. Consequently, the rotation of 180° about the 1-axis involved in $\mathscr{G}$ reflects $|\pi^0\rangle$, i.e., $\exp(i\pi I_1)|\pi^0\rangle = -|\pi^0\rangle$. Hence, $\mathscr{G}|\pi^0\rangle = -|\pi^0\rangle$, and therefore the *G-parity of a pion is* -1, no matter what its charge:

$$G_\pi = -1. \tag{43}$$

The same argument determines the G-parity of any vector meson isomultiplet whose neutral member V^0 has the quantum numbers of the photon, for then $C|V^0\rangle = -|V^0\rangle$. If V^0 is an isoscalar, it is unaltered by the rotation $\exp(i\pi I_1)$, and its G-parity is -1. On the other hand, if V^0 belongs to an isotriplet, the discussion concerning π^0 tells us that V^0 is reflected by $\exp(i\pi I_1)$, giving it G-parity $+1$. We therefore conclude that the mesons produced in e^+e^- annihilation have the following G-parities:

$$\rho: G = 1 \tag{44}$$

$$\omega, \phi, J/\psi, \psi', \Upsilon: G = -1. \tag{45}$$

By the same argument, any isoscalar state that is an eigenstate of C, with eigenvalues ± 1, has that same eigenvalue as its G-parity. This remark tells us that the three charmonium P-states called $\chi(3556, 3510, 3415)$, as well as the ^1S charmonium ground state η_c, which are all fed by one-photon transitions from ψ', have the opposite C-signature to ψ', and therefore:

$$\chi, \eta_c: G = 1. \tag{46}$$

These assignments (43)–(46), and the conservation law (42), lead us to the selection rules listed in Table 1. All the data, most of which are listed in Appendix I, are consistent with these selection rules. The most striking evidence comes from J/ψ, which has a large number of observed multi-meson decay modes. One finds that the $5, 7,$ and 9π modes have branching fractions an order of magnitude larger than the 4 and 6π modes. As we will see shortly, the latter forbidden decays can be ascribed to the electromagnetic interaction. One also observes $J/\psi \to \rho\pi$, ρA_2, $\omega 2\pi$, and $\omega 4\pi$ with branching fractions of $\sim 1\%$, but not $\rho 2\pi$, πA_2, or $\omega 3\pi$. (The G-parity of A_2 is -1, as shown below.) The χ-states are observed to decay into $2, 4,$ and 6 pions, but $3, 5,$ and 7π decays are not seen.

TABLE 1
G-parity selection rules

Allowed	Forbidden
$\rho \to 2\pi$	$\rho \to 3\pi$
$\omega, \phi \to 3\pi$	$\omega, \phi \to 2\pi$
$A_2 \to \omega\pi\pi, \eta\pi$	$A_2 \to \rho\pi\pi, \eta\pi\pi$
$\psi, \psi' \to$ odd # π's	$\psi, \psi' \to$ even # π's
$\to \rho\pi, \rho 3\pi,$	$\to \rho 2\pi, \rho 4\pi$
$\to \omega 2\pi, \omega 4\pi, \rho A_2$	$\to \omega 3\pi, \pi A_2$
$\chi \to$ even # π's	$\chi \to$ odd # π's

Further insight into the concept of G-parity can be obtained by examining the isoscalar η meson. One decay mode of η is $\eta \to 2\gamma$. As with $\pi^0 \to 2\gamma$, this establishes that $C|\eta\rangle = |\eta\rangle$, and since the isospin rotation that differentiates C from $\mathcal{G}$ does not affect an isoscalar, we have

$$G_\eta = 1. \tag{47}$$

This implies that the strong interaction permits $\eta \to 2\pi$, but forbids $\eta \to 3\pi$. Since η has $J^P = 0^-$, in $\eta \to 2\pi$ the pions must be in an S-state, which has even parity. Therefore, $\eta \to 2\pi$ is forbidden to both the strong and the electromagnetic interaction. In view of its low mass of 550 MeV, we conclude that η has no open strong interaction decay channels, which explains its remarkably narrow width of ~0.9 keV.

As a last application, consider the $I = 1, J^P = 2^+$ meson $A_2(1318)$, shown in Fig. 10. Its dominant decay mode is $A_2 \to \rho\pi$, and therefore $G_{A_2} = -1$. The decays $\omega\pi\pi$ and $\eta\pi$ should therefore be copious, but $\rho\pi\pi$ and $\eta\pi\pi$ are forbidden by G-parity. Once more the data are in accord with these expectations.

Since G-parity is not conserved by the electromagnetic interaction, we must expect decays forbidden by G-parity to exist, but with rates comparable to electromagnetic decays. This is exemplified by $\eta \to 3\pi$, which does occur with a branching fraction of 56%, i.e., comparable to the electromagnetic decay $\eta \to 2\gamma$ with 39%.

The preceding discussion of G-parity made no reference to the quark structure of hadrons. By and large the G-parity assignments were based on an observed electromagnetic process, as in the case of π, the vector mesons, η, and the χ states in charmonium; in the case of A_2, the hadronic decay $A_2 \to \rho\pi$ was used to assign $G_{A_2} = -1$. By exploiting the quark–antiquark composition of mesons one can, however, derive these G-parities. From our study of positronium, we know that a fermion-antifermion state of total spin S and orbital angular momentum L has the signature $(-1)^{L+S}$ under C [see Eq. II.C(38)]. The operator $\exp(i\pi I_1)$ gives a factor $(-1)^I$, as we learned in deriving Eq. (43). Consequently,

$$G = (-1)^{I+L+S} \qquad (48)$$

for any uncharged quark–antiquark state that carries no flavor beyond isospin. The G-parity assignments that we derived on phenomenological grounds now follow from (48): η and η_c are $I = 0$ ^1S-states, so $S = L = 0$, and $G = 1$; the vector mesons are ^3S-states, and therefore $G = -(-1)^I$; the χ's are $I = 0$ ^3P-states, so $G = 1$; and A_2 is an $I = 1$ $^3P_2$-state, whence $G = -1$.

To summarize, the accord of the G-parity selection rules with the data provides significant corroboration for the isospin invariance of the strong interactions, and the quark model's ability to explain all G-parity assignments lends further support to the hypothesis that mesons are quark–antiquark bound states.

B. MESONS COMPOSED OF HEAVY QUARK–ANTIQUARK PAIRS

In §I.E.6(e) we already learned that mesons having the composition $c\bar{c}$ and $b\bar{b}$ possess quite rich excitation spectra containing a variety of very narrow (i.e., quasi-stable) levels. This is to be contrasted with mesons built from the quarks u, d, and s, where only the 0^- ground state is narrow. This special feature of the $c\bar{c}$ and $b\bar{b}$ spectra is due to the large masses of the c- and b-quark, which are of order 1.5 and 4.5 GeV, respectively. These masses are very large compared to the momenta of the quarks within these hadrons. *Hence the* $c\bar{c}$ *and* $b\bar{b}$ *systems are nonrelativistic*. This is not the case for systems made up of light quarks, such as u, d, s. That this is so for the $c\bar{c}$ and $b\bar{b}$ systems is most readily seen by observing that their characteristic excitation energies are small compared to the ground-state mass (see Fig. I.8), a statement that does not hold for the mesons composed of u, d, and s. Nor does it hold for mesons having one light and one heavy quark, such as $c\bar{u}$ or $b\bar{d}$, because in such a system the heavy member is essentially at rest, and the spectrum is determined by the motion of the light member.

In view of this distinction between (u, d, s) on the one hand, and (c, b) on the other, we introduce the generic symbol q for light quarks, and Q for heavy quarks. The antiquarks are denoted by $\bar{q}$ and $\bar{Q}$.

Relativistic corrections are of order $(v/c)^2$. The detailed analysis to be described below shows that $(v/c)^2$ is about 0.2 in the $c\bar{c}$ system, and less than 0.1 in the $b\bar{b}$ system. To first approximation, therefore, the internal motion can be described by an ordinary Schrödinger equation with a potential V. This potential, and the spectrum that follows from the Schrödinger equation, was already shown in Figs. I.18 and 19.

The nonrelativistic nature of the $Q\bar{Q}$ system has an importance that transcends our familiarity with such problems. When a system is relativistic, as is the case of $q\bar{q}'$ and $Q\bar{q}$ mesons, the full machinery of quantum field theory is needed for an adequate description. That is to say, the interaction must be described by propagating quanta (gluons, in our case), and fermion-antifermion pairs can appear and disappear with appreciable probability amplitudes. In short, a relativistic bound state is not a system with a definite number of degrees of freedom.

In view of their simplicity, the $Q\bar{Q}$ mesons have a singularly important role, and, with only slight exaggeration, they have been called the "hydrogen atoms of hadronic spectroscopy." To be more precise, the $Q\bar{Q}$

spectra are very similar to those of positronium, $e\bar{e}$, or of "true muonium," $\mu\bar{\mu}$, which are also bound nonrelativistic fermion-antifermion systems.[1] Both the lepton-antilepton and $Q\bar{Q}$ systems have finite lifetimes because their constituents can annihilate into the quanta of their binding field.

1. $e\bar{e} \to \mu\bar{\mu}$ revisited

The simplest way of producing the $Q\bar{Q}$ mesons is by $e\bar{e}$ annihilation. It is easiest to acquaint oneself with the creation of bound states by $e\bar{e}$ annihilation if one first considers the purely electromagnetic process $e\bar{e} \to \mu\bar{\mu}$. We therefore begin by expanding on our discussion of this process in §II.C.2.

In our earlier treatment, we described the scattering amplitude for $e\bar{e} \to \mu\bar{\mu}$ by the Feynman graph

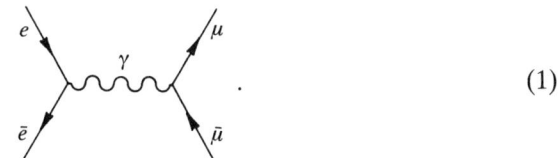

(1)

Supposedly, this is the first term in a power series in α, and so one might be tempted to conclude that it always provides an adequate description of the reaction as long as a precision of better than about 1% is not desired. But this cannot be universally true: The $\mu\bar{\mu}$ pair interacts via the ordinary Coulomb interaction, and this interaction cannot be neglected at will, for among other things it produces the whole Bohr set of $\mu\bar{\mu}$ bound states! The point is that an expansion of powers in α is only possible if the energy W of the incoming electron pair (which equals that of the produced $\mu\bar{\mu}$ pair) is so high that we can neglect the Coulomb interaction between the muon pair. The latter approximation is valid when the kinetic energy of the muons is large compared to their interaction, which is of order one Rydberg, $\frac{1}{2}m_\mu\alpha^2$, where m_μ is the mass of the muon. This is only true if $(W - 2m_\mu) \gg m_\mu\alpha^2$. If W is near $2m_\mu$, the Coulomb interaction is essential, and an expansion in powers of α is not permissible.[2]

What does in fact happen? Obviously the Coulomb interaction between the slow muons should be incorporated, so we must add to (1) the graphs

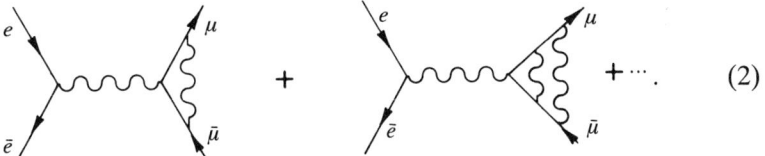

(2)

---

[1] For that reason the $c\bar{c}$ system is often referred to as "charmonium," and any $Q\bar{Q}$ system as "quarkonium."

[2] Indeed the expansion parameter is $\alpha^* = e^2/4\pi\hbar v$, where v is the velocity of the muons. If $v \ll 1$ (nonrelativistic), α^* is not small.

From our discussion of bound states in §II.C.4 we recognize that what (2) does is to replace the plane wave that describes the relative motion of almost free μ an $\bar\mu$ in the final state of (1) by an $\mu\bar\mu$ eigenfunction ψ in the presence of the Coulomb field. These eigenfunctions fall into two classes: If the total $e\bar e$ energy W exceeds $2m_\mu$, we are above threshold for the process $e\bar e \to \mu\bar\mu$, and ψ is a scattering wave function of two muons interacting via their Coulomb attraction; if, however, $W < 2m_\mu$, ψ is a bound state wave function. The masses of these $\mu\bar\mu$ bound states are given by the Bohr formula,

$$M_n = 2m_\mu\left(1 - \frac{\alpha^2}{8n^2}\right), \qquad n = 1, 2, \ldots. \tag{3}$$

These $\mu\bar\mu$ bound states have energies that are less than $2m_\mu$, i.e., they lie below threshold for the reaction $e\bar e \to \mu\bar\mu$. Nevertheless, when $W \simeq M_n$, the nth bound state is formed, but it does not live forever: It decays back into an $e\bar e$ pair and thereby appears as a resonance in $e\bar e$ elastic scattering, or it decays into three photons,[3] and is seen as a resonance in $e\bar e \to 3\gamma$. These reactions can be depicted by

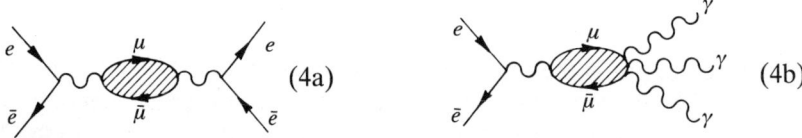

where the shaded loop represents a bound $\mu\bar\mu$ state. The lifetimes Γ^{-1} of these states are extremely long compared to the periods of the motion within the bound states ($\Gamma \ll m_\mu \alpha^2$). Because of the finite lifetime, W need not be exactly equal to the values (3). The small value of Γ causes very sharp resonances in elastic $e\bar e$ scattering and in $e\bar e \to 3\gamma$ slightly below $W = 2m_\mu$, as sketched in Fig. 16.

We can estimate the order of magnitude of the widths of these resonances by observing that the probability amplitude for the process $(\mu\bar\mu)_{\text{bound}} \to \gamma \to e\bar e$ must be proportional to the amplitude for finding the μ pair at zero separation. The latter is given by the wave function evaluated at $\mathbf{r} = 0$, $\psi_n(0)$. There must also be two factors of e, one for the vertex in the process $\mu\bar\mu \to \gamma$, another for $\gamma \to e\bar e$. The decay probability is therefore proportional to $e^4|\psi_n(0)|^2$, and a detailed calculation along the lines discussed in §II.C.7 gives[4]

$$\Gamma_n = \frac{16\pi}{3}\frac{\alpha^2}{M_n^2}|\psi_n(0)|^2, \tag{5}$$

---

[3] The bound state is formed by virtual one-photon annihilation [see (4)], and is therefore odd under charge conjugation (recall §II.C.7). Hence, it can only decay into an odd number of photons.

[4] This is only the width for $e\bar e$ decay. Other processes also contribute to Γ: $\mu\bar\mu \to 3\gamma$, weak decays, and radiative transitions to lower bound states, e.g., $5^3S_1 \to 3^3P_1 + \gamma$. The first two are negligible compared to (4a), while radiative decay is only comparable to direct $e\bar e$ annihilation for n^3S_1 states with $n \geq 3$.

1. $e\bar{e} \to \mu\bar{\mu}$ REVISITED

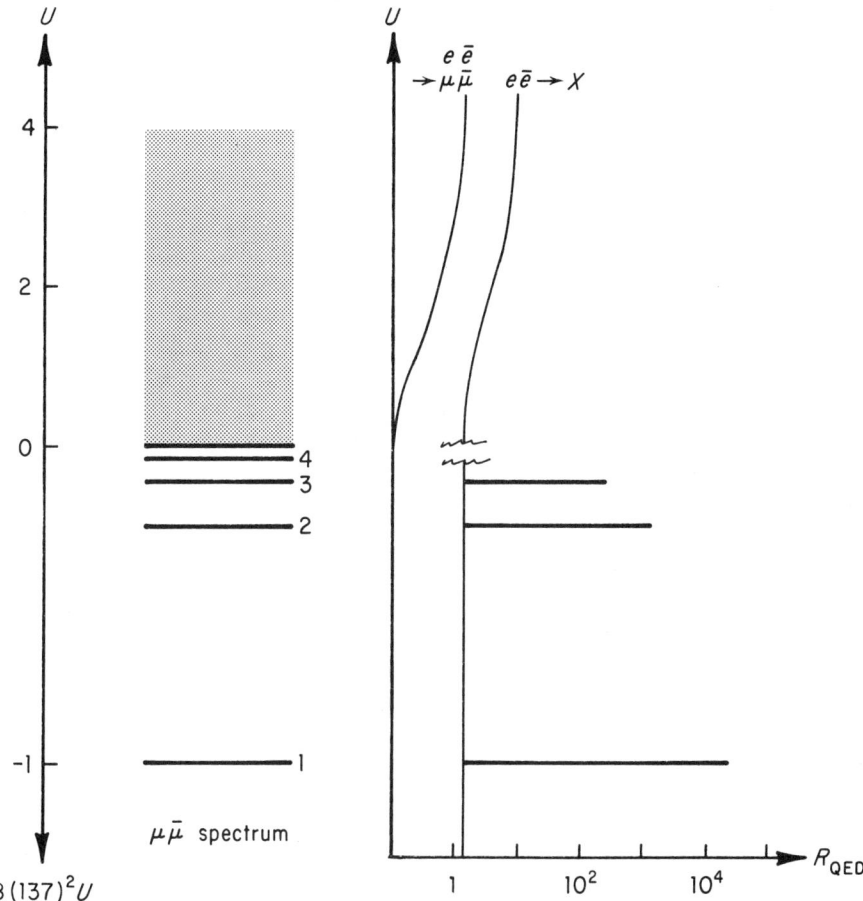

FIG. 16. The $\mu\bar{\mu}$ spectrum and the $e\bar{e}$ cross section in the neighborhood of the threshold for $e\bar{e} \to \mu\bar{\mu}$. The quantity $U \equiv (W/2m_\mu) - 1$ measures energy with respect to the threshold at $U = 0$. For $U < 0$, the scale is vastly expanded by the factor $8/\alpha^2$ so that the bound states can be clearly seen. The $\mu\bar{\mu}$ bound states are labeled by the principal quantum number. The shaded region above $U = 0$ is the $\mu\bar{\mu}$ continuum. $R_{\rm QED}$ is the total $e\bar{e}$ cross section divided by $4\pi\alpha^2/3W^2$, the latter being the $e\bar{e} \to \mu\bar{\mu}$ cross section when W is large enough so that m_μ is negligible. For $U < 0$, $R_{\rm QED}$ includes $e\bar{e}$ elastic scattering and $e\bar{e} \to 2\gamma$. Between resonances these are of comparable magnitude, but at resonance elastic scattering dominates. The resonances are so narrow that their widths cannot be seen on this scale. The region just below threshold is not shown because $U = 0$ is an accumulation point of resonances.

where M_n is the mass of the nth state (which is $2m_\mu$ to good approximation). We can estimate $\psi_n(0)$, because a wave function has the dimension (length)$^{-\frac{3}{2}}$. The characteristic size of the nth state is $na_0 = 2n/m_\mu\alpha$, where a_0 is the Bohr radius. Thus, $|\psi_n(0)|^2 \sim (m_\mu\alpha)^3/n^3$, while a detailed calculation gives

$$\Gamma_n = \frac{m_\mu\alpha^5}{6n^3} = \tau_n^{-1}.$$

We may compare the lifetime τ_n with the characteristic period of the bound state motion, τ_n^0, which is given by $1/\Delta M_n$, where ΔM_n is the level spacing. According to (3), $\Delta M_n \sim m_\mu \alpha^2/n^3$, and therefore

$$\frac{\Gamma_n}{\Delta M_n} = \frac{\tau_n}{\tau_n^0} \sim \frac{1}{6\alpha^3} \sim 10^{-7}.$$

Hence, the resonances due to the $\mu\bar{\mu}$ bound states are exceedingly narrow; their observation would be extremely difficult. Nevertheless, there is such a close analogy between these Coulomb resonances, and those seen in $e\bar{e} \to$ hadrons, that we carry our analysis of the $\mu\bar{\mu}$ resonances somewhat further.

Not all the $\mu\bar{\mu}$ Coulomb bound states show up as resonances in $e\bar{e}$ scattering. From Fig. 17, and also from the Feynman diagrams (4) that give the correct quantum mechanical description of the process, we see that the transitions $e\bar{e} \to \mu\bar{\mu} \to e\bar{e}$ occur via one-photon intermediate states. Therefore, the $\mu\bar{\mu}$ bound states that can be fed in resonant scattering must have the spin-parity of the photon, 1^-. As we learned in §II.C.7, these are the states $^3S_1$ and $^3D_1$, with any principal quantum number. But there is a further restriction, because the virtual process $\gamma \to \mu\bar{\mu}$ can only occur at vanishing $\mu\bar{\mu}$ separation. Consequently, only the ^3S states will be observed as resonances, since the D-state wave functions vanish when $r \to 0$.

It is easy to see that the resonant cross section is far larger than the nonresonant background. Recall from Eq. A(20) that a $J = 1$ resonant cross section in $e\bar{e}$ scattering has the magnitude[5]

$$\sigma_{\text{res}}^{e\bar{e}} = \frac{12\pi}{W^2}, \tag{6}$$

because the relative momentum k is just the energy $\tfrac{1}{2}W$ of either incident particle in our ultrarelativistic situation. The off-resonant cross section is found from the one-photon exchange $e\bar{e}$ cross section as given by Eq. II.C(20). It has a distribution that is strongly forward peaked, and the observed total nonresonant cross section σ_{non} is therefore rather sensitive to the angular acceptance of the detector. Nevertheless, its order of magnitude is obviously $\alpha^2 \mathscr{A}/W^2$, where $\mathscr{A}$ is a detector-dependent quantity of order unity. Thus, the peak-to-background ratio, $\sigma_{\text{res}}/\sigma_{\text{non}}$, is of order $(137)^2$.

As soon as the total c.o.m. energy W crosses the value $2m_\mu$, free muons can be produced. In the threshold region where the muon velocity v_μ is small, there are still important effects due to the Coulomb attraction between the produced particles, but this does not result in any resonances (we are in the continuum region of the $\mu\bar{\mu}$ system). Once W is large enough for v_μ to approach one, lowest order perturbation theory [i.e., the one-photon annihilation graph (1)] becomes valid. The total cross section is

---

[5] Equation (6) is valid if the decay of the resonance is dominated by the channel in which it is produced, i.e., if elastic scattering dominates.

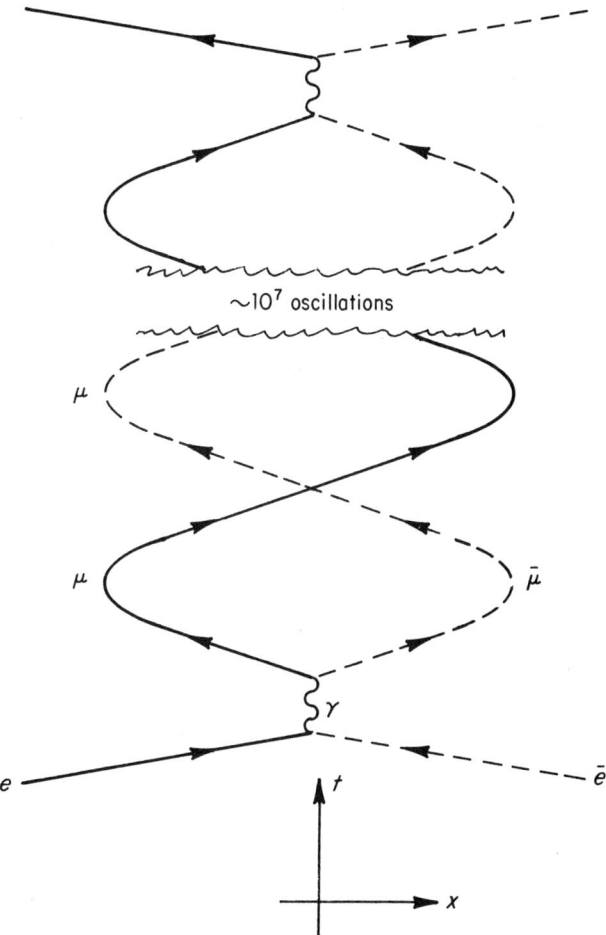

FIG. 17. A crude classical space-time depiction of resonant $e\bar{e}$ scattering through a $\mu\bar{\mu}$ bound state. It is the counterpart of the Feynman diagram (4a), but here the arrows indicate the direction of motion for particles (solid lines) and antiparticles (broken lines). The units of x and t are chosen so that the e and $\bar{e}$, which move at the speed of light, are not on trajectories at 45°; otherwise, the μ and $\bar{\mu}$ trajectories would not show up clearly. Also note that there are, on average, of order 10^7 $\mu\bar{\mu}$ oscillations in this process.

then found by integrating Eq. II.C(16) over angles; this yields

$$\sigma(e^+e^- \to \mu^+\mu^-) = \frac{4\pi}{3}\frac{\alpha^2}{W^2} \equiv \sigma_{\mu\mu} \quad \text{for } (W - 2m_\mu) \gg \alpha^2 m_\mu. \quad (7)$$

2. Narrow hadron resonances in $e\bar{e}$-hadron production

Following the transition of $e\bar{e}$ to a virtual photon, the latter can convert to a quark–antiquark pair, instead of to a lepton pair, with an amplitude proportional to the quark's charge [recall Eq. I.E(10)]. An example of such a process at relatively low energy, $e\bar{e} \to K\bar{K}$, is depicted in Fig. I.23(b), where the intermediate quark pair is $s\bar{s}$. The cross section, shown in Fig. 14(c), has a sharp resonance at $W = 1020$ MeV, just above the $K\bar{K}$ threshold $2m_K$. This resonance was already mentioned in §A.8; it is the vector meson ϕ. There are two other rather light vector mesons, ρ^0 and ω, having a mass near 770 MeV, which are seen in $e\bar{e}$ annihilation, and in many other processes.

In §A we discussed these (and other) hadrons from a purely phenomenological point of view, without invoking their quark structure. But as we learned in §I.E.6, ρ^0 and ω are $I = 1$ and $I = 0$ combinations of $u\bar{u}$ and $d\bar{d}$, respectively, while ϕ is $s\bar{s}$, and all are nodeless $^3S_1$ states. In short, they are the lowest states of these $q\bar{q}$ systems having the photon's quantum numbers. Radially excited states—having nodes but the same $^3S_1$ structure—surely exist, but they lie far above the threshold for decay into multimeson channels, and are therefore very broad and correspondingly difficult to isolate. As a consequence, the total cross section for $e\bar{e} \to$ hadrons above the ϕ-resonance is rather featureless.

A most dramatic change occurs for W above 3 GeV, as seen at a glance in Fig. 18. What is plotted here is the cross section and the ratio

$$R \equiv \frac{\sigma(e^+e^- \to \text{had})}{\sigma(e^+e^- \to \mu^+\mu^-)} \equiv \frac{\sigma_{\text{had}}}{\sigma_{\mu\mu}}, \tag{8}$$

where the denominator is the $\mu^+\mu^-$ production cross section at energy W, as given by (7). As we saw in §I.E.8(c), greater insight is achieved if one presents the data in terms of this ratio R, rather than the hadronic cross section itself. R goes through two enormous resonances at $W = 3.097$ and 3.685 GeV; the lower one is shown in detail in Fig. 18(b)–(d). Both resonances have a width smaller than the energy spread in the beams, which is only several MeV. The background value of R between these resonances is not detectably different from its value below the first resonance. At $W = 3.770$ GeV there is another prominent resonance having a width of about 25 MeV [see Fig. 18(e)]. A very rapid rise of R sets in just below 4 GeV, then a set of prominent but broad resonances are observed between 4 and 5 GeV, while at still higher energies R settles down to a value that is roughly twice as large as the nonresonant background below the 3.685-GeV resonance [see Fig. 18(a) and (e)].

While we have only shown the data for $e^+e^- \to$ hadrons, we emphasize that these resonances are also observed in the channels $e^+e^- \to e^+e^-$ and $e^+e^- \to \mu^+\mu^-$. We draw on some of these data below. Furthermore, purely hadronic collisions also can lead to the production of the 3.097 and 3.685 GeV states. They are then detected by the method described in §A.8, in particular by the decay modes e^+e^- or $\mu^+\mu^-$. Indeed, the 3.097 GeV state was discovered simultaneously in e^+e^- annihilation, and called ψ; and in the reaction p + nucleus $\to e^+e^-$ + hadrons, where it was designated by

J. In the latter reaction, the observed number of events was plotted as a function of e^+e^- mass, and an enormous peak at 3.1 GeV was found. This has led to the rather awkward symbol J/ψ for this object. The higher peaks are designated by the symbols ψ', ψ'', etc., or by giving their mass, as in $\psi' = \psi(3685)$. For the sake of brevity we often use the symbol ψ for J/ψ.

On comparing the data on $e^+e^- \to$ hadrons (Fig. 18) with Fig. 16, which depicts e^+e^- annihilation near the $\mu^+\mu^-$ threshold, one cannot avoid being struck by the similarity. It is only natural to guess that the two very sharp resonances are bound states of some particle-antiparticle pair, with a dynamics similar to the $\mu^+\mu^-$ system, and that the threshold for producing these particles lies just above the sharp resonances.

Before exploring this idea, we quickly survey some data which shed further light on the nature of J/ψ and ψ'.

First, there is direct experimental evidence that the resonances at 3.097 and 3.685 GeV have the quantum numbers of the photon. This is established by demonstrating an interference between the nonresonant amplitude for $e^+e^- \to \mu^+\mu^-$ and the amplitude for $e^+e^- \to$ resonance $\to \mu^+\mu^-$. In terms of graphs, these amplitudes are

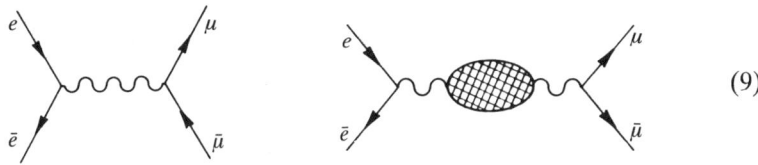 (9)

where the shaded loop depicts the hadronic resonance. As we know, the first of these amplitudes only feeds $\mu^+\mu^-$ states having the photon's spin, parity, and charge conjugation. There can be an interference with the second amplitude if, and only if, the resonance feeds the same $\mu^+\mu^-$ state. Furthermore the nonresonant amplitude does not vary over the width of the resonance, whereas the resonant amplitude, being of the form $(W - M + \frac{1}{2}i\Gamma)^{-1}$, changes sign as W goes from well below to well above the resonance. Thus the interference is destructive on one side of the resonance, constructive on the other, and produces an asymmetric distortion of the resonance in the channel $e^+e^- \to \mu^+\mu^-$. This distortion is observed for both J/ψ and ψ', and establishes that these objects have $J^P = 1^-$.

The second point we discuss concerns the widths of J/ψ and ψ'. As we have said, they are narrower than the energy spread in the beams,[6] and therefore not directly measurable. Nevertheless, it is possible to extract these widths from the data by using the resonance formula, Eq. A(20). The leptonic widths $\Gamma_{e\bar{e}}$ are of the same size as those of the other vector mesons ρ, ω, and ϕ, but the total widths are astonishingly small. The results are

---

[6] The spread of beam energy is due to quantum mechanics, and cannot be avoided. Electrons moving in circular orbits radiate copiously (synchroton radiation), and photons of different frequencies and directions give the electrons different recoil momenta.

III.B. MESONS OF QUARK-ANTIQUARK PAIRS

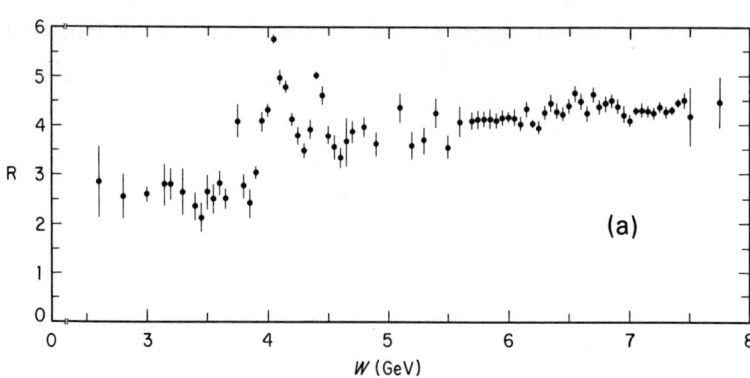

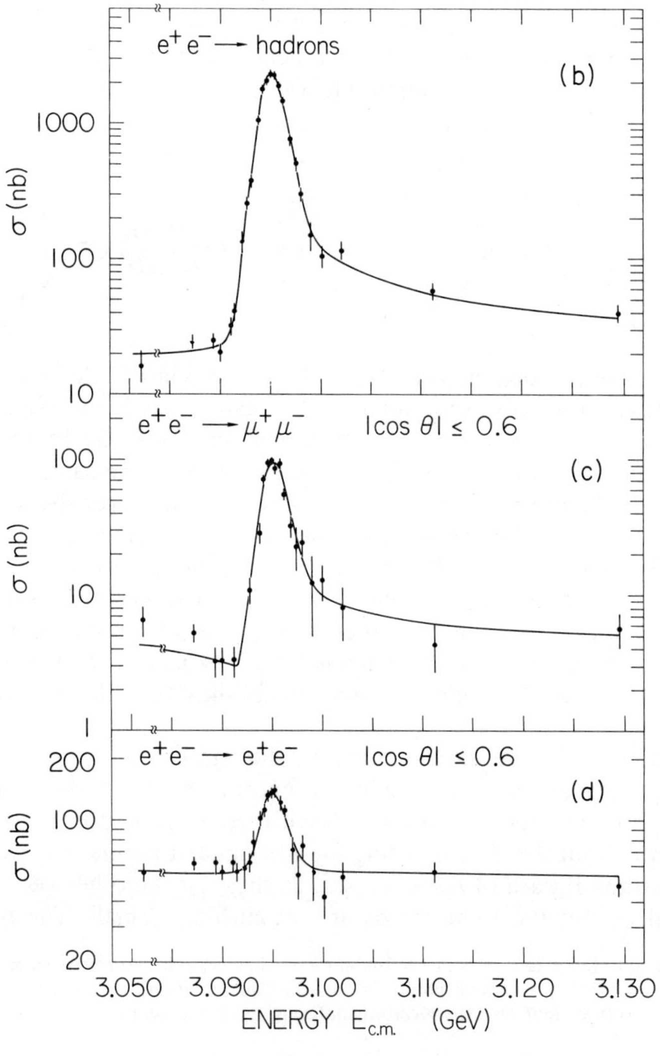

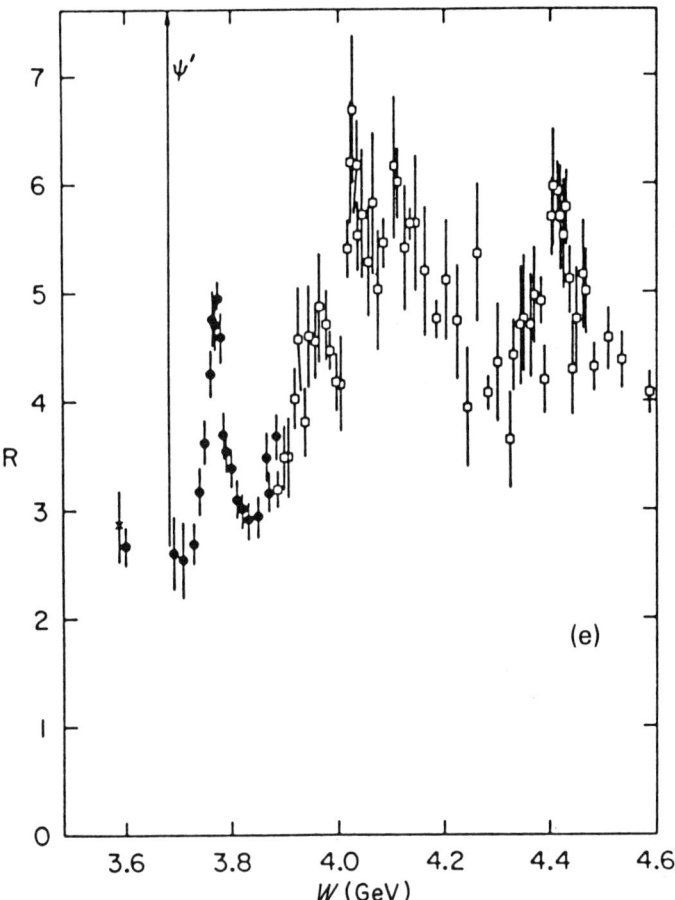

FIG. 18. $e\bar{e}$ annihilation for c.o.m. energies W between 3 and 8 GeV, as observed at the SPEAR storage ring at SLAC.

(a) is a global view of R [defined in Eq. (8)] in this entire energy range [Siegrist et al. (1982)]. This figure does not show the sharp resonances J/ψ and ψ' at 3.1 and 3.7 GeV, respectively; as one sees from (b) and (e), they are $\delta =$ functions on the scale shown here.

(b)–(d) show the J/ψ resonance in detail [Boyarski et al. (1975)]; (b) is the total cross section for $e\bar{e} \to$ hadrons in the angular interval accepted by the detector, while (c) and (d) are the cross sections for the final states $\mu\bar{\mu}$ and $e\bar{e}$. Note from (b) that the peak value of R is of order 10^3; hence, the width and height of J/ψ could not be shown on (a).

(e) shows the resonances above the $D\bar{D}$ threshold at $W = 3.73$ GeV [Barbaro-Galtieri (1977); Rapidis et al. (1977)]. The $^3D_1$ level $\psi(3770)$ is shown in detail, but ψ' is too narrow and high to be plotted on this scale.

tabulated below in keV (not MeV!):

	$J/\psi \equiv \psi(3097)$	$\psi' \equiv \psi(3685)$	
$\Gamma_{e\bar{e}}$	4.4 ± 0.4	1.9 ± 0.2	(10)
Γ	63 ± 9	215 ± 40	

[This extraction proceeds as follows. Let f be some particular final state (e.g., $\mu^+\mu^-$, $\pi^+\pi^-\pi^0 K^+K^-$). The cross section for $e^+e^- \to \psi \to f$ then reads

$$\sigma(e^+e^- \to f) = \frac{3\pi}{M^2} \frac{\Gamma(\psi \to e^+e^-)\Gamma(\psi \to f)}{(W-M)^2 + \frac{1}{4}\Gamma^2},$$

where M is the mass of ψ, and Γ the sought-after total width. The factor $(3\pi/M^2)$ is determined by the requirement that the general resonance formula [A(20)] gives the value $12\pi/W^2$ [recall Eq. (7)] at the peak of an e^+e^- elastic resonance without any inelastic channels; that is to say, while inelastic scattering actually occurs, $\sigma(e^+e^- \to f)$ must reduce to Eq. (7) at $W = M$ if the width only comes from $\Gamma(\psi \to e^+e^-)$. If the experiment's energy resolution is large compared to Γ, one actually measures

$$S_f = \int \sigma(e^+e^- \to f)\, dW \simeq \frac{6\pi^2}{M^2} \frac{\Gamma(\psi \to e^+e^-)\Gamma(\psi \to f)}{\Gamma}.$$

The experiments determine the total integrated cross section, $S = \sum_f S_f$, and the integrated elastic cross section $S_{e\bar{e}}$. But

$$S = \frac{6\pi^2}{M^2}\Gamma(\psi \to e^+e^-), \qquad S_{e\bar{e}} = \frac{6\pi^2}{M^2}\frac{[\Gamma(\psi \to e^+e^-)]^2}{\Gamma},$$

and therefore these poor-resolution measurements do determine the tabulated widths.]

These total widths are so incredibly small compared to mesons of lower mass, such as the ρ or ω, that one could ask whether ψ and ψ' are hadrons at all, and secondly, if they are hadrons, whether their decay is only due to the weak and/or electromagnetic interactions. There is a great deal of compelling evidence that ψ and ψ' are hadrons. If they are hadrons, and their decays are due to the strong interaction, then isospin must be conserved in their hadronic decays. The large body of data on ψ and ψ' hadronic decays is all consistent with the assignment $I = 0$ to both of these objects, and with the conservation of isospin in their decay. Furthermore, by analyzing the photoproduction of ψ in complex nuclei, where the produced ψ scatters from nucleons during its escape from the nucleus, one can estimate the ψ-nucleon total cross section. This yields a cross section of about 0.1 $(\text{fm})^2$, which is far larger than expected for a particle that does not interact strongly. That ψ and ψ' are intimately related hadrons is also indicated by the decay $\psi' \to \psi + \pi\pi$. In fact, about 55% of all ψ' decays lead to ψ. This makes it plausible to assume that ψ' is an excited state of ψ.

That the decays of ψ into hadrons are due to the strong, and not the weak interaction, comes from two observations. First, as just mentioned, isospin is conserved in the decays, as is G-parity (recall Table 1). Second, in the decays $\psi \to \mu^+\mu^-$ and $\psi' \to \mu^+\mu^-$ the fore-aft asymmetry is consistent with zero, as required by parity conservation; on the other hand, if the hadronic decays were due to the weak interaction, ψ and ψ' would not be eigenstates of parity. Thus we conclude that ψ and ψ' are intimately related hadrons, and that their decay into light hadrons is due to the strong interaction.

Another set of narrow hadronic resonances, the Y-family, are found in $e\bar{e} \to$ hadrons in the region of 9.5–10.5 GeV. Three resonances of the ψ-type, called Y, Y', and Y", are seen (see Fig. 19). Their leptonic widths are comparable to those of the ψ's, but their total widths are smaller still.

As we already indicated in §I.E.6(e), these resonances can be understood in detail as being due to a mechanism bearing a very close analogy to that responsible for the $\mu\bar{\mu}$ resonances discussed in §1. In particular, the $\mu\bar{\mu}$ pair is replaced by a $Q\bar{Q}$ pair, where $Q = c$ or b in the case of ψ- and Y-spectra, respectively. The reaction $e\bar{e} \to (\psi, Y) \to l\bar{l}$ can be depicted by a diagram that looks just like (4a):

$$\tag{11}$$

where $l = e$ or μ in the case of ψ, and e, μ, or τ in the case of Y (recall that $2m_\tau \sim 4$ GeV, so well below $m_Y \sim 9.5$ GeV). The processes $e\bar{e} \to (\psi, Y) \to$ hadrons are more complicated than (4b). What happens is schematically indicated in the graph (12) below. The $Q\bar{Q}$ pair annihilates into a color-zero combination of gluons. Two gluons can combine to zero color, but such states are even under charge conjugation, whereas any state produced via one virtual photon must be odd. So the minimum number of gluons is three. This combination, however, cannot propagate as independent gluons to infinity because of color confinement. They must convert into hadrons, glue-balls, or light quark pairs, and eventually materialize as a complicated multihadron state. The shaded box in (12) contains the complicated processes whereby 3 gluons convert into n hadrons:

$$\tag{12}$$

We have more to say about this in §4.

3. The $Q\bar{Q}$ spectrum

The $Q\bar{Q}$ interaction potential $V(r)$ has two asymptotes that we believe to know from general theoretical considerations. The asymptotic freedom of QCD, as discussed in §IV.C, tells us that as the $Q\bar{Q}$ separation r tends to

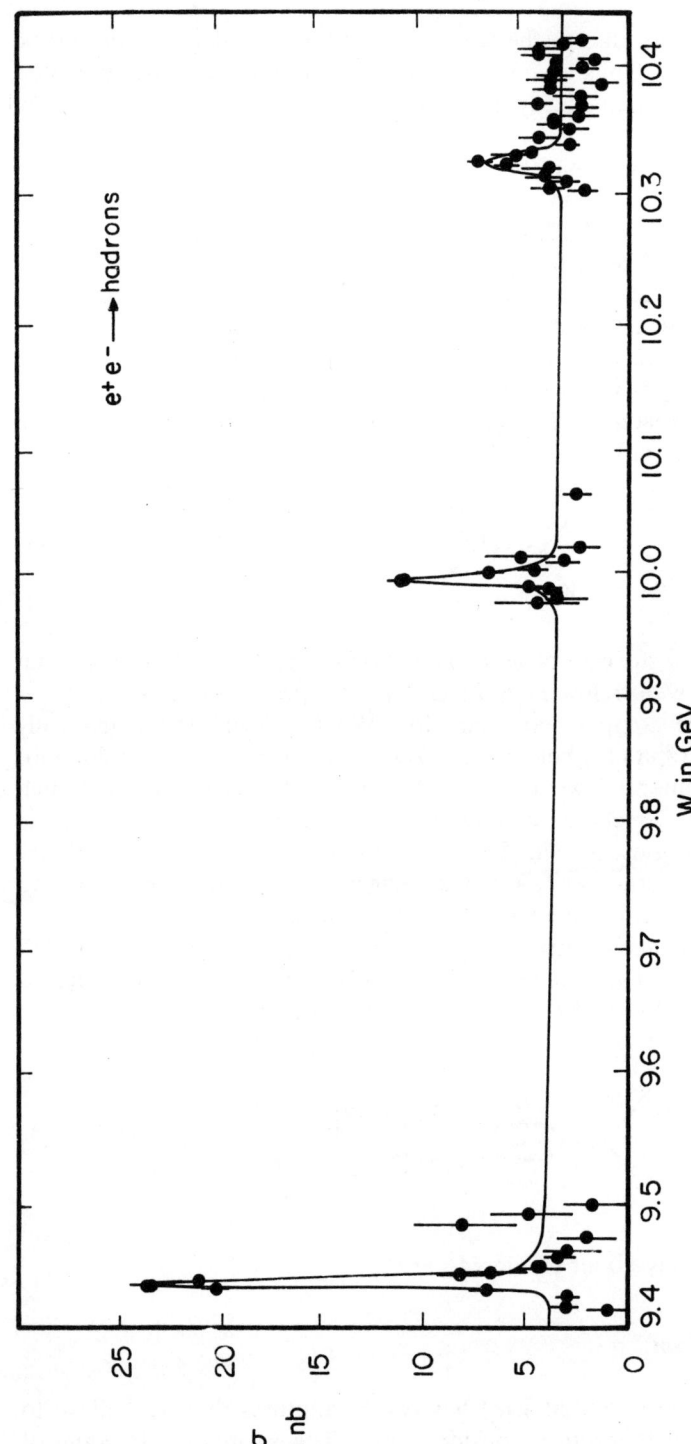

FIG. 19. The cross section for $e\bar{e} \to$ hadrons (in nanobarns) in the Y region, as measured by the CLEO detector at the Cornell electron-positron storage ring CESR. Three sharp resonances, Y, Y', and Y'', are in evidence. These are the lowest three $^3S_1$ states of the $b\bar{b}$ system. A fourth broad resonance is also observed at ~10.59 GeV; its width shows that it is above the "flavor" threshold for production of $B\bar{B}$ pairs, where B is a $b\bar{u}$ or $b\bar{d}$ meson.

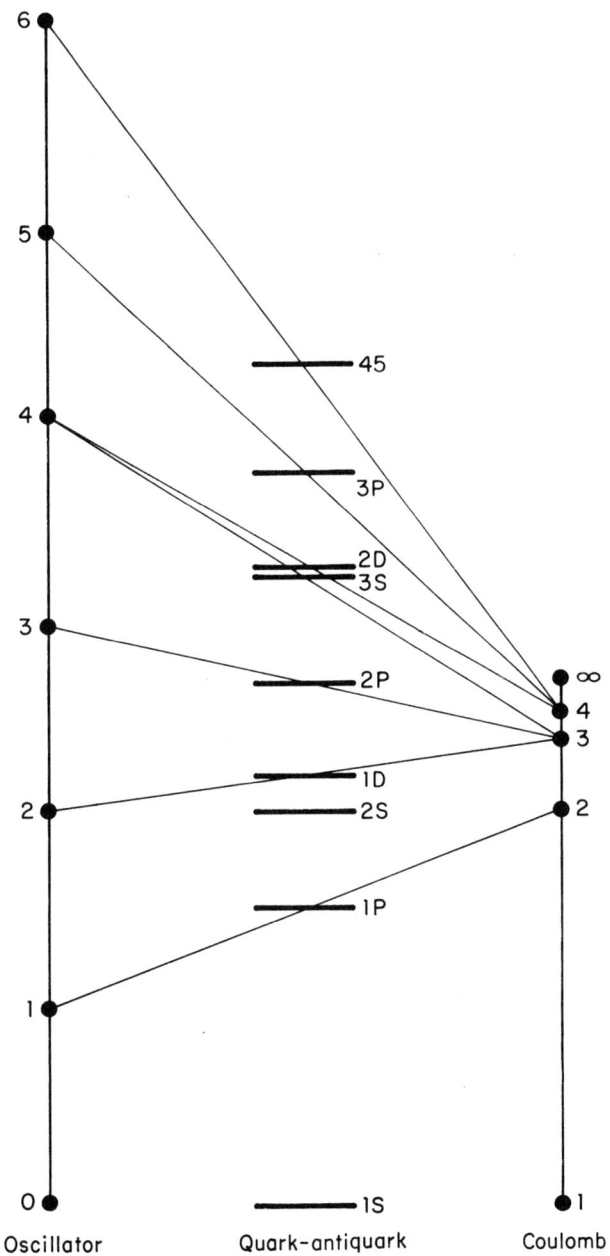

FIG. 20. The spectrum of low-lying states for the Coulomb and three-dimensional harmonic oscillator potentials, as well as for a potential that confines quark–antiquark pairs into hadrons (see Fig. 21). The parameters of these potentials were chosen so that all have the same 1S-2S splittings. The "quark–antiquark" scheme then follows by visual interpolation.

zero, V is logarithmically weaker than the Coulomb $1/r$ law. Color confinement tells us that $V \to \infty$ as $r \to \infty$, and there are good reasons to conclude (see §IV.D) that V grows linearly with r as $r \to \infty$. This linear behavior also follows from the flux tube picture of §I.E.7.

A simple potential that incorporates these two limits is sketched in Fig. I.18, and the corresponding analytic expression (called the Richardson potential) is given in Eq. IV.C(82). One cannot solve the Schrödinger equation analytically with this potential, but one can gain a surprisingly good understanding of the spectrum from a simple argument. To this end, consider Fig. 20. On the right- and left-hand sides of this figure, we show the familiar spectra of the Coulomb potential and the three-dimensional harmonic oscillator. We purposely choose the parameters so that in each case the $2S$ levels have the same excitation energy. The Coulomb potential falls to zero at large r, and therefore has both bound and unbound states. The harmonic potential grows like r^2, however, and is even more confining than the conjectured quark–antiquark interaction $V(r)$. Imagine now a continuous deformation of the potential from the Coulomb field to the oscillator; en route, this deformation will pass through the desired potential (see Fig. 21). Consequently, one can infer qualitative features of the $Q\bar{Q}$-spectrum by interpolation between the Coulomb and the oscillator spectra. This interpolation is shown in Fig. 20: The levels marked $Q\bar{Q}$ constitute the expected $c\bar{c}$- or $b\bar{b}$-level scheme when all spin-dependent forces are ignored.

As in atomic systems, such as positronium, we must expect that there are also spin-dependent forces that remove the degeneracies of Fig. 20. That is, spin–orbit and spin–spin couplings will lift the degeneracy of the states of different J in the multiplet ^3P_J, and spin–spin interactions will lead to a hyperfine splitting of ^3S and ^1S. Only a detailed dynamical theory (such as QED) is capable of predicting these splittings; at this time, our understanding of strong interactions is still unequal to this task. So all that we can say is that in analogy with QED, we expect splittings of the multiplets that are smaller than the gross structure splitting by a factor of order v/c, but we cannot predict the detailed multiplet patterns. The final spectrum has the form shown in Fig. 22.

Let us now turn to the data, and in particular, those for the ψ-family. We assign J/ψ to 1^3S, and ψ' to 2^3S. The photon spectrum observed in $e\bar{e} \to \psi' \to \gamma X$, where X is an unobserved multihadron state, is shown in Fig. 23. It reveals three monochromatic lines due to the transitions $2^3S \to {}^3P_J + \gamma$, with $J = 2, 1$, and 0. As these transitions involve a change of parity, and $\Delta J \leq 1$, they are electric dipoles. These angular momenta and the multipolarity have been verified by studying the angular distribution of the γ-rays, and angular correlations between the γ-ray and two-body decays such as $^3P_J \to K\bar{K}$. The mass and spin assignments shown in Fig. 22 summarize these data. The photon spectrum also shows a fourth well-defined maximum; this is due to two transitions $^3P \to 1^3S + \gamma$ where the different J's overlap because there is a significant Doppler broadening due to the recoil motion of the intermediate P-state.

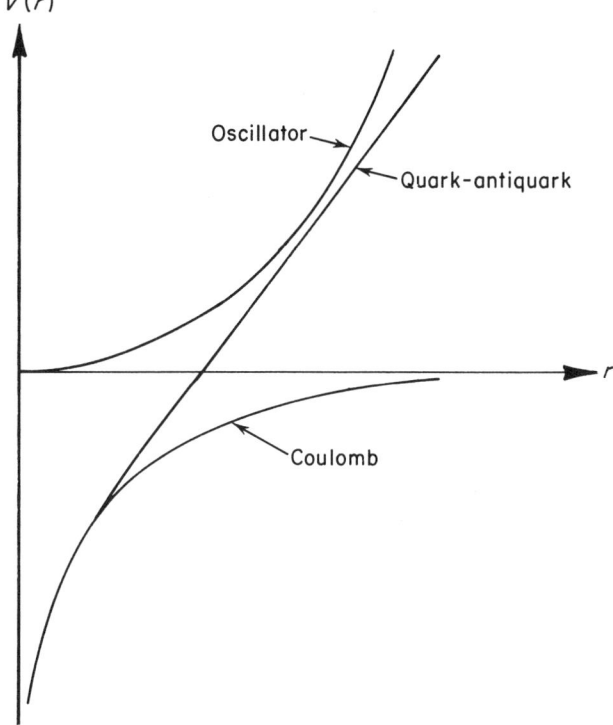

FIG. 21. The three potentials whose low-lying levels are shown in Fig. 20.

In addition to the $E1$ transitions just discussed, there are also three observed $M1$ transitions: $2^3S \to 2^1S + \gamma$, $1^3S \to 1^1S + \gamma$, and $2^3S \to 1^1S + \gamma$. Figure 24 shows the data for the two transitions to the $c\bar{c}$ ground state, 1^1S, also called η_c, because it is the charmed counterpart of the η at 550 MeV. Hence, all states below $\psi' = 2^3S$, with the exception of $^1P_1$, have been observed. The latter should not be seen, because charge conjugation forbids the transition $2^3S \to {}^1P_1 + \gamma$, as one sees from Eq. II.C(38), while the cascade $2^3S \to 2^1S + \gamma \to {}^1P_1 + 2\gamma$ has too low a combined branching fraction to be observable.

As Fig. 18(e) shows, there are a number of resonances in $e\bar{e} \to$ hadrons above ψ'. Of these the first one, at $W = 3770$ MeV, is especially interesting. Since it is observed directly in $e\bar{e}$ annihilation, its spin parity is 1^-. On the other hand, the results of the potential model, to be discussed shortly, make it clear that this is not the second radially excited 3^3S state, for it is much too close in energy to 2^3S. As we remarked before, the state $^3D_1$ also has the quantum numbers of the photon, and all evidence is in accord with the assignment of $^3D_1$ to $\psi(3770)$. At first sight, this assignment may be puzzling, because the D-state wave function vanishes at $r = 0$, so that it cannot couple to the virtual photon. But we must remember that orbital angular momentum L is not a rigorously conserved quantum number, only

III.B. MESONS OF QUARK–ANTIQUARK PAIRS

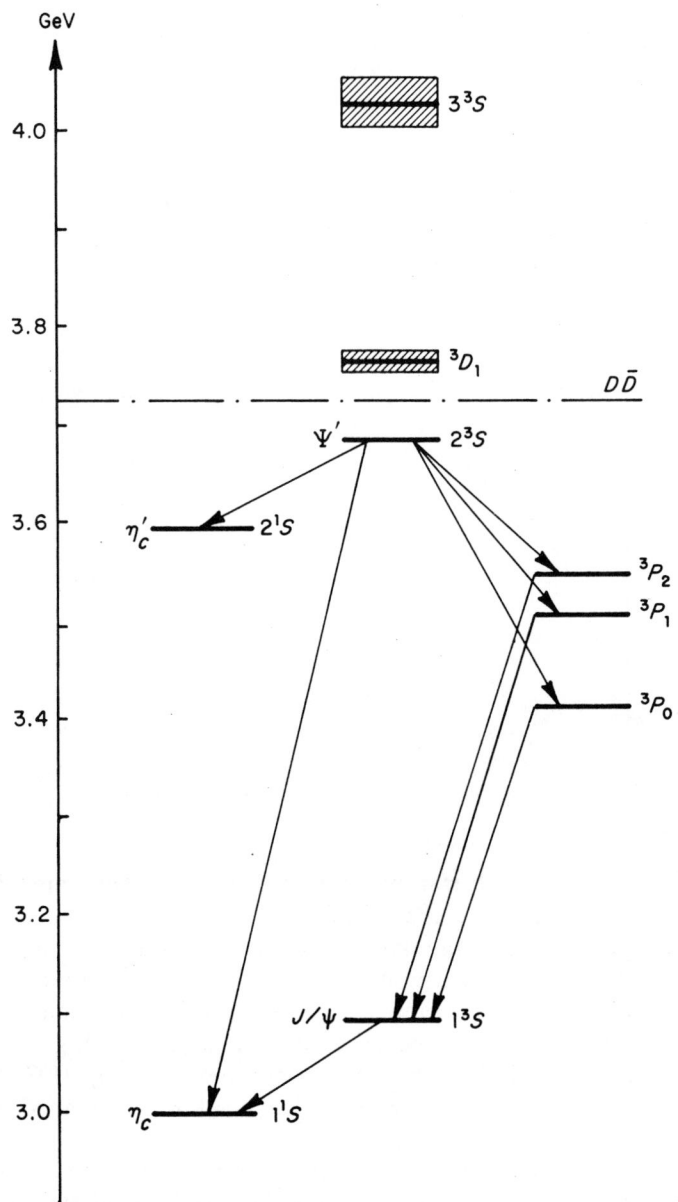

FIG. 22. The observed $c\bar{c}$ (charmonium) spectrum. The evidence for the ^3S levels, and $^3D_1$, is shown in Fig. 18, while levels revealed by the indicated γ-transitions are determined from the data in Figs. 23 and 24, as well as a considerable number of other experiments. The horizontal dot-dash line is the flavor threshold for charm; levels above this decay into $D\bar{D}$ pairs (or, if energetically allowed, heavier charmed mesons, as in $3^3S \to D^*\bar{D}^*$). Levels below this threshold have widths too narrow to be seen on this plot, whereas those above threshold have the widths indicated by shading. All γ-transitions involving the ^3P states are electric dipoles; the others are magnetic dipoles. The levels 1^1S and 2^1S are frequently called η_c and η'_c, respectively, while the ^3P_J states are often denoted by χ_J.

3. THE $Q\bar{Q}$ SPECTRUM

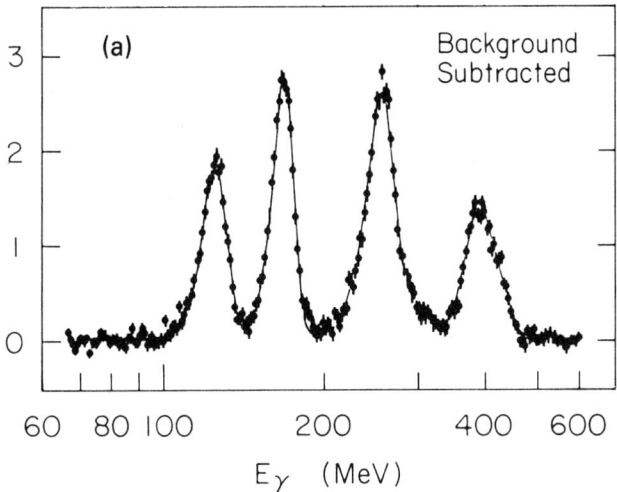

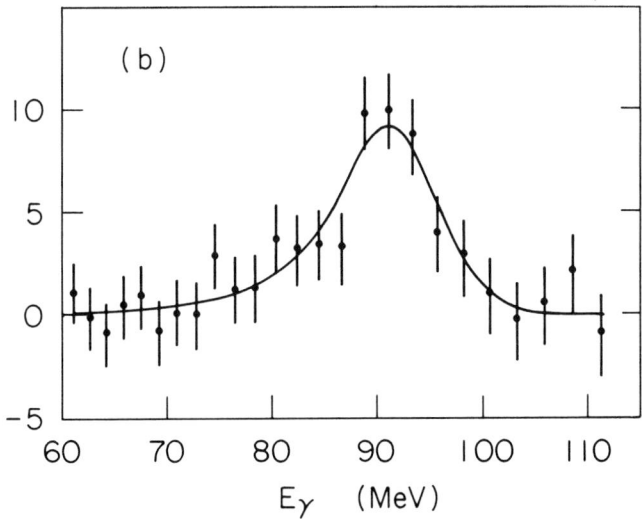

FIG. 23. The photon spectrum observed in $\psi' \to \gamma X$. Reading from left to right, the first three maxima are due to the transitions from ψ' to $^3P_2$, $^3P_1$, and $^3P_0$, respectively. The broad peak at $E_\gamma \sim 400$ MeV is due to the two cascade transitions $^3P_1 \to \psi\gamma$ and $^3P_2 \to \psi\gamma$; these are Doppler broadened because the parent P-state, produced in the preceding ψ'-decay, is in motion. The cascade decay $^3P_0 \to \psi\gamma$ is not visible in these data because $^3P_0$ has a rather sizeable total width (~ 1 MeV), so that its γ-branching fraction is small; this transition has been seen in other experiments, however. The $M1$ hyperfine transition $2^3S_1 \to 2^1S_0 + \gamma$ can be seen by a careful examination of the soft portion of this spectrum, as (b) demonstrates; this determines the mass of $\eta_c' = 2^1S$ as 3592 ± 5 MeV. [The data shown here are from the Crystal Ball Collaboration at SLAC, Partridge et al. (1980), and Edwards et al. (1982).]

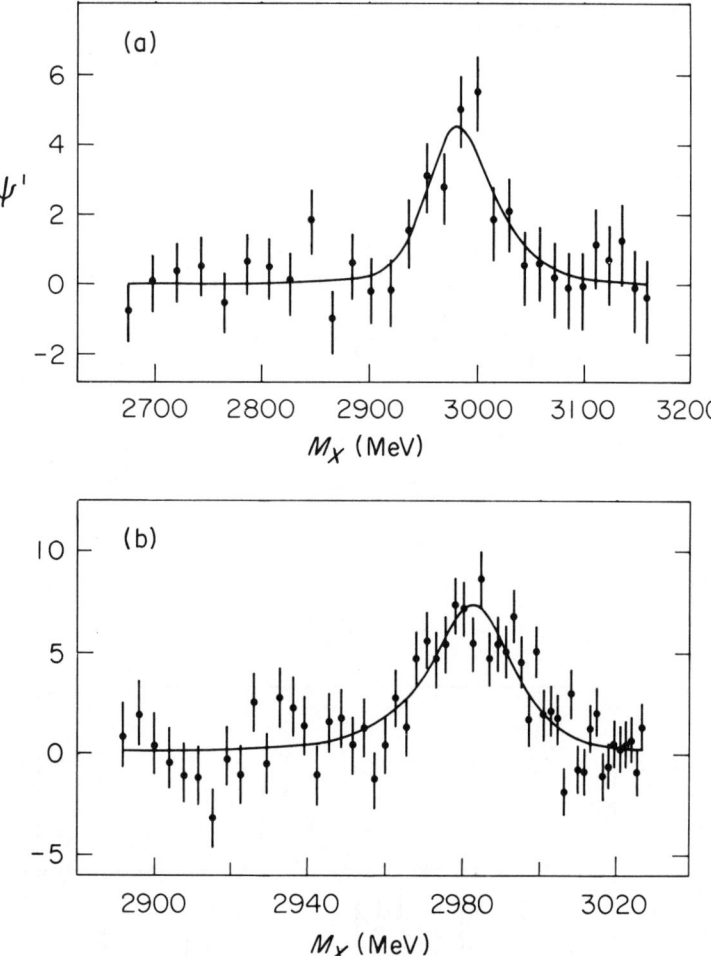

FIG. 24. Evidence for the $c\bar{c}$ ground state, $\eta_c = 1^1S$, at 2981 MeV. In (a) the photon spectrum measured in $\psi' \to \gamma X$ is plotted against the mass of X. This is an expanded view of the hard portion of the spectrum shown in Fig. 23(a); the peak is at $E_\gamma \approx 640$ MeV. Fig. (b) shows the transition $1^3S \to 1^1S + \gamma$ as observed in the decay of ψ itself, i.e., in $\psi \to \gamma X$. The data are from the references cited in Fig. 23.

total angular momentum. There are several known mechanisms[7] that can cause virtual transitions between ^3S and $^3D_1$ states, so that the true eigenstates become linear combinations of these L-eigenstates. One such combination, which is dominantly 2^3S, is $\psi' = \psi(3685)$; the orthogonal

---

[7] One mechanism is familiar to us from atomic physics: The magnetic interaction between two dipoles. This leads to the so-called tensor force, which is proportional to $T = (\boldsymbol{\sigma}_1 \cdot \hat{\mathbf{r}}) \times (\boldsymbol{\sigma}_2 \cdot \hat{\mathbf{r}}) - \frac{1}{3}\boldsymbol{\sigma}_1 \cdot \boldsymbol{\sigma}_2$, where $\boldsymbol{\sigma}_i$ is the spin of particle i, and $\hat{\mathbf{r}}$ a unit vector from one to the other. Though T is a scalar, it is not invariant under rotation of $\hat{\mathbf{r}}$ alone, and therefore it does not conserve L. The mechanism just described is not believed to be the dominant one in producing the $^3S_1$-$^3D_1$ mixing, however. See §4, page 336.

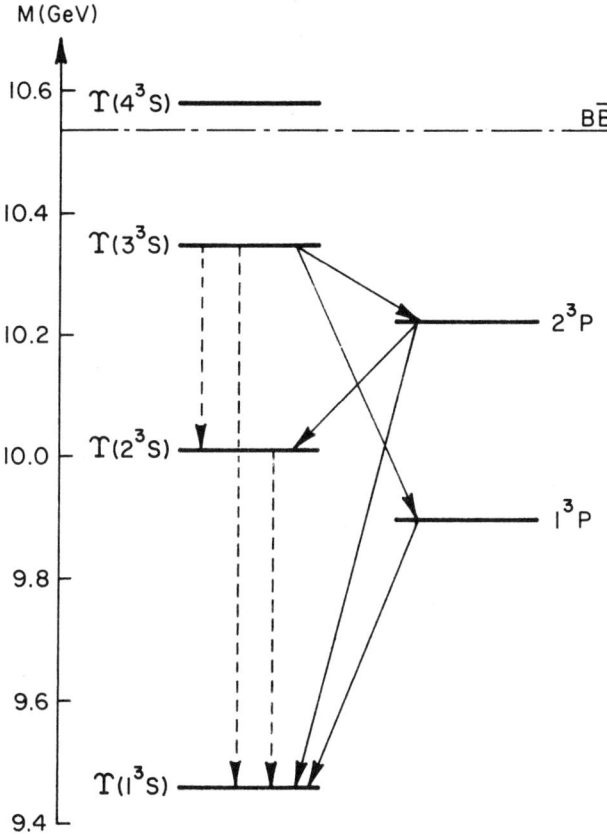

FIG. 25. The currently observed $b\bar{b}$ spectrum. The ^3S levels are those observed directly in $e\bar{e}$ annihilation (see Fig. 19), with 4^3S being above the $B\bar{B}$ threshold at 10.55 GeV [Behrends et al. (1983)], which is indicated. The solid lines show photon transitions to and from ^3P states; the photon spectra indicate that the states fed by transitions from $\Upsilon(3^3S)$ are multiplets, but do not yet give reliable values for the splittings. The dashed lines show $\pi\pi$ transitions, which are discussed in §4. [These data come from the compilations of Berkelman (1983), Franzini (1983), and Silverman (1984)].

combination, which is dominantly $^3D_1$, is $\psi(3770)$, and it is made in $e\bar{e}$ annihilation because the small 2^3S admixture has a nonzero amplitude at $r = 0$.

Finally, we compare the extensive data on the $c\bar{c}$ spectrum, and the data on the $b\bar{b}$ spectrum (see Fig. 25), with the energy eigenvalues found from solving the Schrödinger equation with the Richardson potential, Eq. IV.C(82). Three measured masses are used as inputs: $J/\psi(3097)$, $\psi(3685)$, and $\Upsilon(9460)$, assuming that these three states have the quantum numbers $1S$, $2S$, and $1S$, respectively. The singlet-triplet split is neglected. The first two determine the mass of the c-quark, and the adjustable parameter Λ_1 in the potential, while the third fixes the mass of b. The position of all other levels is then predicted. A comparison with the data is shown in Table 2.

Table 2
Level spacing ratios[a]

Level	ψ Ex	ψ Th	Y Ex	Y Th
2S	input		0.95	0.94
3S	1.59	1.70	1.59	1.60
4S			1.99	2.09
1P	0.76	0.71	0.78	0.83
$^3D_1$	1.14	1.20		
2P			1.41	1.45

[a] All entries except Y(2S) are the ratio of the level spacings between the state in question and the 1S state, divided by the 2S − 1S spacing of the same sytem. The Y(2S) entry is $[m(Y') - m(Y)]/[m(\psi') - m(\psi)]$. In the ψ spectrum, when possible, the center of gravity of multiplets is used, i.e., $\bar{m}_L = N_L^{-1} \sum_J (2J + 1) m_J(L)$, where N_L is the total multiplicity ($N_L = 4$ and 9 for $L = 0$ and 1). This removes the unknown spin-dependent forces from the comparison in the ψ spectrum. The Y data do not yet permit this, but one believes that the spin-splittings are inversely proportional to quark mass (see §IV.D.4), and a straightforward comparison of the potential model with the observed ^3S levels should be adequate in the $b\bar{b}$ system. In the case of $\psi(3770)$, the comparison is perforce only possible with the $J = 1$ member of the ^3D multiplet. The levels $\psi(3S)$ and Y(4S) lie above the flavor threshold, as does $\psi(3770)$, and there are, therefore, effects that the potential model ignores that are expected to shift the levels (see §4). The term *flavor threshold* is defined in connection with Eq. I.E(15), and also on p. 334. The Y data is from Silverman (1984).

One sees that the model is remarkably successful. In particular, the same potential parameter Λ_1 gives a good rendition of the ψ- and Y-spectra, even though the quark mass has tripled. As mentioned in §I.E.7, this is the best evidence we now have for the flavor-independence of the strong interaction.[8]

The wave functions permit further comparisons between theory and experiment. They can be used to compute the rates for photon emission, and to compare these with the data in Figs. 23 and 24. For the $E1$ transitions, there is only good qualitative agreement, but in the case of the $M1$ transition, the agreement is excellent. This is to be expected because $E1$ transitions involve the radial dependence of the charge distributions, and are therefore model-sensitive, whereas $M1$ transition between hyperfine partners merely involve a spin-flip, and an overlap integral that is essentially unity because the radial wave functions are, to a good approximation, identical. This is not true for the transition $2^3S \to 1^1S + \gamma$ in the ψ

---

[8] We emphasize that the potential parameter Λ_1 does not have a known rigorous relationship to the parameters Λ and Λ_0 that appear in the running coupling constant [cf. §IV.C.4(e)], though they are expected to be of comparable magnitude.

spectrum, however, because $2S$ and $1S$ have orthogonal radial wave functions. The rate differs from zero because of the next term in the multipole expansion, which gives a result in agreement with the data in Fig. 24.

To summarize, there is a large body of data that can be understood simply and precisely with the nonrelativistic $Q\bar{Q}$ model. This is one of the most remarkable successes of the quark model of hadronic structure.

4. Hadronic decays of $Q\bar{Q}$ states[9]

The hadronic decays of $Q\bar{Q}$ states fall into three distinct categories:
  (i) decays wherein the heavy pair annihilates, as in $\psi \to 5\pi$ or $\psi' \to p\bar{p}\pi\pi$;
  (ii) decays wherein the heavy quarks separate, and form mesons bearing the flavor of Q, as in $\psi(3770) \to D\bar{D}$;
  (iii) cascade decays, as in $\psi' \to \psi\pi\pi$.

The annihilation decay is presumably due to the 3-gluon mechanism shown in (12). In the case of ψ-decay, there is no direct evidence that this mechanism is actually at work, but in Y-decay there is, because the energy release of nearly 10 GeV is large enough so that the three gluons tend to materialize as three rather ill-defined jets.[10] An analysis of the multihadron final states in Y-decay shows that the events are roughly planar, as one would expect for a 3-jet configuration, and the sharing of energy between the jets is consistent with the theoretical expectation.

The rate for the decay of Y into all hadronic final states provides the least ambiguous determination of the QCD coupling constant, α_s, currently available.[11] While the rate for decay into some particular final state is proportional to the probability that the three virtual gluons in (12) will convert into that state, the total rate for Y$\to$ hadrons, Γ_{had}, is a sum over all these conversion probabilities. Hence, Γ_{had} can be computed as if the gluons could actually escape to infinity. As m_Y is large, the gluons carry large momenta, and their emission involves the QCD running coupling constant $\alpha_s(q^2)$ in the large q^2 regime where asymptotic freedom holds. Once this is the case, Γ_{had} can be computed from the diagram

$$\text{(13)}$$

This amplitude depends on $\psi_Y(0)$, which is a model dependent quantity.

---

[9] This section presupposes some acquaintance with the contents of Chap. IV on QCD.
[10] In this connection, recall §I.E.7(d).
[11] The ratio of $(e\bar{e} \to 3 \text{ jets})/(e\bar{e} \to 2 \text{ jets})$, which should measure the amplitude for gluon bremsstrahlung [recall §I.E.7(d)], is unfortunately very sensitive to the models that must be used to describe the fragmentation of the jets into hadrons.

However, $\psi_Y(0)$ disappears from the ratio $\Gamma(Y \to \gamma gg)/\Gamma(Y \to ggg)$, where the numerator is described by (13) with one gluon replaced by a photon. This ratio is nothing but the branching fraction for $Y \to \gamma X$, where X is a 2-jet multihadron state arising from the underlying process $Y \to \gamma gg$. Such an analysis leads to the evaluation [Mackenzie (1981)] of the running coupling constant at $q^2 \simeq 4m_b^2 \simeq m_Y^2$:

$$\alpha_s = 0.16 \pm 0.02. \tag{14}$$

As we show in §IV.C4(e), an evaluation of $\alpha_s(q^2)$ is equivalent to a determination of the fundamental QCD scale parameter Λ_0; Eq. (14) yields

$$\Lambda_0 = 160 \pm 80 \, \text{MeV}. \tag{15}$$

The relatively large error in Λ_0 is due to the logarithmic relationship between α_s and Λ_0 [see Eq. IV.C(71)]. The small scaling violations observed in deep inelastic scattering (§V.8) yield an even poorer determination of Λ_0 that is consistent with (15).

The *flavor threshold*, W_Q, is defined as twice the mass of the lightest meson carrying the flavor of Q. When $Q = s$, c, or b, these mesons are K, D, or B, respectively, and the corresponding thresholds are at 0.987, 3.730, and 10.54 GeV, respectively. Any $Q\bar{Q}$ level above its flavor threshold will decay by the "fission" mechanism

$$Q\bar{Q} \to Q\bar{q} + \bar{Q}q, \tag{16}$$

already familiar to us from §I.E.7, and as exemplified by $\phi \to K\bar{K}$. In (16) the light $q\bar{q}$ pair is made by soft virtual gluons, so that the QCD coupling constant is large. Hence, the rate for (16) is large, and cannot be computed from perturbation theory. As Fig. 26 shows, there is therefore a dramatic difference between the 3-gluon annihilation rate, and the "fission" rate: the former is very small compared to "typical" hadronic widths, which are all due to "fission." This difference is illustrated by the two $c\bar{c}$ states, $\psi(3685)$ and $\psi(3770)$; the former is 45 MeV below, the latter 40 MeV above the charm threshold. The width $\psi(3685) \to$ (light hadrons) is 44 keV, whereas $\psi(3770)$, which decays dominantly into $D\bar{D}$, has a width of 25 MeV, that is, a factor $\sim 10^3$ larger, even though its mass is barely larger than that of $\psi(3685)$.

[The fission mechanism can also mix $Q\bar{Q}$ states, as shown in Fig. 27 in the important example of the $c\bar{c}$ levels 2^2S and $^3D_1$ discussed in §3. In this instance, the presence of the charm threshold *between* the two unperturbed $c\bar{c}$ levels leads to exceptionally large mixing of these states, and mocks the effect of a large tensor force. A detailed investigation [Eichten (1980)] shows that the process depicted in Fig. 27 accounts for both the total width of $\psi(3770)$ and the admixture of 2^3S as determined from the coupling of $\psi(3770)$ to the $e\bar{e}$ channel.]

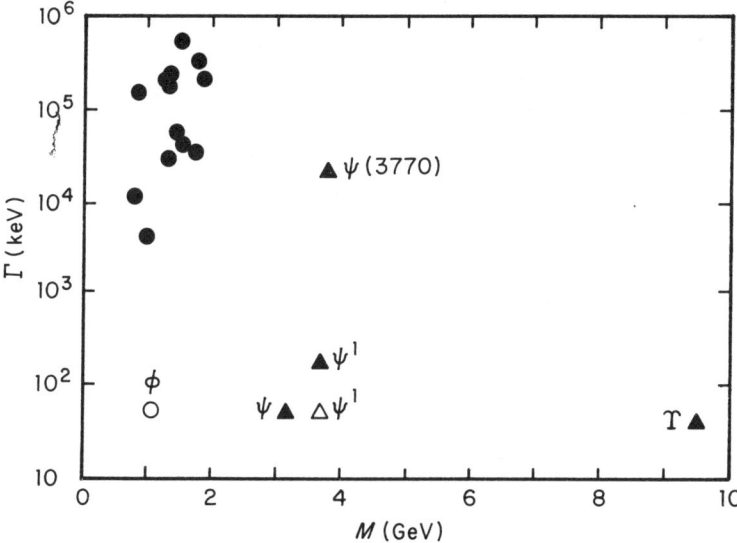

FIG. 26. The widths Γ of mesons, plotted versus their mass. The dots are "light" mesons composed of a $q\bar{q}$ pair, the triangles "heavy" mesons composed of a $Q\bar{Q}$ pair. The solid dots, and $\psi(3770)$, decay by the "fission" mechanism. The open circle marked ϕ gives the width for $\phi \to \pi^+\pi^-\pi^0$, which is due to $s\bar{s}$ annihilation; it is seen to be comparable to the widths of ψ and Υ. (ϕ-decay is dominated by the fission process $\phi \to K\bar{K}$.) $\Gamma(\psi')$ is anomalously large because of $\psi' \to \psi\pi\pi$, and $\psi' \to \gamma\chi$, which do not require $c\bar{c}$ annihilation. The open triangle marked ψ' is due to ψ'-decay into light hadrons, i.e., decay via $c\bar{c}$ annihilation.

Finally, we turn to hadronic cascades within a $Q\bar{Q}$-family, as exemplified by $\psi' \to \psi\pi\pi$ and $\Upsilon' \to \Upsilon\pi\pi$. The essential point here is that the emitter, $Q\bar{Q}$, is small compared to the wavelength of the radiated mesons. This is reminiscent of the emission of light by an atom, where the wavelength of the radiated light is also long compared to atomic dimensions. As we learned in §I.B.6, this circumstance implies that the radiation is accurately described by the lowest multipole field permitted by the conservation of angular momentum and parity. By the same token, the emission of hadrons having a wavelength long compared to the size of the $Q\bar{Q}$ system can also be described to good approximation by a multipole expansion. In the problem of interest here, it is the color gauge field due to the $Q\bar{Q}$ system, and not the electromagnetic field, that is to be approximated by the lowest possible multipole. Naturally, the analogy with radiation by an atom is rather forced, because photons can propagate to infinity, whereas gluons cannot. For that reason a better analogy to our problem is the emission of a positronium state in a nuclear transition,

$$A^* \to A + \gamma \to A + e\bar{e}, \qquad (17)$$

as depicted in Fig. 28(a). The multipole field that describes the virtual photon in (17) is fixed by the spins and parities of the nuclear states A^* and

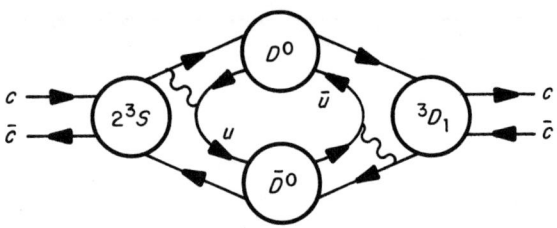

FIG. 27. The 2^3S–$^3D_1$ mixing mechanism, due to virtual transitions to charmed meson pairs. The mixing amplitude is a coherent superposition of amplitudes involving $D\bar{D}$, $D^*\bar{D}^*$, $D\bar{D}^*$, and $D^*\bar{D}$ intermediate states. Because of the mass difference between the pseudoscalar meson D and the vector meson D^*, this mechanism acts like a rather strong tensor force.

A. The rate for $e\bar{e}$ emission is proportional to the probability for the emission of that field, and also to that for the conversion of the field into a positronium state. The latter is to be viewed as the analogue of the hadronic state emitted in a decay such as $\psi' \to \psi\pi\pi$.

In the process of interest to us, the obvious analogue to (17) is

$$\psi_1(Q\bar{Q}) \to \psi_2(Q\bar{Q}) + g \to \psi_2(Q\bar{Q}) + \varphi(q\bar{q}), \qquad (18)$$

where $\psi_{1,2}$ and φ are hadronic wave functions. But (18) is forbidden, because the gluon g carries color, and therefore ψ_1 and ψ_2 cannot both be color singlets, as they must be if they are to describe hadrons. Hence, the lowest order process possible is

$$\psi_1(Q\bar{Q}) \to \psi_2(Q\bar{Q}) + gg \to \psi_2(Q\bar{Q}) + \varphi(q\bar{q}). \qquad (19)$$

This is a 2-gluon emission process [see Fig. 28(b)]. The gluon field is a vector field, so the leading multipole is similar to the familiar electric dipole field, but since the gluon field carries color, what is involved is actually the color-electric dipole moment. This quantity is a vector under a spatial rotation, and simultaneously a color octet. One of its spatial components, say along the z-direction, is therefore

$$\vec{d}_z = g \sum_i \vec{\lambda}_i z_i, \qquad (20)$$

where g is the coupling constant of QCD, $\vec{\lambda}_i$ is the octet of 3×3 color matrices for the ith heavy quark, whose z-coordinate is z_i (see §IV.A.2).

The *amplitude* for (19) is therefore proportional to the *square* of the color-electric dipole moment. As each dipole carries off at most one unit of angular momentum, and produces a change of parity, the selection rule for an allowed transition that exploits the mechanism (19) is $\Delta J \leq 2$, with no change of parity, where these quantum numbers refer, as always, to the emitter, i.e., to the $Q\bar{Q}$ system. The transitions $\psi' \to \psi\pi\pi$ and $\Upsilon' \to \Upsilon\pi\pi$

4. HADRONIC DECAYS OF $Q\bar{Q}$ STATES

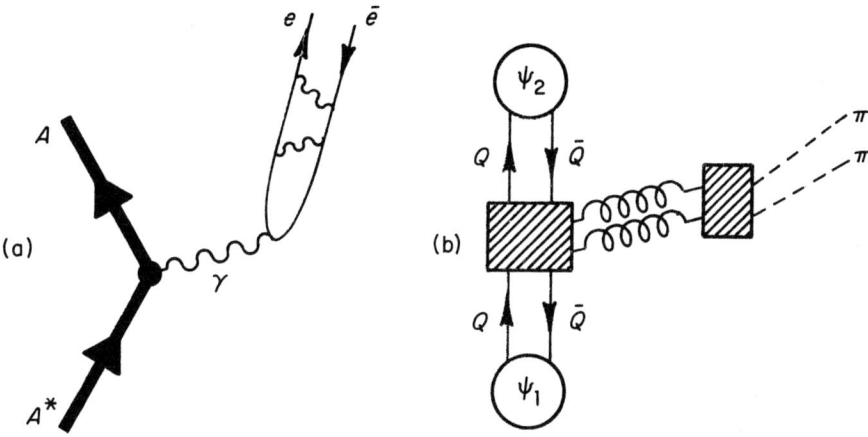

FIG. 28. In (a) we show the Feynman diagram corresponding to Eq. (17), where $e\bar{e}$ are emitted as some 1^- state of positronium. In practice, nuclear emission of positronium is too rare to be seen, but emission of unbound pairs is observed (e.g., $^{16}O^* \to {}^{16}Oe\bar{e}$). In the hadronic analogue (b) the $Q\bar{Q}$ pair emits a gluon field, which then converts into hadrons, by complex processes that are simply represented by the shaded boxes. As the exchanged gluons are soft, their coupling is strong, and one cannot use perturbation theory, as in the electromagnetic case (a). Nevertheless, the multipole expansion can still be carried through, and the result given in Eq. (21) does not depend on perturbation theory [see Yan (1980)].

both satisfy this selection rule, since all these ψ- and Y-states have the quantum numbers ^3S. The rates for these processes are therefore proportional to the fourth power of the color-electric dipole moment, and in consequence to the fourth power of the radius of the emitter. As a result, there is a simple relation between $\pi\pi$ emission rates for transitions between corresponding states in the ψ- and Y-family; in particular [Gottfried (1978)],

$$\frac{\Gamma(\psi' \to \psi\pi\pi)}{\Gamma(Y' \to Y\pi\pi)} \simeq \left(\frac{r_\psi}{r_Y}\right)^4 F, \qquad (21)$$

where F is a phase space correction, which is close to unity in this instance because the energy release is almost the same in the two decays. These radii can be computed from the $Q\bar{Q}$ wave functions, and they give a result in very good agreement with the observed $\pi\pi$ rates [Kuang (1981)].

As we saw, the derivation of (21) relied on two essential properties of the gluon: that it is not a color singlet, whence it cannot be emitted singly in a hadronic transition; and that it carries the same spin-parity as the photon, so that the leading amplitude has the dipole form. The success of the prediction (21) therefore lends credence to these assumed properties of the strong interaction field.

IV

QUANTUM CHROMODYNAMICS

In §I.E.6 color was introduced as a device for averting a serious conflict between the quark model and the Pauli principle, while §I.E.7 provides a brief presentation of the theory of strong interactions based on color as the source of a field whose quanta are called gluons. This theory is called quantum chromodynamics (QCD). The present Chapter is devoted to a more detailed treatment of that theory.

At the outset we should point out that while there is now a widespread acceptance of QCD as *the* theory of strong interactions, much room for skepticism remains. QCD is conceptually elegant. At a qualitative level it successfully accounts for all phenomena, and its radiative corrections are even better behaved than those of QED (asymptotic freedom). But at the moment, some of its major theoretical goals—such an explanation of quark confinement—are not fully achieved, and the growing evidence for the existence of gluons remains rather circumstantial.

A. COLOR

1. The color variable

In §I.E.6(c) we learned that all the low-lying states of the baryon are, but for one serious flaw, elegantly described as bound states of three quarks. The fly in the ointment is that their wave functions appear to violate the Pauli principle. This is most easily seen in the case of the spin-$\frac{3}{2}$ objects Δ^{++}, Δ^-, and Ω^-, with $J_z = \frac{3}{2}$, which are composed of three $u, d,$ and s quarks, respectively, all having the same spin projection and the same spatial wave function. All other baryon wave functions[1] have this unacceptable property. In principle one can avoid this "Pauli catastrophe" by postulating a spatial wave function that is antisymmetric, but this leads to other problems. The most damaging is that such wave functions have nodes, and consequently possess electric form factors that have zeroes at momentum transfers easily accessible to present-day measurement. The data (Fig. III.1) show no hint of such zeros.

As outlined in §I.E.6(c), the Pauli catastrophe can be evaded if one asserts that quarks have a degree of freedom, dubbed "color," beyond the conventional ones of space and spin. That is to say, a single quark has a wave function $\psi(\mathbf{x}\lambda\gamma)$, where $\mathbf{x}$ are any three spatial degrees of freedom (momenta, coordinates, etc.), λ is a spin projection, and γ is an eigenvalue of the new color degree of freedom. A 2-quark wave function satisfying the exclusion principle is then

$$\frac{1}{\sqrt{2}}[\psi_a(\mathbf{x}_1\lambda_1\gamma_1)\psi_b(\mathbf{x}_2\lambda_2\gamma_2) - \psi_b(\mathbf{x}_1\lambda_1\gamma_1)\psi_a(\mathbf{x}_2\lambda_2\gamma_2)],$$

i.e., antisymmetric in *all* the degrees of freedom. States with larger numbers of quarks are constructed in a similar fashion. For example, the aforementioned state of the Ω^- becomes

$$\Psi_{\Omega^-}(J_z = \tfrac{3}{2}) = \Phi(\mathbf{x}_1\mathbf{x}_2\mathbf{x}_3; \uparrow\uparrow\uparrow)\chi(\gamma_1\gamma_2\gamma_3), \tag{1}$$

where Φ is a completely symmetric space-spin function of three s quarks, all with spin "up," and χ is an antisymmetric function of the γ's. Clearly, this is

---

[1] Thus the proton wave function has components with two u quarks in the same spin state; cf. Eq. I.E8(a).

1. THE COLOR VARIABLE

only possible if the eigenvalue γ can assume *at least* three distinct values. As we see, the introduction of color provides a "trivial" escape from the Pauli catastrophe.

Naturally, one cannot go about enlarging the degrees of freedom heedlessly—such a drastic step must have other consequences. One thing is obvious: increasing the number of degrees of freedom inevitably enlarges the number of linearly independent states.[2] This could cause serious obstacles, because the naive quark model, without color and completely symmetric states, was said to have just the correct number of observed low-lying baryon states. But this is not really true: the quark model without color also has an infinity of states that we do not observe: states with $1, 2, 4, 5, 7, \ldots$ quarks. Indeed, we shall now turn this shortcoming to our advantage by presenting other arguments for the introduction of color that are designed to banish the unobserved states from the physical spectrum. This argument is based on a generalization of the hypothesis of quark confinement, which, in turn, leads rather naturally to a theory of the underlying forces between quarks.

As already indicated in §I.E.6, color provides the incisive distinction between existing hadrons—the members of the physical spectrum—and the multitude of quark configurations that are not observed in nature. This was done by assigning an additive degree of freedom to quarks, called color, that has the property that whenever a multiquark state has total color zero (i.e., is color "neutral"), the aggregate is a hadron, or a set of hadrons, and therefore in the physical spectrum; whereas, if the total color is nonzero, the state has infinite energy. For this to be true, the forces coupled to color must be such as to produce this energy spectrum.

We shall assume immediately that the number N of distinct color eigenstates attributable to one quark is 3. They are designated by $|\gamma_\alpha\rangle$, where $\gamma_1, \gamma_2, \gamma_3$ are labels that distinguish them. These labels are often given the names of three distinct colors. Any N other than 3 would lead to a nonsensical set of low-lying baryons, and to other less spectacular failures.

According to the superposition principle, if $|a\rangle, |b\rangle$, and $|c\rangle$ are three possible states of a system, so is any linear combination $A|a\rangle + B|b\rangle + C|c\rangle$, where A, B, and C are arbitrary *complex* numbers satisfying $|A|^2 + |B|^2 + |C|^2 = 1$. Consequently, the three color eigenstates $|\gamma_\alpha\rangle$ associated with any quark generate a state space $\mathscr{C}_3$, the set of all *complex* 3-vectors

$$V = \begin{pmatrix} \zeta_1 \\ \zeta_2 \\ \zeta_3 \end{pmatrix} = \sum_{\alpha=1}^{3} \zeta_\alpha |\gamma_\alpha\rangle, \qquad (2)$$

---

[2] An example will illustrate this, The states of two spin-0 particles obeying the exclusion principle are $P, F, \ldots$, because $S, D, \ldots$ are symmetric. For two spin-$\frac{1}{2}$ particles, however, the states allowed by the exclusion principle are $^1S, ^3P, ^1D, \ldots$. As we see, this last set is even larger than that of *all* the wave functions—both symmetric and antisymmetric—of two spin-0 particles.

where the complex numbers ζ_α satisfy

$$|\zeta_1|^2 + |\zeta_2|^2 + |\zeta_3|^2 = 1. \tag{3}$$

Occasionally we will call $\mathscr{C}_3$ the color space of one quark. (Recall that $\mathscr{C}_3$ is a complex 3-space, not a Euclidean 3-space; cf. §I.B.3(c)). The general principles of quantum mechanics also tell us that the physical observables associated with the color of a quark are the Hermitian operators in $\mathscr{C}_3$. These can be represented by 3×3 Hermitian matrices. One is the 3×3 unit matrix I. The others will be called λ_a, where a is a label that distinguishes them. Later we will show that $a = 1, 2, \ldots, 8$. Hence, the most general possible nontrivial Hermitian operator in $\mathscr{C}_3$ for one quark is $\sum_a \rho_a \lambda_a$, where the ρ_a are arbitrary *real* numbers so as to keep the sum Hermitian. If we are dealing with several quarks, each possesses a set of operators $\lambda_a^{(i)}$, i being the label that distinguishes quarks. By hypothesis, color, like flavor, baryon number, or angular momentum, is additive, so the most general operator for an assembly of n quarks is

$$K = \sum_{i=1}^{n} \sum_a \rho_a^{(i)} \lambda_a^{(i)}. \tag{4}$$

In the following subsection, we construct the λ_a, but for now some brief remarks will suffice. Mathematically speaking, the space $\mathscr{C}_3$, and its operators λ_a, are a simple generalization of the two-dimensional state space $\mathscr{C}_2$ defined by the possible spin orientations of one quark.[3] If $|\mu_\alpha\rangle$, with $\mu_1 = \frac{1}{2}, \mu_2 = -\frac{1}{2}$, are spin states, and η_1 and η_2 are complex numbers, the spinors

$$\begin{pmatrix} \eta_1 \\ \eta_2 \end{pmatrix} = \sum_{\alpha=1}^{2} \eta_\alpha |\mu_\alpha\rangle, \tag{5}$$

with

$$|\eta_1|^2 + |\eta_2|^2 = 1, \tag{5'}$$

define $\mathscr{C}_2$. The linearly independent Hermitian operators in $\mathscr{C}_2$ are the 2×2 unit matrix and the three Pauli matrices σ_a ($a = 1, 2, 3$). The λ_a are the counterparts of the Pauli matrices in the three-dimensional space $\mathscr{C}_3$.

We can now define the crucial notion that "a state has total color zero" in a precise manner: we say that a state $|\Phi\rangle$ representing an assembly of n colored quarks has no total color if

$$\sum_{i=1}^{n} \lambda_a^{(i)} |\Phi\rangle = 0 \tag{6}$$

for all a. Note that this is analogous to the statement that a state $|\Psi\rangle$

---

[3] Or, equivalently, the isospin of the u and d quarks, or of the proton and neutron.

1. THE COLOR VARIABLE

representing an assembly of n spins has no net angular momentum J, for that requires $\sum_i \sigma_a^{(i)}|\Psi\rangle$ to vanish for all $a = 1, 2, 3$. An important clue is provided by the observation that a $J = 0$ state is invariant under arbitrary spatial rotations, or equivalently, under arbitrary unitary transformations of the spinor (5). In this last form, the statement generalizes to $\mathscr{C}_3$, as we shall soon demonstrate, to wit, if (6) is fulfilled, the state $|\Phi\rangle$ is invariant under unitary transformation of the complex color vectors (2) that represent each quark in the assembly. For reasons that will become clear in §2, we also call a state that has no color a *color singlet*.

Let us now spell this out in detail. The argument will be more transparent if we first go through the analogous exercise for a multispin state $|\Psi\rangle$. An arbitrary unitary transformation W that leaves (5') invariant can be written in the form

$$W = e^{i\chi}U, \tag{7}$$

where χ is real, and U is a 2×2 unitary matrix with $\det U = 1$. The expression for U in terms of Euler angles was given in Eq. I.B(15). The overall phase χ has nothing to do with rotations or angular momentum, so we set $\chi = 0$ henceforth. Equation (7) tells us how the state of one spin-$\frac{1}{2}$ transforms; amongst other things, it is not an invariant. The simplest state that is invariant is the $J = 0$ combination of two spin-$\frac{1}{2}$'s. Let $\eta_\alpha^{(1)}$ and $\eta_\beta^{(2)}$ be the components of the spinors of the two constituents. According to Eq. I.B.(46), the $J = 0$ state is the antisymmetric combination

$$\Psi_A = \frac{1}{\sqrt{2}}[\eta_1^{(1)}\eta_2^{(2)} - \eta_2^{(1)}\eta_1^{(2)}]. \tag{8}$$

If we perform the rotation (7) on *both* spinors (setting $\chi = 0$), we find

$$\Psi_A \to \frac{1}{\sqrt{2}}(U_{1\alpha}U_{2\beta} - U_{2\alpha}U_{1\beta})\eta_\alpha^{(1)}\eta_\beta^{(2)}.$$

Since the expression in parentheses is antisymmetric in α and β, this is also

$$\Psi_A \to (U_{11}U_{22} - U_{12}U_{21})\Psi_A = (\det U)\Psi_A = \Psi_A, \tag{9}$$

and Ψ_A is invariant as expected.

We now return to color. Let $\zeta_\alpha^{(i)}$ be the components of the color 3-vector for quark i. As in any quantum mechanical situation, we may change our "representation" to any other by carrying out a unitary transformation, in this instance via a unitary 3×3 matrix U. As in the spin example, we restrict ourselves to the U's with $\det U = 1$. Then

$$\zeta_\alpha^{(i)} \to U_{\alpha\beta}\zeta_\beta^{(i)}. \tag{10}$$

We must find a colorless multiquark state, $\zeta_\alpha^{(1)}\zeta_\beta^{(2)}\cdots$, which is invariant under (10), i.e., the analog of the $J = 0$ state (8). The simplest state with this property is the totally antisymmetric 3-quark state [cf. Eq. I.E(5)]

$$\chi_B = \frac{1}{\sqrt{6}} \varepsilon_{\alpha\beta\gamma} \zeta_\alpha^{(1)} \zeta_\beta^{(2)} \zeta_\gamma^{(3)}, \tag{11}$$

where $\varepsilon_{\alpha\beta\gamma}$ is the totally antisymmetric symbol ($\varepsilon_{123} = 1$, $\varepsilon_{213} = -1$, $\varepsilon_{223} = 0$, etc.). Upon carrying out the transformation (10) on (11) as in (8), one finds after rearrangement that

$$\chi_B \to (\det U)\chi_B = \chi_B. \tag{12}$$

Hence χ_B is invariant—a color singlet; it carries the subscript B because it represents a baryon.

We must also construct a $q\bar{q}$ color-singlet wave function χ_M to represent mesons. χ_M must be a combination of products of two color wave functions, one factor for q, the other for $\bar{q}$. But in all field operators antiparticles appear with the complex conjugate wave functions of the corresponding particle [cf. Eq. III.A(27)]. Therefore, $\bar{q}$ is represented by a complex 3-vector ($\zeta_1^*, \zeta_2^*, \zeta_3^*$), where $\zeta_\alpha^* \to U_{\alpha\beta}^* \zeta_\beta^*$ under a transformation in $\mathscr{C}_3$. [A detailed proof of this will be given in §2(b).] As a consequence

$$\chi_M = \frac{1}{\sqrt{3}} (\zeta_1 \zeta_1^* + \zeta_2 \zeta_2^* + \zeta_3 \zeta_3^*) \tag{13}$$

is invariant—a color singlet. χ_M is the desired color wave function for mesons.

Several theorems will now complete the picture:[4]
  I. χ_B is the only invariant qqq state;
  II. χ_M is the only invariant $q\bar{q}$ state;
  III. the only other invariant multiquark states contain $3, 6, 9, \ldots, q$'s and arbitrary numbers of $q\bar{q}$ pairs.

It follows from I and the Pauli principle that the 3-quark states that do *not* have symmetric space-spin wave functions must be in states that have nonzero color; i.e., the operators $\sum_i \lambda_a^{(i)}$ do not give zero when acting on them.

This completes our demonstration that all observed hadrons are color singlets. As we saw in §I.E.6, this leads naturally to the basic hypothesis of QCD: Color is confined; the only objects occurring in nature have vanishing total color—are color singlets. The strong force must be such that colored portions (e.g., quarks) of a color singlet aggregate (a hadron) cannot be separated from each other to infinity because this causes the energy of the system to grow indefinitely.

---

[4] These theorems are either proved, or made highly plausible, in §2. For complete proofs, see Carruthers (1966).

We close this section with two remarks. First, there are no long-range confinement forces between color singlets. The "nuclear force" between n and p is the residual interaction that survives between these singlets because the quarks contained in a nucleon are not always at the same point in space, and therefore allow local departures from color neutrality, even though the overall color vanishes. Thus nuclear forces are similar to the Van der Waals interactions between *neutral* atoms;[5] these, too, fall off much more rapidly at large distances than the underlying Coulomb force. Indeed, our current view is that all the forces observed in hadronic interactions, as for example in πN scattering, are of this Van der Waals character.

The second remark concerns aggregates of $6, 9, 12, \ldots$ quarks with baryon number $B = 2, 3, 4, \ldots$. To be specific, focus on the $B = 2$ states with $3u$ and $3d$ quarks. These can be grouped into two nucleons $(uud)_A = p$ and $(ddu)_A = n$, where $(\cdots)_A$ means that the color variables are antisymmetrized into a singlet. Such a 6-quark system (e.g., the deuteron) can be separated into two color singlets. But there are other color-singlet states of the $B = 2$ system (and also for all $B > 2$ systems) that cannot be divided into portions that can then be separated by large distances. As B increases, the number of linearly independent color singlets increases. Those "exotic" $B \geq 2$ states that are not aggregates of nucleons presumably have a higher mass than ordinary nuclei, and do not appear to play a significant role in conventional nuclear physics. They probably did in the early universe when matter was at an energy density far higher than in ordinary nuclei.

Finally, a word about leptons and photons. Like hadrons, these are color singlets, but in contrast to hadrons, they have no local departures from color neutrality whatsoever. Therefore leptons and photons[6] have no coupling at all to the strong force.

2. The group SU(3)

At the moment, our understanding of the color variable is limited to the following: We know that it defines a three-dimensional complex vector space $\mathscr{C}_3$, and we know how to construct two invariant states, the baryon state χ_B of Eq. (11) and the meson state χ_M of Eq. (13). To proceed further we must enlarge our knowledge. It is therefore necessary to interject a mathematical digression at this point.

The material of this section goes under the general rubric of *group theory*. The language, concepts, and techniques of group theory have gained a permanent place in particle physics, and one purpose of this section is to introduce the reader to these. Some elementary aspects of group theory have already been used in Vol. I, §I.E.6, and in the preceding section. Others are of central importance in the formulation of both QCD and the electroweak interaction. Others still are used in hadronic spectroscopy.

---

[5] At this time, however, there is no explicit theory of nuclear forces based on QCD.
[6] This also holds for the bosons $W^\pm$ and Z^0 that mediate the weak interaction.

Throughout this section we lean heavily on analogies between $\mathscr{C}_2$, the two-dimensional complex vector space defined by spinors, and $\mathscr{C}_3$. Spinors, the theory of angular momentum, and the associated rotations, were discussed in some detail[7] in §I.B. As we shall see, the color operators, such as λ_a of Eq. (4), are involved in a group of transformations of $\mathscr{C}_3$, and the analogies alluded to hold because this group can be viewed as a generalization of the familiar rotation group.

Nothing is easier than to lose oneself in group theory. For that reason we begin with a trail guide. Our first objective is to convince the reader that he or she already knows something about groups; subsection (a) is devoted to codifying this knowledge, and providing some indications of its relevance to particle physics. Subsection (b) contains a description of the group-theoretic material that is essential for QCD; the treatment consists largely of analogies with the theory of rotations, and other plausibility arguments. A convenient summary of this information is provided by Table 1 on p. 359). Subsection (c) goes somewhat further into the mathematical theory, but those who wish to acquire a mastery of this material will have to consult the monographs referred to below.[8] Those who wish to minimize this mathematical interruption should browse through (a) and (b), examine Table 1, and proceed directly to §B.

(a) Definition of groups. SU(2) and SU(3)

$\mathscr{C}_2$ is the space of all complex 2-vectors of unit length, as defined by Eq. (5). According to Eq. (7), any pair of these vectors is related by a 2×2 unitary matrix W. Let $W_1, W_2, W_3, \ldots$ be unitary matrices. They satisfy the following properties:

1. The product $W_1 W_2$ is again a unitary matrix, which relates other pairs of vectors in $\mathscr{C}_2$.
2. To any W there is a unique inverse, W^{-1}, such that $W^{-1}W = WW^{-1} = 1$ is the unit matrix. Since W is unitary, $W^{-1} = W^\dagger$.
3. Matrix multiplication is associative: $(W_1 W_2)W_3 = W_1(W_2 W_3)$.

Any set of objects $\{W_i\}$ that satisfy these three conditions is called a *group*, in our case the *unitary group* in two dimensions, called $U(2)$. The members of a group are called *elements*; thus any 2×2 unitary matrix is an element of $U(2)$. As already noted in Eq. (7), $W = e^{i\chi}U$, where the unitary matrix U satisfies $\det U = 1$. Such a matrix U is called *unimodular*. Observe that the phase factors $e^{i\chi}$ themselves form a group:

$$e^{i\chi_1}e^{i\chi_2} = e^{i(\chi_1+\chi_2)}, \qquad e^{i\chi}e^{-i\chi} = 1;$$

this is called $U(1)$, the one-dimensional unitary group. Further, the unimodular matrices $U_1, U_2, \ldots$, separately form a group, since $\det(U_1 U_2) = (\det U_1)(\det U_2) = 1$; this group is called $SU(2)$, the two-dimensional *special unitary group*. Finally, note that $e^{i\chi}U = Ue^{i\chi}$.

---
[7] The mathematically identical problem of isospin was treated in §I.D.2.
[8] See van der Waerden (1974), Wybourne (1974), Georgi (1982), and Cahn (1984).

2. THE GROUP $SU(3)$

If $\mathfrak{G}$ is a group of transformations, each element of which can be factored into two *commuting* transformations that separately constitute the groups $\mathfrak{G}_1$ and $\mathfrak{G}_2$, one expresses this by

$$\mathfrak{G} = \mathfrak{G}_1 \otimes \mathfrak{G}_2. \tag{14}$$

$\mathfrak{G}_1 \otimes \mathfrak{G}_2$ is called the direct product. As an example, we have seen that

$$U(2) = U(1) \otimes SU(2). \tag{15}$$

Reconsider now the paragraph beginning with "$\mathscr{C}_2$ is the space of all 2-vectors..." (p. 348). Nothing of substance said between that sentence and Eq. (15) depends on the dimensionality of $\mathscr{C}_2$: The $N \times N$ unitary matrices form a group, called $U(N)$. This is the group of linear transformations on the N complex variables $(\zeta_1, \ldots, \zeta_N)$ that leaves invariant the Hermitian form

$$\sum_{\alpha=1}^{N} |\zeta_\alpha|^2 = \text{constant}. \tag{16}$$

Further, any such unitary matrix can be written as $e^{i\chi}U$, with $\det U = 1$, where the unimodular matrices themselves form a group, called $SU(N)$. Thus

$$U(N) = U(1) \otimes SU(N). \tag{17}$$

In particular, the group associated with $\mathscr{C}_3$ is

$$U(3) = U(1) \otimes SU(3). \tag{18}$$

A group $\mathfrak{G}$, all of whose elements commute with each other, is said to be *Abelian*; otherwise, it is called *non-Abelian*. As we see, $U(1)$ is Abelian, but $SU(N)$ is not.

The direct product notation of Eq. (14) is often used as a compact description of all the independent transformations that a physical system can undergo. It therefore gives a succinct account of the independent dynamical variables, and *if* the Hamiltonian has appropriate symmetries, of the constants of motion. We illustrate this with several examples. Consider a quark of definite flavor fixed at a point. It has no spatial degrees of freedom, but its spin and color degrees of freedom are independent variables, and can be "rotated" separately. Consequently, the transformation group is $SU_s(2) \otimes SU_c(3)$, where the subscripts s and c indicate that the factors refer to spin and color.[9] If the quark is not fixed, the group is enlarged to $\mathfrak{P} \otimes SU_c(3)$, where $\mathfrak{P}$ is called the Poincaré group, the group

---

[9] One may wonder why the group is not $U_s(2) \otimes U_c(3)$. The difference is an overall phase factor, which plays no role in QCD, but is important in the electroweak interaction (see §VI.B).

of space-time translations and Lorentz transformations. (The latter already contain spatial rotations.) As a last example, assume that this quark can be in either of the light flavor states u or d. These two states define an isospin space $\mathscr{C}_2^f$, which is distinct from spin or color. Furthermore, color and isospin (more generally, any flavor) are unrelated variables, so now the group is larger still: $\mathscr{P} \otimes SU_c(3) \otimes SU_f(2)$, where f stands for flavor.

(b) An SU(3) primer

As a basis in $\mathscr{C}_3$, we choose the orthonormal vectors

$$V_1 = \begin{pmatrix} 1 \\ 0 \\ 0 \end{pmatrix}, \quad V_2 = \begin{pmatrix} 0 \\ 1 \\ 0 \end{pmatrix}, \quad V_3 = \begin{pmatrix} 0 \\ 0 \\ 1 \end{pmatrix}. \tag{19}$$

As we already saw in Eq. (2), any vector in $\mathscr{C}_3$ is

$$V = \begin{pmatrix} \zeta_1 \\ \zeta_2 \\ \zeta_3 \end{pmatrix} = \sum_\alpha \zeta_\alpha V_\alpha. \tag{20}$$

We denote *any* set of three orthogonal complex vectors of this type by **3**.

Our first task is the construction of the matrices λ_a introduced in Eq. (4). An arbitrary operator in $\mathscr{C}_3$ is some 3×3 matrix, and can be written as a linear combination of nine matrices $I_{\alpha\beta}$ ($\alpha, \beta = 1, 2, 3$), where all elements of $I_{\alpha\beta}$ vanish except the element in the βth column and αth row, which has the value of one; for example,

$$I_{12} = \begin{pmatrix} 0 & 1 & 0 \\ 0 & 0 & 0 \\ 0 & 0 & 0 \end{pmatrix}, \quad I_{33} = \begin{pmatrix} 0 & 0 & 0 \\ 0 & 0 & 0 \\ 0 & 0 & 1 \end{pmatrix}. \tag{21}$$

Of the nine $I_{\alpha\beta}$, three are diagonal. But $\sum_\alpha I_{\alpha\alpha} = \mathbf{1}$, the unit matrix. Hence there are only eight nontrivial matrices, two of which are linear combinations of the diagonal matrices $I_{\alpha\alpha}$. Call these diagonal matrices λ_3 and λ_8. We can, without any loss of generality, choose λ_3 and λ_8 as traceless.[10]

An infinitesimal transformation of $SU(3)$ is described by a matrix of the form $U = \mathbf{1} + i\Delta$, where Δ is an infinitesimal 3×3 Hermitian matrix. If $\text{Det } U = 1$, $\text{Tr } \Delta = 0$. Hence Δ does not contain $\mathbf{1}$, and is some linear combination of the two diagonal matrices λ_3 and λ_8, and the 6 off-diagonal matrices $I_{12}, I_{13}, \ldots$ The latter 6 can be combined into 6 Hermitian matrices

---

[10] To prove this, consider $M = \sum_\alpha a_\alpha I_{\alpha\alpha}$, with $\text{Tr } M = \bar{a}$. Define $a_\alpha - \frac{1}{3}\bar{a} \equiv \delta_\alpha$; since $\sum_\alpha \delta_\alpha = 0$, only two of these are independent. Thus, for example, $M = \frac{1}{3}\bar{a} + \delta_1(I_{11} - I_{22}) + \delta_3(I_{33} - I_{22})$. The coefficients of δ_1 and δ_3 are traceless matrices.

2. THE GROUP $SU(3)$

in many ways. A standard choice is

$$\lambda_1 = I_{12} + I_{21}, \qquad \lambda_2 = i(I_{21} - I_{12}),$$
$$\lambda_4 = I_{13} + I_{31}, \qquad \lambda_5 = i(I_{31} - I_{13}), \qquad (22)$$
$$\lambda_6 = I_{23} + I_{32}, \qquad \lambda_7 = i(I_{32} - I_{23}).$$

Therefore Δ is a linear combination of the λ_α's. It is convenient to write any infinitesimal element of $SU(3)$ in the form

$$U = 1 - \tfrac{1}{2} i \omega_a \lambda_a. \qquad (23)$$

Since Δ is Hermitian, the ω_a are eight real, infinitesimal parameters—the analogues in $SU(3)$ of the three rotation angles of $SU(2)$ [see Eq. I.B(40)]. Further, the eight matrices λ_a are the counterparts of the three Pauli matrices σ_a; they are the *generators* of infinitesimal transformations in $\mathscr{C}_3$.

The existence of two commuting generators λ_3 and λ_8 shows that $SU(3)$ is very different from $SU(2)$: two eigenvalues must be specified if one wants to single out a member of an $SU(3)$ multiplet, whereas in $SU(2)$ just one, the spin projection, suffices. Let us look at this more closely. Naturally, there is a great deal of arbitrariness in choosing the diagonal generators, just as σ_3 is diagonal only by convention. The traditional choice is

$$\lambda_3 = \begin{pmatrix} 1 & 0 & 0 \\ 0 & -1 & 0 \\ 0 & 0 & 0 \end{pmatrix}, \qquad \lambda_8 = \frac{1}{\sqrt{3}} \begin{pmatrix} 1 & 0 & 0 \\ 0 & 1 & 0 \\ 0 & 0 & -2 \end{pmatrix}. \qquad (24)$$

λ_3 is just σ_3 in the $\mathscr{C}_2$-subspace $\zeta_3 = 0$. [The coefficient $1/\sqrt{3}$ in λ_8 is explained in the discussion after Eq. (56)].

Let λ_3' and λ_8' be the eigenvalues of λ_3 and λ_8. For the basis vectors (19), these eigenvalues are

	λ_3'	λ_8'
V_1	1	$\dfrac{1}{\sqrt{3}}$
V_2	-1	$\dfrac{1}{\sqrt{3}}$
V_3	0	$-\dfrac{2}{\sqrt{3}}$

(25)

λ_3 therefore distinguishes[11] V_1 from V_2, which have the same eigenvalue of λ_8.

---

[11] Here one is entitled to demur: λ_3', by itself, distinguishes all three vectors (19). But this is an accident. In larger multiplets, that is no longer true: see, for example, **8** in Fig. 4.

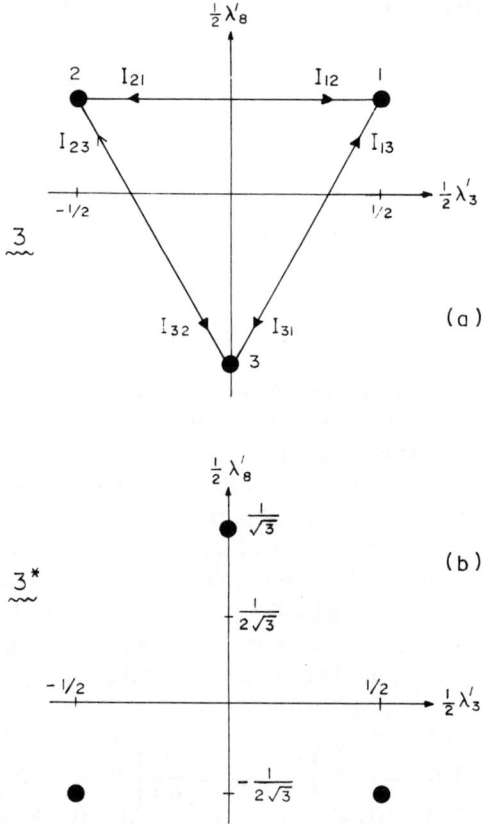

FIG. 1(a) Weight diagram for the fundamental triplet **3** of $SU(3)$. The dots carry the same labels as the vectors in Eq. (19). The shift operators $I_{\alpha\beta}$, defined in Eq. (21), are shown. The scales on the axes are chosen as $\frac{1}{2}\lambda_3'$ and $\frac{1}{2}\lambda_8'$, because $\frac{1}{2}\lambda_a$ is the quantity that generalizes naturally to larger multiplets, not λ_a itself.
(b) Weight diagram of the complex conjugate (or antiquark) triplet **3***.

As we shall see, considerable insight is gained if one draws a diagram of **3** in a plane where λ_3' and λ_8' are Cartesian coordinates. Each of the 3-vectors V_α is represented by a dot at the point given by its eigenvalues (25). Obviously, **3** is then a triangle, and because of the factor $1/\sqrt{3}$ in the definition of λ_8, it is equilateral, as shown in Fig. 1(a). Diagrams of this type are called *weight diagrams*.

The action of the operators $I_{\alpha\beta}$, defined in (21), can also be shown on the weight diagram. Consider, for example,

$$I_{13} = \begin{pmatrix} 0 & 0 & 1 \\ 0 & 0 & 0 \\ 0 & 0 & 0 \end{pmatrix} = \tfrac{1}{2}(\lambda_4 + i\lambda_5). \tag{26}$$

2. THE GROUP $SU(3)$

Obviously $I_{13}V_3 = V_1$, and quite generally $I_{\alpha\beta}V_\beta = V_\alpha$, where, for once, there is no sum on the repeated index. Hence for $\alpha \neq \beta$, the $I_{\alpha\beta}$ are shift operators in the weight diagram, similar to the operators $J_1 \pm iJ_2$ in angular momentum theory. These shift operators are also shown in Fig. 1(a).

At this point we briefly return to physics, because the alert reader will have noted that diagrams like Fig. 1(a) already appeared in §I.E.6: Fig. I.10, showing the u, d, and s quarks, is just the weight diagram of **3**. Indeed, $\tfrac{1}{2}\lambda_3$ corresponds to the third component of isospin, and $(-\tfrac{1}{3} + \lambda_B/\sqrt{3})$ to strangeness, and the group $SU(3)$ is the mathematical structure that underlies much of the discussion of meson and baryon spectroscopy in §I.E.6. But one must recognize that the trivalued physical variable that was under discussion there was the *flavor* of the three light quarks, whereas here the trivalued variable is color! In the notation defined just before Eq. (19), if we are concerned with a colored quark that can assume any of the three light flavors (u, d, s), the transformation group of the "internal" variables is

$$\mathfrak{G}_{\text{int}} = SU_f(3) \otimes SU_c(3). \tag{27}$$

The mathematics of both $SU(3)$ factors is identical, of course, but their physical significance must never be confused. For example, the weak transitions change flavor, not color, and the operators that describe these transitions are shifts on the *flavor* $SU(3)$ weight diagram. Hence, an understanding of $SU(3)$ is not only essential to QCD, it is also very useful in the analysis of all flavor-dependent phenomena that involve the three light quarks. We are now ready to return to our study of $SU(3)$, with our eyes set on QCD, but on occasion we will allude to facets of flavor $SU(3)$ that were actually discussed in §I.E.6.

In the theory of angular momentum, the single most important mathematical relationship is the commutation rule; virtually everything else follows from it. These rules are the same for any and all angular momenta, and in particular, for the Pauli matrices:

$$[\tfrac{1}{2}\sigma_a, \tfrac{1}{2}\sigma_b] = i\varepsilon_{abc}\tfrac{1}{2}\sigma_c, \tag{28}$$

where $a, b, c = 1, 2, 3$ and ε_{abc} was defined after Eq. (11). What is the analogue of (28) for the λ_a? This can be answered by direct computation with the simple matrices $I_{\alpha\beta}$; from (21), (22) and (24) one finds that the basic commutation rule of $SU(3)$ is

$$[\tfrac{1}{2}\lambda_a, \tfrac{1}{2}\lambda_b] = if_{abc}\tfrac{1}{2}\lambda_c. \tag{29}$$

We do not need the actual values[12] of f_{abc}, the analogue of ε_{abc}; for our purpose it suffices to know that f_{abc} is real and totally antisymmetric in all its

---

[12] The independent elements of f_{abc} are listed in Carruthers (1966), p. 31.

8-valued indices. We note, however, that for $a = 1, 2, 3$

$$\lambda_a = \begin{pmatrix} \sigma_a & 0 \\ 0 & 0 \end{pmatrix}, \tag{30}$$

and therefore f_{abc} is just ε_{abc} for $a, b, c = 1, 2,$ or 3.

We have denoted the triplet of Pauli matrices by $\boldsymbol{\sigma}$, and treated this as an ordinary 3-vector. Why is this valid? The answer leads to a useful analogue of vector algebra for $SU(3)$. First, we recall that if $|\psi\rangle$ is any state, X an observable, and U a rotation operator, the rotation of the system can be described by either rotating the state, $|\psi\rangle \to U|\psi\rangle$, or by rotating the observable, $X \to U^\dagger X U$. Consider, for example, an infinitesimal rotation of σ_1 through the angle $\delta\alpha$ about the 3-axis. From Eq. I.B(36) we find

$$\sigma_1 \to (1 + i\tfrac{1}{2}\sigma_3\delta\alpha)\sigma_1(1 - i\tfrac{1}{2}\sigma_3\delta\alpha)$$
$$\simeq \sigma_1 + \tfrac{1}{2}i\delta\alpha[\sigma_3, \sigma_1] = \sigma_1 - \sigma_2\delta\alpha. \tag{31}$$

The important point here is that even though the half-angle occurs in the 2×2 matrix $(1 - \tfrac{1}{2}i\sigma_3\delta\alpha)$, the actual rotation angle $\delta\alpha$ appears in (31), as it must for a vector.[13] If we repeat this for an arbitrary rotation, with 2×2 matrix $(1 - \tfrac{1}{2}i\theta_a\sigma_a)$, where θ_a are three infinitesimal rotation angles, we find from (28) that the change of σ_a is

$$\delta\sigma_a = -\tfrac{1}{2}i[\sigma_a, \sigma_b]\theta_b = \varepsilon_{abc}\theta_b\sigma_c. \tag{32}$$

Since the cross-product of two 3-vectors $\mathbf{A}$ and $\mathbf{B}$ is $(\mathbf{A} \times \mathbf{B})_a = \varepsilon_{abc}A_bB_c$, (32) is

$$\delta\boldsymbol{\sigma} = \boldsymbol{\theta} \times \boldsymbol{\sigma}. \tag{33}$$

Since the change $\delta\mathbf{A}$ of any Euclidean 3-vector under an infinitesimal rotation equals $\boldsymbol{\theta} \times \mathbf{A}$, we have demonstrated that $\boldsymbol{\sigma}$ transforms in the required manner.

We repeat exercise (31) for $SU(3)$. From (23)

$$\lambda_a \to (1 + \tfrac{1}{2}i\omega_b\lambda_b)\lambda_a(1 - \tfrac{1}{2}i\omega_b\lambda_b) \simeq \lambda_a - \tfrac{1}{2}i[\lambda_a, \lambda_b]\omega_b.$$

Upon using (29) we find that the change of λ_a is

$$\delta\lambda_a = f_{abc}\omega_b\lambda_c. \tag{34}$$

On comparing with (32), we see that the eight λ_a undergo a linear transformation having a structure analogous to the rotation of the σ_a. The transformation law (32) serves to characterize the rotational behavior of all operators that are Euclidean 3-vectors. In the same sense, *any* set of eight

---
[13] The reader should check that for a finite angle α, $\sigma_1 \to \sigma_1 \cos \alpha - \sigma_2 \sin \alpha$.

2. THE GROUP $SU(3)$

quantities $\{E_a\}$, with $a = 1, 2, \ldots, 8$, that transforms under $SU(3)$ like the λ_a, will be denoted by $\vec{E}$, and simply called an *octet*.[14]

As we now see, such $SU(3)$ octets obey algebraic relations that are very similar to those of Euclidean 3-vectors. First we define the analogue of the Euclidean cross-product. If $\vec{D}$ and $\vec{E}$ are octets, we define another octet by their $SU(3)$ cross-product:[15]

$$(\vec{D} \times \vec{E})_a \equiv f_{abc} D_b E_c. \tag{35}$$

With this notation, the $SU(3)$ transformation law (34) for any octet looks very familiar:

$$\delta \vec{E} = \vec{\omega} \times \vec{E}. \tag{36}$$

This formula plays a central role in QCD, where it will be used to demonstrate that the strong interaction field is itself an $SU(3)$ octet. Secondly, we define the scalar product of two octets by

$$\vec{D} \cdot \vec{E} = D_a E_a; \tag{37}$$

this formula will also see continuous service in QCD. Combining (35) and (37), and recalling the antisymmetry of f_{abc}, we have the familiar-looking identity

$$(\vec{C} \times \vec{D}) \cdot \vec{E} = -(\vec{C} \times \vec{E}) \cdot \vec{D}. \tag{38}$$

With its help we establish that (37) is indeed an invariant:

$$\delta(\vec{D} \cdot \vec{E}) = \delta \vec{D} \cdot \vec{E} + \vec{D} \cdot \delta \vec{E} = (\vec{\omega} \times \vec{D}) \cdot \vec{E} - \vec{D} \cdot (\vec{\omega} \times \vec{E}) = 0.$$

Lastly, we need to know something about the $SU(3)$ analogue of angular momentum addition. (We have already worked out two special cases of such an "addition" where the colors "add" to zero: the wave functions χ_B and χ_M of §1.) Recall first that if we multiply two $SU(2)$ spinors together, each of the four resulting bilinear expressions can be written as a sum of two eigenfunctions of total angular momentum J, one with $J = 0$, the other with

---

[14] The term "octet" is used in two different, though intimately related, contexts. Here the term refers to *any set of eight matrices* that satisfy the commutation rule (29), and therefore transform like (34). In the case of the $\{\lambda_a\}$, these are 3×3 matrices, but matrices of higher dimension having the same algebraic properties exist, as we shall see below. [The analogue in $SU(2)$ is that not only the 2×2 Pauli matrices, but any set of three $(2j + 1)$-dimensional matrices V_i that satisfy $[J_1, V_2] = iV_3$, etc., constitute a 3-vector operator **V**.] In the second usage of the term "octet," one refers to *any set of eight states* in an eight-dimensional Hilbert space, which, under infinitesimal $SU(3)$ transformations, also undergo the change (34). When we refer to an octet of gluons, we are using the term in this second sense, while the gluon field operators are an octet in the first sense. The intimate connection between these two usages is familiar to us from rotations, i.e., from $SU(2)$; there the three angular momentum operators J_i form a 3-vector, as do the spherical harmonics of degree $l = 1$. In the same connection, see the entries under "generator diagram" and "corresponding multiplet" in Table 1 below.

[15] The proof that (35) is actually an octet is given after Eq. (52).

$J = 1$ [cf. Eq. I.B(45)]. It is customary to write this in the following shorthand:

$$2 \otimes 2 = 1 \oplus 3, \tag{39}$$

where each angular momentum multiplet is denoted by $2 \mathbf{J} + 1$, its multiplicity. Further, a $j = \frac{1}{2}$ doublet multiplied by a $j = 1$ triplet can be decomposed into multiplets of total angular momentum $J = \frac{1}{2}$ and $J = \frac{3}{2}$, or $2 \otimes 3 = 2 \oplus 4$, giving

$$2 \otimes 2 \otimes 2 = 2 \oplus 2 \oplus 4. \tag{40}$$

These formulas always hold if one replaces the $\otimes$ and $\oplus$ symbols by $\times$ and $+$, as they must, because they also keep track of the number of linearly independent states.

We seek the $SU(3)$ analogues of (39) and (40), for that will tell us the color content of di- and triquark states. In addressing this question we encounter an immediate "complication" that we already saw in §I.E.5: there are two inequivalent triplets in $SU(3)$! There is no analogue to this in $SU(2)$—there is just one $j = \frac{1}{2}$ doublet, **2**. This novel feature of $SU(3)$ is absolutely essential for our purpose, because we must be able to build color singlets—the mesons—from a quark q and an antiquark $\bar{q}$, and *no* color singlet out of qq. Hence $\bar{q}$ and q must be different $SU(3)$ multiplets.

It is therefore necessary to understand the behavior of color under particle-antiparticle interchange (or "charge" conjugation), an operation usually denoted by C. We assume that the color eigenvalues, like other internal quantum numbers such as all the flavors, change sign under C. If these internal quantum numbers belong to $SU(2)$, as in the case of isospin, the set of eigenvalues belonging to the multiplet is unchanged by C. This can be seen with any $SU(2)$ multiplet, for example **2**, with isospin eigenvalues $(\frac{1}{2}, -\frac{1}{2})$ which go into $(-\frac{1}{2}, \frac{1}{2})$ under C. But from (25) it is obvious that $\lambda_3' \to -\lambda_3'$, $\lambda_8' \to -\lambda_8'$ does *not* map **3** into itself, because λ_8' is not symmetrically distributed about zero. Indeed, as we see from Fig. 1(a), C turns the triangular weight diagram of **3** upside-down into Fig. 1(b). This inverted triangle for the antiquarks was already shown in Fig. I.10(b), but there it depicted the flavor quantum numbers of the $\bar{q}$'s, not their color.

We now show that the color vectors that correspond to the vertices of the inverted triangle in Fig. 1(b) are of the form $V^* = \zeta_\alpha^* V_\alpha$ in the notation of Eq. (20), and $\zeta_\alpha^* \to U_{\alpha\beta}^* \zeta_\beta^*$ under $SU(3)$. We denote any set of three orthogonal vectors of the type V^* by $\mathbf{3}^*$. To show that $\{V^*\}$ is associated with the inverted triangle, consider the special transformation:

$$V^* \to [1 - \tfrac{1}{2}i(-\omega_3 \lambda_3 - \omega_8 \lambda_8)] V^*. \tag{41}$$

The generators F_a of an $SU(3)$ transformation are, by definition, obtained from the formula $U = 1 - i\omega_a F_a$ for *any* multiplet of $SU(3)$. Thus the $\frac{1}{2}\lambda_a$ are generators for **3**. According to (41), $-\frac{1}{2}\lambda_3$ and $-\frac{1}{2}\lambda_8$ are the diagonal

2. THE GROUP $SU(3)$

generators of $\mathbf{3}^*$, and as promised, their eigenvalues are those of the inverted triangle. On the other hand, in $SU(2)$, $\mathbf{2}$ and $\mathbf{2}^*$ have the same diagram, and are equivalent.

In the preceding section we already learned how to construct the $SU(3)$ bilinear singlet (13) from $\mathbf{3}$ and $\mathbf{3}^*$. This leaves eight other bilinear combinations of $\mathbf{3} \otimes \mathbf{3}^*$. These form an octet, that is designated by $\mathbf{8}$. The explicit construction of this octet was already given for *flavor* $SU(3)$ in §I.E.6, and the corresponding weight diagram is Fig. I.16. Thus we have

$$\mathbf{3} \otimes \mathbf{3}^* = \mathbf{1} \oplus \mathbf{8}. \tag{42}$$

This is the $SU(3)$ analogue of (39). Note that in both (39) and (42) there is a singlet, and a multiplet that has the same transformation properties as the generators of the group. This is no accident.

The final analogue is that of (40):

$$\mathbf{3} \otimes \mathbf{3} \otimes \mathbf{3} = \mathbf{1} \oplus \mathbf{8} \oplus \mathbf{8} \oplus \mathbf{10}. \tag{43}$$

The singlet is just the antisymmetric color state of the baryon, Eq. (11). The decouplet $\mathbf{10}$ was already constructed for flavor $SU(3)$ in §I.E.6. Its weight diagram is shown in Fig. I.15; as we recall, the spin-$\frac{3}{2}$ baryons belong to this *flavor* $\mathbf{10}$.

Equations (42) and (43) establish the theorems quoted in §1: χ_B and χ_M are the only qqq and $q\bar{q}$ color singlets.

Some further remarks concerning the "true meaning" of Eqs. (39)–(43) are in order here. As always, we begin with familiar ground, $SU(2)$: a rotation of the coordinate system changes the orientation of the angular momentum, but leaves its magnitude invariant. As a consequence a rotation never mixes the multiplets on the right-hand sides of Eqs. (39) and (40). Stated more elegantly, an equation like $\mathbf{2} \otimes \mathbf{3} = \mathbf{2} \oplus \mathbf{4}$ means that the six-dimensional space of states spanned by the six products of $\mathbf{2} \otimes \mathbf{3}$ can be decomposed into two rotationally invariant subspaces, one of dimension 2, the other of dimension 4. The $SU(3)$ equations have the analogous meaning: according to (43), for example, the 27-dimensional space of states spanned by the triquark wave functions $\mathbf{3} \otimes \mathbf{3} \otimes \mathbf{3}$ can be decomposed into one one-, one ten-, and two eight-dimensional subspaces that cannot be connected to each other by any transformation of $SU(3)$. This has an important consequence, which is seen by forming the octet "total color" operator, already introduced in Eq. (6):

$$\vec{F} = \tfrac{1}{2}[\vec{\lambda}^{(1)} + \vec{\lambda}^{(2)} + \vec{\lambda}^{(3)}]. \tag{44}$$

The superscript on $\vec{\lambda}^{(i)}$ refers to the separate quarks in $\mathbf{3} \otimes \mathbf{3} \otimes \mathbf{3}$, and consequently $\vec{\lambda}^{(i)}$ and $\vec{\lambda}^{(j)}$ commute with each other if $i \neq j$. An infinitesimal transformation of $\mathbf{3} \otimes \mathbf{3} \otimes \mathbf{3}$ is of the form

$$\prod_{i=1}^{3} [1 - \tfrac{1}{2} i \vec{\omega} \cdot \vec{\lambda}^{(i)}] \simeq 1 - i \vec{\omega} \cdot \vec{F}. \tag{45}$$

In view of the preceding statements about the invariance of the subspaces listed on the right-hand side of (43), the 27-dimensional matrix $\vec{F}$ must have the block-diagonal form

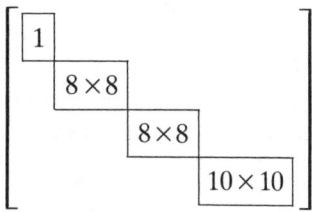

where all the blanks are zero. In short, when $\vec{F}$ acts on any one of the $SU(3)$ multiplets listed on the right-hand side of of Eqs. (42) and (43), it never produces an admixture of any other multiplet. Thus, the total color operator shares this property with the total angular momentum. In particular, if a state Ψ has the eigenvalue zero of $\vec{F}$, Eq. (45) tells us that Ψ is invariant, and therefore a singlet. This concept was already exploited extensively in §1.1.

We have now accumulated all the properties of $SU(3)$ needed for QCD. A convenient overview giving the analogies between $SU(2)$ and $SU(3)$ is provided by Table 1. A few of the entries in this table refer to material in the mathematical discussion that now follows.

(c) Further aspects of $SU(3)$

[The $2j + 1$ members of an arbitrary angular momentum multiplet can be constructed by forming appropriate products of the fundamental 2-component spinors. In $SU(3)$, the complex 3-vectors (20) are the fundamental building blocks. Let $(\zeta_1^{(i)}, \zeta_2^{(i)}, \zeta_3^{(i)})$ be the components of such a 3-vector; the products

$$\zeta_{\alpha_1}^{(1)} \zeta_{\alpha_2}^{(2)} \cdots \zeta_{\alpha_n}^{(n)} \equiv P^{(n)}, \tag{46}$$

for arbitrary n, will then generate all possible multiplets of $SU(3)$. In the preceding paragraph, we discussed the special case $n = 3$ of (46). Under an infinitesimal transformation, each factor in (46) suffers the change (23):

$$\delta \zeta_\alpha^{(i)} = -\tfrac{1}{2} i \omega_a (\lambda_a)_{\alpha\beta} \zeta_\beta^{(i)}; \tag{47}$$

observe that, as in (45), the eight parameters ω_a are the *same* for *all* factors in (46). Consequently, the change of $P^{(n)}$ is given by

$$\delta P^{(n)} = -i\omega_a F_a P^{(n)}, \tag{48}$$

where

$$F_a = \tfrac{1}{2} \sum_{i=1}^{n} \lambda_a^{(i)}. \tag{49}$$

This is the obvious generalization of (44). It shows that for *any* multiplet, the

2. THE GROUP SU(3)

TABLE 1
Résumé of SU(2) and SU(3)

Group: $U(N)$	$U(2) = U(1) \otimes SU(2)$	$U(3) = U(1) \otimes SU(3)$
Defining invariant	$|\eta_1|^2 + |\eta_2|^2$	$|\zeta_1|^2 + |\zeta_2|^2 + |\zeta_3|^2$
Generators	$\tfrac{1}{2}\sigma_a; \quad a = 1, 2, 3$	$\tfrac{1}{2}\lambda_a; \quad a = 1, \ldots, 8$
Diagonal generators	σ_3	λ_3, λ_8
Commutation rules	$[\tfrac{1}{2}\sigma_a, \tfrac{1}{2}\sigma_b] = i\varepsilon_{abc}\tfrac{1}{2}\sigma_c$	$[\tfrac{1}{2}\lambda_a, \tfrac{1}{2}\lambda_b] = if_{abc}\tfrac{1}{2}\lambda_c$
N	•——•	△
Generator diagram	$J_- \xleftarrow{\;J_3\;} J_+$	✶
Corresponding multiplet	•—•—•	⬡ (hexagon with center)
N*	•——•	▽
N ⊗ N*	$1 \oplus 3$	$1 \oplus 8$
N ⊗ N ⊗ N	$2 \oplus 2 \oplus 4$	$1 \oplus 8 \oplus 8 \oplus 10$

generators F_a have the commutation rules (29) of $\tfrac{1}{2}\lambda_a$, viz.,

$$[F_a, F_b] = if_{abc}F_c. \tag{50}$$

Thus in $SU(3)$, as in $SU(2)$, the commutation rule of the generators is universal: it does not depend on the multiplet.

Since F_a is a sum of λ_a's, it is clear that $\{F_a\}$ forms an octet $\vec{f}$. Further, if $\{G_a\}$ is an arbitrary set of eight operators satisfying[16]

$$[F_a, G_b] = if_{abc}G_c, \tag{51}$$

then they also form an octet $\vec{G}$. To prove this, one repeats the calculation of

---

[16] This general definition of an octet parallels that of a Euclidean 3-vector: if the objects W_a, where $a = 1, 2, 3$, satisfy $[J_a, W_b] = i\varepsilon_{abc}W_c$, where J_a are the components of the angular momentum, then $\{W_a\}$ forms a Euclidean 3-vector.

Eq. (34) with the unitary matrix $1 - i\vec{\omega}\cdot\vec{F}$. This leads to

$$\delta G_a = -i[G_a, \vec{\omega}\cdot\vec{F}], \qquad (52)$$

and when (51) is used it yields $\delta\vec{G} = \vec{\omega}\times\vec{G}$, as desired.[17]

Since the commutation rules (50) are universal, we know that in any $SU(3)$ multiplet two generates commute, and can be diagonalized simultaneously. As before, we choose these as F_3 and F_8, and call their eigenvalues F'_3 and F'_8. As with the basic **3**, the weight diagram of any multiplet is a set of contiguous dots in the $F'_3 - F'_8$ plane. There are two properties of these diagrams that follow immediately from what we already know. First, all λ_a have vanishing traces; but F_a is a sum of $\lambda_a^{(i)}$'s and therefore the sums of eigenvalues, $\sum F'_3$ and $\sum F'_8$, vanish for any multiplet. Consequently, the "center-of-gravity" of all weight diagrams is at the origin of the $F'_3 - F'_8$ plane.[18] Secondly, because of (49), the six operators

$$I_\pm = F_1 \pm iF_2,$$
$$V = F_4 \mp iF_5, \qquad (53)$$
$$U_\pm = F_6 \pm iF_7,$$

are sums of shift operators $I^{(i)}_{\alpha\beta}$, $\alpha \neq \beta$. From (22) one finds that $I_+ = \sum_i I^{(i)}_{12}$, $V_- = \sum_i I^{(i)}_{13}$, and $U_- = \sum_i I^{(i)}_{32}$, etc.). Hence, on any $SU(3)$ weight diagram, the generators F_a, when combined as in (53), will produce the same shifts as already shown for **3** in Fig. 1(a). To make this clearer, we depict the actions of the generators themselves in Fig. 2.

The structure of all $SU(3)$ weight diagrams can be surmised by reexamining angular momentum addition with $SU(2)$ weight diagrams. In $SU(2)$, only $\frac{1}{2}\sigma_3$ can be diagonalized, and for the spinor **2** the weight diagram is just two dots at $\pm\frac{1}{2}$ on a line giving the J_3 eigenvalues (see Table 1). From this diagram for **2**, we can, by replication, lay down an infinite line of equally spaced dots, and the weight diagram of any angular momentum [or $SU(2)$] multiplet is a contiguous set of $2j + 1$ dots on this line centered on the origin. The reasons behind this are: (1) any $SU(2)$ multiplet can be formed as a product of spinors; (2) the J_3 eigenvalue of the resultant is the *sum* of the J_3-eigenvalues of the factors; and (3) $J_1 \pm iJ_2$ produce unit shifts on the weight diagrams. But according to (46), (49), and Fig. 2, the analogues are true of $SU(3)$, and in particular, the eigenvalues of F_3 and F_8 are sums of the eigenvalues of the factors in (46). Hence, the $SU(3)$ analogue of the construction of the infinite $SU(2)$ line is just the following: one lays down a triangular lattice whose unit cells are the triangles **3** and **3***.

The weight diagram of any $SU(3)$ multiplet must then be a set of contiguous dots on this triangular lattice.[19] Figure 3 shows several examples: **8**, **10**, and **6**.

---

[17] We can now indicate how one demonstrates that $\delta(\vec{D}\times\vec{E}) = \vec{\omega}\times(\vec{D}\times\vec{E})$, as required by our claim that $\vec{D}\times\vec{E}$ is an octet. From (52), $\delta(D_b E_d) = \delta D_b E_c + D_b \delta E_c = -i[D_b, \vec{\omega}\cdot\vec{F}]E_c - iD_b[E_c, \vec{\omega}\cdot\vec{F}]$. These commutators are then evaluated with (51), and after multiplication by f_{abc}, one obtains the desired result.

[18] In Vol. I (see Figs. I.15 and 16) some weight diagrams are not centered in this way because the vertical axis was taken as strangeness, $S = (2/\sqrt{3})F_8 - B$, where B is the baryon number.

[19] Obviously, this sentence does not tell one which contiguous dots belong to a given multiplet. Some examples of multiplet construction were given in §I.E.6. For a complete treatment, see Carruthers (1966).

2. THE GROUP $SU(3)$

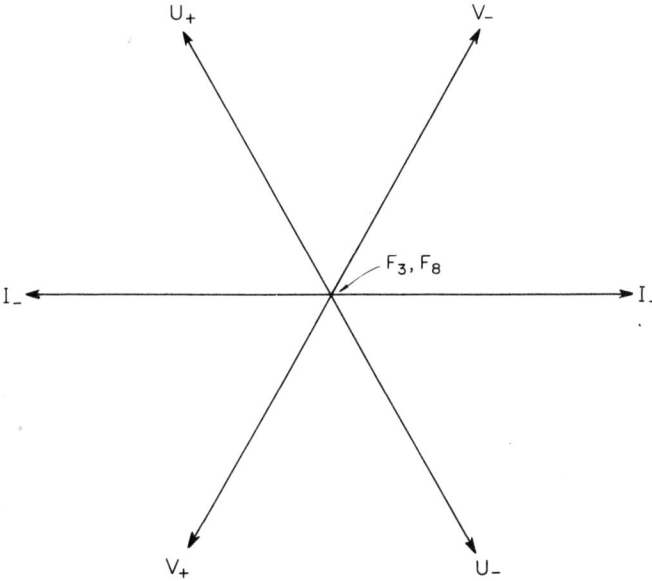

FIG. 2. Generator diagram for $SU(3)$. The shift operators are defined in Eq. (53). The operators F_3 and F_8 can be diagonalized simultaneously, and do not change states. On this diagram they therefore lie at the center.

The multiplet **6** arises when one combines two **3**'s: $\mathbf{3} \otimes \mathbf{3} = \mathbf{3}^* \oplus \mathbf{6}$; furthermore, $\mathbf{3} \otimes \mathbf{6} = \mathbf{10} \oplus \mathbf{8}$, which explains (43).

One should note the striking resemblance between the **8** in Fig. 3 and the generator diagram in Fig. 2. This same resemblance occurs in $SU(2)$: there is a one-to-one correspondence between the generator diagram for J_+, J_3, and J_-, and the weight diagram for $j = 1, m = 1, 0, -1$ (see Table 1). The reason is that both the angular momentum and the $j = 1$ triplet transform like a Euclidean 3-vector. The analogue occurs in $SU(3)$: if $\psi_a, a = 1, 2, \ldots, 8$, are the members of the multiplet **8**, they transform precisely like the F_a, i.e., $\delta\psi_a = f_{abc}\omega_b\psi_c$, and that is the underlying reason for the resemblance between the generator diagram and the weight diagram for **8**.

In our later work we shall need some properties of the λ-matrices for a quark–antiquark system. From (49) for $n = 2$, we have

$$F^2 \equiv \vec{F} \cdot \vec{F} = \tfrac{1}{4}[\vec{\lambda}^{(1)} \cdot \vec{\lambda}^{(1)} + \vec{\lambda}^{(2)} \cdot \vec{\lambda}^{(2)} + 2\vec{\lambda}^{(1)} \cdot \vec{\lambda}^{(2)}]. \tag{54}$$

Like any 3×3 matrix, $\tfrac{1}{4}\vec{\lambda}^{(i)} \cdot \vec{\lambda}^{(i)}$ is a linear combination of I and the λ_a, but since it is a scalar, it must be proportional to I:

$$\tfrac{1}{4}\vec{\lambda} \cdot \vec{\lambda} = \mathrm{I}k,$$

where we drop the common label i for now. We now take the trace over the

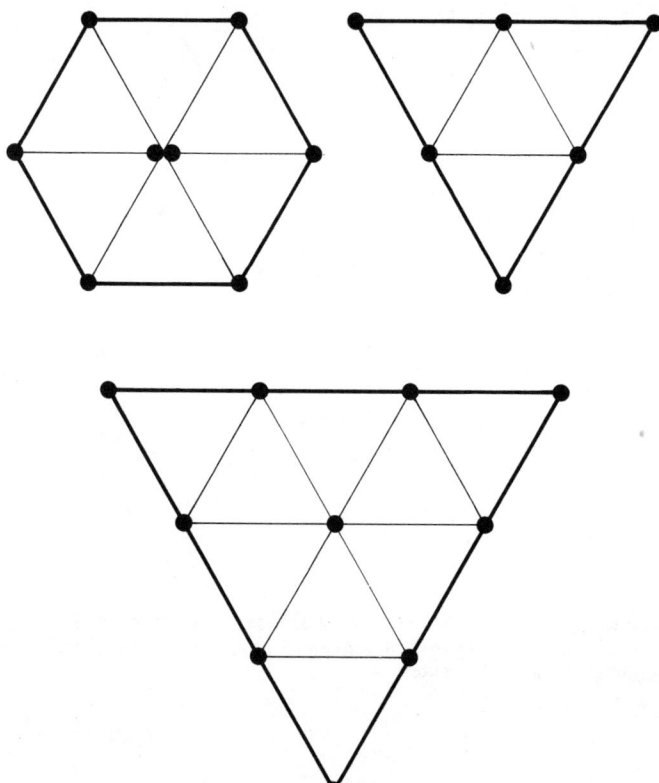

FIG. 3. The $SU(3)$ multiplets **6**, **8** and **10** formed by combining the fundamental triplets **3** and **3***. The triangular lattice described in the text is shown.

members of **3**:

$$\tfrac{1}{4}\sum_{a=1}^{8} \operatorname{Tr} \lambda_a^2 = 3k. \tag{55}$$

The trace of any matrix is invariant under all unitary transformations. Using this freedom, we can choose new axes in $\mathscr{C}_3$ so that any term $\operatorname{Tr} \lambda_a^2$ in (55) is "turned into" any other, with a different index; in particular, all terms can be turned into one of the diagonal ones. Thus,

$$3k = \tfrac{8}{4}\operatorname{Tr}\lambda_3^2 = \tfrac{8}{4}\operatorname{Tr}\lambda_8^2 = 2\sum \lambda_3'^2 = 2\sum \lambda_8'^2, \tag{56}$$

where the sums run over the eigenvalues listed in Eq. (25). This argument shows why the factor $1/\sqrt{3}$ must appear in λ_8 [cf. Eq. (24)]. Further, it leads to $3k = 2 \times 2$, or $k = \tfrac{4}{3}$, which shows that

$$\vec{\lambda}^{(i)} \cdot \vec{\lambda}^{(i)} = \tfrac{16}{3} \tag{57}$$

in **3**, and since $|\lambda_3'|$ and $|\lambda_8'|$ have the same values in **3***, (57) also applies to the antiquarks in **3***.

2. THE GROUP SU(3)

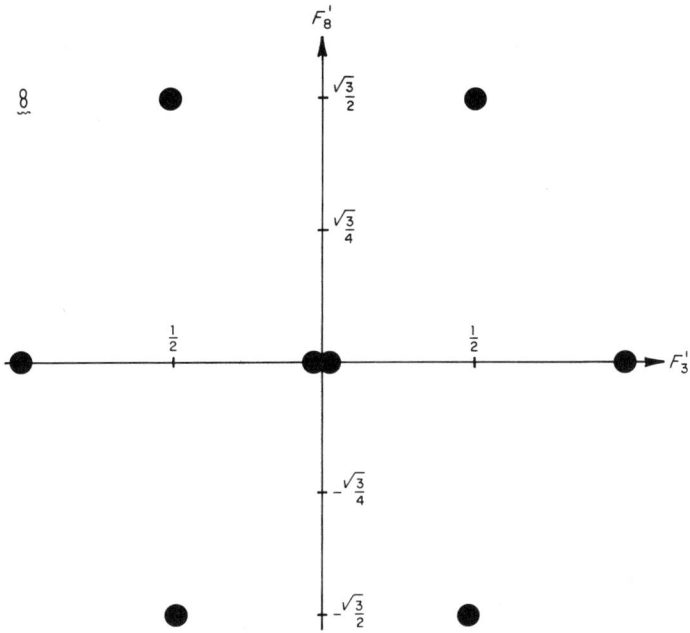

FIG. 4. Weight diagram for the octet **8**. The construction of this diagram for flavor $SU(3)$ was already given in §I.E.6 (see Figs. I.11 and 16). The axes give the eigenvalues F_3' and F_8' of the diagonal generators F_3 and F_8 (these generators are $\frac{1}{2}\lambda_a$ in the case of **3**). There are two linearly independent members of **8** with $F_3' = F_8' = 0$, which we call ψ_3 and ψ_8. In the case of the baryon octet shown in Fig. I.16, they correspond to Σ^0 and Λ, respectively. Hence $(F_1 \pm iF_2)\psi_3 \neq 0$, whereas $(F_1 \pm iF_2)\psi_8 = 0$.

The same argument, together with the detailed weight diagram (see Fig. 4) for **8**, leads to the result $F^2 = 3$ for an octet. Since $F^2 = 0$ in a singlet, we conclude that for a $q\bar{q}$ system

$$\tfrac{1}{4}\vec{\lambda}^{(1)} \cdot \vec{\lambda}^{(2)} = \begin{cases} -\tfrac{4}{3} & \mathbf{1} \\ \tfrac{1}{6} & \mathbf{8} \end{cases} \tag{58}$$

Because the gluons form an octet, and two gluons can be combined into a color singlet, we shall also need the analogue of (58) for this case. As before, we set $\vec{F} \cdot \vec{F} = 1k$, where $\vec{F}$ is the generator in any *one* octet. Then $\text{Tr}\,\vec{F} \cdot \vec{F} = 8k = 8\,\text{Tr}\,F_3^2$. The eigenvalues of F_3 are shown in Fig. 4 (they are also those of I_3 in Fig. I.16). Hence, $\text{Tr}\,F_3^2 = 4(\tfrac{1}{2})^2 + 2(1)^2 = 3$, and therefore $F^2 = 3$. Now we combine *two* independent octets into a singlet, $(\vec{F}^{(1)} + \vec{F}^{(2)})^2 = 0$, whence

$$\vec{F}^{(1)} \cdot \vec{F}^{(2)} = -3. \tag{59}$$

B. GAUGE FIELDS[1]

1. Global vs. local symmetries

All the symmetries that we have studied in this book up to now are called *global*. By that one means that the symmetry transformation is the same for all observers, no matter where they may be located in space-time. We recall several examples. Whenever we spoke of a rotation of the coordinate system, we took it for granted that the angles specifying this rotation did not vary from place to place. Indeed, one usually thinks of just one family of rectilinear coordinate frames, with a common origin, when one considers rotations. A second example was the isospin transformation that mixes the two components of the nucleon doublet (p, n), or of the quark doublet (u, d). Here, too, we tacitly assumed that the $SU(2)$ transformation did not vary from place to place. And the same went for the $SU(3)$ transformations that intermingle the three color states of any quark: in Eq. A(23) for an infinitesimal $SU(3)$ transformation,

$$U = 1 - \tfrac{1}{2}i\vec{\omega} \cdot \vec{\lambda}, \tag{1}$$

the eight parameters $\vec{\omega}$ were, without discussion, assumed to be independent of the space-time coordinates. Thus (1) is a global $SU(3)$ transformation.

If one reflects on the color transformations, one may well wonder why these transformations should be global. The mass of a quark does not depend on its color, and its electromagnetic and weak interactions are oblivious to color. In view of this, it is not even clear how two observers,

---

[1] Readers who want to study the $SU(2)$ gauge theory of the electroweak interaction (Chap. VI) before QCD, and who have not yet read the preceding material on $SU(3)$, can read what follows by using the following dictionary. Wherever you see Ψ, instead of calling this a quark color triplet, call it a quark or lepton left-handed flavor doublet (e.g., $\psi_1 = u_L$, $\psi_2 = d'_L$, or $\psi_1 = \nu_{eL}$, $\psi_2 = e^-_L$, etc.); our eight 3×3 matrices λ_a are to be replaced by the three 2×2 Pauli matrices τ_a which act on the weak isospin (*not* the spin) variable, and all 3×3 matrices, such as U, are then 2×2; any object $\vec{O}$, which we call an $SU(3)$ octet, you call a 3-component vector in the weak isospin space—examples are the matrices $\vec{\tau}$, the infinitesimal parameters $\vec{\omega}$, and the gauge fields $\vec{V}_i$, of which there are three in $SU(2)$, in contrast to the eight of $SU(3)$; formulas that look like familiar expressions of three-dimensional vector algebra, such as $\vec{\lambda} \cdot (\vec{\omega} \times \vec{V}_i)$, are to be replaced by $\vec{\tau} \cdot (\vec{\omega} \times \vec{V}_i)$, and handled without fear—in $SU(2)$, they *are* just the formulas of vector algebra. In $SU(3)$ these formulas are a concise notation for somewhat different expressions [see §A.2(b and c)].

one, say, at Fermilab, the other at CERN, are to know that they must carry out all color transformations in lock-step.[2] Nevertheless, if we stipulate (1) as the symmetry transformation, all observers must redefine their coordinate frame in $\mathscr{C}_3$ in the same way.

Indeed, all continuous global symmetries are open to the same critique: They violate the spirit, if not the letter, of the principles of locality and causality. This deep insight was first recognized and exploited by Einstein, and led him to the General Theory of Relativity. There are significant analogies between the train of thought that leads from a global "internal" symmetry [such as color $SU(3)$] to the associated local symmetry. Such a symmetry requires the existence of interactions mediated by "gauge" fields. There is now reason to hope that all the interactions of nature, not just gravity, stem from local symmetry principles. A brief sketch of some basic concepts of the General Theory of Relativity is therefore in order.

In Special Relativity we consider two arbitrary inertial frames, K_1 and K_2. Each K_i has a clock at its origin O_i, and a family F_i of observers spread throughout space who are all at rest with respect to O_i, and who have synchronized their clocks with the one at O_i. The transformation from K_1 to K_2 is a *global* Lorentz transformation, because *any* observer in F_1 has the *same* relative velocity with respect to *any* observer in F_2, no matter where they may be located in space-time. It is the basic premise of the Special Theory that the laws of physics do not change their form under such a global Lorentz transformation.

In contrast to this, the General Theory only uses local concepts. Once again, one imagines two families of observers, F_1 and F_2, spread throughout space-time, but one does *not* demand that all members of one family are at rest with respect to each other, use synchronized clocks, and orient their spatial coordinates in the same way. Instead, one only requires that all those observers in one family that are in the immediate neighborhood of an arbitrary space-time point are at rest with respect to one another, have synchronized clocks, etc. As the space-time separation between observers belonging to *one* family, say F_1, grows, their relative velocity, relative clock rates, and relative orientation of spatial coordinates are allowed to depart smoothly from zero in an arbitrary fashion. Now focus on any point in space-time, and one of the members of F_1 and F_2 in the neighborhood of that point: obviously, their relative velocity $\mathbf{v}_{12}$ will depend on the chosen point. Consequently, the Lorentz transformation that relates them varies from point to point—it is a *local* Lorentz transformation. The General Theory requires that the laws of physics do not distinguish the two families of observers F_1 and F_2. This requirement compels one to introduce a field related to the accelerations that are inevitably associated with a space-dependent velocity. The Equivalence Principle asserts that this is a

---

[2] This sentence borders on demagoguery. If color confinement holds, all objects larger than 10^{-13} cm are color singlets. Since singlets are invariant, two observers separated by distances large compared to this can choose $\mathscr{C}_3$ coordinates independently even if the Hamiltonian is only invariant under global transformations.

gravitational field. In short, the imposition of a local, as compared to a global, symmetry requires the existence of force fields.

We now try to impose this General Principle on a physical law that already conforms with the Special Principle of Relativity. For the sake of concreteness, consider Maxwell's equations. When we carry out a global Lorentz transformation on the coordinates and electromagnetic fields, these equations retain their form because the velocity **v** that appears in the transformations is a constant. If **v** is an arbitrary function of **x** and t, however, the transformed equations will contain expressions like $\partial \mathbf{v}/\partial t$ and $\partial v_i/\partial x_j$ arising from the time and space derivatives in Maxwell's equations. Hence, the covariance is lost, and the transformed equations contain arbitrary space-time dependent functions. This arbitrariness is removed by introducing the gravitational field, with its own dynamics. Once it is introduced, and properly coupled to the electromagnetic field, the modified Maxwell equations satisfy the General Principle of Relativity. Above all, the new equations contain new phenomena, such as the bending of light in a gravitational field.

In the case of "internal" symmetries, the requirement of local, as compared to global invariance also requires the introduction of new degrees of freedom. These are the gauge fields. They have their own dynamics, and an interaction with their sources, such as quarks, that has a form that is prescribed by the local symmetry principle. As a consequence, new phenomena are expected that cannot be ascribed to quarks alone.

2. The gauge field

We now try to implement the requirement that a system of colored quarks satisfy physical laws that are invariant under local $SU(3)$ transformations. For this purpose it will suffice to begin with heavy nonrelativistic quarks Q; nothing essentially new arises when one goes to a relativistic theory of light quarks. If this is granted, we can also ignore the quark spin, which only becomes important when velocities comparable to c are considered.

Consider first a single quark Q, with Schrödinger wave function $\psi_\alpha(\mathbf{r}t)$. The index $\alpha = 1, 2, 3$ is the color variable. A *local SU(3)* transformation is given by

$$\psi_\alpha(\mathbf{r}t) \to U_{\alpha\beta}(\mathbf{r}t)\psi_\beta(\mathbf{r}t), \tag{2}$$

where, for an infinitesimal transformation,[3] the 3×3 matrix U is just the obvious generalization of (1):

$$U_{\alpha\beta}(\mathbf{r}t) = \delta_{\alpha\beta} - \tfrac{1}{2}i\vec{\omega}(\mathbf{r}t) \cdot \vec{\lambda}_{\alpha\beta}. \tag{3}$$

[3] It can readily be shown that invariance under infinitesimal transformations implies invariance under arbitrary finite transformations, because the latter can always be achieved as a product of infinitesimal ones.

2. THE GAUGE FIELD

As in the Pauli theory of electron spin, we introduce a more compact notation:

$$\Psi(\mathbf{r}t) = \begin{pmatrix} \psi_1(\mathbf{r}t) \\ \psi_2(\mathbf{r}t) \\ \psi_3(\mathbf{r}t) \end{pmatrix}. \tag{4}$$

Then (2) reads

$$\Psi(\mathbf{r}t) \to [1 - \tfrac{1}{2}i\vec{\omega}(\mathbf{r}t) \cdot \vec{\lambda}]\Psi(\mathbf{r}t) \equiv \Psi'(\mathbf{r}t). \tag{5}$$

The essential new feature is that the octet of infinitesimal parameters now depends on location.

We shall now follow the line of attack described in §1 when we discussed the imposition of the General Principle of Relativity on Maxwell's equations: we ask whether the Schrödinger equation of one isolated quark can be made invariant under (5), and if not, what modifications invariance requires. The equation is

$$-\frac{\nabla^2}{2m}\Psi = i\frac{\partial \Psi}{\partial t}. \tag{6}$$

Obviously, this is covariant if $\vec{\omega}$ is not a function of t and $\vec{r} = (x_1, x_2, x_3)$. But when $\vec{\omega}$ varies, we have

$$i\frac{\partial \Psi'}{\partial t} = iU\frac{\partial \Psi}{\partial t} + \frac{1}{2}\left(\frac{\partial \vec{\omega}}{\partial t} \cdot \vec{\lambda}\right)\Psi, \tag{7}$$

$$\frac{1}{i}\frac{\partial \Psi'}{\partial x_i} = U\frac{1}{i}\frac{\partial \Psi}{\partial x_i} - \frac{1}{2}\left(\frac{\partial \vec{\omega}}{\partial x_i} \cdot \vec{\lambda}\right)\Psi. \tag{8}$$

The second terms in (7) and (8) contain arbitrary space-time functions. As in the example of §1, where we applied a space-time dependent Lorentz transformation to Maxwell's equations, these terms spoil the invariance of the equations. New dynamical fields must be added to the system, and coupled to our quark, if the symmetry principle is to be upheld.

The way in which this is to be accomplished can be inferred from electrodynamics. If e_Q is the quark's charge, the Schrödinger equation in the presence of an electromagnetic field is

$$\frac{1}{2m}\left(\frac{1}{i}\nabla + e_Q\mathbf{A}\right)^2\psi = \left(i\frac{\partial}{\partial t} + e_Q A_0\right)\psi, \tag{9}$$

where $\mathbf{A}$ and A_0 are the vector and scalar potentials, respectively. The electric and magnetic fields are then

$$\mathbf{E} = -\frac{\partial \mathbf{A}}{\partial t} - \nabla A_0, \tag{10}$$

$$\mathbf{B} = \nabla \times \mathbf{A}. \tag{11}$$

As we know, only **E** and **B** appear in Maxwell's equations and the Lorentz force law; potentials that give the same fields are physically indistinguishable. From (10) and (11) we see that **E** and **B** are unaltered if the potentials undergo *the electromagnetic gauge transformation*

$$\mathbf{A} \to \mathbf{A} + \frac{1}{e_Q}\nabla\chi \equiv \mathbf{A}', \qquad A_0 \to A_0 - \frac{1}{e_Q}\frac{\partial \chi}{\partial t} \equiv A_0', \tag{12}$$

where $\chi(\mathbf{r}t)$ is an arbitrary (smooth) scalar function. While (12) leaves Maxwell's equations unaltered, it appears to alter the Schrödinger equation (9). But now we exploit the complex nature of ψ, and redefine its phase as follows:

$$\psi \to e^{-i\chi}\psi \equiv \psi'. \tag{13}$$

On combining (12) and (13), we see that

$$\left(\frac{1}{i}\frac{\partial}{\partial x_i} + e_Q A_i'\right)\psi' = e^{-i\chi}\left(\frac{1}{i}\frac{\partial}{\partial x_i} + e_Q A_i\right)\psi, \tag{14}$$

$$\left(i\frac{\partial}{\partial t} + e_Q A_0'\right)\psi' = e^{-i\chi}\left(i\frac{\partial}{\partial t} + e_Q A_0\right)\psi. \tag{15}$$

These equations show that if the two transformations (12) and (13) are applied together, they leave the Schrödinger equation (9) invariant, because the space-time dependent phase $e^{-i\chi}$ is now common to all terms in the Schrödinger equation, and cancels out.

That electrodynamics provides the clue to the escape from the difficulty encountered in Eqs. (6)–(8) is best seen by considering an infinitesimal electromagnetic gauge function χ, in which case (13) is replaced by

$$\psi \to [1 - i\chi(\mathbf{r}t)]\psi \equiv \psi'. \tag{16}$$

This has the same structure as the local $SU(3)$ transformation (5). The lesson to be learned from (14) and (15) is that the local transformation (16) is only a symmetry if the momentum and energy operators in Schrödinger's equation are modified by adding couplings to other fields that also change under that transformation. The form of this modification can be guessed by repeating our electromagnetic exercise in reverse, as if we knew that we want (16) to be a symmetry, but had not yet heard of electromagnetism, for

2. THE GAUGE FIELD

that is precisely the situation we are in with $SU(3)$. So we apply the momentum operator to (16) and find

$$\frac{1}{i}\frac{\partial \psi'}{\partial x_i} = (1 - i\chi)\frac{1}{i}\frac{\partial \psi}{\partial x_i} - \frac{\partial \chi}{\partial x_i}\psi. \tag{17}$$

This looks about as bad as (8); but we know already how to repair (17): we replace the momentum by

$$\frac{1}{i}\frac{\partial}{\partial x_i} \rightarrow \frac{1}{i}\frac{\partial}{\partial x_i} + e_Q A_i, \tag{18}$$

and require A_i to transform as in (12).

We now try the same cure for $SU(3)$. First, (8) has an octet of arbitrary vector fields, $\partial \vec{\omega}/\partial x_i$ that must be removed, whereas in (17) there is just one, $\partial \chi/\partial x_i$. So the analogue of the electromagnetic vector potential A_i must be an octet of vector potentials, $\vec{V}_i$, where $i(=1, 2, 3)$ labels the components of each of the eight vector fields in everyday 3-space. Similarly, the temporal equation (7) implies an octet of scalar fields, $\vec{V}_0$, that are the counterpart of the electromagnetic scalar potential A_0. Instead of (18) we try the replacement

$$\frac{1}{i}\frac{\partial}{\partial x_i} \rightarrow \left(\frac{1}{i}\frac{\partial}{\partial x_i} + \tfrac{1}{2}g\vec{\lambda} \cdot \vec{V}_i\right) \equiv \frac{1}{i}D_i, \tag{19}$$

and its temporal partner

$$i\frac{\partial}{\partial t} \rightarrow \left(i\frac{\partial}{\partial t} + \tfrac{1}{2}g\vec{\lambda} \cdot \vec{V}_0\right) \equiv iD_0, \tag{20}$$

where g, *a pure number, is the color coupling constant*—the analogue of e in quantum electrodynamics. As our modified Schrödinger equation, we try

$$-\frac{1}{2m}D_i D_i \Psi = iD_0 \Psi. \tag{21}$$

Note that the so-called *covariant derivatives*, D_0 and D_i, are 3×3 matrices—the derivative terms, such as $i\partial/\partial t$, are understood to contain the 3×3 unit matrix that is not shown.

How do we know when we have succeeded? As always, by demonstrating the invariance of the equation of motion (21) when the wave function Ψ and the potentials $\vec{V}_i$ are transformed. The transformation law for the source Ψ is $\Psi \rightarrow U\Psi$, which is our basic assumption. But how does $\vec{V}_i$ transform?

That transformation is found by requiring the analogues of (14) and (15),

$$D_i'\Psi' = U(D_i\Psi), \tag{22}$$

$$D_0'\Psi' = U(D_0\Psi),$$

where $D_{i,0}'$ is $D_{i,0}$, are as given by (19) and (20), but with $\vec{V}_{i,0}$ replaced by the transformed field, $\vec{V}_{i,0}'$. Since $\Psi' = U\Psi$, (22) implies

$$UD_i U^{-1} = D_i', \qquad UD_0 U^{-1} = D_0', \tag{23}$$

and these, in turn, guarantee that the Schrödinger equation transforms properly under the local color transformation, i.e., that (21) and (23) imply $D_i' D_i' \Psi' = -2mi D_0' \Psi'$.

Equation (23) also provides the gauge transformation law for the fields. Consider, for example, D_i; from (19) and (23) we have

$$\frac{1}{i} U \frac{\partial}{\partial x_i} U^{-1} + \frac{1}{2} g \vec{\lambda}' \cdot \vec{V}_i \equiv \frac{1}{i} \frac{\partial}{\partial x_i} + \frac{1}{2} g \vec{\lambda} \cdot \vec{V}_i', \tag{24}$$

where

$$\vec{\lambda}' \equiv U \vec{\lambda} U^{-1} = \vec{\lambda} - \vec{\omega} \times \vec{\lambda}, \tag{24'}$$

according to Eqs. A(34) and A(35). To first order in $\vec{\omega}$,

$$\frac{1}{i} U \frac{\partial}{\partial x_i} U^{-1} = \frac{1}{i} \frac{\partial}{\partial x_i} + \frac{1}{2} \vec{\lambda} \cdot \frac{\partial \vec{\omega}}{\partial x_i}.$$

We now collect coefficients of $\vec{\lambda}$ on both sides of (24), use $(\vec{\omega} \times \vec{\lambda}) \cdot \vec{V} = -\vec{\lambda} \cdot (\vec{\omega} \times \vec{V})$, and find

$$\vec{V}_i' = \vec{V}_i + \frac{1}{g} \frac{\partial \vec{\omega}}{\partial x_i} + \vec{\omega} \times \vec{V}_i. \tag{25}$$

The same calculation for D_0 leads to

$$\vec{V}_0' = \vec{V}_0 - \frac{1}{g} \frac{\partial \vec{\omega}}{\partial t} + \vec{\omega} \times \vec{V}_0. \tag{26}$$

These are the $SU(3)$ counterparts of the electromagnetic gauge transformation (12). We note there are two terms: the derivatives of $\vec{\omega}$ have the same structure as in electrodynamics—they are only there if the transformation is local; the $\vec{\omega} \times$ terms are present even if the transformation is global—they confirm that $\vec{V}_i$ and $\vec{V}_0$ are octets. The former is present in any gauge transformation; the latter arises because in contrast to electrodynamics we are now dealing with a *non-Abelian* group of gauge transformations.

3. Quanta of the gauge field and coupling to quarks

The quantization of the free gauge field, and the coupling between the field and quarks, are patterned after QED.

The steps that turn the classical gauge field into a quantum mechanical operator are just those of §II.A (and Appendix II). One first decomposes $\vec{V}_i(\mathbf{r}t)$ into a sum of plane waves, as in Eq. II.A(37). The complex amplitudes are then replaced by operators—the QCD counterparts of the photon destruction and creation operators $a_h^\dagger(\mathbf{k})$ and $a_h(\mathbf{k})$. They must carry degrees of freedom beyond helicity h and momentum $\mathbf{k}$ so as to account for the color quantum numbers. Since there is a linear relation between the field, on the one hand, and the creation and destruction operators, on the other, and as the field is an $SU(3)$ octet, the creation and destruction operators are also octets. We call them $\vec{c}_h^\dagger(\mathbf{k})$ and $\vec{c}_h(\mathbf{k})$, respectively, where $\rightarrow$ indicates that for each $(h, \mathbf{k})$ there are eight distinct operators, $c_h^a(\mathbf{k})$, with $a = 1, \ldots, 8$. As always, the commutation rules are those of independent harmonic oscillators:

$$[c_h^a(\mathbf{k}), c_h^b(\mathbf{k}')] = \delta_{ab}\, \delta_{hh'}\, \delta_{\mathbf{k}\mathbf{k}'}. \tag{27}$$

The complete expression for the quantum field $\vec{V}_i$ is then the obvious generalization of Eq. II.A(40) and (41):

$$\vec{V}_i(\mathbf{r}t) = \sum_{h\mathbf{k}} \frac{1}{\sqrt{2V\omega}} [\vec{c}_h(\mathbf{k})\hat{\varepsilon}_h(\mathbf{k})_i e^{i(\mathbf{k}\cdot\mathbf{r}-\omega t)} + \text{h.c.}]. \tag{28}$$

Except for the dispersion law, the relation between ω and $\mathbf{k}$, $\vec{V}_i$ is now completely specified.

In electrodynamics $\omega = |\mathbf{k}|$, and this implies that the photon is massless. What is the underlying reason for this? If the photon had a mass μ, the vector potential $\mathbf{A}$ would satisfy the Klein–Gordon equation, which would then impose the energy-momentum relation $\omega^2 = k^2 + \mu^2$. Naturally, the electromagnetic energy density $\frac{1}{2}(E^2 + B^2)$ must also be altered if $\mu \neq 0$, and a dimensional argument suffices to show that the additional term is proportional to $\mu^2 A_i A_i$. But this is not a gauge invariant expression, so $\mu = 0$. This argument applies without modification to the color gauge field: if ω in (28) is to be $\sqrt{k^2 + \mu^2}$, the energy density of the $SU(3)$ gauge field must contain a term $\propto \mu^2 \vec{V}_i \cdot \vec{V}_i$, which is not invariant under the local transformation (25), so $\mu = 0$ again.

Hence quantization leads[4] to a color SU(3) *octet of massless spin one quanta. These are called "gluons."*

---

[4] Gauge bosons only acquire a mass if the vacuum state is not invariant under gauge transformations. This phenomenon does not occur in QCD, but it is a crucial ingredient in the gauge theory of the electroweak interaction (see Chap. VI). We might add that the current experimental upper limit on the photon mass is $\sim 0.5 \times 10^{-21} m_e$.

The gluons, being members of an octet, are obviously not color singlets. If our color confinement hypothesis is correct, it must be that *isolated gluons have infinite energy, and, like quarks, cannot be observed directly*. Presumably they materialize into hadrons in a fashion similar to the "hadronization" of quarks. In short, the interactions in QCD must play an overriding role. This shows that despite the mathematical and conceptual similarities between QED and QCD, the gluon-photon analogy must be treated with great caution.

The *interaction between quarks and gluons* has essentially the same form as that between photons and electrons. In the Feynman diagrams of QCD, this coupling is depicted by the vertex

(29)

where α and β are the color quantum numbers of the incoming and outgoing quarks, and a is the octet label of the gluon, which is represented by the coiled line as shown. The vertex carries a factor in the mathematical expression for the diagram that is not quite the same as in QED. To see what it is, it suffices to note that the Schrödinger equation (21) for a quark in the presence of a color field becomes that for the motion of a particle carrying a charge e, in an electromagnetic field A_i, if one makes the substitution $\frac{1}{2}g\vec{\lambda} \cdot \vec{V}_i \to eA_i$. Consequently, in QCD the vertex in (29) carries a factor $\frac{1}{2}g(\lambda_a)_{\alpha\beta}$ in addition to the Dirac matrix familiar from QED.

We now turn to several important consequences of the basic symmetry principles.

Local symmetry—gauge invariance—contains global symmetry as a special case. As we already saw in the case of isospin (§I.D.2) the generators of a global symmetry transformation are constants of motion. *Consequently, the total color is exactly conserved in QCD, and in particular, a color singlet (e.g., a photon, lepton, or hadron) is strictly forbidden from transforming into a colored state (e.g., a quark or gluon).*

Finally we come to a remarkable consequence of *local SU(3)* symmetry: *the strength g of the coupling of the color gauge field cannot depend on quark flavor—QCD is characterized by just one dimensionless "charge" g.*

The significance of this assertion can only be appreciated if one recognizes that electrodynamics contains no principle[5] that forbids different particles from having arbitrary charges. Naturally, all transitions (e.g., decays) that violate charge conservation would be forbidden; but that does not prevent particles that are not connected by transitions from having arbitrarily

---

[5] Here we assume that magnetic monopoles, which have never been seen, do not exist, because Dirac demonstrated in 1931 that such objects would force all charges to be a multiple of a basic unit ("charge quantization"). In this connection, recall §I.E.13(a).

different charges. The ultimate reason for this is that the electromagnetic field carries no charge. In a non-Abelian gauge theory, however, the "charge" (such as color) flows from sources into the field, and vice versa. This establishes a channel of communication between different source species, and implies that all matter fields have color charges that are uniquely determined by the color charge of the gauge mesons.

⟦The preceding sentence provides a complete physical explanation of why QCD requires all quark flavors to have the same coupling constant to the color field. For a formal proof, imagine a system of two nonrelativistic quarks of different flavor, having coordinates r_1 and r_2, and color matrices $\vec{\lambda}^{(1)}$ and $\vec{\lambda}^{(2)}$. It suffices to consider a product wave function, because any state is a linear superposition of such functions. The right-hand side of the Schrödinger equation (21) then generalizes to

$$\left[i\frac{\partial}{\partial t} + \frac{1}{2}g_1\vec{\lambda}^{(1)}\cdot\vec{V}_0(r_1 t) + \frac{1}{2}g_2\vec{\lambda}^{(2)}\cdot\vec{V}_0(r_2 t)\right]\Psi_1(r_1 t)\Psi_2(r_2 t).$$

By demanding that this be covariant [in the sense of Eq. (22)] when $\vec{V}_0$ undergoes the gauge transformation (26), one is led to the conclusion that $g_1 = g_2 = g$.⟧

This remarkable property of non-Abelian gauge theories appears to be of profound importance. It means that if all of physics could be described by a local symmetry based on a group $\mathfrak{G}$ that cannot be factored into a direct product $\mathfrak{G}_1 \otimes \mathfrak{G}_2$, there would be just one basic constant that describes all interactions. The Grand Unified Theories described in §I.E.13 are of this character. They only have one coupling constant, to which the coupling constants of the weak, electromagnetic, and strong interactions are related by algebraic expressions dictated by group theory.

C. GAUGE FIELD DYNAMICS

This Part is devoted to somewhat more sophisticated aspects of QCD.[1] Our first task is really an exercise in classical field theory: Construction of the Yang-Mills equations that govern the dynamics of the gauge field. We then come to a subtle and most remarkable property of the quantum mechanical radiative corrections: asymptotic freedom. In essence, asymptotic freedom is the statement that at very short distances the interaction between quarks is *weaker* than the $1/r$ law that one would expect classically when the forces are mediated by a massless vector field. Asymptotic freedom is a phenomenon that *only* occurs in non-Abelian gauge theories. As we learned in §II.D.3, the opposite occurs in QED, i.e., the interaction is stronger than $1/r$ at short distances. Asymptotic freedom offers an explanation of why quarks appear to behave as if they were almost free in deep inelastic scattering. This chapter will close with a discussion of color confinement, which is widely believed to be a consequence of QCD, but which has not yet been rigorously demonstrated.

1. Color analogues of the electromagnetic field strengths

The construction of the equations that govern the evolution of the gauge fields is patterned on the construction of Maxwell's equations. The quanta of the gauge fields are the gluons. But there is one important difference: *non-Abelian gauge invariance demands couplings of the gauge fields to themselves*. That is, even in the absence of colored sources, such as quarks, there is no such thing as a gauge invariant theory of noninteracting gluons. The reason is quite simple: as we know, the gluons carry color. Furthermore, gauge invariance means that the gauge field is universally coupled to color. Consequently, there must be self-couplings of the gauge field. Here again we see an analogy to Einstein's Theory of Gravitation: the gravitational field is coupled to all energy, whether in the form of mass or of fields; but the gravitational field itself carries energy, so it must be coupled to itself. In contrast to this, the electromagnetic field [with the Abelian gauge group $U(1)$] does not carry any electromagnetic charge itself, and there is therefore no electromagnetic self-coupling. In more physical terms this

---

[1] For a treatment of QCD at a "professional" level, consult Cheng (1984).

1. FIELD STRENGTHS

means that there is no scattering of light by light *in vacuo* in Maxwell's classical theory.[2] Such scattering does occur in QED thanks to vacuum polarization (cf. §II.D.3), i.e., because of quantum fluctuations in the electron-positron density. Even in the quantum theory, therefore, there would be no photon–photon scattering if charged particles did not exist. Non-Abelian gauge theories are fundamentally different: in an imaginary world where nothing but gluons exists, there is already gluon–gluon scattering, because the classical field equations are nonlinear; such scattering does not require vacuum fluctuations of any sort for its existence.

As we have said, we shall try to copy the construction of Maxwell's equations as far as that will take us. From Eqs. B(10) and (11) we recall how the electric and magnetic fields **E** and **B** are constructed from the potentials A_0, **A**. The fields are gauge invariant. From **E** and **B** one can form gauge invariant expressions for a host of physically important quantities, in particular the energy density

$$u = \tfrac{1}{2}(E^2 + B^2). \tag{1}$$

We begin by constructing the curl of $\vec{V}_i$, in the hope that it will be the analogue of the magnetic field **B**:

$$\vec{b}_i = \frac{\partial}{\partial x_j}\vec{V}_k - \frac{\partial}{\partial x_k}\vec{V}_j, \tag{2}$$

where (i, j, k) is a cyclic permutation of the spatial indices 1, 2, 3. Here we see that our candidate $\vec{b}_i$ for the "color-magnetic" field appears to be an $SU(3)$ octet, because it is merely a spatial derivative of the octet $\vec{V}_i$. Consequently, these field strengths cannot be gauge invariant! Although this conclusion is astonishing when viewed from the vantage point of electrodynamics, it is only to be expected, since the field carries color. All that we can expect is a gauge invariant energy density—some analogue of (1). This analogy will actually hold if our color magnetic field $\vec{b}_i$ really transforms like an octet, because the magnetic energy density could then be taken to be

$$\tfrac{1}{2}\sum_{a=1}^{8}\sum_{i=1}^{3} b_{ia} b_{ia} \equiv \tfrac{1}{2}\vec{b}_i \cdot \vec{b}_i, \tag{3}$$

which we know to be an $SU(3)$ invariant. So we must examine the transformation law of $\vec{b}_i$. For that purpose, we merely substitute the transformation law B(25) into our definition (2). As in electrodynamics, curl removes the "gauge shift" $\partial \vec{\omega}/\partial x_i$; but these spatial derivatives act also

---

[2] For a discussion of experimental limits on nonlinearities in classical electrodynamics, see Jackson (1975), pp. 10–13.

on $\vec{\omega}$ in $\vec{\omega} \times \vec{V}_i$, and lead to

$$\vec{b}_i' = \vec{b}_i + \vec{\omega} \times \vec{b}_i + \left(\frac{\partial \vec{\omega}}{\partial x_j} \times \vec{V}_k - \frac{\partial \vec{\omega}}{\partial x_k} \times \vec{V}_j\right). \tag{4}$$

The last term shows that $\vec{b}_i$ is *not* an octet for *local* gauge transformations, and we have actually broken our rules by using the notation $\vec{b}_i$ for it. For a global transformation, the last term is absent, and the transformation law has the desired form.

Having seen why this mindless copying fails, it is not difficult to repair the damage. From the discussion pertaining to Eq. B(8), we recall that the presence of the unwanted quantity $\partial \vec{\omega}/\partial x_i$ in an unsuccessful transformation could be removed by introducing a gauge field $\vec{V}_i$ in the object of interest; the "gauge shift" of $\vec{V}_i$ then removed the unwanted term. In the light of this remark we define *the color magnetic field* by

$$\vec{B}_i = \frac{\partial \vec{V}_k}{\partial x_j} - \frac{\partial \vec{V}_j}{\partial x_k} - g\vec{V}_j \times \vec{V}_k. \tag{5}$$

The transformation law for the curl term is just (4), whereas that of the second term, to first order in $\vec{\omega}$, is

$$g\vec{V}_j' \times \vec{V}_k' = g\vec{V}_j \times \vec{V}_k + \left(\frac{\partial \vec{\omega}}{\partial x_j} \times \vec{V}_k + \vec{V}_j \times \frac{\partial \vec{\omega}}{\partial x_k}\right)$$

$$+ g[(\vec{\omega} \times \vec{V}_j) \times \vec{V}_k + \vec{V}_j \times (\vec{\omega} \times \vec{V}_k)].$$

The term inside ( ) cancels the unwanted last term of (4), and the expression inside [ ] is just $\vec{\omega} \times (\vec{V}_j \times \vec{V}_k)$. Consequently, the transformation law for $\vec{B}_i$ is

$$\vec{B}_i \to \vec{B}_i' = \vec{B}_i + \vec{\omega} \times \vec{B}_i \tag{6}$$

for *any local* gauge transformation. Thus (5) is the desired expression for the octet of color magnetic fields. The identical argument also yields *an octet of color electric fields*:

$$\vec{E}_i = -\frac{\partial \vec{V}_i}{\partial t} - \frac{\partial \vec{V}_0}{\partial x_i} + g\vec{V}_i \times \vec{V}_0, \tag{7}$$

where the linear portion is again the obvious generalization from electromagnetism [cf. Eq. B(10)]. The transformation law for $\vec{E}_i$ is also

$$\vec{E}_i \to \vec{E}_i' = \vec{E}_i + \vec{\omega} \times \vec{E}_i. \tag{8}$$

We have now seen that the requirement that the field strengths $\vec{B}_i$ and $\vec{E}_i$ transform as octets under local $SU(3)$ transformations forces the presence of the nonlinear terms $g\vec{V}_j \times \vec{V}_k$ and $g\vec{V}_i \times \vec{V}_0$. The structure of these terms is fixed by the symmetry principle.

2. Self-interactions of the gauge field

The gauge invariant energy density u can be written down by direct analogy with electrodynamics:[3]

$$u = \tfrac{1}{2}(\vec{E}_i \cdot \vec{E}_i + \vec{B}_i \cdot \vec{B}_i). \tag{9}$$

It contains terms that are quadratic, cubic, and quartic in the gauge fields. The quadratic terms lead to freely traveling plane waves, and the higher order terms to interactions between these waves. The cubic and quartic interactions due to the "electric" part of u are

$$u_3^E = -g\left(\frac{\partial \vec{V}_i}{\partial t} + \frac{\partial \vec{V}_0}{\partial x_i}\right) \cdot (\vec{V}_i \times \vec{V}_0), \tag{10}$$

$$u_4^E = \tfrac{1}{2}g^2(\vec{V}_i \times \vec{V}_0) \cdot (\vec{V}_i \times \vec{V}_0), \tag{11}$$

and the "magnetic" self-interactions are of a similar form.

These interactions are represented by vertices in the Feynman diagrams of QCD that have no QED counterpart. The cubic terms are depicted by the vertex shown in Fig. I.20(b), the quartic by that of Fig. I.20(c). From (9) and (10) we see that these are of order g and g^2, respectively. The cubic terms describe transitions from states with one gluon to states where there are two, and vice versa. The quartic term permits gluon–gluon scattering in the very first Born approximation, but second-order perturbation theory using the cubic interaction also contributes to this same order in g (i.e., g^2).

3. The Yang–Mills field equations

We are now ready to derive the equations of motion for the color gauge field. If we try a naive generalization of Maxwell's equations, say

$$\frac{\partial \vec{E}_i}{\partial x_j} - \frac{\partial \vec{E}_j}{\partial x_i} + \frac{\partial \vec{B}_k}{\partial t} = 0,$$

where (i, j, k) is again a cyclic permutation of $(1, 2, 3)$, we shall fail. Once more the reason is that an expression like $\partial \vec{E}_i / \partial x_j$ does not transform like an

---

[3] Admittedly one could conceive of adding terms that are of higher order in V, and one therefore calls this the minimal theory. Nonminimal theories are not renormalizable, and are therefore rejected. This ambiguity also arises in gravitation, where Einstein's field equations are merely the simplest structures that obey the General Principle of Relativity.

octet under a local $SU(3)$ transformation, even though $\vec{E}_i$ does, because

$$\frac{\partial \vec{E}'_i}{\partial x_j} = \frac{\partial \vec{E}_i}{\partial x_j} + \vec{\omega} \times \frac{\partial \vec{E}_i}{\partial x_j} + \frac{\partial \vec{\omega}}{\partial x_j} \times \vec{E}_i. \tag{12}$$

The last term, which spoils the transformation, can be removed by defining "covariant derivatives" for the gauge fields:

$$\begin{aligned} \mathscr{D}_i &= \frac{\partial}{\partial x_i} - g\vec{V}_i \times, \\ \mathscr{D}_0 &= \frac{\partial}{\partial t} + g\vec{V}_0 \times. \end{aligned} \tag{13}$$

These covariant derivatives[4] are the analogues for the V-fields of the operators D_i and D_0 for the quark fields introduced in §B.2. By using Eqs. (8) and B(25) and (26) one readily demonstrates that

$$\mathscr{D}'_j \vec{E}'_i = \mathscr{D}_j \vec{E}_i + \vec{\omega} \times (\mathscr{D}_j \vec{E}_i), \tag{14}$$

etc. That is, $\mathscr{D}_j \vec{E}_i$, $\mathscr{D}_0 \vec{B}_k$, etc., are octets under local $SU(3)$ transformations. In terms of these derivatives, the *Yang-Mills field equations* look just like Maxwell's:

$$\mathscr{D}_i \vec{E}_j - \mathscr{D}_j \vec{E}_i + \mathscr{D}_0 \vec{B}_k = 0, \tag{15}$$

$$\mathscr{D}_i \vec{B}_i = 0, \tag{16}$$

$$\mathscr{D}_i \vec{E}_i = g\vec{\rho}, \tag{17}$$

$$\mathscr{D}_i \vec{B}_j - \mathscr{D}_j \vec{B}_i = \mathscr{D}_0 \vec{E}_k + g\vec{j}_k. \tag{18}$$

Here $\vec{\rho}$ and $\vec{j}_k$ are the density and current of color carried by objects *other* than the gauge field; these, to our knowledge, are totally due to quarks. For massive nonrelativistic quarks, $\vec{\rho}$ and $\vec{j}_k$ follow from the Schrödinger equation B(21) in the usual manner:

$$\vec{\rho} = (\Psi^* \tfrac{1}{2} \vec{\lambda} \Psi), \tag{19}$$

and

$$\vec{j}_k = \frac{1}{im} \operatorname{Im} \left(\Psi^* \tfrac{1}{2} \vec{\lambda} D_k \Psi \right), \tag{20}$$

---

[4] If $\{\vec{F}\}$ are the $SU(3)$ generators for an arbitrary representation of $SU(3)$, the covariant derivatives for fields belonging to that representation are $i(\partial/\partial t) + g\vec{F} \cdot \vec{V}_0$, etc. In the case of **8**, this becomes the cross product of Eq. (13), because the 8×8 matrices F_a are $(F_a)_{bc} = -if_{abc}$ (see Carruthers (1966), p. 50), and therefore

$$(\vec{V}_0 \cdot \vec{F})_{ab} E_{kb} = -iV_{0c} f_{cab} E_{kb} = i(\vec{V}_0 \times \vec{E}_k)_a.$$

3. THE YANG–MILLS FIELD EQUATIONS

where D_k was defined in B(19). The matrix $\frac{1}{2}\vec{\lambda}$ appears here because that is the $SU(3)$-generator for quarks. [In (19) and (20) we use a shorthand where the 3-valued quark color index is suppressed, i.e., $\rho_a = \frac{1}{2}\Psi_\alpha^*(\lambda_a)_{\alpha\beta}\Psi_\beta$, etc.] Both $\vec{\rho}$ and $\vec{j}_k$ are octets under local $SU(3)$ transformations.[5]

⟦If a relativistic description of the quarks is required, as is always the case for the u, d, and s quarks, one must replace Ψ in these expressions by the appropriate Dirac fields. That is, one combines $\vec{\rho}$ and $\vec{j}_k$ into a 4-current of color $\vec{j}_\mu$, where $\mu = 0, 1, 2, 3$; then

$$\vec{j}_\mu = \sum_f \bar{\psi}_f \gamma_\mu \tfrac{1}{2}\vec{\lambda}\psi_f, \tag{21}$$

where ψ_f is an $SU(3)$ triplet of Dirac fields for quarks of flavor f. Equation (21) is actually a highly abbreviated expression; in detail it reads

$$j_\mu^a = \sum_f \sum_{\alpha\beta} \sum_{rs} \bar{\psi}_{fr\alpha}(\gamma_\mu)_{rs}(\tfrac{1}{2}\lambda_a)_{\alpha\beta}\psi_{fs\beta}, \tag{21'}$$

where r and s are the usual 4-valued Dirac spinor indices, and (α, β) are 3-valued color labels.⟧

The energy of interaction between quarks and the transverse (or radiation) part of the gauge field is then

$$H_{\text{int}} = g \int d^3 r\, \vec{j}_i(\mathbf{r}t) \cdot \vec{V}_i(\mathbf{r}t), \tag{22}$$

which has the same form as in QED (cf. §II.B.1). This explains why the vertex Eq. B(29) has the structure described in §B.3. From Eq. (21) onward, we have also used the fact that there is but one coupling constant in QCD (recall §B.3).

As expected, the field equations are nonlinear even when the "matter" densities $\vec{\rho}$ and $\vec{j}_i$ *vanish*. For example, from (13) we see that "Poisson's equation" (17) reads as follows in such spatial regions:

$$\frac{\partial \vec{E}_i}{\partial x_i} = g\vec{V}_i \times \vec{E}_i. \tag{23}$$

One can therefore call $g\vec{V}_i \times \vec{E}_i$ the *color charge density of the gluon field*. Where there is a matter source, (23) becomes

$$\frac{\partial \vec{E}_i}{\partial x_i} = g\vec{V}_i \times \vec{E}_i + g\vec{\rho}. \tag{24}$$

---

[5] One verifies this as follows: $\vec{j}_k^{\prime t} = (1/im)\,\text{Im}\,(\Psi^* U^{-1}\tfrac{1}{2}\vec{\lambda}UD_k U^{-1}U\Psi)$ in virtue of Eq. B(23), and Eq. B(24') then shows that $\delta \vec{j}_k = \vec{\omega} \times \vec{j}_k$, as required.

As in electrodynamics, the divergence of the electric field, when integrated over all of space, is a constant of motion. In our case, there is an octet of such conserved quantities. This octet is, apart from the overall factor of g, the total color of the entire system:[6]

$$\vec{\mathcal{F}} = \int (\vec{V}_i \times \vec{E}_i)\, d^3r + \int \vec{\rho}\, d^3r. \qquad (25)$$

While $\vec{\mathcal{F}}$ is a constant of motion, the two portions of (25) are not conserved separately. As is clear from everything that has been said—and as demonstrated explicitly by (24)—color flows from quarks into the gauge field, and vice versa. As a consequence, even "color electrostatics" is a complex and rich subject, in contrast to conventional electrostatics. This can be seen very easily by considering a fixed point source of color—an infinitely massive quark. The color electric field arising from it has two distinct contributions: a Coulomb field and a field due to a color charge distribution that must exist throughout space because of the presence of a color electric field. As we will see below, the latter nonlinear contribution is responsible for "asymptotic freedom"—the softening of hadronic interaction at short distances. There are also compelling arguments that this nonlinearity produces color confinement.

4. Asymptotic freedom

We now address ourselves to the following question: What is the energy of interaction $W(r)$ of two stationary point sources of color when their separation **r** is small? The answer to this question determines the amplitude A for scattering of quarks at very large momentum transfers, and is therefore of crucial importance in deep inelastic scattering.

(a) Comparison with QED

Let q^2 be the square of the 4-momentum transfer, and $A(q^2)$ the aforementioned amplitude. As we shall see, $W(r)$ is weaker than the Coulomb interactions when r tends to zero. This implies that $q^2 A(q^2) \to 0$ as $q^2 \to \infty$, that is, the amplitude is *smaller* than the elementary point-particle scattering amplitude. This phenomenon is called *asymptotic freedom*; the word "asymptotic" refers to the limit $q^2 \to \infty$, or equivalently, to $r \to 0$.

To gain some appreciation of the significance of asymptotic freedom, consider the analogous problem in quantum electrodynamics—the interaction between two stationary ("external") point charges. We recall from

---

[6] $\vec{\mathcal{F}}$, as given by (25), satisfies the $SU(3)$ commutation rules [i.e., Eq. A(29), or equivalently, Eq. A(50)]. We use the notation $\vec{\mathcal{F}}$, instead of $\vec{F}$, because $\vec{\mathcal{F}}$ is the specific QCD operator (25). By using the operator B(28) for $\vec{V}_i$, and the standard Bose–Einstein commutation rule B(27), one can show that this requirement is met.

4. ASYMPTOTIC FREEDOM

§II.D.3 that virtual electron-positron pairs produce a "polarization of the vacuum" that modifies Coulomb's law at distances small compared to the electron's Compton wavelength. From §II.D.3 we know that vacuum polarization leads to a *stronger* interaction at short distances than that given by Coulomb's law. That is, *QED is not asymptotically free*: electromagnetic scattering of two leptons through large momentum transfers is *more* probable than what one would naively expect if one assumed an unlimited validity of the Coulomb law.[7] As we shall soon see, this difference between QED and QCD is entirely due to the nonlinear character of the non-Abelian gauge field.

Why does vacuum polarization produce a strengthening of the interaction at short distances? This was explained in detail in §II.D.3, so we can be brief here. The charge distribution due to virtual pairs *screens* an external charge. By definition, the total charge is that seen by a distant observer. When one approaches the external charge, one leaves the screening cloud behind, and so one sees an evergrowing charge. It is this growth that causes the interaction of two charges to increase more rapidly than $1/r$ as $r \to 0$.

Vacuum polarization is also present in QCD. Indeed, it is larger than in QED, because QCD has many more objects that can appear as virtual pairs: eight gluons, as well as three colors of quarks of various flavors. If QCD is to be asymptotically free, there must be another mechanism that produces an "antiscreening" effect that overcomes the screening due to all this vacuum polarization. As we shall now see, the Yang-Mills charge density $\vec{E}_i \times \vec{V}_i$ is responsible for such a mechanism.

(b) Antiscreening—A qualitative discussion

If we are to proceed further, we shall have to improve our understanding of the gauge field when sources are present. We recall that in electrodynamics the presence of charges requires one to separate the electric field into transverse ($\mathbf{E}_T$) and longitudinal ($\mathbf{E}_L$) parts that satisfy $\nabla \cdot \mathbf{E}_L \neq 0$, $\nabla \times \mathbf{E}_T \neq 0$, and $\nabla \cdot \mathbf{E}_T = 0$, $\nabla \times \mathbf{E}_L = 0$. These statements assume a more intuitive form if one decomposes $\mathbf{E}$ into plane waves: in any Fourier component, $\mathbf{E}_L$ points along the direction of propagation, whereas $\mathbf{E}_T$ is transverse to that direction. $\mathbf{E}_L$ is determined completely by the positions of the charges, and is not a dynamical variable of the electromagnetic field. On the other hand, $\mathbf{E}_T$ is such a dynamical variable. This is brought out very clearly by the expression for the total energy of the electromagnetic field in the presence of stationary point charges $\{e_i\}$ at positions $\{\mathbf{r}_i\}$:

$$H = \frac{1}{2} \int (E_T^2 + B^2) \, d^3r + \frac{1}{2} \sum_{ij} \frac{e_i e_j}{4\pi |\mathbf{r}_i - \mathbf{r}_j|}. \tag{26}$$

---

[7] Because α is so small, this effect cannot be seen in QED tests involving large momentum-transfer collisions of leptons (as in $e^+e^- \to \mu^+\mu^-$, cf. §II.C.3). On the other hand, the spectroscopic measurements that are sensitive to vacuum polarization (cf. §II.D.3) demonstrate that the electromagnetic interaction is stronger than Coulombic at short distances.

Since **B** is always transverse ($\nabla \cdot \mathbf{B} = 0$), all fields that appear in the Hamiltonian are transverse.[8]

We decompose the color electric field in the same way:

$$\vec{E}_i = \vec{E}_{Ti} + \vec{E}_{Li}, \tag{27}$$

where

$$\nabla_i \vec{E}_{Ti} = 0. \tag{28}$$

As in electrodynamics,

$$\vec{E}_{Ti} = -\frac{\partial \vec{V}_i}{\partial t}. \tag{29}$$

The longitudinal field $\vec{E}_{Li}$ is not an independent variable; like the electrostatic field, it is determined by the sources of color, that is, the quarks *and* gluons. The gauge field color density can be split into two physically distinct terms by means of (27) and (29):

$$\vec{V}_i \times \vec{E}_i = \vec{V}_i \times \vec{E}_{Li} - \vec{V}_i \times \frac{\partial \vec{V}_i}{\partial t} \tag{30}$$

If we recall Eq. B(28), we see that the second term of (3) is a bilinear expression in the gluon creation and destruction operators; therefore, it can create gluon pairs when acting on the gauge field vacuum state, and it is ultimately responsible for the gluons' contribution to vacuum polarization. As we are not concerned with vacuum polarization at the moment, we can drop this second term of (30) for now. Once this is done, the "Poisson" equation (24) becomes

$$\nabla_i \vec{E}_{Li} = g \vec{V}_j \times \vec{E}_{Lj} + g \vec{\rho}, \tag{31}$$

where, for simplicity, we place one quark at the origin:

$$\vec{\rho} = \tfrac{1}{2} \langle \vec{\lambda} \rangle \, \delta^3(\mathbf{r}), \tag{32}$$

with $\langle \ \rangle$ indicating an expectation value in its color state.[9]

Our goal is to find the color charge density of the field induced by the presence of this quark. That is, we wish to determine

$$\vec{\rho}_{\text{ind}} \equiv \vec{V}_j \times \vec{E}_{Lj}. \tag{33}$$

Here the field $\vec{V}$ will arise from the vacuum fluctuations, which are always

---

[8] This is known as the "Coulomb gauge" formulation of electrodynamics. Other, equivalent, formulations exist, but are ill-suited to our needs. If the sources move, one must add to (26) the familiar expression $\sum_i e_i \mathbf{v}_i \cdot \mathbf{A}(\mathbf{r}_i t)$, where **A** is also transverse.

[9] For an arbitrary state, $\langle \vec{\lambda} \rangle = \sum_{\alpha\beta} \zeta_\alpha^* \vec{\lambda}_{\alpha\beta} \zeta_\beta$, in the notation of Eq. A(20).

present. Let us first solve (31) for the longitudinal fields. Since we wish to know how the field modifies the color density of a quark, we only need the expectation value of the operator $\vec{\rho}_{\text{ind}}$ in the lowest state containing one quark; insofar as the gluons are concerned, this is the vacuum state $|\Omega\rangle$.

While an exact solution of (31) for $\vec{E}_{Li}$ is not known, it is not difficult to find a solution when g is small. This will suffice for our purpose.[10] We therefore expand the longitudinal field in powers of g:

$$\vec{E}_{Li} = g\vec{E}_{Li}^{(1)} + g^2\vec{E}_{Li}^{(2)} + \cdots. \tag{34}$$

The induced density has a related expansion:

$$\vec{\rho}_{\text{ind}} = g\vec{\rho}_{\text{ind}}^{(1)} + g^2\vec{\rho}_{\text{ind}}^{(2)} + \cdots. \tag{35}$$

Since $\vec{V}$ is given by the vacuum fluctuations, which are independent of g, we have

$$\vec{\rho}_{\text{ind}}^{(n)} = \vec{V}_i \times \vec{E}_{Li}^{(n)}. \tag{36}$$

After substituting (34) into (31), and equating coefficients of g and g^2 on both sides, we find

$$\nabla_i \vec{E}_{Li}^{(1)} = \vec{\rho}, \tag{37}$$

$$\nabla_i \vec{E}_{Li}^{(2)} = \vec{V}_i \times \vec{E}_{Li}^{(1)} = \vec{\rho}_{\text{ind}}^{(1)}. \tag{38}$$

$\vec{E}_{Li}^{(1)}$ is the Coulomb field of $\vec{\rho}$; $\vec{E}_{Li}^{(2)}$ is the Coulomb field of $\rho_{\text{ind}}^{(1)}$. Equations (37) and (38) are just a pair of Poisson equations: in (37) the source is pointlike, whereas it is distributed in (38). The solution of equations of this type is straightforward: for any source of color $\vec{f}$, the solution of $\nabla_i \vec{E}_i = \vec{f}$ is given by Coulomb's law:

$$\vec{E}_i(\mathbf{r}) = \int \frac{(\mathbf{r} - \mathbf{r}')_i}{4\pi |\mathbf{r} - \mathbf{r}'|^3} \vec{f}(\mathbf{r}') \, d^3r'. \tag{39}$$

In particular,

$$\vec{E}_{Li}^{(1)} = \frac{r_i}{4\pi r^3} \frac{1}{2} \langle \vec{\lambda} \rangle. \tag{40}$$

The construction of $\vec{\rho}_{\text{ind}}$ therefore proceeds as follows: for the quark at $\mathbf{r} = 0$, one constructs a point Coulomb field (40). This is then used to construct the first-order induced density $\vec{\rho}_{\text{ind}}^{(1)}$ via (36); the latter is used as a static source distribution for $\vec{E}_{Li}^{(2)}$; and so forth.

What about the field $\vec{V}$, which also enters into the construction of $\vec{\rho}_{\text{ind}}$?

---

[10] This is so because of asymptotic freedom: g becomes progressively smaller as r decreases; see Subsection (d).

We are only concerned with the vacuum expectation value $\langle\Omega|\vec{\rho}_{\text{ind}}|\Omega\rangle$. As $\vec{V}_i$ is a linear superposition of creation and destruction operators, the only terms in (35) that matter are those that have an even number of gauge fields. Clearly, $\vec{\rho}_{\text{ind}}^{(1)}$ is of the first degree in $\vec{V}_i$, and the leading contribution is therefore

$$\langle\Omega|\vec{\rho}_{\text{ind}}(\mathbf{r})|\Omega\rangle = g^2 \langle\Omega|\vec{V}_i(\mathbf{r}) \times \vec{E}_{Li}^{(2)}(\mathbf{r})|\Omega\rangle. \tag{41}$$

According to (39),

$$\vec{E}_{Li}^{(2)}(\mathbf{r}) = \int \frac{(\mathbf{r}-\mathbf{r}')_i}{4\pi |\mathbf{r}-\mathbf{r}'|^3} \vec{\rho}_{\text{ind}}^{(1)}(\mathbf{r}') \, d^3r', \tag{42}$$

where $\vec{\rho}_{\text{ind}}^{(1)}$ is given by (38).

Since the gauge field in (41) describes vacuum fluctuations, it can assume all possible color orientations and wavelengths (or momenta). As we have said, the field appears twice in (41); once explicitly as $\vec{V}_i(\mathbf{r})$, and once implicitly as a factor $\vec{V}_j(\mathbf{r}')$ in $\vec{\rho}_{\text{ind}}^{(1)}$ [cf. (38)]. But whatever $\vec{V}_j(\mathbf{r}')$ creates out of the vacuum must be destroyed by $\vec{V}_i(\mathbf{r})$, and therefore the wavelength and color labels of these two field fluctuations must be identical.

Now we shall give a geometrical demonstration[11] that (41) is an antiscreening distribution when the color group is $SU(2)$. The argument for $SU(3)$, though entirely similar, is more cumbersome and not as enlightening. In $SU(2)$ the quark color is 2-valued, all objects bearing the symbol $\to$ are 3-vectors [in contrast to the "8-vectors" of $SU(3)$], and cross-products in color space are conventional cross-products.

Consider a quark at $\mathbf{r}=0$ for which $\langle\vec{\lambda}\rangle$ is in the 1-direction of color space. According to (40), this will produce a Coulomb field $\vec{E}_{Li}^{(1)}$ of color label 1 everywhere in ordinary space, as shown in Fig. 5(a). Now we turn to the gauge field. The result obtained for one mode must then be summed appropriately over all of them. We choose $\vec{V}$ in the 3-direction of color, and oriented in space at the point $\mathbf{r}'$ as shown in Fig. 5. To compute the first-order induced density $\rho_{\text{ind}}^{(1)}(\mathbf{r}')$, we must take the scalar product in ordinary space and the cross-product in color space of the two fields already depicted. For the acute angle shown, the former is positive; the color cross-product, for the assumed colors of $\langle\vec{\lambda}\rangle$ and $\vec{V}(\mathbf{r}')$, results in a positive 2-component for $\vec{\rho}_{\text{ind}}^{(1)}$, as shown in Fig. 5(b). Like any charge, $\vec{\rho}_{\text{ind}}^{(1)}$ has an associated electric field $\vec{E}_{Li}^{(2)}$ [cf. (38)], which must be in the 2-direction of color [cf. (42)], and which is also shown in Fig. 5(b). The induced color density $\vec{\rho}_{\text{ind}}^{(2)}$ of actual interest to us is then found [cf. (41)] by once again forming the spatial scalar-product and color cross-product of the induced electric field $\vec{E}_{Li}^{(2)}$ with the gauge field $\vec{V}_i$. The latter, as explained, must also have the label 3, so the resultant color of $\vec{V}_i \times \vec{E}_{Li}^{(2)}$ is *negative* along the 1-direction. This negative sign is the crux of the matter; it is a consequence

---

[11] We owe this demonstration to M. Peskin (private communication).

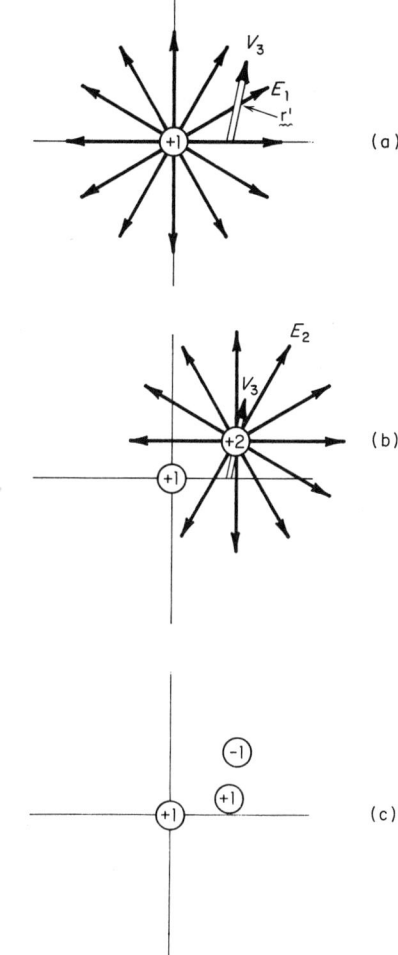

FIG. 5. The induced color density. (a) A quark is placed at the origin. Its density $\langle \vec{\lambda} \rangle$ is taken to be positive in the 1-direction of color, and this is indicated by the circle marked +1. Associated with it there is a Coulomb field $\vec{E}^{(1)}$ of the same color [cf. Eq. (40)], whose field lines are depicted by the outward arrows. A gauge field vacuum fluctuation, of color 3, is shown at the spatial point $\mathbf{r}'$ as a double arrow.

(b) At $\mathbf{r}'$ the fields shown in (a) induce a positive color-2 density, as given by Eq. (38). [Here, and in (c), we consider only the gauge group $SU(2)$.] The figure shows this induced density as a circle marked +2, the associated Coulomb field $\vec{E}^{(2)}$, and the gauge field fluctuation.

(c) The electric and gauge fields in (b) produce the second-order induced density $\vec{\rho}^{(2)}$, given by Eqs. (36) and (41). This induced density is shown at the tip and base of the gauge field arrow, where it is assumed that these points are within one wavelength or so of $\mathbf{r}'$. The sign change from -1 to $+1$ of color is because, at these points, $\vec{V}_3$ and $\vec{E}_2$ are parallel, and antiparallel, respectively.

of two successive color cross-products. Finally, we must still take care of the spatial scalar-product in forming $\vec{\rho}_{\text{ind}}^{(2)}$; Fig. 5(b) shows that this is positive or negative, depending on whether the chosen points are closer to, or farther from, the quark than the point **r**′ at which the density $\vec{\rho}_{\text{ind}}^{(1)}$ was evaluated. The upshot is then an induced dipole shown in Fig. 5(c); it is an antiscreening dipole, having its +1 color piece closer to the +1 quark than its −1 piece. Actually, it is a configuration of opposite charge densities—a "spread out dipole"—with the positive portion nearer the quark. One can repeat this exercise for arbitrary choices of the color label of $\vec{V}$ and arbitrary spatial orientations of $\vec{V}$: one always finds such an antiscreening charge distribution. In the preceding discussion we assumed, for the sake of simplicity, that the wavelength of the gauge field $\vec{V}_i$ is large compared to the distance from the origin **r**′. The conclusions are similar for any wavelength, but the precise spatial configuration of the induced dipole distribution depends on that wavelength.

This argument suffices to show that vacuum fluctuations of the Yang–Mills field induce an antiscreening charge distribution, but it does not tell us whether this effect is large enough to overcome the screening dipole distribution due to conventional vacuum polarization.

(c) Antiscreening by an electromagnetic analogy

There is another way to see that a theory containing charge-carrying field quanta (gluons) will give rise to antiscreening [Nielsen (1981)]. Moreover, this approach shows that the antiscreening is much stronger than the screening due to the gluon vacuum polarization.

We describe the vacuum by the zero-point fluctuations of the gluon field. For the moment we ignore the effect of the zero-point fluctuations of the quarks, and regard the vacuum as the state in which each mode has a probability $\frac{1}{2}$ to be occupied by a free gluon, because the zero-point fluctuations carry half a quantum. Furthermore, we rely on the analogy between QED and QCD, and replace the gluon field by a massless charged boson field with spin 1 and charge e. The scalar charged boson field is treated in Appendix II.2(b). There it is shown that it gives rise to particles of a given charge—say, e_0—and their antiparticles. If we generalize these considerations to a vector field, the same method of quantization gives rise to spin-1 particles with a gyromagnetic ratio[12] of $\gamma = 2$.

Our argument exploits a relation between the dielectric "coefficient" ε and the magnetic permeability v. Since the velocity of light in a medium is given by $(\varepsilon v)^{-1/2}$, the relation

$$\varepsilon v = 1 \qquad (43)$$

holds in the vacuum. Let us introduce the electric and magnetic

---

[12] This is often called the g factor; our notation avoids confusion with the coupling constant g.

4. ASYMPTOTIC FREEDOM

susceptibilities:

$$\varepsilon = 1 + \chi_e, \qquad \nu = 1 + \chi_m. \tag{44}$$

In the limit of small deviations of ε and ν from unity, (43) implies that

$$\chi_e \simeq -\chi_m. \tag{45}$$

The magnetic properties of an assembly of gluons are easier to evaluate than the electric, but we can use (45) to go from one to the other.

A medium with $\nu > 1$ is paramagnetic; when $\nu < 1$, it is diamagnetic. The corresponding electric properties do not have such names, but we introduce the terms paraelectric and diaelectric for $\varepsilon > 1$, and $\varepsilon < 1$, respectively. Ordinary materials containing molecules with electric dipoles have $\varepsilon > 1$. A paraelectric medium "screens" charges, and the interaction between two charges in the medium is reduced by a factor of ε^{-1}. On the other hand, diaelectric media are "antiscreening"—the interaction is increased. We show that the gluonic vacuum is *paramagnetic*, and therefore diaelectric.

The energy of interaction $W(r)$ between two equal charges, Q, separated by a distance r, placed in a medium, has the form

$$W(r) = Q^2 \int e^{i\mathbf{q}\cdot\mathbf{r}} \frac{1}{q^2 \varepsilon(q)} \frac{d^3q}{(2\pi)^3}. \tag{46}$$

If ε is a constant, this reduces to $Q^2/4\pi r \varepsilon$. There is a dependence of ε on q because the response of the medium changes when the spatial extent $\sim 1/q$ of an externally applied field is varied. Loosely speaking, one can therefore write[13]

$$W(r) \simeq \frac{Q^2}{4\pi r \varepsilon(r^{-1})}. \tag{47}$$

Actually an assembly of free particles with spin and charge exhibit both para- and diamagnetism. The former comes from the contribution of the magnetic moments, the latter from the change of trajectories caused by the magnetic field. In a state where each mode is occupied by one particle, the spins add to zero, but the magnetic moments do not! This is because of the relativistic relationship between the magnetic moment μ and the spin s:

$$\mu = \frac{e_0}{2E} \gamma s, \tag{48}$$

---

[13] The counterpart of $\varepsilon(q)$ in coordinate space is its Fourier transform $\tilde{\varepsilon}(r)$, which relates the displacement vector to the electric field in a nonlocal fashion: $\mathbf{D}(\mathbf{r}) = \int \tilde{\varepsilon}(\mathbf{r} - \mathbf{r}') \mathbf{E}(\mathbf{r}') d^3r'$. There is no local relationship between $W(r)$, the Coulomb potential, and $\tilde{\varepsilon}(r)$, however. Equation (47) is a crude approximation to the actual nonlocal relationship.

where E is the energy of the particle. This is a generalization of the nonrelativistic relation $\mu = e_0 \gamma s / 2m$. When a magnetic field B is present, the energy E is changed by $\mp \mu B$, depending on the directions of the spin relative to B. Thus, the total magnetic moment of a pair of particles of the same momentum $\mathbf{p}$ with opposite spin does not vanish: the energy of the spin-up particle is reduced and its magnetic moment increased; the opposite is true for the other particle. Since the magnetic moments point in opposite directions, the resultant magnetic moment $\mu(\mathbf{p})$ is

$$\mu(\mathbf{p}) = \frac{e_0 \gamma s}{2(E - \mu B)} - \frac{e_0 \gamma s}{2(E + \mu B)} \simeq e_0 \gamma s \frac{\mu B}{E^2} = \frac{e_0^2 \gamma^2 s^2}{2E^3} B, \tag{49}$$

and it points in the direction of the field, neglecting terms of order B^2. In short, this is a paramagnetic contribution.

The magnetization $\mathcal{M}^{\text{sp}}$ per unit volume due to spins is obtained by integrating (49) over all momenta (note that $E = |\mathbf{p}|$):

$$\mathcal{M}^{\text{sp}} = \frac{1}{2} \cdot 2 \int \frac{d^3 p}{8\pi^3} \mu(E) = \frac{e_0^2 \gamma^2 s^2}{4\pi^2} B \ln \frac{K}{\sqrt{e_0 B}}. \tag{50}$$

As zero-point fluctuations correspond to 50% occupancy, (50) has a factor of $\frac{1}{2}$, which is compensated by a factor 2 for particles and antiparticles. K is a cutoff momentum, and $\sqrt{e_0 B}$ is the lower limit of integration because the approximation (49) ceases to be valid when $E < \mu B$, or $E^2 < e_0 \gamma s B / 2 = e_0 B$, with $\gamma = 2$, $s = 1$. The magnetic susceptibility χ_m^{sp} corresponding to $\mathcal{M}^{\text{sp}}$ is

$$\chi_m^{\text{sp}} = \mathcal{M}^{\text{sp}} / B = \frac{e_0^2}{\pi^2} \ln \frac{K}{\sqrt{e_0 B}}. \tag{51}$$

The diamagnetic effect is harder to determine. A detailed calculation shows (see Appendix VI) that the induced diamagnetic moment (opposed to the field) is smaller by the ratio $(3\gamma^2 s^2)^{-1}$ than the above-described paramagnetic effect. For example, it is $\frac{1}{12}$ for vector particles ($s = 1$, $\gamma = 2$) and $\frac{1}{3}$ for spinor particles ($s = \frac{1}{2}$, $\gamma = 2$). This ratio also holds for a system of nonrelativistic free particles; for example, the ratio between diamagnetism and paramagnetism in a free electron gas is $1:3$.[14]

Our final result for the magnetic susceptibility of the vacuum state of a

---

[14] One could wrongly infer from this that the vacuum of quantum electrodynamics is antiscreening. The QED vacuum can be thought of as consisting of electrons in negative energy states, from which one might conclude that the diaelectric effects should be three times larger than the paraelectric ones. But the vacuum electrons are in *negative* energy states, which turns the situation around, as Eq. (49) shows. The induced momentum is proportional to E^3, which changes sign for negative energies. Indeed, this argument can be used to show that the QED vacuum screens (i.e., $\chi_e > 0$).

4. ASYMPTOTIC FREEDOM

charged spin-1 field, including the diamagnetic effect, is

$$\chi_m = \chi_m^{sp}\left(1 - \frac{1}{12}\right) = \frac{11}{12}\frac{e_0^2}{\pi^2}\ln\frac{K}{\sqrt{e_0 B}}. \tag{52}$$

This is positive, and therefore paramagnetic.

We now make use of relation (45) to show that the vacuum is "diaelectric", i.e., antiscreening. The paraelectric part is only $\frac{1}{12}$ of the diaelectric. This calculation makes it plausible that the antiscreening of the gluon vacuum is 12 times stronger than the screening effect. The paraelectric part, corresponding to the diamagnetic orbital effect, can be considered as being due to the vacuum polarization of the gluon field.

How do we interpret the logarithmic dependence of χ_m on the field B? It expresses the dependence of the susceptibility on the momentum transfer $q \sim r^{-1}$ of the processes considered. This is made plausible as follows. The field B transmits a momentum $q \sim e_0 Br$ to a particle moving with $v \sim c$ over a distance $r \sim 1/q$; consequently, $q \sim \sqrt{e_0 B}$. This dielectric "coefficient" ε_v of a spin 1 field vacuum then follows from (43), (44), and (52):

$$\frac{1}{\varepsilon_v(q)} = 1 + \chi_m = 1 + \frac{11}{12}\frac{e_0^2}{2\pi^2}\ln\left(\frac{K^2}{q^2}\right). \tag{53}$$

We now define a *running coupling constant*, $\alpha(q^2)$, by considering two test charges of charge $Q = e_0$, the charge carried by the field quanta. This is done by rewriting (46) as

$$W(r) = \int e^{i\mathbf{q}\cdot\mathbf{r}}\frac{\alpha(q^2)}{q^2}\frac{d^3q}{2\pi^2}. \tag{54}$$

Thus $\alpha(q^2) = e_0^2[4\pi\varepsilon_v(q)]^{-1}$, or

$$\alpha(q^2) = \alpha_0\left[1 + \frac{22}{12\pi}\alpha_0\ln\left(\frac{K^2}{q^2}\right)\right], \tag{55}$$

where $\alpha_0 = e_0^2/4\pi$, i.e., the value of $\alpha(q^2)$ at $q = K$.

This is the running coupling constant caused by the vacuum polarization of a field of electrically charged bosons of spin 1. The positive sign in (51) and (52) indicates that a charge e_0 is immersed in an antiscreening medium: the effective coupling constant increases with decreasing q. If we use the loose terminology of an r-dependent ε, as indicated in (47), we may say that $\varepsilon(r^{-1})$ decreases and the coupling increases with increasing r. This is equivalent to an induced charge distribution surrounding e_0 of the same sign as e_0.

As we report in the next subsection, the exact QCD calculation gives results very similar to (55). Therefore, we can use the charged vector boson model to describe the effect of the QCD vacuum. The situation can be described as follows: At a momentum transfer $K \sim 1/r^*$, $\alpha = \alpha_0$, and we measure the "bare charge." In QED the bare charge is larger than the charge measured at distances $r > r^*$. It is surrounded by a cloud of *opposite* charge. In QCD the bare color charge g_0 is surrounded by a "charge" cloud of the same sign. Thus, as the distance decreases, the effective charge decreases (as long as $r > r^*$), since more of the "charge" cloud remains outside. This explains the origin of asymptotic freedom.

There is another essential difference between QCD and QED. The cloud of opposite charge in QED lies within a distance $r \sim 1/m_e$. This is why we measure a definite charge as $r \to \infty$. But in QCD the gluons are massless, so there is no outer limit to the cloud. However, our perturbative calculation is no longer valid for distances beyond r_0 at which the effective coupling constant approaches unity.[15] The problem of large distances has not yet been solved completely, but there are compelling indications that the effective coupling constant increases without limit as r increases. This leads to the confinement phenomenon, which is discussed in §5.

(d) The results of an exact perturbation calculation

The interaction of two fixed quarks at sufficiently small separation can be evaluated exactly by perturbative QCD. Such calculations surpass the level of this book, and we therefore restrict ourselves to an account of the results. They bear out the conclusions of the preceding Subsections (b) and (c).

There is a fundamental difficulty with the application of the field concept to QCD. In QED one can, at least in principle, probe the charge distribution surrounding an external charge by using a test charge that is arbitrarily small compared to the external charge, so that it does not disturb the system of interest. We have tacitly thought about the induced color density surrounding a quark in the same way. A mythical observer can "see" a color charge only if he or she has a test charge. But in QCD one cannot have a test charge arbitrarily smaller than the color charge of a quark! As we learned in §B.3, the color charges of all entities are simply related to g, and a bit of group theory shows that the smallest total color charge is carried by the smallest $SU(3)$ multiplets, **3** and **3***, to which Q and $\bar{Q}$ themselves belong. In short, the least upsetting test charge we could use is as big as the one we are probing, and this raises the question of whether the charge induced in the field by one quark has any well-defined physical meaning. We avoid this difficulty by considering instead the energy W of a massive $Q\bar{Q}$ pair in a state of total color zero. This is a gauge invariant quantity, and therefore has an unambiguous physical significance.

The energy W is also important in its own right; indeed, we defined asymptotic freedom by the behavior of $W(r)$ as the $Q\bar{Q}$ separation r tends

---

[15] Note that the effective coupling varies only logarithmically with distance, both in QED and QCD, as long as perturbation theory is valid. This is the case for $r < r_0$ in QCD, and practically at all distances in QED (see §II.D.3).

4. ASYMPTOTIC FREEDOM

to zero. In a more practical context, the short distance $Q\bar{Q}$ interaction plays a central role in the theory of deep inelastic scattering and the spectroscopy of very massive quarks.

The Hamiltonian of a set of massive (stationary) quarks in interaction with the color field is of the same form as in electrodynamics [recall Eq. (26)]:

$$H = \frac{1}{2}\int (\vec{E}_{Ti} \cdot \vec{E}_{Ti} + \vec{B}_i \cdot \vec{B}_i)\, d^3r + \frac{g^2}{2}\int \frac{\vec{R}(\mathbf{r}) \cdot \vec{R}(\mathbf{r}')}{4\pi |\mathbf{r} - \mathbf{r}'|}\, d^3r\, d^3r'. \quad (56)$$

Here $\vec{R}$ is the *total* color density. For a $Q\bar{Q}$ pair placed at $\mathbf{r}_1$ and $\mathbf{r}_2$, respectively,

$$\vec{R}(\mathbf{r}) = \tfrac{1}{2}\vec{\lambda}_1\, \delta^3(\mathbf{r} - \mathbf{r}_1) + \tfrac{1}{2}\vec{\lambda}_2\, \delta^3(\mathbf{r} - \mathbf{r}_2) + \vec{\rho}_T + \vec{\rho}_{\text{ind}}, \quad (57)$$

where $\vec{\rho}_T = \vec{V}_i \times \vec{E}_{Ti}$ is the color density of free gluons [cf. Eq. (30) et seq.], and $\vec{\rho}_{\text{ind}}$ is the induced field density just analyzed [cf. Eq. (33) et seq.]. The matrix $\vec{\lambda}_1$ is just our standard $\vec{\lambda}$, and acts on the color variable of Q, while $\vec{\lambda}_2$ is the matrix $-\vec{\lambda}^*$, which acts on those of $\bar{Q}$. As in QED, the first integral in (56) contains the energy of free gluons, but as we learned in §2, it also describes interactions between gluons. The nonlinear character of the gauge field also shows up in the second term of (56), because even when there are no quarks, $\vec{R}$ is nonzero.

By substituting (57) into (56), one sees that the energy is composed of a variety of interactions.[16] First there is the color analogue of the ordinary Coulomb interaction, which is depicted in Fig. 6(a):

$$H_c = \frac{1}{4}\vec{\lambda}_1 \cdot \vec{\lambda}_2\, \frac{\alpha_{s0}}{|\mathbf{r}_1 - \mathbf{r}_2|}, \quad (58)$$

where $\alpha_{s0} \equiv g_0^2/4\pi$. The color dependence of this interaction is given by $\vec{\lambda}_1 \cdot \vec{\lambda}_2$, the analogue of the spin–spin and isospin–isospin interaction familiar from atomic and nuclear physics [cf. Eq. I.B(48)]. As we showed in Eq. A(58), for a $Q\bar{Q}$ system $\tfrac{1}{4}\vec{\lambda}_1 \cdot \vec{\lambda}_2$ is $-\tfrac{4}{3}$ or $\tfrac{1}{6}$, depending on whether its state is a color singlet or octet, respectively.

Next we turn to the interaction of one point source with the induced color charge surrounding the other source. Here the point source acts like a test charge. From Eqs. (56) and (57) we see that this piece of H is

$$H_{\text{ind}}^{(1)} = \alpha_{s0}\int \left(\frac{\tfrac{1}{2}\vec{\lambda}_1 \cdot \vec{\rho}_{\text{ind},2}(\mathbf{r})}{|\mathbf{r} - \mathbf{r}_1|} + \frac{\tfrac{1}{2}\vec{\lambda}_2 \cdot \vec{\rho}_{\text{ind},1}(\mathbf{r})}{|\mathbf{r} - \mathbf{r}_2|} \right) d^3r, \quad (59)$$

where $\vec{\rho}_{\text{ind},i}$ is the induced color density associated with the ith quark. If we

---

[16] We only consider those terms that depend on the $Q\bar{Q}$ separation. There are also terms that have no relation to where both quarks are, as well as terms that only depend on one quark at a time. These give the self-energy of gluons and quarks, and are removed by renormalization.

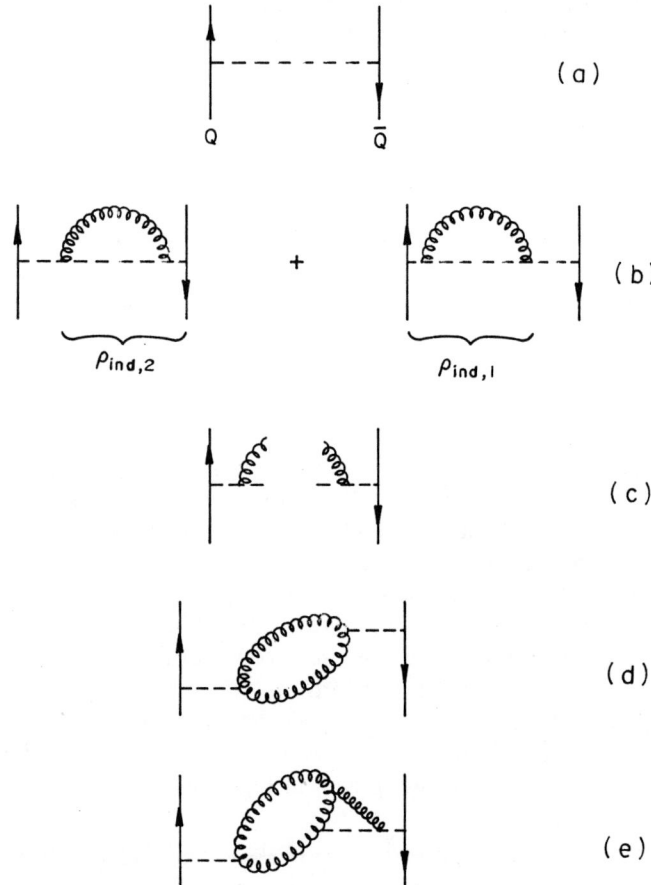

FIG. 6. Contributions to the $Q\bar{Q}$ interaction in perturbation theory. Here solid vertical lines represent stationary quarks, dashed horizontal lines the instantaneous Coulomb interaction $1/4\pi\,|\mathbf{r}_1 - \mathbf{r}_2|$, and coiled lines transverse gluons. Each vertex carries a factor of g.

(a) The elementary instantaneous interaction of Eq. (58).

(b) The interaction between a point quark (or antiquark) and the gauge field color density induced by its partner, as given by Eq. (59). In the first diagram Q interacts with $\vec{\rho}_{\text{ind}}$ due to $\bar{Q}$, the latter being represented by the portion of the diagram indicated as $\vec{\rho}_{\text{ind},2}$. In the second diagram the roles of Q and $\bar{Q}$ are interchanged.

(c) The interaction between the induced color densities of Q and $\bar{Q}$. The left- and right-hand halves of the diagram represent the operators $\vec{\rho}_{\text{ind},1}$ and $\vec{\rho}_{\text{ind},2}$ appearing in Eq. (60). Each of these operators is linear in the gauge field $\vec{V}$ [cf. Eq. (38)]; one creates a gluon, the other destroys it.

(d) The contribution of gluons to vacuum polarization. A similar diagram with a quark loop, one per flavor, also contributes to $W(r)$ in the same order (g^4).

(e) An order g^6 contribution to $W(r)$ representing the interaction between a vacuum fluctuation of a gluon pair produced by Q, and the induced charge due to $\bar{Q}$. Note that the triple-gluon vertex of Fig. I.20(b) appears here.

For the *cognoscenti* we remark that the gluons in (d) and (e) actually propagate, as described by Feynman propagators, whereas those in (b) and (c) do not propagate because their creation and destruction occurs at the same instant.

recall the expansion (35) and the remarks following it, we see that the leading contribution to the vacuum expectation value $\langle \Omega | H_{\text{ind}}^{(1)} | \Omega \rangle$ comes from the term (41) of second order in the gauge field. This contribution of $H_{\text{ind}}^{(1)}$ is depicted graphically in Fig. 6(b). The qualitative behavior of the contribution of $\langle \Omega | H_{\text{ind}}^{(1)} | \Omega \rangle$ to the interaction energy can be surmised from our earlier considerations.

There is also a contribution to the energy when the induced charge $\vec{\rho}_{\text{ind},1}$ associated with quark "one" interacts with that belonging to quark "two", $\vec{\rho}_{\text{ind},2}$:

$$H_{\text{ind}}^{(2)} = \alpha_{s0} \int \frac{\vec{\rho}_{\text{ind},1}(\mathbf{r}) \cdot \vec{\rho}_{\text{ind},2}(\mathbf{r}')}{|\mathbf{r} - \mathbf{r}'|} d^3r \, d^3r'. \tag{60}$$

This term actually gives a contribution to the energy that is exactly half as big as (59). Why these two contributions are so simply related can be seen from Fig. 6(b) and (c).

Finally,[17] we come to the interaction between the primitive point densities of the quarks and the color density ρ_T of the transverse gauge (i.e., of "real, physical" gluons):

$$H_T = \alpha_{s0} \int \left(\frac{\frac{1}{2}\vec{\lambda}_1}{|\mathbf{r}_1 - \mathbf{r}|} + \frac{\frac{1}{2}\vec{\lambda}_2}{|\mathbf{r}_2 - \mathbf{r}|} \right) \cdot \vec{\rho}_T(\mathbf{r}) \, d^3r. \tag{61}$$

When acting on the gauge field vacuum $|\Omega\rangle$, $\vec{\rho}_T$ can only do one thing: create a gluon pair. Consequently, $\langle \Omega | H_T | \Omega \rangle = 0$. But H_T does contribute in second-order perturbation theory, for then a virtual gluon pair created by one quark's Coulomb field can propagate before annihilating into the field of the other, as shown in Fig. 6(d). This is the vacuum polarization due to gluons, and as already stated, it produces a strengthening of the interaction $W(r)$ as $r \to 0$, which has the same form as in QED (cf. §II.D.3).

A detailed evaluation of all these contributions to the energy yields the following expression

$$W(r) = \alpha_{s0} \pi \vec{\lambda}_1 \cdot \vec{\lambda}_2 \int \frac{d^3q}{(2\pi)^3} \frac{e^{i\mathbf{q}\cdot\mathbf{r}}}{q^2} \left[1 + b\alpha_{s0} \ln \frac{K^2}{q^2} \right]; \tag{62}$$

K is a very large momentum cutoff that must be introduced to render the integral finite, as was done in Eq. (50). Just as in QED, the radiative corrections are not finite, but renormalization allows one to express everything in terms of finite and physically well-defined quantities, as we see in the next subsection.

The constant b is given by

$$b = \frac{1}{12\pi}(33 - 2N_f), \tag{63}$$

---

[17] By "finally" we actually mean that there is nothing else to order g^4. A more precise perturbative calculation of the energy would include other types of effects. For example, one quark can cause a vacuum fluctuation of gluon pairs (via H_T), which then interacts with $\vec{\rho}_{\text{ind}}$ due to the other quark; this effect first enters in order g^6, as shown in Fig. 6(e).

where N_f is the number of quark flavors satisfying $m_q r \ll 1$. The numbers in (63) have the following origin: 24 and 12 come from the two Yang–Mills terms involving $\tilde{\rho}_{\text{ind}}$, as given by Eqs. (59) and (60) [or diagram 6(b) and (c)], respectively; gluon vacuum polarization [Fig. 6(d)] has the coefficient -3; fermion vacuum polarization (not shown in Fig. 6) has coefficient $-2N_f$. Expression (62) is valid as long as an expansion in powers of α_{s0} is applicable, that is, as long as the addition to unity in the square bracket is small.

When the bracket [ ] in Eq. (62) is approximated by 1, this is just the Fourier representation of the Coulomb potential. The term of order α_{s0}^2 leads to a term in the potential of the form

$$-\alpha_{s0}^2 b \frac{\ln Kr}{r}, \tag{64}$$

which is less singular than the Coulomb potential.

The sign of b is decisive for the behavior of the potential at small distances. It is positive if the number of flavors is less than 17. We already know that there are five flavors, and a sixth is likely to exist. If b is positive, the interaction is asymptotically free, that is, the radiative correction *reduces* the interaction. This can be seen most clearly by recalling that the scattering amplitude, in Born approximation, is given by the Fourier transform of the potential. Hence the $Q\bar{Q}$ scattering amplitude for momentum transfer $\mathbf{q}$ is

$$A(q^2) = \frac{1}{4}\vec{\lambda}_1 \cdot \vec{\lambda}_2 \frac{\alpha_s(q^2)}{q^2},$$

$$\alpha_s(q^2) \equiv \alpha_{s0}\left[1 + \alpha_{s0} b \ln \frac{K^2}{q^2}\right]. \tag{65}$$

The coefficient $\alpha_s(q^2)$ provides a measure of the interaction strength relative to that of a pure Coulomb field; it is a *decreasing* function of q^2 as long as b is positive. Hence, the QCD scattering amplitude $A(q^2)$, as given by (65), is characteristic of an interaction that is weaker than Coulombic at large momentum transfers, or equivalently, at short distances.

It is interesting to compare the expression (62) with the effective coupling constant (55) calculated with the electromagnetic analogue to the gluonic vacuum. $\alpha(q^2)$ should be compared with the coefficient of the Coulombic interaction in (62). Expression (55) would give a value $b = 22/12\pi$, which is not very far from $33/12\pi$, the part due to the gluonic field alone. The negative contribution $-2N_f/(12\pi)$ to (63) comes from the quark vacuum polarization, and was not included in our model of a field of spin 1 bosons.

(e) Renormalization of the coupling constant. Asymptotic freedom

The preceding discussion still leaves the following question unanswered: Can the cutoff K be eliminated by defining a coupling constant in terms of some physically well-defined quantity, as we did in QED?

4. ASYMPTOTIC FREEDOM

In QED our definition of the coupling constant was also based on the interaction $\phi(r)$ between two point sources (cf. §II.D.3). Let us quickly summarize the argument. For simplicity, we introduce a cutoff distance $r^* \ll m^{-1}$ at which these sources have a "bare" charge e_0. This charge equals the charge that couples electrons to the electromagnetic field (i.e., $H_{\text{int}} = e_0 \int \mathbf{j} \cdot \mathbf{A}\, d^3r$). We computed $\phi(r)$ as a perturbation series in powers of e_0. At large distances $(r \gg m^{-1})$, $\phi(r)$ becomes Coulombic, but the coefficient $e^2/4\pi$ of $1/r$ was found to be much smaller than $e_0^2/4\pi$. The quantity e is the observed (or renormalized) charge. The potential $\phi(r)$, and its deviations from the Coulombic form $e^2/4\pi r$ for distance $r^* < r < m^{-1}$, can be expressed as in (46):

$$\phi(r) = \int \frac{d^3q}{(2\pi)^3} e^{i\mathbf{q}\cdot\mathbf{r}} \frac{e^2(q)}{q^2}. \tag{66}$$

The elementary Coulomb interaction corresponds to $e^2(q) \equiv e^2$, while the function $e(q)$ incorporates vacuum polarization. The conventional charge, previously defined by $\phi(r)$ as $r \to \infty$, is $e(q = 0)$. But *any* $q \neq 0$ will define a basic coupling constant equally well, because the difference $e(q) - e(0)$ is finite. Hence, one can switch from one definition, $e(q_1)$, to another, $e(q_2)$, at will. If q is of order m_e, $e(q) - e(0)$ is minute, but this difference grows significant once $\log(q/m_e)$ is appreciable.[18] This increase in strength is what we have learned to expect from the screening due to vacuum polarization (cf. §II.D.3), because $e(q)$ is the charge inside a sphere around the electron having a radius $\sim 1/q$.

Unfortunately, this lesson from QED cannot be applied directly to QCD: the increase of the effective charge (color confinement) prevents the separation of quarks to large distances, and so there is no QCD analogue to the Coulomb definition of the renormalized charge. By the same token, there is no other large-distance or low-energy phenomenon[19] involving quarks and gluons that can be used for this purpose. What we need is a definition of the coupling constant that does not involve large distances. In principle, any short-distance phenomenon for which we know the perturbation expansion, *and also* have the experimental data, will do, though some phenomena are more convenient for this purpose than others.

We therefore define the renormalized coupling constant of QCD by the $Q\bar{Q}$ interaction at a distance r that is short enough so that the effective charge is less than unity, and the considerations of the last section are applicable. Let r_0 be the distance at which the effective charge is unity; r_0 is

---

[18] To illustrate this, consider $q \simeq 100$ GeV, which is of order the mass of the Z^0- and W-bosons. In collisions having such momentum transfers (e.g., $e^+e^- \to \mu^+\mu^-$), the weak interaction is comparable to the electromagnetic. For such values of q, $e^2(q)/4\pi$ has the value $\frac{1}{129}$, an ~6% increase from $\frac{1}{137}$.

[19] In QED, many low-energy phenomena can be used to give an equivalent definition of the renormalized charge; for example, low-energy photon–electron scattering (Thomson scattering).

approximately 10^{-13} cm. Then any $r < r_0$ will do. There is a simple relation between coupling constants defined at any two distances inside r_0. These are small distances compared to typical hadronic dimensions, for as we know from hadronic spectroscopy and deep inelastic scattering, that is where hadrons behave as if they were made of weakly coupled quarks and gluons.

A precise formulation of these notions begins with the $Q\bar{Q}$ interaction $W(r)$, as given by Eq. (62), which we write in the same form as (66):

$$W(r) = \pi \vec{\lambda}_1 \cdot \vec{\lambda}_2 \int \frac{d^3q}{(2\pi)^3} e^{i\mathbf{q}\cdot\mathbf{r}} \frac{\alpha_s(q^2)}{q^2}, \tag{67}$$

where

$$\alpha_s(q^2) = \alpha_{s0}\left[1 + b\alpha_{s0} \ln \frac{K^2}{q^2}\right]. \tag{68}$$

Our goal is to eliminate the cutoff K, and the input coupling constant $\alpha_{s0} = g_0^2/4\pi$, where g_0 is the analogue of e_0, in favor of the coupling constant defined at some momentum Λ, such that the length Λ^{-1} falls inside our short-distance regime. For this purpose we rewrite (68) as

$$\frac{1}{\alpha_s(q^2)} = \frac{1}{\alpha_{s0}}\left[1 - b\alpha_{s0} \ln \frac{K^2}{q^2}\right],$$

because our perturbation expansion assumes that the bracket in (68) is near unity. Then

$$\frac{1}{\alpha_s(q^2)} - \frac{1}{\alpha_s(\Lambda^2)} = b \ln\left(\frac{q^2}{\Lambda^2}\right), \tag{69}$$

or equivalently,[20]

$$\alpha_s(q^2) = \frac{\alpha_s(\Lambda^2)}{1 + b\alpha_s(\Lambda^2) \ln (q^2/\Lambda^2)}, \tag{70}$$

which no longer involves the cutoff, or the coupling constant α_{s0} that appears in the Hamiltonian. Equation (70) is a finite expression for the running coupling constant of QCD in terms of its value at some calibration distance Λ^{-1}. Since $b > 0$, $\alpha_s(q^2) < \alpha_s(\Lambda^2)$, if $q^2 > \Lambda^2$; this is asymptotic freedom. *Because of this property, one can always find a regime of short enough distances where perturbation theory in QCD becomes as accurate as one pleases!*

A convenient parametrization of the running coupling constant can be

---

[20] Straightforward lowest order perturbation theory does not distinguish (70) from $\alpha_s(\Lambda^2)[1 - b\alpha_s(\Lambda^2) \ln (q^2/\Lambda^2)]$. However, the theory of the renormalization group shows that $\alpha_s(q^2)$ has the form given by (70).

obtained if one treats Eq. (70) as if it were valid for values of q^2, where $\alpha_s(q^2)$ is not small. That being done, one can define a calibration momentum Λ_0 such that $\alpha(\Lambda_0^2) \equiv 1$. The distance Λ_0^{-1} corresponds roughly to the distance r_0 introduced previously (see p. 395). In terms of Λ_0, Eq. (70) reads

$$\alpha_s(q^2) = \frac{1}{1 + b \ln(q^2/\Lambda_0^2)} ; \qquad (71)$$

this equation is only valid if $q^2 \gg \Lambda_0^2$.

5. Confinement

(a) The QCD vacuum

What can be said about large distances or small momentum transfers? We have introduced a length r_0 (or a momentum transfer Λ_0) above which (or below which) the running coupling constant becomes large because of antiscreening. At present, there is no analytic technique for solving the field equations in this region. Unfortunately, this is a very important regime; it deals with the forces felt by quarks within hadrons, since r_0 is of the order of hadronic dimensions. The structure of hadrons, their spectra and interactions, all fall within this regime. In the last analysis, the nuclear force is also a consequence of QCD effects at large distances.

At this time only analytic arguments of a qualitative nature are available.[21] On the other hand, extensive numerical calculations confirm the general features of these qualitative considerations.

We give a qualitative description of some consequences of QCD at large distances, which may or may not be substantiated by future calculations.[22] QCD reliably predicts antiscreening as long as the effective coupling constant is less than unity, i.e., for distances small compared to r_0, or momentum transfers much higher than Λ_0. In this regime the effective charge increases with increasing distance from the source. We now make the plausible guess that *the effective coupling constant α_s rises rapidly towards infinity with increasing $r(=q^{-1})$ when $r > r_0$*. Using the notation of Eq. (47), we express this by a dielectric "coefficient" $\varepsilon'(r^{-1})$. Here we normalize this coefficient so that[23] $\varepsilon'(r^{-1}) = 1$ when $r = r_0$:

$$\alpha_s(r^{-1}) = \alpha_s(r_0^{-1})/\varepsilon'(r^{-1}) \approx 1/\varepsilon'(r^{-1}). \qquad (72)$$

Our guess regarding the behavior for $r > r_0$ is equivalent to a function $\varepsilon'(r^{-1})$ that decreases strongly with increasing r and approaches zero for $r \gg r_0$ (see Fig. 7).

---

[21] These considerations are due to Hansson et al. (1982).
[22] While the conclusions we shall draw are in conformity with detailed numerical calculations, what is not clear is whether the intuitive picture we use is correct in detail.
[23] We use the symbol ε' in order to distinguish it from ε_v used in (53) which is unity for $r^{-1} \simeq K$.

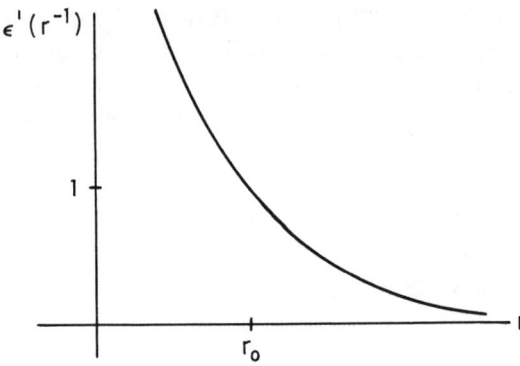

FIG. 7. A sketch of the dielectric "coefficient" defined by the rough relationship $\varepsilon'(r^{-1}) \simeq 1/\alpha_s(r^{-1})$; see Eq. (72).

We now discuss the consequences of this conjecture by using the electromagnetic analogy to QCD. The electric field energy density u is given by

$$u = \tfrac{1}{2}\mathbf{E} \cdot \mathbf{D}, \qquad \mathbf{D} = \varepsilon'\mathbf{E}. \tag{73a}$$

According to the Maxwell equation div $\mathbf{D} = \rho$; $\mathbf{D}$, not $\mathbf{E}$, is determined by the charge density ρ. We write (73a) in the form

$$u = \frac{1}{2}\frac{D^2}{\varepsilon'}. \tag{73b}$$

If we place a charge at the origin, we see that the field energy density would become infinite in the region outside r_0. Thus, the electric field would not be able to penetrate beyond r_0, and would be confined. This is equivalent to the statement that an isolated charge cannot exist, since Gauss's Law requires the field to extend to infinity. However, any assembly of charges that has no net charge can exist, as long as the constituents are contained within a volume of linear dimensions r_0. Such an uncharged assembly does not require a field extending to infinity.

In QCD only groups of three quarks, and quark–antiquark pairs, can have a total color charge zero. Hence, only such groups, or multiples thereof, can exist. Furthermore, they will be confined in a volume of linear dimensions r_0. Hence, our qualitative picture of QCD for $r > r_0$ gives an explanation of the most important features of quark systems. (In addition, there are also states of zero color that contain "constituent" gluons; see below.)

⟦Let us now discuss the consequences of our picture for the properties of the vacuum itself. Here we again concentrate on the effects of the gluon field, as they are more important than those of the quark field.

5. CONFINEMENT

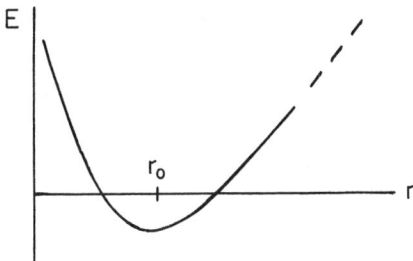

FIG. 8. The energy of a pair of gluons in a color singlet state of zero spin as a function of their separation. This curve is merely an educated guess.

We begin by considering the "empty" vacuum, that is, one where all gluon occupation numbers vanish. Naturally, there are still the zero-point fluctuations. Now let us add to the "empty" vacuum a pair of gluons of opposite color "charge" (i.e., a color singlet), and opposite spin, with an average separation r. In this qualitative argument we neglect the difference in character (**8** vs. **3**) of the gluon vs. the quark charges.

The kinetic energy of such a massless pair is $A/4\pi r$, where A is a constant. The potential energy is $-C\alpha_s(r^{-1})/4\pi r$ as long as $r \lesssim r_0$, where C is a numerical coefficient. The value of C should be relatively large since the potential energy contains the color-electric attraction between gluons, and also the attraction between their color-magnetic moments; furthermore, the large multiplicity of gluons will amplify C. The energy is then

$$E(r) = \frac{A}{4\pi r} - \frac{C\alpha_s(r^{-1})}{4\pi r} \qquad (r \lesssim r_0). \tag{74}$$

For $r \ll r_0$, α_s is very small and $E(r)$ is positive. But in the neighborhood of r_0, $E(r)$ probably becomes negative! When r grows larger than r_0, expression (74) is no longer valid; we then expect $E(r)$ to increase rapidly as the gluons become more and more separated, since colored objects cannot exist in isolation. (A qualitative sketch of $E(r)$ is given in Fig. 8.) It therefore seems to pay to create a pair of gluons with opposite spin and color out of the "empty" vacuum, in a state having linear dimensions of order r_0, to take advantage of the negative energy when $r \sim r_0$.

Hence we are led to describe the "true" vacuum as follows. The "empty", or naive, vacuum is unstable. There is a state of lower energy that consists of cells, each containing a gluon pair in a color- and spin-singlet state. The size of these cells is of order r_0. We may speak of a "liquid" vacuum, a random distribution of such cells [see Fig. 9(a)].

This picture also lends itself to a crude description of how an assembly of quarks having a total color charge zero is immersed into the gluon vacuum. The gluonic cells will be displaced by the quarks, and therefore the quarks will find themselves in a bubble or "bag" of "empty" vacuum [see Fig. 9(b)]. This occurs because the field of the quarks, when added to that of the gluon pairs, yields an energy density that is not favorable to having gluon pairs and quarks in the same region.

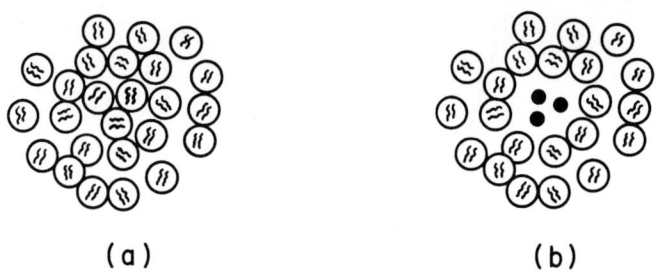

FIG. 9. The QCD vacuum state is depicted in (a). It is a random distribution of cells that contain a gluon pair in a color and spin singlet state. Quarks (in a color singlet configuration) displace these cells, creating a region (or "bag") of "empty" vacuum, as shown in (b).

We expect the bag size to be of the order r_0, since the fields of the quarks cannot reach beyond such distances. If the bag is to describe hadrons, r_0 must be of order hadronic dimensions, i.e., $r_0 \sim 1$ fm. The corresponding momentum Λ_0 is therefore ~ 200 MeV. The value of Λ_0 can also be determined by means of (71) from measurements of $\alpha_s(q^2)$ by analyzing processes at high momentum transfer. This, too, gives values of Λ_0 of order 100 MeV.]

We now consider the energy of the quark-confining "bag." Since the "true" vacuum has a lower energy density than the naive or empty vacuum, the bag has a positive energy E_B, relative to the true vacuum, which is proportional to the bag's volume:

$$E_B = BV. \tag{75}$$

Here B is a constant, having dimension (energy)^4, determined by the mechanism of "cell" formation in the "true" vacuum. It is not yet possible to calculate the value B from QCD.

Equation (75) is equivalent to the statement that the true vacuum exerts a pressure upon the bag, or bubble. This pressure is counteracted by the kinetic energy of the quarks. In §D we show that such a picture describes the observed size, structure, and low-lying spectra of hadrons reasonably well, if one chooses $B^{\frac{1}{4}} \approx 150$ MeV. The pressure B of the true vacuum on the bag therefore has the value $\sim 10^{23}$ atmospheres.

(b) The long-distance interaction

The force between quarks at large separation can also be inferred from this picture, as we already saw in §I.E.7(b). Let us compare the field of two opposite electric charges with the field of two opposite color charges at separation larger than r_0 (e.g., a quark and an antiquark, or a quark and a pair of quarks in a color 3* inside a baryon). In the electromagnetic case [Fig. I.21(a)], the field lines spread further and further when the separation grows. The number of field lines crossing a unit area decreases, which is why the force between the charges decreases like $1/r^2$. In QCD [Fig.

I.21(b)] the pressure of the "true" vacuum compresses the field lines into a tube of diameter $\sim r_0$. Hence, when $r \gg r_0$, the number of field lines per unit area within this tube remains constant, leading to a constant force, or to a linear potential

$$W(r) = ar, \qquad r \gg r_0, \tag{76}$$

where a is a constant. Its value can be extracted from an analysis of the $c\bar{c}$- and $b\bar{b}$-spectra, and is found to be ~ 0.14 GeV^2. A theoretical estimate of a is given in §D.4.

We therefore conclude that there is a constant force acting on a quark if one tries to remove it from a hadron to distances larger than r_0. One would need an infinite energy to remove the quark to infinite distances. This conforms with the model we used in the spectroscopy of heavy quark–antiquark mesons (§III.B).

(c) The semiempirical quark–antiquark interaction

In §4(d), Eq. (64), we showed that the quark–antiquark interaction $W(r)$ at short distances ($r \ll r_0$) is essentially Coulombic, with a singularity as $r \to 0$ that is softened by a logarithmic factor. As we have just seen, at large distances $W(r)$ rises linearly. In Fig. I.18 we showed a potential that interpolates between these asymptotes. This potential can be expressed as a one-parameter function of r that we now construct.

First, we write the Fourier transform $V(q)$ of the potential (67) for a quark–antiquark color singlet by using Eq. A(58):

$$V(q) = -\frac{16\pi}{3} \frac{\alpha_s(q^2)}{q^2}. \tag{77}$$

According to (71), at very short distances ($q^2 \to \infty$), $\alpha_s(q^2)$ tends to $[b \ln (q^2/\Lambda_0^2)]^{-1}$, where b is given by (63). The number N_f is the number of quark flavors that contribute to vacuum polarization in the energy range of interest. For the $c\bar{c}$ and $b\bar{b}$ families, the u, d, and s quarks satisfy this requirement; we shall ignore the small effect of $c\bar{c}$ vacuum polarization in the $b\bar{b}$ system. Hence, $N_f = 3$ and $b = 27/12\pi$, so that

$$\alpha_s(q^2) \xrightarrow[q^2 \to \infty]{} \frac{4\pi}{9 \ln (q^2/\Lambda_0^2)}. \tag{78}$$

It is our intention to introduce a functional form for $\alpha_s(q^2)$, so that (77) describes the interaction for long as well as short distances (i.e., for small as well as large q^2). This requires $V(q)$ to become the Fourier transform of the linear confinement potential (76) as $q^2 \to 0$. Since the Fourier transform of the linear potential (76) is $-8\pi a/q^4$, this leads us to set

$$\alpha_s(q^2) \xrightarrow[q^2 \to 0]{} \frac{3a}{2q^2}. \tag{79}$$

A simple semiempirical *Ansatz* that has the desired functional form at the two asymptotes (78) and (79) is

$$\alpha_s(q^2) = \frac{4\pi}{9 \ln[1 + (q^2/\Lambda_1^2)]}, \tag{80}$$

where Λ_1 is a parameter to be determined by spectroscopic data. For $q^2 \to \infty$, this is designed to give (78); as $q^2 \to 0$, (80) becomes $4\pi\Lambda_1^2/9q^2$, so that

$$\frac{\Lambda_1^2}{a} = \frac{27}{8\pi}. \tag{81}$$

Combining (80) with (77) gives the so-called Richardson potential

$$V(q^2) = -\frac{64\pi^2}{27} \frac{1}{q^2 \ln[1 + (q^2/\Lambda_1^2)]}. \tag{82}$$

The Fourier transform of (82) is shown in Fig. I.18 when $\Lambda_1 \simeq 400$ MeV, as determined from the $c\bar{c}$ spectrum.

This parameter Λ_1 can be compared with two other parameters of hadronic physics: (1) the momentum Λ_0, defined by Eq. (71), which is measured in deep inelastic phenomena; and (2) the slope of the Regge trajectories m_J^2/J, as shown in Figs. I.13 and 17. As we mentioned, Λ_0 is of order 100 MeV, but is not well-determined at present. From the semiempirical argument that led to $V(q^2)$, we can only expect Λ_1 to be of the same magnitude as Λ_0, and this is indeed the case.[24] The Regge slope is related to the linear restoring force in §D.4, where it is shown that $m_J^2/J = 2\pi a$. With $\Lambda_1 = 400$ MeV, it follows from (81) that $2\pi a = 0.9$ GeV^2, which agrees to about 20% with the observed slopes shown in Figs. I.13 and 17. If the Richardson potential is taken at face value, the Regge slope and the c and b quark masses suffice to determine the $c\bar{c}$- and $b\bar{b}$-spectra.

It is probably fortuitous that the preceding sentence is quantitatively correct. Nevertheless, if QCD is the correct theory of the strong interaction, there should be just one parameter (any one of the lengths Λ_0^{-1}, r_0, or $a^{-\frac{1}{2}}$) that, together with the quark masses, determines all aspects of hadronic physics. The success of the potential (82) indicates that such an economical description of hadronic physics may be within our grasp.

Finally, we note that there is an evergrowing number of numerical calculations of the $Q\bar{Q}$ interaction. These do not rely on perturbation theory in any shape or form, and they are in very satisfactory agreement with

---

[24] One should not expect quantitative agreement between Λ_0 and Λ_1. At this time, the values of q^2 measured in deep inelastic phenomena are much too small to permit the approximation (78). Furthermore, the large q^2 (small distance) regime is only of secondary importance in the $c\bar{c}$- and $b\bar{b}$-spectra, as one sees from Fig. I.18.

heuristic and empirically successful potentials, such as the one of Eq. (82). These calculations are based on a discrete version of QCD, wherein the space-time continuum is replaced by a four-dimensional lattice, but strict gauge invariance is maintained. A large variety of problems in hadronic spectroscopy are now being examined within the framework of lattice QCD. There is good reason to hope that this investigation will provide a firm bridge between theory and experiment, so that the basic assumptions of the theory can be examined without an intervening swamp of questionable approximations.[25]

---

[25] For an introduction to lattice QCD, see Cheng (1984), §10.5.

D. THE BAG MODEL

1. The primitive bag model

The bag model provides a simple semiquantitative approach that relates quark confinement to the low-lying, stationary states of hadrons. It is based on the tentative conclusion, reached in the last section, that quarks are confined to a bubble of radius R of "empty" vacuum, upon which the "true" vacuum exerts a pressure B, as given by Eq. C(75) (see Fig. 10). The radius R is not fixed; it adjusts itself so that the outward pressure of the quarks is balanced by the inward pressure of the true vacuum.

In its primitive form, the bag model assumes that the quarks move within the sphere of radius R as free particles. The improved bag model considers the interactions between the quarks due to their color fields. These interactions will be treated by perturbation theory. Such a method can only be semiquantitative, since the distances between the quarks within hadrons are not very small compared to the length Λ_0^{-1}. Indeed, the relevant α_s will turn out to be about 2.

The nonrelativistic treatment of this problem is elementary. A state of a free quark of mass m confined to a sphere of radius R has a wave function with the boundary condition

$$\psi(R) = 0. \tag{1}$$

The lowest state of a free particle inside R is $r\psi = A \sin(\pi r/R)$, and has a total energy

$$\varepsilon = \frac{\pi^2}{2mR^2} + m, \tag{2}$$

where m is the mass of the quark. Let us apply this to the nucleon with three quarks of the same mass. The total energy, the proton mass M, would be three times (2) plus the volume energy Eq. C(75):

$$M = 3m + \eta, \qquad \eta = \frac{3\pi^2}{2mR^2} + \frac{4\pi}{3} BR^3. \tag{3}$$

The radius R of the bag must be adjusted to minimize M, which yields

$$R_{\min} = \left(\frac{3\pi}{4mB}\right)^{1/5}.$$

1. THE PRIMITIVE BAG MODEL

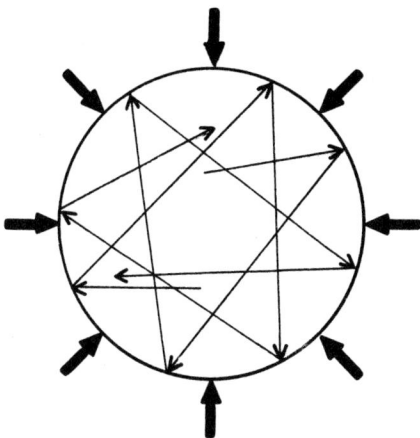

FIG. 10. Pictorial representation of the bag model. The pressure of the vacuum outside the bag is balanced by the counterpressure of the kinetic energy of the quarks inside.

The corresponding value of η is

$$\eta = \frac{5}{3} \cdot \frac{3\pi^2}{2mR_{\min}^2},$$

that is $\frac{5}{3}$ times the kinetic energy.

But this nonrelativistic treatment is not internally consistent, for it would require that the quark momenta, $\sim \pi/R$, be small compared to m, and that M be well approximated by $3m$. This gives the condition

$$R \gg \frac{3\pi}{M} \simeq 2 \times 10^{-13} \text{ cm}.$$

This is an unreasonably large radius. Furthermore, there are many indications that the masses of the u- and d-quarks are very much smaller than $M/3$.

That is why we use the opposite approach, and consider the u- and d-quark masses to be considerably smaller than all the energies in the problem, and approximate them by zero. The masses of s-quarks cannot be neglected, but their motion is still relativistic. Therefore, we must use the Dirac equations for the quarks in the bag.

What are the boundary conditions at $|\mathbf{r}| = R$? We cannot set ψ to zero there because the Dirac equations are first-order differential equations, and that would make ψ vanish everywhere. Since ψ must be zero for $r > R$, we conclude that at least some of the four components of ψ are not continuous at $r = R$.

We can determine the relativistic wave functions by considering the bag as equivalent to an infinitely deep potential well:

$$V(r) = 0 \quad \text{for} \quad r < R,$$
$$V(r) = V_0 \quad \text{for} \quad r > R, \quad (V_0 \to \infty). \tag{4}$$

$V(r)$ cannot be the time component of a 4-vector, like the electrostatic potential, since it must confine both particles and antiparticles. Thus, V must be a scalar potential that appears in the Dirac equation as an addition to the mass term. In other words, we consider the masses of the quarks to be small or zero for $r < R$, and to be so large for $r > R$ that the quarks cannot penetrate into that region.

The Dirac equations for a scalar potential can be solved easily, as shown in Appendix VII. As expected, $\psi \neq 0$ at $r = R$. The correct boundary condition is a simple relation between the components of ψ,

$$(\hat{\mathbf{r}} \cdot \boldsymbol{\gamma})\psi = -i\psi \quad \text{at} \quad |\mathbf{r}| = R, \tag{5}$$

where $\hat{\mathbf{r}}$ is the unit vector in the radial direction, and $\boldsymbol{\gamma}$ is the 3-vector of Dirac matrices that appears in the current density $\mathbf{j} = \bar{\psi}\boldsymbol{\gamma}\psi$ [cf. Eq. II.A(34)]. While ψ itself does not vanish at $r = R$, the current normal to the bag surface vanishes there, as expected, as does $\bar{\psi}\psi$.

The lowest energy eigenvalue ε for the scalar potential (4) is

$$\varepsilon = \left(m^2 + \frac{\beta^2(mR)}{R^2}\right)^{1/2}, \tag{6}$$

where m is the quark mass inside the bag, and $\beta(mR)$ is the function shown in Fig. 11. For $m = 0$, (6) takes the simple form

$$\varepsilon = \beta_0/R, \tag{7}$$

with $\beta_0 = 2.04$. For $mR \to \infty$, (6) becomes identical with the nonrelativistic energy (2). The form (7) is understandable since no other length but R appears in the $m = 0$ Dirac equation. Figure 12 shows the charge density and the azimuthal current density (perpendicular to $\hat{\mathbf{r}}$ and the spin direction) as functions of r. They are finite at $r = R$ and drop abruptly to zero there.

We apply the primitive bag model only to hadrons made of u and d quarks ($m = 0$). The mass M of a hadron consisting of n massless quarks, all in the lowest state, is then

$$M = \frac{n\beta_0}{R} + \frac{4\pi}{3}BR^3, \tag{8}$$

where the second term represents the volume energy. The bag adjusts itself so as to minimize M. Minimization of (8) with respect to R yields $R = (n\beta/4\pi B)^{1/4}$, while the virial theorem in this case states that the

1. THE PRIMITIVE BAG MODEL

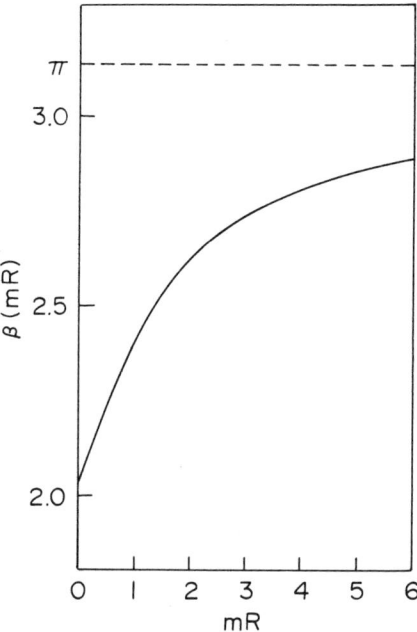

FIG. 11. The factor $\beta(mR)$, as defined in Eq. (6), for the lowest quark mode in a spherical cavity of radius R.

volume energy must be a quarter of the total. This gives the following relations between M, R, and B:

$$R = \frac{4}{3}\frac{n\beta_0}{M}, \tag{9}$$

$$M = \frac{16\pi}{3} BR^3 = \frac{4}{3}(4\pi)^{1/4}(n\beta_0)^{3/4} B^{1/4}. \tag{10}$$

Since B, the pressure of the vacuum, is a universal constant, we may determine it with the help of (10) from the mass M of the proton. Setting $n = 3$ gives

$$B^{1/4} = 0.102 M = 96 \text{ MeV}. \tag{11}$$

The radius of the proton bag is 1.71 fm from (9). The root-mean-square radius of the quark distribution can be calculated from the eigenfunction ψ_0 belonging to the eigenvalue (7):

$$\langle r^2 \rangle^{1/2} = 0.74 R. \tag{12}$$

This would give 1.3 fm for the proton, whereas the experimental value is only 0.8 fm. The refinement of the theory reduces the predicted value.

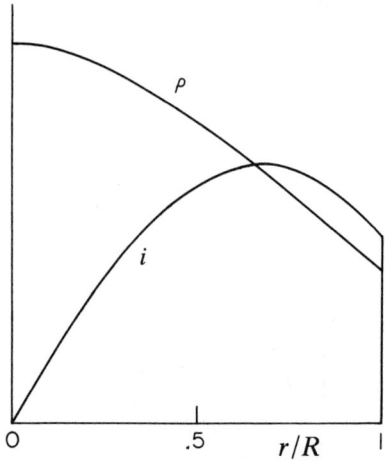

FIG. 12. The charge density ρ and the azimuthal current density i as a function of r in the lowest state of a massless quark in a bag of radius R. By "azimuthal" we mean the component perpendicular to the spin of the quark. Both quantities are finite at R, and then drop abruptly to zero.

Knowing ψ_0, one can calculate the magnetic moment. The moment μ_q of one quark with charge e_q is given by

$$\boldsymbol{\mu}_q = -\tfrac{1}{2}e_q \int (\bar{\psi}_0 \boldsymbol{\gamma} \psi_0) \times \mathbf{r}\, d^3x. \tag{13}$$

For dimensional reasons, (13) must be proportional to $e_q R$. The result is $\mu_q = 0.20 e_q R$, or using (9),

$$\mu_q = 3.3 e_q \frac{1}{2M}, \tag{14}$$

where M is the proton mass. The magnetic moments μ_p and μ_n of the nucleons can be determined from (14) by the technique described in §I.E.8(b), with the results

$$\mu_p = 3.3 \frac{e}{2M}, \qquad \mu_n = -2.2 \frac{e}{2M}. \tag{15}$$

These values are a little too high, but satisfactory in view of the crudity of the model.

Let us now apply the model to a meson. In the scalar potential (4), an antiquark has the same energy eigenvalue as a quark. Hence the meson mass M_m will be

$$M_m = 2\varepsilon(R) + \frac{4\pi}{3} BR^3. \tag{16}$$

2. THE IMPROVED BAG MODEL

For mesons made up of u and d quarks, we may use (8) with $n = 2$. Since B is fixed by (11), we find from (10) that the hadron masses are proportional to $n^{3/4}$, so that

$$M_m = (\tfrac{2}{3})^{3/4} M = 692 \text{ MeV}. \tag{17}$$

The primitive model cannot account for the dependence of hadron masses on the spin orientation of their constituent quarks. Thus Eq. (17) gives the same mass for the π and ρ mesons, and should be compared to the spin-weighted average mass

$$\tfrac{3}{4} m_\rho + \tfrac{1}{4} m_\pi \simeq 600 \text{ MeV}.$$

By the same token, (9) should really have been used to determine R from the spin-weighted average mass of N and Δ, which is $\simeq 1140$ MeV. These refinements will be considered in the next section.

A serious deficiency of the primitive model is revealed by considering more massless quarks in a bag. The Pauli principle allows up to twelve u and d quarks to be in the lowest state, considering both color and spin. The energy of an n-quark system in the ground state is proportional to $n^{3/4}$. This leads to a contradiction: $3f$ quarks coalesced into *one* bag would have a lower mass than f nucleons, that is, f 3-quark bags! A deuteron, for example, would have a state with a binding energy of $(2 - 2^{3/4})M \simeq 300$ MeV when all six quarks were in one bag, and an α-particle would have a state with 1100 MeV of binding. Such collapsed "one-bag" nuclei do not exist. This difficulty will disappear in the more realistic model that we now discuss.

2. The improved bag model

The model can be improved by considering the following effects, which were ignored in the primitive model:
(a) The finite mass of the s quark;
(b) The interactions between the quarks;
(c) The zero-point energy of the quantum fields in the bag;
(d) A correct treatment of the center of mass.

The mass m_s of the s quark enters into its kinetic energy ε_s. We use Eq. (6) instead of (7), and write it in the form

$$\varepsilon_s = (\kappa_s^2 + \beta_s^2)^{1/2}/R, \qquad \kappa_s = m_s R, \tag{18}$$

where β_s is the value of $\beta(mR)$ for the mass m_s of the s quark (see Fig. 11). As we see below, κ_s also appears in the color-magnetic interaction energy.

The interactions between the quarks are carried by the color fields. The

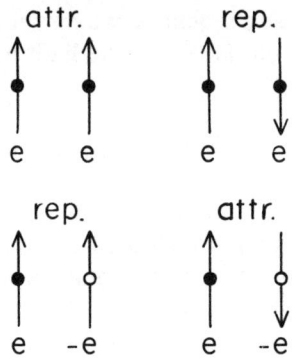

Fig. 13. The sign of the magnetic interaction for different relative spin directions of quark pairs or quark–antiquark pairs.

color-magnetic effects are especially important because they lift the degeneracy between states that only differ in the orientation of quark spins, such as π and ρ, or N and Δ. The color-electric interaction, on the other hand, gives rise to an overall energy shift, but no splittings, since it depends only on the spatial wave function of the single-particle states; in our approximation, the latter is unaffected by the interactions.

Let us start with a qualitative discussion of the color-magnetic interaction resulting from the color-magnetic moments. We exploit the close analogy between QED and QCD, and recall the magnetic interaction of electrically charged particles of spin $\frac{1}{2}$. The interaction between two equally charged particles with parallel spin is attractive, because parallel currents attract each other. By the same token, opposite charges with parallel spin repel, and opposite charges with opposite spin attract (Fig. 13). In mesons, which are quark–antiquark systems, the parallel spin configuration will therefore have higher energy than the antiparallel configuration. Thus the mass eigenvalues (16) for the mesons composed of u and d quarks will be split, and the vector mesons will be more massive than the scalar mesons. This is indeed true: ρ is heavier than π, and K^* is heavier than K.

A similar but smaller split is expected for the baryons. Each quark is in a color triplet and faces a pair of quarks that together form an antitriplet. Hence, there will be a repulsion if all spins are parallel (the decuplet). In the octet we have one pair of parallel spins and two pairs that are antiparallel, giving a net attraction. Indeed, the Δ lies higher than N, and Σ^* is higher than Σ, but the splittings are smaller than in the mesons. It is remarkable that the dependence of the masses on the relative spins is so easily understood.

Now let us be more quantitative. The calculation of the spin-spin interaction is somewhat more complicated than in the electromagnetic case, because the color fields are restricted to the bag. The boundary conditions

2. THE IMPROVED BAG MODEL

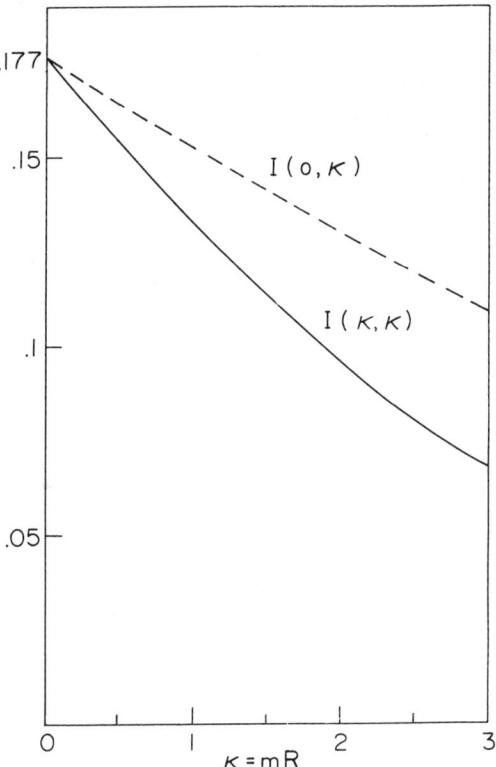

FIG. 14. The factors $I(\kappa_i, \kappa_j)$ appearing in the color-magnetic interaction energy, Eq. (19), of two quarks as functions of $\kappa_i = m_i R$. The solid line refers to equal-mass quarks, the dashed to the case of one massless and one massive quark.

imply that on the bag surface the color-magnetic field is tangential, while the color-electric field is normal.[1]

We do not reproduce the calculation [DeGrand (1975)], and only give the resulting expression for the magnetic interaction energy ΔE_m between two quarks i and j, both being in the lowest state in the bag:

$$\Delta E_m = -\frac{1}{4}\alpha_s \sum_{i<j} (\vec{\lambda}^{(i)} \cdot \vec{\lambda}^{(j)})(\sigma^{(i)} \cdot \sigma^{(j)}) \frac{I(\kappa_i, \kappa_j)}{R}. \tag{19}$$

This is the interaction to lowest order in α_s, the strong coupling constant. The sum is taken over all quark pairs in the bag, $\vec{\lambda}^{(i)}$ is the octet of 3 × 3 color matrices of quark i, $\sigma^{(i)}$ is its spin, and $I(\kappa_i, \kappa_j)$ is a function of the $\kappa_i = m_i R$, which is plotted in Fig. 14. $I(0, 0) = 0.177$ for massless quarks,

---

[1] The dielectric coefficient of the vacuum outside the bag is zero. Therefore the magnetic susceptibility μ must be infinite [see §C.4(c)]. So the true vacuum outside acts magnetically like a perfect conductor does electrically: the corresponding field must be perpendicular to the surface.

and decreases as the masses increase. This effect comes about because the wave functions are more concentrated toward the center for higher mass, and therefore lead to circular currents of smaller average radius.

Equation (19) results from the following Feynman diagrams:

$$\text{(20)}$$

The appearance of the spin product $\boldsymbol{\sigma}^{(i)} \cdot \boldsymbol{\sigma}^{(j)}$ weighted by the color product $\vec{\lambda}^{(i)} \cdot \vec{\lambda}^{(j)}$ is expected in a spin-spin interaction between two objects in a color triplet.

Since α_s is not small for separations as large as R, Eq. (19) can only serve as a rough approximation. The factor R^{-1} is the dominant R-dependence. There is a weaker R-dependence through $I(\kappa_i, \kappa_j)$, since the variables κ are proportional to R. The weak R-dependence of α_s as a running coupling constant will be neglected.

For a quark–antiquark pair the color product was evaluated to be $-\frac{16}{3}$ in Eq. A(58). In the case of baryons, one obtains $\vec{\lambda}^{(i)} \cdot \vec{\lambda}^{(k)} = -\frac{8}{3}$ by using $\vec{\lambda}^{(1)} + \vec{\lambda}^{(2)} + \vec{\lambda}^{(3)} = 0$.

The spin products for two-fermion systems (mesons) were evaluated in Eq. I.B(49). The total spin S is either 0 or 1 and $\boldsymbol{\sigma}^{(1)} \cdot \boldsymbol{\sigma}^{(2)} = -3$ or 1, respectively. For three-fermion systems (baryons), $S = \frac{1}{2}$ or $\frac{3}{2}$; a similar evaluation gives $\sum_{i>j} \boldsymbol{\sigma}^{(i)} \cdot \boldsymbol{\sigma}^{(j)} = -3$ or 3, respectively.

We apply the bag model only to systems composed of u, d, and s quarks. In general, ΔE_m consists of three terms corresponding to the interactions of massless pairs, of pairs with one s quark, and of two s quarks:

$$\Delta E_m = \frac{2}{3}\frac{\alpha_s}{R}[a_{00}I(0, 0) + a_{0s}I(0, \kappa_s) + a_{ss}I(\kappa_s, \kappa_s)]. \tag{21}$$

The coefficients a_{ik} contain the sums over the color and spin products. The index 0 refers to u and d quarks, and s refers to the s quark. Table 2 gives the values of these coefficients for the different hadrons:

TABLE 2
Coefficients in the magnetic energy

	N	Δ	π	ρ, ω	Λ	Σ	Σ*	Ξ	Ξ*	K	K*	Ω	φ
a_{00}	−3	3	−6	2	−4	1	1	0	0	0	0	0	0
a_{0s}	0	0	0	0	1	−4	2	−4	2	−6	2	0	0
a_{ss}	0	0	0	0	0	0	0	1	1	0	0	3	2

We now turn to the color-electric interaction ΔE_e, which is essentially the Coulomb energy between the quarks and antiquarks. It is of less importance

2. THE IMPROVED BAG MODEL

than the magnetic interaction, because it depends only on the color-charge distribution, which is the same for all hadrons in which the quarks are in the lowest single-particle state. Therefore, we treat it in a very cursory way. We neglect the influence of the mass of the strange quark, so that no length other than R enters. Then ΔE_e must be proportional to $1/R$, with a dimensionless coefficient that does not depend on R. This coefficient will depend on the number of constituents, but it cannot be very different for mesons and baryons. In the former case, two particles with opposite color charges interact; in the latter case, the three colors add up to zero. Since we expect this Coulomb energy to be negative, we make the simple assumption

$$\Delta E_e \approx -\frac{z}{R}, \tag{22}$$

where z is a positive constant, which we adjust to provide a best fit.[2]

We now come to the center-of-mass problem. In the bag model the total momentum of the quarks is zero only on the average. The expectation value $\langle(\sum \mathbf{p}_i)^2\rangle$ does not vanish, as it should for a hadron at rest. The bag model determines the energy E_h of the system, which is related to the mass M_h by $E_h = \sqrt{(\sum \mathbf{p}_i)^2 + M_h^2}$. We therefore approximate the mass M_h by the expression

$$M_h \approx \sqrt{E_h^2 - \langle(\sum \mathbf{p}_i)^2\rangle}.$$

Since the quarks are treated as independent particles in the bag, we obtain $\langle(\sum \mathbf{p}_i)^2\rangle = \sum \langle \mathbf{p}_i^2 \rangle$. For massless quarks the momentum is equal to the energy (7): $\langle \mathbf{p}_i^2 \rangle = (\beta_0/R)^2$. For an s quark we get $\langle \mathbf{p}_i^2 \rangle = (\beta_s/R)^2$, so that the sum over the momentum squares becomes:

$$\sum_i \langle \mathbf{p}_i^2 \rangle = n_0(\beta_0/R)^2 + n_s(\beta_s/R)^2, \tag{23}$$

where n_0 is the number of u and d quarks, and n_s is the number of s quarks. The mass M_h is therefore

$$M_h = \sqrt{E_h^2 - \sum \langle \mathbf{p}_i^2 \rangle} = \sqrt{E_h^2 - n_0(\beta_0/R)^2 - n_s(\beta_s/R)^2}. \tag{24}$$

We now collect the various improvements. The energy of a hadron made up of u, d, and s quarks is

$$E_h = \frac{A}{R} + \frac{4\pi}{3} B R^3, \tag{25}$$

---

[2] We do not compute the value of z because we expect other factors of the same form as (22) to contribute to ΔE_e, such as the vacuum self-energy of the color field enclosed in the bag. Again there is no other magnitude involved but R. This remark takes care of point (c) on p. 409.

with

$$A = n_0\beta_0 + n_s\sqrt{\beta_s^2 + \kappa_s^2} + b - z. \tag{26}$$

The first and second terms refer to the kinetic energies of the quarks [see (7) and (18)]; the third term to ΔE_m [see (21)]:

$$b = \tfrac{2}{3}\alpha_s[a_{00}I(0, 0) + a_{0s}I(0, \kappa_s) + a_{ss}I(\kappa_s, \kappa_s)]; \tag{27}$$

the fourth term in (26) refers to ΔE_e, as given by (22).

For hadrons consisting of only u and d quarks, A is independent of R, and by minimizing (25) we obtain

$$R = \left(\frac{A}{4\pi B}\right)^{1/4}. \tag{28}$$

When s quarks are present, A depends on R through κ_s. We obtain a good approximation to the minimum of (25) by choosing a self-consistent value of R in (26) and (27) that reproduces (28). The resulting expression for the energy is

$$E_h = \frac{4A}{3R}. \tag{29}$$

Finally, we use (24) to determine the mass M_h from the energy (29). Recall that all this is valid only for hadrons in which the quarks are in the lowest single particle state of the bag.

3. Quantitative test of the model

There are four adjustable parameters in this formulation: B, α_s, z, and m_s. They can be determined by fitting to the masses of N, Δ, and Ω, and by requiring that the π mass be close to zero. Although all four parameters appear in these input data, B is mainly determined by the mass of N, m_s by the mass of Ω, and α_s by the requirement $m_\pi = 0$. The value of z enters in all masses. The values that fit these conditions are

$$B^{1/4} = 0.135 \text{ GeV}, \qquad \alpha_s = 2.0$$
$$m_s = 0.270 \text{ GeV}, \qquad z = 0.50. \tag{30}$$

These are of the expected order of magnitude. We point out that the mass steps between members of the baryon decuplet, which differ by one s quark, are somewhat smaller than m_s because of the relativistic motion of the s quark.

3. QUANTITATIVE TEST OF THE MODEL

TABLE 3
Hadron masses and bag radii[a]

Name	Experimental mass (MeV)	Theoretical mass (MeV)	Bag radius (GeV^{-1})
N	939		5.86
Δ	1232		6.24
Λ	1116	1100	6.10
Σ	1193	1163	6.18
Ξ	1318	1342	6.44
Ω	1672		6.89
Σ^*	1384	1386	6.47
Ξ^*	1533	1537	6.68
π	137	93	4.77
ρ, ω	769, 783	818	5.58
K	496	548	5.28
K^*	892	964	5.85
ϕ	1019	1115	6.09

[a] The parameters in (30) have been rounded off; as a consequence (29) will not reproduce the exact values of the input masses of N, Δ, and Ω. Fewer significant figures are given for z and α_s because they enter into relatively small contributions to the masses, and for that reason m_π is not zero. The theoretical values are calculated by the simple method described in the preceding pages, and differ slightly from those published in the literature.

Table 3 presents the calculated and experimental values of the masses, and the bag radii, of the most important low-lying hadrons.

The agreement is better than one would expect from such a simplistic approach. Only the π and the ϕ masses disagree by more than 10%. The mass of π is so close to zero that in this case the whole method is not reliable;[3] the ϕ is assumed to be pure $s\bar{s}$, but it may have admixtures of $u\bar{u}$ and $d\bar{d}$ that would reduce the theortical value. The transition from the primitive to the improved model is shown schematically in Fig. 15.

The model gets into difficulty only with the masses of η and η'. According to the footnote on p. 83 of Vol. I, η is the antiparallel spin combination $\eta = 6^{-1/2}(u\bar{u} + d\bar{d} - 2s\bar{s})$, whereas $\eta' = 3^{-1/2}(u\bar{u} + d\bar{d} + s\bar{s})$. That would imply that η should have the mass $m_\eta = \frac{1}{3}m_\pi + \frac{2}{3}m'$, where m' is the mass of a hadron made of an s and an $\bar{s}$ quark with opposite spin. Equation (25) would give $m' = 0.72$ GeV, so that $m_\eta = 0.53$ GeV, which is in adequate agreement with the observed value 0.55 GeV. However, η' would be lower

[3] For massless u and d quarks, QCD is invariant under *separate* $SU(3)$ transformations of the left- and right-handed u and d fields. (This is called *chiral symmetry*, and it has the same origin as the symmetries of the electroweak theory when the quarks and leptons are massless.) If this symmetry were exact, the pion mass would vanish identically. The anomalously small value of m_π is presumably due to the small values of m_u and m_d. Unfortunately, the boundary condition on the wave function imposed in the bag model spoils this symmetry even when $m_d = m_u \equiv 0$, since these conditions are equivalent to assigning a very high mass to the quarks for $r > R$. For that reason the model should not be expected to yield a reliable value of m_π.

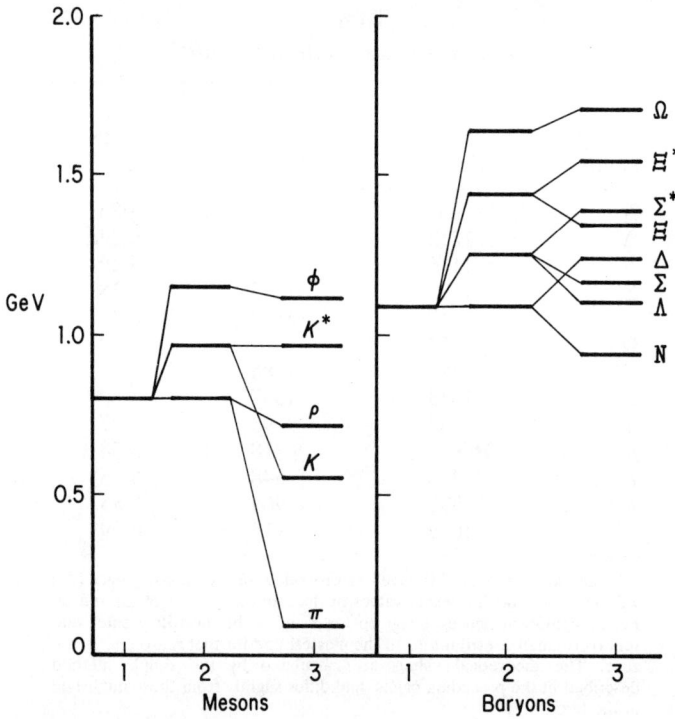

FIG. 15. The calculated masses of mesons and baryons. Columns 1 correspond to the primitive bag model with all quark masses zero; columns 2 take the finite mass of the s quark into account; columns 3 correspond to the improved model.

than η, because it contains less of $s\bar{s}$: $m_{\eta'} = \frac{2}{3}m_\pi + \frac{1}{3}m' = 0.25$ GeV. This failure is not yet understood.

We now turn to other properties of these low-lying states. The root-mean-square radius of the proton is considerably improved compared to that of the primitive model. Equation (12) gives $\langle r^2 \rangle^{1/2} = 0.86$ fm, which is rather close to the observed value 0.8 fm. This reduction is caused by the attraction of the color-electric and color-magnetic forces, which reduces the kinetic pressure of the quarks on the bag.

To evaluate the magnetic moments of the proton and the neutron, we agin use (13) since the wave function ψ_0 is the same in the improved model. On inserting the bag radius from Table 2 into $\mu_q = 0.20 e_q R$, one obtains $\mu_q = 2.2 e_q/2M$ for the magnetic moments of the u and d quarks, as compared to the result $3.3 e_q/2M$ found in the primitive model. This leads to the following magnetic moments of the nucleons:

$$\mu_p = 2.2 e/2M, \qquad \mu_n = -1.5 e/2M. \tag{31}$$

These results are too small, in contrast to the results (15) of the primitive model, which were too big. This discrepancy probably arises because the current density of our model wave function ψ_0 drops abruptly at the surface (as one sees from Fig. 12). Actually, of course, there is no discontinuity; the current spills out beyond $r = R$, and gives rise to a higher magnetic moment than (31).

The improved model also explains why nuclei do not collapse into one-bag states. This collapse is prevented by the color-magnetic interaction ΔE_m between the quarks having parallel spins. The expression for ΔE_m when $3f$ massless quarks are enclosed in a bag is[4]

$$\Delta E_m = \frac{2}{3}\frac{\alpha_s}{R} I(0,0) \frac{1}{2}[9f(f-2) + J(J+1) + 3I(I+1)], \qquad (32)$$

where J is the total angular momentum and I the total isospin, and $I(0, 0)$ is the value of the function plotted in Fig. 14 when both its arguments vanish. ΔE_m becomes nonnegative for $f \geq 2$, whereas it is negative for the nucleon ($f = 1$). For six u and d quarks ($f = 2$), the eigenvalues (J, I) allowed by the Pauli principle, which give (32) its lowest value, are $J = 1, I = 0$. The corresponding mass, according to (25), is 2.08 GeV, which is higher than twice the nucleon mass, 1.88 GeV. Hence, six quarks in one bag would be unstable and decay into two nucleons. In the case of the α-particle ($f = 4$), ΔE_m is very large. The minimum value 7.09 GeV is attained for $I = J = 0$, which is much larger than 4(0.94) GeV.

4. *The long-range potential and the Regge slope*[5]

[Finally, we apply the bag model to meson states of very high angular momentum. The centrifugal force then leads to a greatly elongated bag whose transverse dimension is of order R, but whose length is such that the quark–antiquark separation r is much larger that R, or equivalently, to the characteristic length $r_0 = \Lambda_0^{-1}$ (see Fig. 16). In §C.5(b) we learned that under this circumstance the quark–antiquark interaction is expected to be a linear potential, $W(r) = ar$, because the color-electric field lines are compressed into a tube with a diameter of the order of r_0. We now discuss this problem and establish a connection between a and the constants B and α_s of the bag model.

The compression of the field lines is resisted by the outward pressure of the color-electric field $\vec{\mathbf{E}}$. The two are in balance if the energy density $\frac{1}{2}\vec{\mathbf{E}} \cdot \vec{\mathbf{E}}$ equals B. We gloss over the octet nature of $\vec{\mathbf{E}}$, and assume that the field in the tube has the same value $\mathscr{E}_0$ everywhere except in the regions very near to the quarks (see Fig. 16). Thus,

$$\mathscr{E}_0^2 = 2B. \qquad (33)$$

The magnitude of $\mathscr{E}_0$ is determined from the requirement that the field flux within the tube must equal the charge of the source. Call A_0 the cross-sectional

---
[4] See DeGrand (1975).
[5] This is a sketch of Johnson (1976).

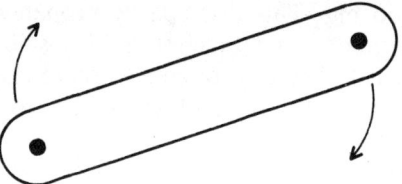

FIG. 16. A schematic picture of a rotating meson elongated by the centrifugal effect.

area of the tube; then the flux is $\mathscr{E}_0 A_0$. The color charge is $[\tfrac{1}{4}(\vec{\lambda} \cdot \vec{\lambda})(4\pi \alpha_s^*)]^{1/2}$, where $\vec{\lambda}$ is the color matrix of a quark and α_s^* is the value of the coupling constant appropriate to typical hadronic dimensions, since the end regions of the tube, where the field originates, are of that size. Hence, we are justified to use the value α_s from (30). Using Eq. A(57) we find for the flux

$$\mathscr{E}_0 A_0 = (\tfrac{4}{3} \cdot 4\pi \alpha_s)^{1/2}. \tag{34}$$

We can interpret a as the energy of the tube per unit length along the axis. This energy consists of the field energy, and the volume energy of the displaced vacuum:

$$a = \tfrac{1}{2}\mathscr{E}_0^2 A_0 + B A_0 = \sqrt{2B}\,\mathscr{E}_0 A_0, \tag{35}$$

where the second equality follows from (33). We have neglected the contributions to the energy of the ends of the tube where the kinetic energies of the particles would enter. It can be shown that they are negligible if the length $2L$ of the tube satisfies $L \gg A_0^{1/2}$. Equations (34) and (35) then yield

$$a = 4\left(\frac{2\pi}{3} B\alpha_s\right)^{1/2}, \tag{36}$$

which gives the value $a = 0.15\,\text{GeV}^2$ with the constants (30), which is in excellent agreement with the value extracted from the $c\bar{c}$- and $b\bar{b}$-spectra.

The same picture also allows one to determine the Regge slope for mesons, i.e., the ratio between the square of the energy and the angular momentum (see Fig. I.13) in the limit of high excitation when the meson is extended into a long tube. We consider rotations around an axis perpendicular to the tube direction (see Fig. 16). Assume the ends of the tube move with the velocity v_0. We again calculate the energy per unit length, which now depends on the speed v of rotation, because of the Lorentz transformation of the fields and of the area A_0:

$$\mathscr{E} = \mathscr{E}_0/\sqrt{1-v^2}, \qquad \mathscr{B} = v\mathscr{E}, \qquad A = A_0\sqrt{1-v^2}, \tag{37}$$

where $\mathscr{E}$ and $\mathscr{B}$ are the absolute values of the color-electric and magnetic fields. Call x the coordinate along the tube axis measured from the center, and $2L$ the length of the tube. We then have $v = v_0 x/L$. The field energy is then

$$E_f = 2\int_0^L \frac{1}{2}(\mathscr{E}^2 + \mathscr{B}^2) A\, dx = \frac{\mathscr{E}_0^2 A_0 L}{v_0} \int_0^{v_0} \frac{1+v^2}{\sqrt{1-v^2}}\, dv,$$

4. THE REGGE SLOPE

and the volume energy is

$$E_v = 2B \int_0^L A\, dx = \frac{2BA_0 L}{v_0} \int_0^{v_0} \frac{1-v^2}{\sqrt{1-v^2}}\, dv.$$

Using (33) and (35), we find $\mathscr{E}_0^2 A_0 = 2BA_0 = a$, and therefore

$$E_f + E_v \equiv E = \frac{aL}{v_0} \int_0^{v_0} \frac{2\, dv}{\sqrt{1-v^2}} = \pi a L f(v_0), \tag{38}$$

where $f(v_0)$ is a function of v_0 that is 1 for $v_0 = 1$. The angular momentum density of the field along the axis of rotation is $x\mathscr{E}\mathscr{B}$, so that the total field angular momentum becomes

$$J = 2 \int_0^L x\mathscr{E}\mathscr{B} A\, dx = \frac{2\mathscr{E}_0^2 A_0 L^2}{v_0} \int_0^{v_0} \frac{v^2}{\sqrt{1-v^2}}\, dv = L^2 a \frac{\pi}{2} g(v_0). \tag{39}$$

Here $g(v_0)$ also equals 1 when $v_0 = 1$.

Equations (38) and (39) represent the energy and angular momentum of the fields only. In principle, the kinetic energy of the rotating quarks near the ends of the tube, and the angular momentum of the quark motion, should be added to (38) and (39), respectively. A simple estimate shows, however, that these contributions are small compared to (38) and (39).

The energy E is proportional to L because of the integration over x, whereas J is proportional to L^2, because x also appears in the expression for J. Hence, the ratio $J/E^2 = (2\pi a)^{-1} g(v_0)/f^2(v_0)$ does not depend on L. For a given value of J, the energy is a minimum[6] for $v_0 = 1$, where $g = f = 1$. Thus the rotational excitations satisfy the following simple relationship:

$$J/E^2 = (2\pi a)^{-1} = 1.1\, \text{GeV}^{-2}. \tag{40}$$

This is expected to hold universally for sufficiently large J, since the field energy was found to dominate the $q\bar{q}$ energy. Equation 40 is called the Regge slope. As we see from Fig. I.13, the linear $J - E^2$ relation already holds for the smallest angular momenta, and our value (40) for the Regge slope, as inferred from the confinement potential, is in good agreement with the data on the rotational excitations of the ρ. The universal nature of the Regge slope is seen in the data on other mesonic rotational bands, and in Fig. I.17 for a baryon trajectory.]

---

[6] The quarks are inside the tube by a distance $\sim A_0^{1/2}$, and therefore move with less than the velocity of light even if $v_0 = 1$.

V

DEEP INELASTIC LEPTON-HADRON SCATTERING

1. Introduction

One of the most important techniques for exploring the internal structure of hadrons (in practice, of nucleons) is the scattering of high-energy electrons, muons, and neutrinos. They are particularly suitable tools for such investigations because we believe that we understand the weak and electromagnetic forces between leptons and the quarks that constitute hadrons. These interactions are feeble in the sense that the first Born approximation provides an accurate description of the collision. As a consequence, the mathematical analysis is straightforward, and the interpretation of the data relatively unambiguous. Neither of these simplifying circumstances holds in hadron-hadron collisions.

The internal structure of hadrons reveals itself most clearly in "deep inelastic scattering," that is, scattering where the momentum transfer is much larger than the relative momenta of the hadron's constituents. The latter momenta serve as a measure of the strong interaction felt by the quarks. The collision is therefore instantaneous in comparison to the time scale that characterizes the internal motions. Thus, deep inelastic scattering provides a "snapshot" of the hadronic constituents.

To clarify the concepts that we shall encounter, we first consider a more familiar setting—the scattering of a beam of electrons by an atomic nucleus. When the energy ε of the electrons is less than the lowest excitation energy of the nucleus, the scattering is *elastic*. The nucleus remains in the ground state, and the electron does not suffer any energy loss in the c.o.m. frame $\mathscr{F}_0$. The angular dependence (or, equivalently, the dependence on the momentum transfer q) gives information about the charge distribution in the nuclear ground state, which is expressed by the form factor $F(q)$, as discussed in §III.A.1.

If the incident energy ε is higher than the lowest excitation energy, and the momentum transfer is not considerably larger than the internal momenta of the nucleons in the nucleus, the process may be called "shallow" inelastic scattering. The energy ε' of the scattered electron is then distributed over many values. In $\mathscr{F}_0$ the scattering cross section for a given ε, as a function of ε', exhibits peaks whenever $\varepsilon - \varepsilon'$ is equal to an excitation energy of the nucleus. Elastic scattering also occurs, but becomes less prominent as more states can be excited. When $\varepsilon - \varepsilon'$ is larger then the energy necessary to remove a nucleon from the nucleus, we find a continuum instead of peaks. The dependence of the cross section on the momentum transfer is complicated, since it depends on the properties of the initial and the final states of the nucleus. We do not discuss it further.

When the momentum transfer is much larger than the average internal momentum of the nucleons in the ground state, the process[1] is called "deep inelastic" scattering. Under these conditions the interaction between the electron and the nucleons is so sudden, and involves such a large change of momentum, that one can neglect the binding forces between the constituents during the collision. In a first approximation, the constituents behave like free particles with a momentum distribution given by the momentum-space wave function of the ground state. This is called the *impulse approximation*: the electron can be considered as being scattered by one of the "free" nucleons. The total scattering cross section is then the sum of the cross sections for each of the constituent "free" nucleons,[2] appropriately weighted by the probability for finding a nucleon in the ground state with a given momentum. Since the scattering of electrons by individual nucleons is known, it is possible to determine the ground-state momentum distribution from an analysis of deep inelastic scattering by the nucleus. The state after the collision consists of the highly energetic nucleon that was struck, and the remnant nucleus. (The nucleon acquired a large momentum compared to the internal ones.) This remnant is a linear superposition of energy eigenstates of a nucleus with $A - 1$ constituents; it corresponds to the original nucleus with a "hole" at "the place" of the ejected nucleon.

We now apply this analysis to the scattering by an isolated nucleon. Again, if the lepton energy is low ($\lesssim 300$ MeV), the first excited state (Δ) cannot be reached, and only elastic scattering takes place.[3] Such measurements are used to determine the form factor of the nucleon (see §III.A.1), which leads to a radius R of $\sim 10^{-13}$ cm. The average internal quark momentum therefore is of the order $R^{-1} \sim 200$ MeV/c. What we called "shallow" inelastic scattering occurs when the lepton energy is in the region of one to several GeV. The cross section will have peaks when the energy loss equals one of the excitation energies of the nucleon (see Fig. I.5). For lepton energies of many GeV, the momentum and energy transfers can be considerably larger than 200 MeV/c. We then have deep inelastic scattering of leptons by nucleons.

Let us exploit the analogy with the scattering of electrons by nuclei. The role of the electrons is played by one of the leptons (electron, muon, or neutrino), that of the nucleus is played by the nucleon, and that of nucleons within the nucleus by quarks. But there is one important difference: quarks cannot be ejected from the nucleon (since we believe that no free quarks exist). However, this neither plays a significant role in the evaluation of the

---

[1] Even in this case there is a small probability for elastic scattering.

[2] The absence of interference terms between amplitudes for scattering by different nucleons is also a consequence of the large difference between the momentum of the struck nucleon and the momenta of all the other, unstruck nucleons. Once $1/q$ is small compared to the mean internucleon spacing, the collision, in effect, uniquely defines the constituent involved in the process, and the quantum-mechanical interference effects become negligible.

[3] This is not quite true, because the reaction $eN \to eN\pi$ has a small cross section for energies below that required for $eN \to e\Delta$.

scattering cross section, nor in the momentum distribution of the scattered leptons. This is so because the momentum acquired by the quark immediately after the collision is much larger than R^{-1}. The wavelength of its subsequent motion is therefore such that it can be localized in a wave packet that is small compared to the nucleon's size. The collision is over before this packet can leave the nucleon, and the cross section for scattering of the lepton (summed over *all* accessible final states of the hadron) is therefore insensitive to the subsequent fate of the quark. On the other hand, the detailed properties of the final states depend sensitively on the mechanisms that come into play when the highly energetic quark attempts to leave the hadron. As we know from the discussion of §I.E.7 (and especially Fig. I.25), quark–antiquark pairs are produced, and these eventually materialize into a jet of mesons moving in the general direction of the struck quark.

Since quark confinement does not, in any appreciable way, influence the scattering of high-energy leptons by quarks, we may use the same description of deep inelastic scattering by hadrons as we used in the case of lepton-nucleus scattering. We neglect the quark binding, and treat the nucleon as an assembly of free quarks with a momentum distribution given by the nucleon ground state. The basis of our analysis, therefore, is the scattering of ultrarelativistic leptons by free pointlike fermions. The experimental results can be used to test the assumption that quarks are pointlike, or at least small compared to the resolving power of the measurement, which is of order q^{-1}. As we shall see, the data are consistent with this assumption. Furthermore, the experimental results provide information about the momentum distribution of quarks within the nucleon.

2. Kinematic variables

Before attacking lepton-quark scattering, we define the variables that describe the scattering of a lepton by a nucleon.

Let k and k' be the 4-momenta of the lepton before and after the scattering, respectively. The corresponding 4-momenta of the nucleon are P and P'. Conservation of energy and momentum gives

$$k + P = k' + P'. \qquad (1)$$

An important quantity is the 4-momentum transfer

$$q = k - k' = P' - P. \qquad (2)$$

In an elastic collision, the target state is unchanged, and $P^2 = P'^2 = M^2$, from which it follows that[4]

---

[4] $q^2 = (P' - P)^2 = 2M^2 - 2P' \cdot P$; it follows from (2) that $P' \cdot P = (P + q) \cdot P = M^2 + P \cdot q$, which gives (3).

$$t \equiv q^2 = -2P \cdot q. \tag{3}$$

Here we introduced the variable $t \equiv q^2$ (see §II.C.1), since q^2 will appear frequently. In an inelastic collision the target is excited into a state that consists of two or more emerging hadrons. As we learned in §III.A.3, any set of particles has an invariant mass W that is related to their total 4-momentum P' by $W = \sqrt{P'^2}$. As the state in question is excited, $W > M$. Instead of (3), we then find[5]

$$-t \equiv -q^2 = 2P \cdot q - (W^2 - M^2). \tag{4}$$

Define

$$x \equiv -q^2/2P \cdot q,$$
$$= 1 - \frac{W^2 - M^2}{2(P \cdot q)}, \tag{5}$$

and

$$\nu \equiv P \cdot q/M^2. \tag{5'}$$

The ratio x is smaller than one, since $W^2 - M^2$ and $2P \cdot q$ are positive. The latter follows from its value in the laboratory frame $\mathscr{F}_L$, where the target is at rest, and the components of P are $(M, \mathbf{0})$. Hence,

$$P \cdot q = M^2 \nu = M(\varepsilon - \varepsilon'), \tag{6}$$

where ε and ε' are the energies of the lepton before and after collision in $\mathscr{F}_L$. Thus, ν is the energy loss in units of M of the lepton in $\mathscr{F}_L$, which is necessarily positive, and therefore the invariant $P \cdot q$ is positive [see (6)]. As for t, we have $q^2 = 2m^2 - 2(\varepsilon\varepsilon' - kk' \cos \theta)$, where θ is the scattering angle in $\mathscr{F}_L$. In deep inelastic scattering $k \simeq \varepsilon, k' \simeq \varepsilon'$, and the term $2m^2$ is negligible, so that

$$q^2 \equiv t = -4\varepsilon\varepsilon' \sin^2 \tfrac{1}{2}\theta. \tag{7}$$

Hence[6] $t \leq 0$, and since $P \cdot q \geq 0$, (5) implies that

$$0 \leq x \leq 1. \tag{8}$$

---

[5] Now $q^2 = (P' - P)^2 = W^2 + M^2 - 2P' \cdot P = W^2 - M^2 - 2P \cdot q$.
[6] That $(k - k')^2 < 0$ is always true in scattering, whether "deep" or not, because one can find a frame in which the energies k_0 and k_0' are equal; i.e., q is a spacelike 4-vector.

The limit $x = 1$ corresponds to elastic scattering, $W^2 = M^2$. In that case, there is a definite relation between the energy loss v of the leptons in $\mathscr{F}_L$ and the momentum transfer q:

$$-q^2 = 2M^2 v \quad \text{(elastic scattering)}. \tag{9a}$$

In general, v and q^2 are not uniquely related, but depend on the variable x:

$$\frac{-q^2}{2M^2 v} = x \quad \text{(inelastic scattering)}. \tag{9b}$$

Henceforth we shall mostly use the notation $q^2 = \mathbf{q}^2$.

3. Deep inelastic electron and muon scattering cross sections

To the best of our knowledge, electrons and muons interact with hadrons only via the electromagnetic and weak interactions. For lepton (laboratory) energies ε up to hundreds of GeV, the electromagnetic interaction dominates, and we can ignore the weak interaction.

There exists a generally valid expression for the cross section of an ultrarelativistic pointlike fermion with charge e scattered electromagnetically by any system N that carries electrical charges and currents, no matter how complex its structure may be. That expression is derived by using only general principles. The general structure of this expression can be inferred from the diagram

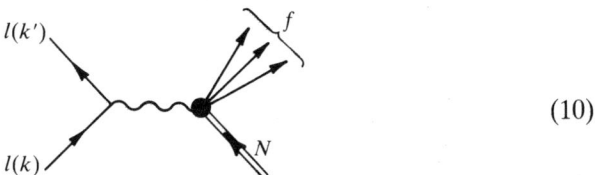

(10)

where f is some multihadron state. The corresponding amplitude has a factor of $1/t$ for the photon propagator, a factor for the lepton-photon vertex identical to that in elastic electron-electron scattering, and a matrix element of the electromagnetic current operator j between the states N and f, $\langle f|j|N\rangle$. The form of $\langle f|j|N\rangle$ is constrained because j is a conserved 4-current, satisfying $\partial j_\mu/\partial x_\mu = 0$. The resulting cross-section in $\mathscr{F}_L$ is

$$\frac{d^2\sigma}{dt\,dv} = \frac{1}{4\pi}\frac{e^4}{t^2}\frac{\varepsilon'}{\varepsilon}\left\{\frac{1}{v}F_2^\gamma(t, v) - \frac{t}{4\varepsilon\varepsilon'}\left[2F_1^\gamma(t, v) - \frac{1}{v}F_2^\gamma(t, v)\right]\right\}. \tag{11}$$

Here the superscript γ indicates that we are dealing with the electromagnetic interaction. It is expressed in terms of variables defined completely by the scattering of the lepton: the lepton's energies ε and ε' before and after scattering, its energy loss v, and the square of its 4-momentum transfer t. The first factor $e^4/(4\pi t^2)$ is essentially the cross section for the scattering of point charges (compare with §II.C.1). The cross section contains two functions F_1^γ and F_2^γ, which depend only on t and v. These so-called *structure functions* characterize the scattering system, and can be determined from the observed lepton scattering by use of (11). As one might expect from our introductory discussion, the structure functions turn out to be related to the momentum distribution of quarks within the nucleon.

We do not provide a detailed derivation of Eq. (11) for a general system because we will show that it takes on a much simpler form when we assume the hadron to be a system of point quarks. As we shall see, this assumption leads to the conclusion that the structure functions $F_{1,2}^\gamma(t, v)$ should depend not on two, but only on *one* variable, the previously defined quantity $x = -t/2M^2 v$. This property of the structure functions is called their *scaling behavior*. Furthermore, in the quark model there is just *one* function $G^\gamma(x)$ that determines *both* structure functions:

$$G^\gamma(x) = \frac{1}{x} F_2^\gamma(x) = 2F_1^\gamma(x). \tag{12}$$

This connection between F_1^γ and F_2^γ is called the *Callan-Gross relation*; it is only valid for spin $\frac{1}{2}$ scatterers. As we shall see, $G^\gamma(x)$ is, in essence, the quark momentum distribution.

4. Deep inelastic electron and muon scattering according to the quark model

We now proceed to calculate the scattering of a lepton from a nucleon by assuming that it consists of quarks, so that the lepton is scattered by just one[7] of the quarks. Furthermore, we use the impulse approximation, as defined on p. 424. In this picture, the final state differs from the nucleon ground state in that one of the quarks has received a 4-momentum transfer q, whereas all other quarks remain untouched. In other words, the energy difference $W - M$ between the final and the initial nucleon states is equal to the energy acquired by the quark that scattered the lepton.

As we have already said, in deep inelastic scattering both particles—the lepton and the nucleon—have initial 3-momenta $\mathbf{k}$ and $\mathbf{P} = -\mathbf{k}$ in the center-of-mass frame $\mathscr{F}_0$ that are large compared to their masses, $|\mathbf{k}| \gg m$, $|\mathbf{P}| \gg M$. Furthermore, the 4-momentum transfer is much larger than M; $|q^2| \gg M^2$. Under these circumstances the nucleon can be treated approximately as if it were an assembly of freely moving quarks. In a frame

---

[7] Double scattering can only occur in the second Born approximation, and is smaller by a factor $\frac{1}{137}$.

4. DEEP INELASTIC SCATTERING

in which the nucleon has a 3-momentum $\mathbf{P}$ very much larger than its mass M, the 3-momenta $\mathbf{p}$ of the constituent quarks are all approximately parallel to $\mathbf{P}$, because the components perpendicular to $\mathbf{P}$, due to the internal motion, are only of order M; it is important to recognize that this circumstance holds in the c.o.m. frame $\mathscr{F}_0$. Hence, we may write

$$\mathbf{p} = x\mathbf{P}, \qquad |\mathbf{p}| < |\mathbf{P}|. \tag{13}$$

We have used the symbol x for the momentum fraction because it will turn out to be identical to the quantity introduced in (5). We expect that

$$\sum_i x_i \leq 1, \tag{14}$$

where x_i is the momentum fraction of the ith quark, and the sum extends over all quarks. This relation says that the sum of the momenta of the charged constituents cannot exceed the momentum of the hadron. The sum (14) may be less than unity because we must entertain the possibility of constituents other than quarks—in particular, the electrically neutral gluons that are the quanta of the interquark force field.

In $\mathscr{F}_0$, we do not expect that a quark could run in the direction opposite to that of the hadron when $P \gg M$, so we assume[8] that $x_i > 0$. From this we also obtain $x_i < 1$ in order to fulfill (14). The restriction $0 < x_i < 1$ is supported more stringently later on.

We assume the deep inelastic scattering cross section of a lepton by a nucleon to be the sum of the elastic scattering cross sections of the lepton with the different quarks within the hadron.[9] The elastic scattering cross section between two charged point-particles of spin $\frac{1}{2}$ was given in §II.C.1 in an invariant form valid in any frame:

$$\frac{d\sigma}{dt} = \frac{e^2 e'^2}{8\pi t^2}\left(1 + \frac{(\mathbf{p}\cdot\mathbf{k}')^2}{(\mathbf{p}\cdot\mathbf{k})^2}\right). \tag{15}$$

Here e and e' are the charges of the two particles, and their masses are put equal to zero since they can be neglected in this kinematic regime. The relation between the quark 4-momentum p and the hadron 4-momentum P follows from the 3-momentum relation (13),

$$p = xP, \tag{16}$$

because, in extreme relativistic motion, the time component is proportional to the spatial component. Therefore, the cross section (15) is independent of the quark's momentum p; it contains only the hadron's momentum P

---

[8] This is only a plausibility argument. The condition $x_i > 0$ also follows from field-theoretical arguments beyond the scope of this presentation; see Cheng (1984), §10.4.
[9] Recall footnote 2 on p. 424, but replace "nucleon" with "quark."

and the momentum transfer:

$$\frac{d\sigma}{dt} = \frac{e^2 e'^2}{8\pi q^4}\left(1 + \frac{(\mathbf{P}\cdot \mathbf{k}')^2}{(\mathbf{P}\cdot \mathbf{k})^2}\right). \tag{17}$$

Since the lepton-quark collision is elastic, the relation (3) must hold, with $\mathbf{P}$ replaced by the quark momentum $\mathbf{p}$:

$$q^2 = -2p \cdot q = -2xP \cdot q; \tag{18}$$

hence,

$$x = \frac{-q^2}{2(P \cdot q)} = -\frac{t}{2M^2\nu}. \tag{19}$$

Thus the variable x, here introduced as the momentum fraction of the quark, is indeed the kinematic variable $x = -q^2/2M^2\nu$, as defined by (5), that depends only on the lepton variables. We can interpret these two definitions of x in the following way: Whenever a lepton is observed to be scattered by a hadron with a momentum transfer q and an energy loss ν in $\mathscr{F}_L$, it must have been scattered by a quark of momentum xP. This unique relation between the lepton's momenta, and the momentum of the scattering quark, only holds in the impulse approximation.

Let us return to the lepton-quark scattering cross section (17), which has been cast into a form where the quark momenta do not appear, only those of the nucleon and the lepton. We now introduce the probability $G_i(x)\,dx$ of finding a quark or antiquark of flavor i, with the charge $\pm e_i$, and the momentum fraction between x and $x + dx$ within the nucleon.[10] This momentum distribution does not refer to a nucleon at rest, but to a frame in which the nucelon has an extreme relativistic momentum ($|\mathbf{P}| \gg M$), since we want to exploit (16). We multiply (17) by $G_i(x)$ and sum over all flavors i. This then gives us the cross section of the hadron for a given value of t and x:

$$\frac{d^2\sigma}{dt\,dx} = \frac{e^4}{8\pi t^2}\left(1 + \frac{(\mathbf{P}\cdot \mathbf{k}')^2}{(\mathbf{P}\cdot \mathbf{k})^2}\right)G^\gamma(x), \tag{20}$$

with

$$G^\gamma(x) = \sum_i Q_i^2 G_i(x), \tag{21}$$

where $Q_i = e_i/e$ is the quark charge in units of e.

---

[10] $G_i(x)\,dx$ is the probability of finding the flavor i, without distinguishing quarks from antiquarks, because the charge enters into the cross section quadratically.

4. DEEP INELASTIC SCATTERING

Equation (19) gives the cross section in terms of quark variables. We now compare it to the generally valid expression (11), which does not rely on any theory of nucleon structure, or the impulse approximation, but only on the first Born approximation. For this purpose we must relate the differential of energy loss, dv, to that of momentum fraction dx. From (19) we see that at constant t

$$\frac{dx}{x} = \left|\frac{dv}{v}\right|. \qquad (22)$$

Our goal, Eq. (11), is expressed in terms of laboratory frame ($\mathscr{F}_L$) variables, whereas (20) is not. But (20) is given in terms of Lorentz-invariant expressions, and it can therefore be evaluated in any frame.[11] In $\mathscr{F}_L$, in particular, $\mathbf{P} \cdot \mathbf{k} = M\varepsilon$ and $\mathbf{P} \cdot \mathbf{k}' = M\varepsilon'$, so that

$$1 + \frac{(\mathbf{P} \cdot \mathbf{k}')^2}{(\mathbf{P} \cdot \mathbf{k})^2} = 1 + \left(\frac{\varepsilon'}{\varepsilon}\right)^2 \equiv B, \qquad (23)$$

an expression that we shall encounter again. Since $v = (\varepsilon - \varepsilon')/M$ and therefore $\varepsilon^2 + \varepsilon'^2 = M^2 v^2 + 2\varepsilon\varepsilon'$, we have

$$B = 2\frac{\varepsilon'}{\varepsilon}\left(1 + \frac{M^2 v^2}{2\varepsilon\varepsilon'}\right) = 2\frac{\varepsilon'}{\varepsilon}\left(1 - \frac{v}{x}\frac{t}{4\varepsilon\varepsilon'}\right), \qquad (24)$$

where we used (19) and (6). Combining (20), (22), and (24) yields the following expression for the deep inelastic electromagnetic scattering of a lepton by a nucleon in the frame $\mathscr{F}_L$:

$$\frac{d^2\sigma}{dt\,dv} = \frac{e^4}{4\pi t^2}\frac{\varepsilon'}{\varepsilon}\left[\frac{x}{v} - \frac{t}{4\varepsilon\varepsilon'}\right]G^\gamma(x). \qquad (25)$$

Having cast the quark model cross section into the format (11), we can identify the structure functions. From the first term in the braces { } we obtain

$$F_2^\gamma = xG^\gamma(x). \qquad (26)$$

The other term yields

$$2F_1^\gamma - \frac{1}{v}F_2^\gamma = G^\gamma. \qquad (27)$$

---

[11] Our derivation of (20) exploits (16), which only holds in frames where $|\mathbf{P}|^2 \gg M^2$, whereas $\mathbf{P} = 0$ in $\mathscr{F}_L$. But the cross section (20) is a Lorentz invariant, and it can therefore be expressed in terms of quantities evaluated in any frame.

According to (26) and (27)

$$2F_1^\gamma = F_2^\gamma\left(\frac{1}{x} + \frac{1}{v}\right). \tag{28}$$

In deep inelastic scattering $v \gg 1$, and since $x \leq 1$, we get the Callan–Gross relation (12). Thus we have shown that F_1^γ and F_2^γ should "scale," i.e., they should be functions of x only.

The scaling property is a consequence of our assumption that quarks are pointlike, so that no quantity having the dimension of a length enters into deep inelastic scattering. That is why the structure functions depend only on the dimensionless variable x, and not separately on the variables t and v. The Callan–Gross relation devolves from the spin $\frac{1}{2}$ character of quarks. The quantity B [see Eqs. (20) and (23)] arises from the helicity structure of the lepton-quark scattering amplitude (cf. §II.C.1); B would not have the form (24) if quarks had another spin, and there then would be a different relationship between F_1^γ and F_2^γ.

Deep inelastic electron and muon scattering has been studied for a wide range of incident momenta, scattering angles, and energy losses. By casting the results in the form (11), where everything but the structure functions F_1^γ and F_2^γ are determined by the energies and momenta of the incident and scattered lepton, one can extract those functions for various values of q^2 and v. The experimental results for $F_2^\gamma(x)$ are plotted in Fig. 1 for protons and neutrons. Figure 2 shows that the Callan–Gross relation (12) is fulfilled. Figure 3 illustrates how well the scaling law is satisfied. As we see, for fixed values of x, F_2 changes somewhat, but only when q^2 undergoes very large variations.

The observed validity of the Callan–Gross relation, and the approximate scaling, support our basic assumption that all *charged* constituents of the nucleon are pointlike spin-$\frac{1}{2}$ objects. As we will learn in §8, the weak scaling violation revealed by Fig. 3 can be understood as being due to a correction to the impulse approximation; it does not imply that quarks are spatially extended systems.

According to (26), $F_2^\gamma(x)$, as shown in Fig. 1, is supposed to be identical with $xG^\gamma(x)$. As Eq. (21) shows, it can be used to determine the momentum distributions $G_i(x)$ of the various quarks.

5. The quark momentum distributions in the nucleon

Let us examine the quark momentum distributions $G_i(x)$ that enter into the expression (25) for the inelastic cross section. We repeat that these distributions refer to a frame in which the nucleon momentum is highly relativistic. It is not easy to relate them to the momentum distributions of quarks in a nucleon at rest. This problem is discussed in §7.

According to the primitive quark model, the nucleon consists of three

5. THE QUARK MOMENTUM DISTRIBUTIONS

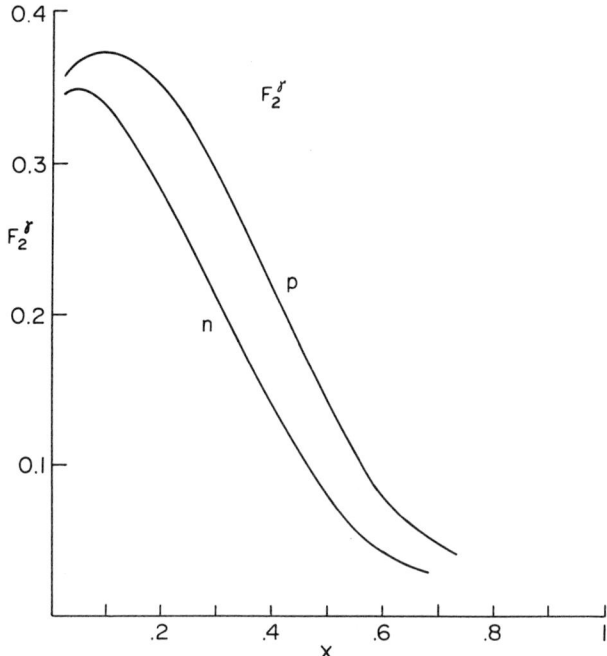

FIG. 1. The structure functions $F_2^\gamma(x)$ of the proton and neutron, measured by means of deep inelastic electron and muon scattering, for momentum transfers around $|Q^2| = (6.5 \text{ GeV}/c)^2$. The data are from SLAC-MIT (electron) and from the European Muon Collaboration (EMC). There is a normalization discrepancy of about 10% between the two groups, which was readjusted by increasing the EMC data by 10%. The neutron data for muons were obtained by multiplying the EMC-proton data by the ratio shown in Fig. 4. [Data from SLAC (M. Barnett, private communication); from EMC (E. Gabathuler, private communication); J. J. Aubert et al., *Phys Lett.* **105B**, 315 (1981).]

quarks of the flavor u and d, the proton being uud, the neutron ddu. It will emerge that this is too simple to explain the observed deep inelastic scattering. Let us therefore anticipate the possibility of finding other quarks, such as s quarks, and therefore also antiquarks, in the nucleon. We call $u(x)\,dx$ the probability of finding a u quark in a nucleon with momentum $x\mathbf{P}$, when x lies between x and $x + dx$, and $|\mathbf{P}| \gg M$ is the momentum of the nucleon. We also introduce the corresponding functions $d(x), s(x), \bar{u}(x), \bar{d}(x), \bar{s}(x)$. Equation (21) then gives

$$G^\gamma(x) = \tfrac{4}{9}[u(x) + \bar{u}(x)] + \tfrac{1}{9}[d(x) + \bar{d}(x) + s(x) + \bar{s}(x)]. \quad (29)$$

We divide the functions $u(x)$, etc., into two parts, one that comes from the three quarks that the primitive quark model requires, the other from contributions of additional effects:

$$u(x) = u_v(x) + u_s(x), \quad \text{etc.} \quad (30)$$

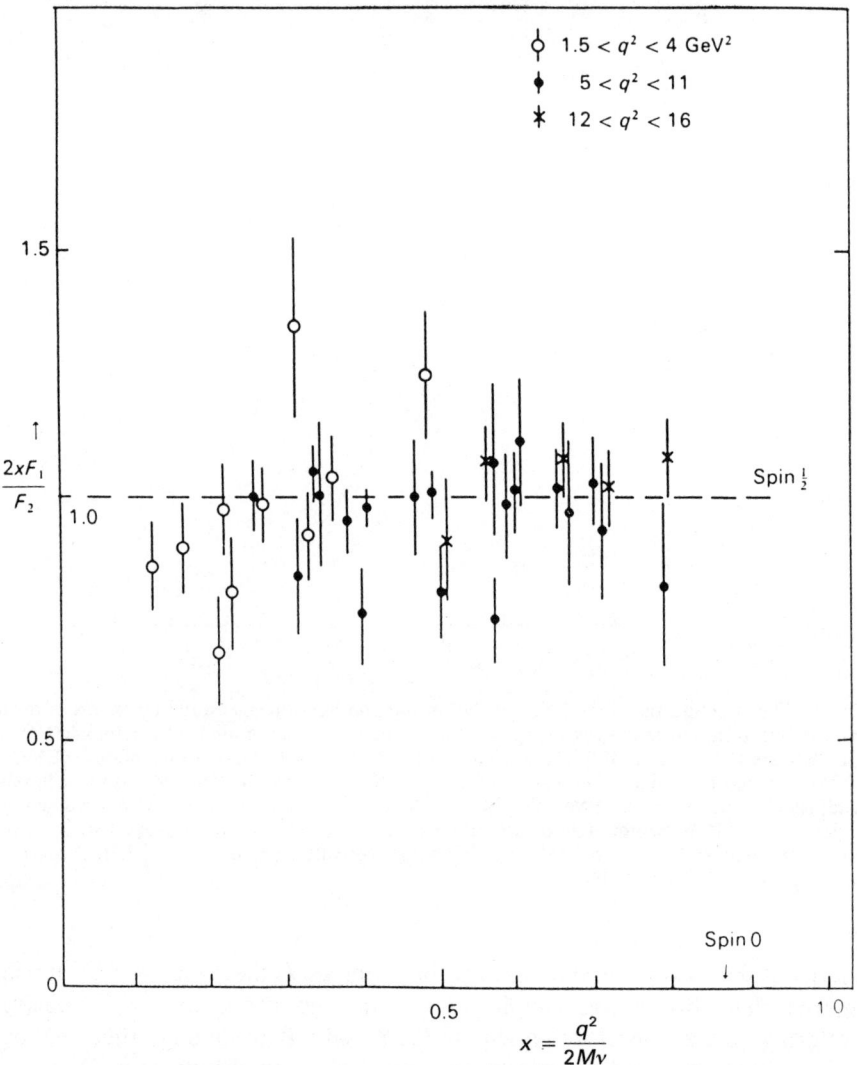

FIG. 2. The ratio of $2xF_1$ to F_2 as a function of x. This shows that the Callan–Gross relation (12) is well fulfilled, in particular for larger momentum transfers. [From Perkins (1982), page 298.]

The subscripts refer to the terms "valence" and "sea." The "valence" quarks are the three quarks of the primitive model; hence, only $u_v(x)$ and $d_v(x)$ differ from zero in the nucleon:

$$s_v(x) = \bar{s}_v(x) = \bar{u}_v(x) = \bar{d}_v(x) = 0. \tag{31}$$

The nucleon, like any other hadron, contains not only the valence quarks

5. THE QUARK MOMENTUM DISTRIBUTIONS

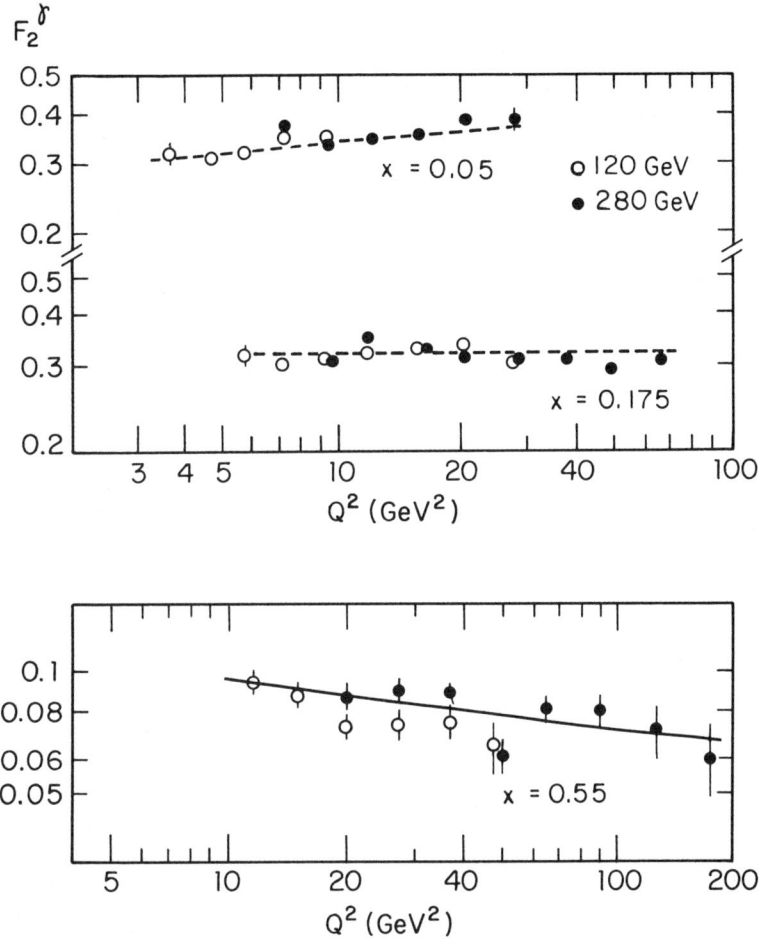

FIG. 3. Test of scaling. F_2^γ is shown, for different x, as function of $Q^2 = -t$. If scaling were exactly fulfilled, the curves would be horizontal lines, but as one sees, there is weak scaling violation. F_2^γ increases slightly with Q^2 for low x and decreases for high x. This behavior is explained in §8.

that determine its low-lying spectrum, but also the force field that binds these quarks. That field is able to create virtual quark pairs, which are referred to as a "sea" through which the valence quarks move. If this picture is valid, the number of quarks in the sea must be equal to the number of antiquarks. Furthermore, since the mass difference between the u and the d quark is small compared to the energies exchanged within the nucleon, we shall assume that the number of $u\bar{u}$ pairs is equal to the number of $d\bar{d}$ pairs:

$$u_s(x) = \bar{u}_s(x) = d_s(x) = \bar{d}_s(x) = c(x); \quad s_s(x) = \bar{s}_s(x) = s(x), \quad (32)$$

where $c(x)$ is the common distribution function for the u and d sea quarks, and $s(x)$ for the strange quarks. (The very heavy quarks c and b give a negligible contribution to the nucleon's sea.) We expect the magnitude of the distribution functions of the sea quarks to be a good deal smaller than those of the valence quarks. They will be concentrated around small x-values, for reasons to be discussed later.

We must differentiate the quark distribution functions in the proton from those in the neutron. The neutron differs from the proton by an interchange of the u's with d's. If we distinguish the distribution functions by a superscript p or n, this implies that

$$u_v^p(x) = d_v^n(x) \equiv 2a(x), \tag{33}$$

$$d_v^p(x) = u_v^n(x) \equiv b(x). \tag{34}$$

Thus $a(x)$ is the distribution of a valence quark of the flavor that is doubly present, and $b(x)$ refers to the quark with the flavor that is singly present. We have

$$\int_0^1 a(x)\,dx = \int_0^1 b(x)\,dx = 1, \tag{35}$$

because these integrals are the total probability that a valence quark is present. One might expect that $a(x) = b(x)$, since the interactions between quarks do not depend on flavor, and therefore the two flavors u and d should have the same distribution. This is indeed true for the four Δ's (uuu, uud, udd, ddd) because in this case the states are completely symmetric, so that the three quarks have parallel spins [see §I.E.6(d)]. In the nucleon, however, the two quarks with the same flavor have parallel spins, whereas the other quark has the opposite spin. The quark–quark forces are spin-dependent, and therefore the two flavors do not necessarily have the same distribution:

$$a(x) \neq b(x). \tag{36}$$

As we will see, the experiments indeed indicate that $a(x)$ differs from $b(x)$. Since the valence quarks carry the entire isospin, and the proton and neutron only differ by an isospin rotation, the sea-quark distributions must be the same for the proton and the neutron.

We call $G^{\gamma p}$ and $G^{\gamma n}$ the function $G^{\gamma}(x)$ for the proton and for the neutron. Equation (29), and the argument of the preceding paragraph, then yields

$$\begin{aligned} G^{\gamma p}(x) &= \tfrac{1}{9}[8a(x) + b(x) + 10c(x) + 2s(x)], \\ G^{\gamma n}(x) &= \tfrac{1}{9}[2a(x) + 4b(x) + 10c(x) + 2s(x)]. \end{aligned} \tag{37}$$

5. THE QUARK MOMENTUM DISTRIBUTIONS

The sum of the two functions is given by

$$G^{\gamma p}(x) + G^{\gamma n}(x) = \tfrac{5}{9}[2a(x) + b(x) + 4c(x) + \tfrac{4}{5}s(x)]. \tag{38}$$

The two functions (37) are determined from the deep inelastic scattering measurements. The experimental results for $xG^{\gamma p,n}$ are plotted in Fig. 1. They give very useful information about the motion of quarks within the nucleon.

Next we determine the total momentum $\mathbf{P}_q$ carried by the quarks. There are, for each value of x, $2a$ valence quarks of the doubly represented flavor, b valence quarks of the singly represented one, $4c$ sea quarks and antiquarks of the d and u flavor, and $2s$ of the s flavor. From (16) we therefore have

$$\mathbf{P}_q = \int x\mathbf{P}[2a(x) + b(x) + 4c(x) + 2s(x)]\,dx,$$

where $\mathbf{P}$ is the 3-momentum of the whole nucleon. On using (38), we obtain

$$\mathbf{P}_q = \tfrac{9}{5}\mathbf{P}\int [xG^{\gamma p}(x) + xG^{\gamma n}(x) + \tfrac{6}{5}xs(x)]\,dx. \tag{39}$$

Therefore

$$\mathbf{P}_q \geq \tfrac{9}{5}\mathbf{P}\int x[G^{\gamma p}(x) + G^{\gamma n}(x)]\,dx = \tfrac{9}{5}\mathbf{P}\int [F_2^{\gamma p}(x) + F_2^{\gamma n}(x)]\,dx, \tag{40}$$

where $F_2^{\gamma p,n}$ are the structure functions of the proton and the neutron, as defined in (26). The third term of (39) is small compared to the others. As we shall learn, the sea contribution is peaked toward small x, which is suppressed in (39). Furthermore, the contribution to the sea of s quarks is considerably smaller than that of the others because the larger mass of s inhibits the creation of $s\bar{s}$ pairs. Consequently, the actual value of $\mathbf{P}_q$ is close to the lower bound (40). If one uses the observed structure functions, and ignores the small correction due to s quarks, one finds

$$\mathbf{P}_q \sim 0.5\,\mathbf{P}. \tag{41}$$

This is a remarkable result: it says that only about half of the nucleon's momentum is carried by quarks! What carries the balance? It is the contribution of the strong interaction field. A field carries energy; when the system is in motion, this field also contributes to the momentum. As we know, the quanta of the strong field are the gluons. Equation (41) therefore says that *about half of the nucleon's momentum is carried by gluons*.

The difference $G^{\gamma p} - G^{\gamma n}$ is also of interest. According to (37), contributions of the sea disappear from this quantity:

$$G^{\gamma p}(x) - G^{\gamma n}(x) = \tfrac{2}{3}a(x) - \tfrac{1}{3}b(x).$$

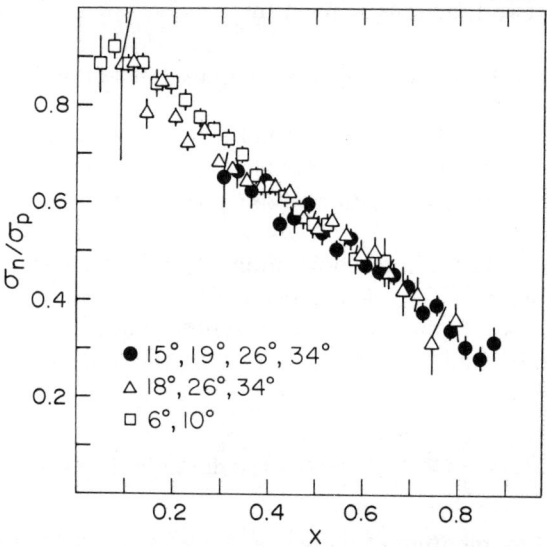

FIG. 4. The ratio of the neutron and proton deep inelastic scattering cross sections as a function of x, measured at different scattering angles (i.e., momentum transfers). Only in the region near $x \sim 0.4$ is the ratio $\simeq \frac{2}{3}$, as one would expect for pure valence quarks and identical momentum distributions for u- and d-quarks. [From A. Bodek et al., Phys. Rev. D20, 1471 (1979).]

By using (35) and (26), this gives

$$\int [G^{\gamma p}(x) - G^{\gamma n}(x)] \, dx = \int [F_2^{\gamma p}(x) - F_2^{\gamma n}(x)] \frac{dx}{x} = \frac{1}{3}. \qquad (42)$$

Unfortunately, it is difficult to verify this relation, since $F_2^{\gamma p}$ and $F_2^{\gamma n}$ are both finite for $x \to 0$, though their difference must vanish. Consequently, the two functions have to be known with great accuracy at low x if (42) is to be evaluated.

The ratio

$$\rho(x) = \frac{G^{\gamma n}(x)}{G^{\gamma p}(x)} \qquad (43)$$

can be measured by simply taking the ratio $d\sigma_n/d\sigma_p$ at the appropriate values of t and ν [see Eq. (20)]. This reveals interesting information about the quark distributions. If one ignores the sea, and simply puts $a = b$, one has $\rho(x) = \frac{2}{3}$. The data (Fig. 4) show that this is roughly correct for intermediate x only. On the other hand, at small x, the data say that $\rho \simeq 1$. This is consistent[12] with sea-quark dominance as $x \to 0$: $a/c \approx b/c = 0$.

---

[12] That result is not unique, because $2a = b$, $c = s = 0$, also gives $\rho = 1$.

As we shall see, the neutrino data unambiguously show that the sea gives a contribution peaked toward $x = 0$. From Fig. 4 we see that $\rho \simeq \frac{1}{4}$ as $x \to 1$. According to (37), $\frac{1}{4}$ is the minimum value that ρ can have, and it is attained when $b = c = s = 0$. This leads us to the conclusion that at x near 1 the dominant contribution comes from $a(x)$. i.e., from the valence quark that is doubly present in the nucleon. This is a surprising result that, at this time, has no rigorous explanation.[13]

6. Charge-changing neutrino scattering

(a) Cross sections for inelastic neutrino scattering

We now modify the considerations of the preceding sections so that they apply to deep inelastic scattering of neutrinos caused by the weak interaction. Here[14] we restrict ourselves to the part of the weak interaction that is transmitted by the charged intermediate bosons W^+ and W^-. Then the scattering mechanism transforms the neutrino into an electron or into a muon, depending upon whether it was an electron neutrino or a muon neutrino. There are important differences[15] between the scattering of charged leptons and of neutrinos because the weak interaction is transmitted by field quanta of very high mass, and also because it violates parity.

We start again by quoting, without derivation, a general expression analogous to (11) for the cross section for scattering of a neutrino or antineutrino by an arbitrary system of mass M whose constituents interact weakly with the neutrinos. The derivation only exploits general principles, and in particular, the 4-vector character of the weak charge-changing current. The resulting differential cross section $d\sigma^{\nu,\bar{\nu}}/dt\,d\nu$ for the scattering of neutrinos (or antineutrinos) through a momentum transfer q ($t \equiv q^2$) with an energy loss ν in units of M is given by[16]

$$\frac{d^2\sigma^{\nu,\bar{\nu}}}{dt\,d\nu} = \frac{G_F^2 \varepsilon'}{2\pi \varepsilon}\left\{\frac{1}{\nu}F_2^{\nu,\bar{\nu}}(t,\nu) - \frac{t}{4\varepsilon\varepsilon'}\right.$$
$$\left. \times \left[2F_1^{\nu,\bar{\nu}}(t,\nu) - \frac{1}{\nu}F_2^{\nu,\bar{\nu}}(t,\nu) \pm \frac{\varepsilon + \varepsilon'}{M\nu}F_3^{\nu,\bar{\nu}}(t,\nu)\right]\right\}. \quad (44)$$

---

[13] A possible explanation is that the known attraction between quarks of opposite spin favors configurations where such quarks are close, not only in ordinary space but also in momentum-, or x-space. If one quark is near $x = 1$, it requires the others to be near $x = 0$ [see (14)], and these should then have opposite spin which, in the case of the nucleon, is equivalent to differing flavor. Hence, the proton should have dominantly a u near $x = 1$, while in the case of the neutron, the d should dominate there.

[14] Deep inelastic neutrino scattering caused by Z^0 exchange is discussed in §VI.B.

[15] There is also a weak interaction, mediated by Z^0, between charged leptons and quarks. These effects are negligible at the energies discussed in the previous section, unless one designs an experiment that is sensitive to parity-violating effects. This has been done, and is discussed in §VI.B.3.

[16] We apologize for the confusing, lamentable, but standard usage ν for "neutrino" when it appears as a superscript, and for the energy loss of the lepton elsewhere!

Here G_F is the so-called Fermi constant (see footnote on p. 475), which is connected with the weak coupling constant g and the mass of the intermediate boson m_W by

$$G_F^2 = \frac{1}{32}\frac{g^4}{m_W^4}. \tag{44a}$$

Equation (44) is only valid for momentum transfers that are small compared to the mass m_W. The upper sign in (44) holds for neutrinos, the lower for antineutrinos; ε and ε' are the incident and the outgoing energies of the lepton, all in the lab-frame, $\mathscr{F}_L$; and $v = (\varepsilon - \varepsilon')/M$. In contrast to (11), this formula contains three structure functions; moreover, they are different for neutrinos and antineutrinos. The lack of parity conservation requires the term containing F_3. In principle, these expressions permit a determination of the six structure functions from the scattering data.

As in the case of electromagnetic scattering, we do not derive Eq. (44), because we show that it takes on a much simpler form when we consider the target to be a system of point quarks. There is again "scaling," that is, the functions $F_i^{\nu,\bar{\nu}}$ depend only on the variable x, and the Callan–Gross relation (12) between $F_1^{\nu,\bar{\nu}}$ and $F_2^{\nu,\bar{\nu}}$ remains valid.

(b) Neutrino-quark cross sections

In order to derive the neutrino scattering cross section, we again make use of the impulse approximation as it is described in §1. For this purpose we need expressions, analogous to (15), for the scattering of a neutrino or antineutrino by a free quark or antiquark. We first write down these expressions, and afterward explain how they follow from what we already know about the weak and electromagnetic interactions. The formulas in question are

$$\frac{d\sigma_1}{dt} = \frac{g^2 g_1^2}{32\pi m_W^4} \qquad \text{(for } vq \text{ and } \bar{v}\bar{q} \text{ scattering)} \tag{45}$$

$$\frac{d\sigma_2}{dt} = \frac{g^2 g_2^2}{32\pi m_W^4} \frac{(p \cdot k')^2}{(p \cdot k)^2} \qquad \text{(for } v\bar{q} \text{ and } \bar{v}q \text{ scattering)} \tag{46}$$

where p is the quark 4-momentum, while k and k' are the 4-momenta of the incident and outgoing leptons. The g_i are effective weak coupling constants for the different quarks:

$$g_1 = \begin{cases} g \cos \theta_C & \text{for } d \text{ and } \bar{d} \\ g \sin \theta_C & \text{for } s \text{ and } \bar{s} \\ 0 & \text{for } u \text{ and } \bar{u} \end{cases} \tag{47a}$$

$$g_2 = \begin{cases} g & \text{for } u \text{ and } \bar{u} \\ 0 & \text{for } d, \bar{d}, s, \text{ and } \bar{s} \end{cases} \tag{47b}$$

where θ_C is the Cabibbo angle, defined in §I.E.9(g).

The coupling constants (47) can be understood by recalling that ν_l changes into l^-, whereas $\bar{\nu}_l$ changes into l^+. Hence, neutrino scattering can only occur with a quark or antiquark that can change into another that has a charge larger by one unit, whereas antineutrino scattering requires a decrease of charge by one unit. Furthermore, we recall from §I.E.9(g) that the weak interaction produces the $\Delta Q = 1$ process $d' \to u$ with amplitude g, where d' is the combination $d \cos \theta_C + s \sin \theta_C$.

The dependence of the cross sections (45) and (46) on the kinematical variables can be understood by noting that the basic electromagnetic and weak processes have essentially the same Feynman diagram:

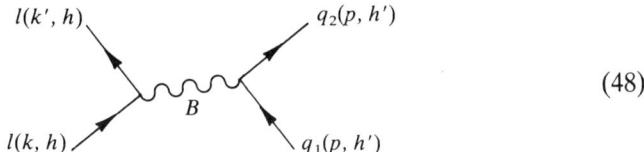

(48)

Here the l's designate leptons or antileptons of momenta k and k' and helicity h; the q's designate quarks or antiquarks of momenta p and p' and helicity h'; while B is the exchanged boson. In the electromagnetic case $B = \gamma$, the propagator is t^{-1}, while charge and flavor conservation imply $l_1 = l_2, q_1 = q_2$. For weak processes, $B = W^\pm$, and the propagator is $(t + m_W^2)^{-1}$, which is m_W^{-2} at the momentum transfers of interest, whereas $l_1 = \nu_l$ or $\bar{\nu}_l$, $l_2 = l$ or $\bar{l}$, and there are corresponding transitions among the quarks. If all momenta are large compared to the fermion masses, then *both interactions conserve fermion helicities*, as already implied by (48). The essential difference[17] between the two interactions (apart from the coupling constant and flavor dependence) is that electromagnetism has equal strength for *both* helicities, whereas the weak interaction *only* involves $h = -\frac{1}{2}$ fermions and $h = \frac{1}{2}$ antifermions [recall §I.E.9(f)]. Hence, in νq or $\bar{\nu}\bar{q}$ scattering, both particles have the same helicity, and no net angular momentum along the incident and outgoing directions in the c.o.m. frame $\mathcal{F}_0$, whereas in $\nu\bar{q}$ and $\bar{\nu}q$ there is one unit of angular momentum along these directions. As we learned in §II.C.1, in the first case the fermion spins do not affect the angular distribution, whereas in the second they contribute the factor $(p \cdot k')^2/(p \cdot k)^2$.

This explains all aspects of (45) and (46): They are related to the formulas

---

[17] Both are interactions between a 4-current carried by fermions with a 4-vector field. Helicity conservation then follows for reasons already given in §II.B.1(b). The detailed structure of the charge-changing weak current is derived in §VI.A.1.

for scattering of ultrarelativistic charged particles, Eqs. II.C(7), by $1/t \to 1/m_W^2$ and the substitution of coupling constants.[18]

(c) Neutrino-nucleon scattering

Equations (45) and (46) tell us how neutrinos are scattered by quarks. We proceed from them in the same way as we did from (17). The expressions (45) and (46) are valid for any quark momentum $p = xP$, where P is the nucleon's 4-momentum, and therefore $(p \cdot k')/(p \cdot k) = (P \cdot k')/(P \cdot k)$. We now multiply (45) with the probabilities of finding a quark or an antiquark of a given flavor with x between x and $x + dx$, and again introduce the distributions $u(x), d(x), s(x)$ and $\bar{u}(x), \bar{d}(x), \bar{s}(x)$ of the three quark and antiquark flavors. The ν and $\bar{\nu}$ cross sections then follow from (45) and (46), and the abbreviation (44a),

$$\frac{d^2\sigma^\nu}{dt\,dx} = \frac{1}{\pi} G_F^2 [\alpha d(x) + \beta s(x) + A\bar{u}(x)], \qquad (49)$$

$$\frac{d^2\sigma^{\bar{\nu}}}{dt\,dx} = \frac{1}{\pi} G_F^2 [\alpha \bar{d}(x) + \beta \bar{s}(x) + Au(x)], \qquad (50)$$

with
$$A = (P \cdot k')^2/(P \cdot k)^2,$$
$$\alpha = \cos^2 \theta_C, \quad \beta = \sin^2 \theta_C. \qquad (51)$$

A is, of course, related to B, as defined in (23), by $A = B - 1$. In the lab frame $\mathscr{F}_L$

$$A = \varepsilon'^2/\varepsilon^2. \qquad (52)$$

Interesting conclusions can be drawn from these formulas if one neglects the contributions of the antiquarks; there are but few of them in the nucleon. Then the term proportional to A drops out of σ^ν, whereas $\sigma^{\bar{\nu}}$ becomes proportional to A. This follows directly from (45) and (46) since, in this approximation, both neutrino and antineutrino scattering are only due to collisions with q's. In short, if sea quarks are neglected, the neutrino cross section, written as a function of t and x, does not depend on the momentum transfer t, only on x; for fixed x it is therefore independent of the lepton energy loss. Antineutrino scattering is proportional to A, however, and according to (52), A is proportional to the square of the energy ε' of the scattered lepton. This characteristic difference is shown by the experimental results of Fig. 5.

It is useful to integrate (49) and (50) over all t. (Such an integration is not

---

[18] The factor $\frac{1}{32}$ instead of $\frac{1}{8}$ comes about because $g/\sqrt{2}$ corresponds to e as the coupling constant; see Eqs. I.E(44) and (45).

6. CHARGE-CHANGING NEUTRINO SCATTERING

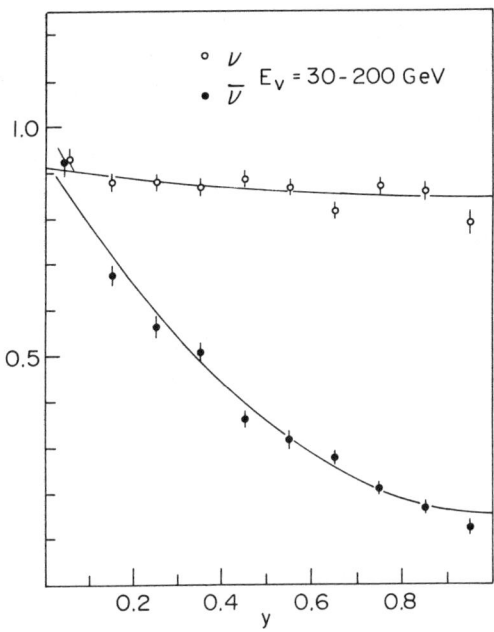

FIG. 5. The deep inelastic cross section of neutrinos and antineutrinos as function of the energy loss $\varepsilon - \varepsilon'$. The cross section is independent of $y \equiv (\varepsilon - \varepsilon')/\varepsilon$ for neutrinos, but quadratic in y for antineutrinos. [From de Groot et al., Z. Phys. C1, 143 (1979).]

possible for electron scattering, because of the familiar t^{-2} singularity at $t = 0$.) From (5) and (2) we get $t = -2x\mathbf{P} \cdot \mathbf{q} = 2x\mathbf{P} \cdot (\mathbf{k}' - \mathbf{k})$. Let $2(\mathbf{P} \cdot \mathbf{k}') \equiv u$ and $2(\mathbf{P} \cdot \mathbf{k}) \equiv s$. Then $t = x(u - s)$ and $dt = x \, du$, because[19] x is the other independent variable, while s, which is the square of the c.o.m. energy, is fixed. The only quantity depending on t in (49) and (50) is A. According to (51), $A = u^2/s^2$. The integrated cross section is then

$$\frac{d\sigma^{\nu,\bar{\nu}}}{dx} = x \int_0^s du \, \frac{d^2\sigma^{\nu,\bar{\nu}}}{dt \, dx}$$

because $s \geq u \geq 0$. Hence,

$$\begin{aligned}\frac{d\sigma^\nu}{dx} &= \frac{1}{\pi} G_F^2 sx[\alpha d(x) + \beta s(x) + \frac{1}{3}\bar{u}(x)], \\ \frac{d\sigma^{\bar{\nu}}}{dx} &= \frac{1}{\pi} G_F^2 sx[\alpha \bar{d}(x) + \beta \bar{s}(x) + \frac{1}{3}u(x)].\end{aligned} \quad (53)$$

[19] Here we again conform to a confusing but standard notation, and one should not confuse the kinematic variables u and s with the quark distribution functions designated by the *same* symbol!

In terms of the lab-frame energy ε, $s = 2M\varepsilon$, and Eq. (53) therefore states that the neutrino-nucleon cross sections are proportional to the neutrino beam energy. As long as $s \ll m_W^2$, this energy dependence is a direct consequence of the assumed pointlike structure of leptons and quarks, as one can see from the neutrino-quark cross sections, (45) and (46). The data are in reasonably good agreement with this expected linear growth of the cross sections (see Fig. 6). *This provides striking confirmation of the assumption that quarks and leptons are pointlike sources of the weak interaction field.*

We can obtain some information about the ratio between neutrino and antineutrino cross sections from (53) by neglecting βs and $\beta \bar{s}$, since β, $\bar{s}$, and s are small, and by approximating $\alpha = \cos^2 \theta_C$ by one:

$$\frac{d\sigma^{\nu,\bar{\nu}}}{dx} \approx \frac{1}{\pi} G_F^2 sx \begin{cases} [d(x) + \tfrac{1}{3}\bar{u}(x)] & \text{for } \nu \\ [\tfrac{1}{3}u(x) + \bar{d}(x)] & \text{for } \bar{\nu} \end{cases}. \quad (54)$$

Consider the sum of the cross sections for protons and neutrons. The latter arises from the former by interchanging u and d. Neglecting the sea contributions, (54) gives

$$\frac{d\sigma^{\nu n}}{dx} + \frac{d\sigma^{\nu p}}{dx} = 3\left(\frac{d\sigma^{\bar{\nu} n}}{dx} + \frac{d\sigma^{\bar{\nu} p}}{dx}\right).$$

Since this relation is approximately valid for all x, we conclude that the neutrino cross section for a nucleus with equal numbers of protons and neutrons is approximately three times the antineutrino cross section. As seen from (54), the sea contribution is three times larger for antineutrinos than for neutrinos. Hence, the ratio 3 is an upper limit, but a reasonably close one:

$$\sigma^{\nu n} + \sigma^{\nu p} \leq 3(\sigma^{\bar{\nu} n} + \sigma^{\bar{\nu} p}). \quad (55)$$

This relation is borne out by the observations shown in Fig. 6.

(d) Structure functions

To extract the structure functions from (49) and (50), we must cast these expressions in the form of Eq. (44). For this purpose we introduce

$$\delta(x) = \alpha d(x) + \beta s(x), \qquad \bar{\delta}(x) = \alpha \bar{d}(x) + \beta \bar{s}(x). \quad (56)$$

Note that $\delta(x) \approx \alpha d(x)$, and that β, $\bar{s}(x)$, and $s(x)$ are small quantities. The expression in the bracket of (49) becomes

$$\delta(x) + A\bar{u}(x) = \tfrac{1}{2}(1+A)(\delta + \bar{u}) + \tfrac{1}{2}(1-A)(\delta - \bar{u}). \quad (57)$$

Recall that $1 + A = B$, where the latter is given by (23). In combination

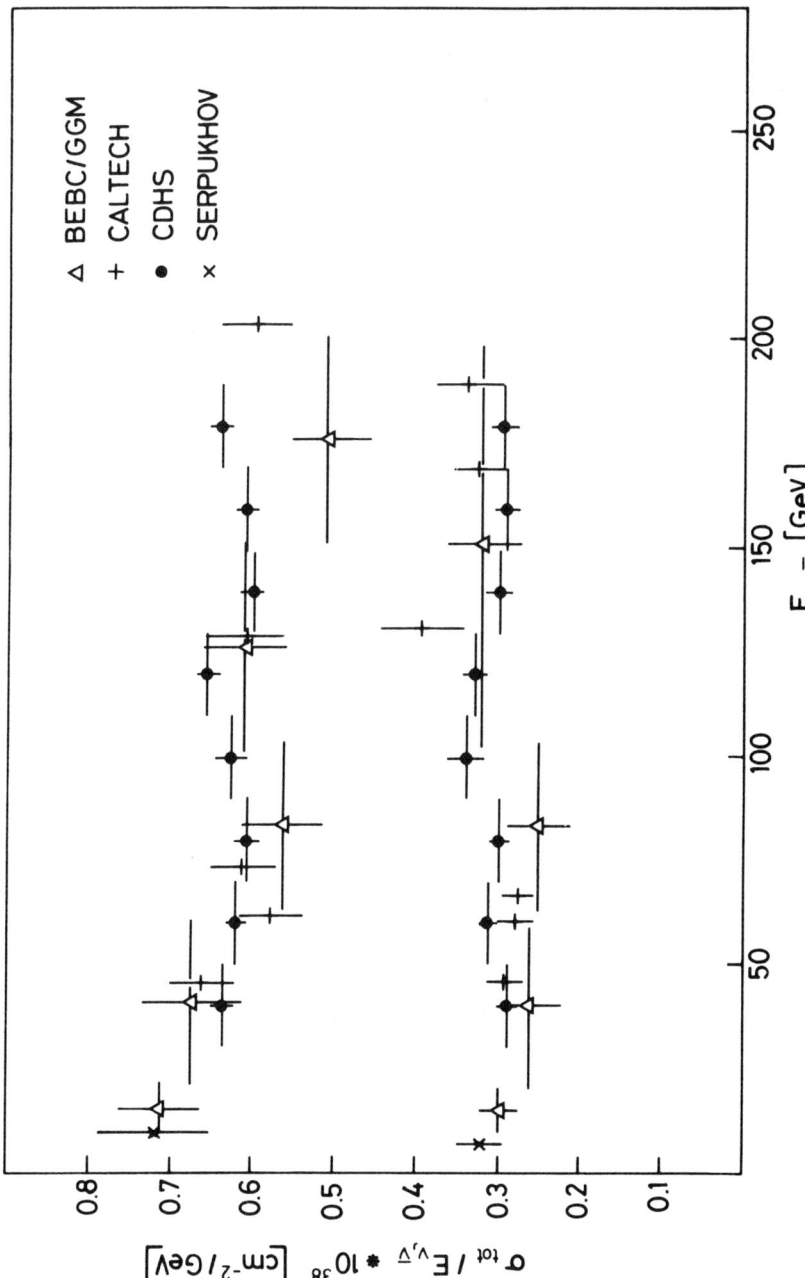

FIG. 6. The total deep inelastic neutrino and antineutrino cross sections per nucleon divided by the incident energy, demonstrating the linear dependence on energy. The ratio between neutrino and antineutrino cross sections is somewhat less than three. (From deGroot et al., loc. cit.)

with (19) and (52) this also yields

$$1 - A = -2\frac{\varepsilon'\varepsilon + \varepsilon'}{\varepsilon}\frac{t}{Mx\,4\varepsilon\varepsilon'}. \tag{57'}$$

Finally, we express the differential dx in terms of dv by using (22), and use (24), (57), and (57') in (49):

$$\frac{d^2\sigma^\nu}{dt\,dv} = \frac{G_F^2\,\varepsilon'}{\pi\,\varepsilon}\left\{\frac{x}{v}(\delta(x) + \bar{u}(x)) - \frac{t}{4\varepsilon\varepsilon'}\left[(\delta(x) + \bar{u}(x))\right.\right.$$
$$\left.\left. + \frac{(\varepsilon + \varepsilon')}{Mv}(\delta(x) - \bar{u}(x))\right]\right\}.$$

The corresponding cross section for antineutrinos is found by replacing $\delta(x)$ by $\bar{\delta}(x)$, and $\bar{u}(x)$ by $u(x)$. Comparing this with (44), we obtain the following expressions for the structure functions:

$$F_2^\nu(x) = 2x(\delta(x) + \bar{u}(x)); \qquad F_2^{\bar\nu}(x) = 2x(\bar\delta(x) + u(x));$$
$$2F_1^\nu(x) - \frac{1}{v}F_2^\nu(x) = 2(\delta(x) + \bar{u}(x));$$
$$2F_1^{\bar\nu}(x) - \frac{1}{v}F_2^{\bar\nu}(x) = 2(\bar\delta(x) + u(x)); \tag{58}$$
$$F_3^\nu(x) = 2(\delta(x) - \bar{u}(x)); \qquad F_3^{\bar\nu}(x) = 2(u(x) - \bar\delta(x)).$$

These are the analogues of the expressions (26)–(29) for the electromagnetic structure functions. As before, the functions F_1, F_2, and F_3 scale—they depend only on x—and the Callan–Gross relation (12) is fulfilled by F_i^ν and $F_i^{\bar\nu}$ ($i = 1, 2$), when $v \gg 1$ is taken into account.

In analogy to (26) and (37), we may express the neutrino structure function of the proton and neutron in terms of the valence and sea-quark distributions, as defined in (32), (33), and (34):

$$F_2^{\nu p} = F_2^{\bar\nu n} \simeq 2x(b + 2c),$$
$$F_2^{\nu n} = F_2^{\bar\nu p} \simeq 4x(a + c),$$
$$F_3^{\nu p} = F_3^{\bar\nu n} \simeq 2b, \tag{59}$$
$$F_3^{\nu n} = F_3^{\bar\nu p} \simeq 4a.$$

Here we used the approximations $\alpha = 1$ and $\beta = 0$.

The neutrino structure functions (58) can be extracted from the data (see Fig. 7) by means of (44); they are related to the different quark

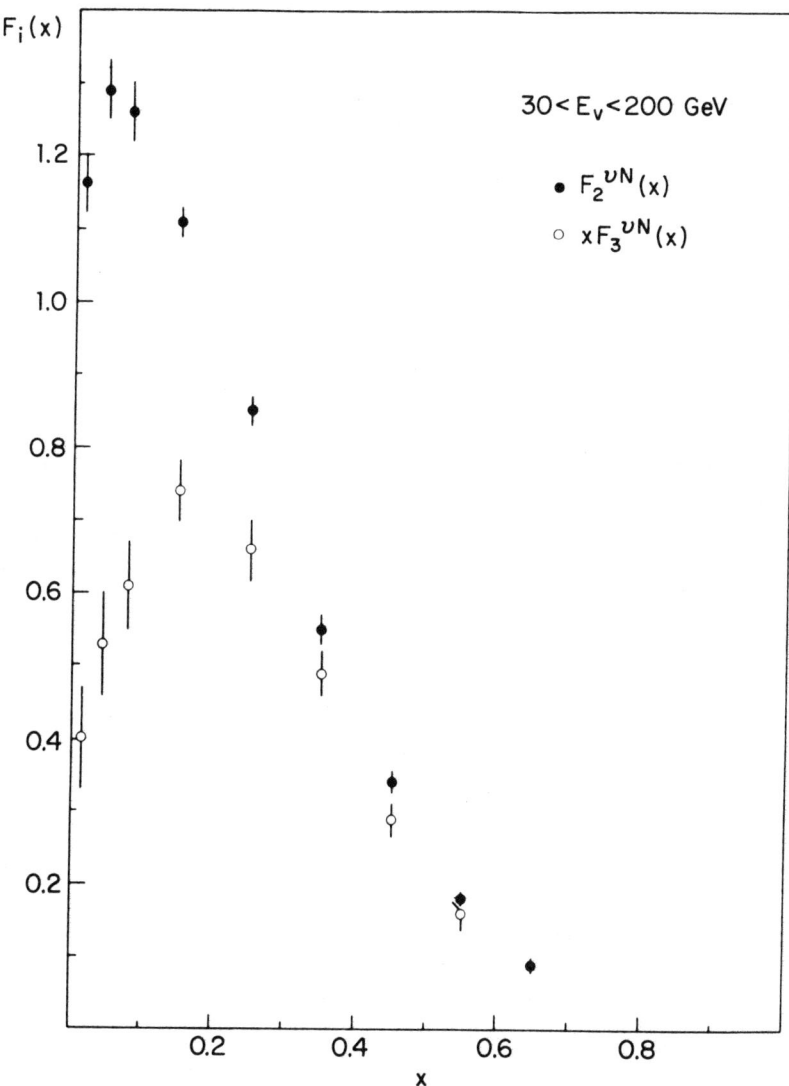

FIG. 7. The structure functions $F_2^{\nu N}$ and $xF_3^{\nu N}$, as observed in deep inelastic scattering of neutrinos by nuclei containing equal numbers of protons and neutrons. These functions are the average of the proton and neutron functions, and are averages over the data in the energy interval $30 < E_\nu < 200$ GeV. (From de Groot et al., loc. cit.)

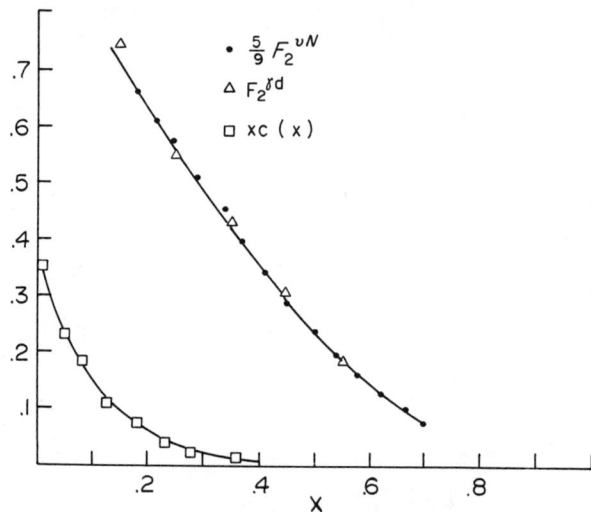

FIG. 8. Comparison of $F_2^{\gamma d}$ and $F_2^{\nu N}$ observed via electromagnetic and weak interactions, respectively. Here $F_2^{\gamma d}$ is the sum of $F_2^{\gamma p}$ and $F_2^{\gamma n}$, because deuterium is the target. The theoretically expected ratio 5/9 is observed to be a good approximation. The figure also contains the distribution function $xc(x)$ of the antiquarks [see Eq. (32)], as determined from the observed structure functions in accordance with Eq. (59). $F_2^{\nu N}$ is defined in the caption of Fig. 7. (Data from SLAC-MIT, private communication; and de Groot et al.)

distributions. In particular, we find

$$\begin{aligned}\delta(x) &\approx \alpha\, d(x) = \tfrac{1}{4}(F_2^\nu(x)/x + F_3^\nu(x));\\ u(x) &= \tfrac{1}{4}(F_2^{\bar\nu}(x)/x + F_3^{\bar\nu}(x));\\ \bar u(x) &= \tfrac{1}{4}(F_2^\nu(x)/x - F_3^\nu(x));\\ \bar\delta(x) &= \tfrac{1}{4}(F_2^{\bar\nu}(x)/x - F_3^{\bar\nu}(x)).\end{aligned} \quad (60)$$

The expressions for $\bar u$ and $\bar\delta$ are of special interest since, in contrast to the F_i^γ, the neutrino structure functions give us direct information about the distribution of antiquarks in the nucleon. Neutrino scattering measurements are difficult, especially for small values of x; therefore, the results are not too accurate. Figure 8 shows the measurement of $c(x)$, which is expected to be identical with $\bar u(x)$ and $\bar d(x)$. These results indicate the $1/x$ dependence for small x which follows from theoretical considerations that will be developed in the next Section.

There is a simple relation between the structure functions determined by electromagnetic and weak interactions. Let us make the approximation $s \approx 0$ in (37), and then add the proton and neutron structure functions F_2 with the help of (26). Then we compare this with (59), and find that

$$\tfrac{9}{5}(F_2^{\gamma p}(x) + F_2^{\gamma n}(x)) = \tfrac{1}{2}(F_2^{\nu p}(x) + F_2^{\nu n}(x)). \quad (61)$$

The sum $F_2^p + F_2^n$ is found by the analysis of deep inelastic scattering from nuclei with the same number of protons and neutrons. Indeed, relation (61) is verified within the accuracy of observations, as shown by Fig. 8. This represents good evidence for the validity of the quark model.

7. Theoretical considerations regarding the quark distributions

⟦The analysis of the deep inelastic scattering enables us to determine experimentally, with a reasonable degree of accuracy, the various distribution functions ($u(x), d(x), s(x), \bar{u}(x)$, etc.) for the different quark flavors in the nucleon. These are *not* the distributions of quark momenta in a nucleon at rest, but of the z-components of the quark momenta in a nucleon that moves with almost the velocity of light in the z-direction. Obviously, the two distributions are connected by a Lorentz transformation. It is not easy to establish this connection, because the quarks are not free particles, but are bound by the strong forces within the nucleus. In order to perform this transformation a knowledge of the bound states is necessary, which is not available today.

In this section we present a crude treatment of this problem, in order to show that the observed shapes of the distribution functions are indeed what one would expect from a plausible distribution in a nucleon at rest. We do this in a rather inexact way by neglecting the binding forces between the quarks, and other crude approximations. Correspondingly, the results serve only as a first orientation, and cannot be considered as quantitatively correct.

(a) The quark momentum distribution in the nucleon

Let us consider the nucleon in the frame $\mathcal{F}_L$ in which it is at rest ($\mathbf{P} = 0$). We begin by taking only the three valence quarks into account. We expect that the quark momenta will be of the order of M or smaller.[20] Let us therefore express them in units of M:

$$p_z = \beta M, \qquad p_x = \zeta_1 M, \qquad p_y = \zeta_2 M.$$

We introduce the following quantities

$$\zeta = (\zeta_1^2 + \zeta_2^2)^{1/2}, \qquad \alpha = (\beta^2 + \zeta^2)^{1/2}, \tag{62}$$

where α is the absolute value of the momentum of the quark in units of M.

Now we define the probability of finding a quark with the values β, ζ_1, ζ_2 in the intervals $d\beta, d\zeta_1, d\zeta_2$:

$$F(\alpha) \, d\beta \, d\zeta_1 \, d\zeta_2. \tag{63}$$

The nucleon is spherically symmetric in $\mathcal{F}_L$, so F will be a function of α only,

---

[20] The momenta are of the order R^{-1}, where R is the radius of the nucleon; $R \sim 10^{-13}$ cm, so $R^{-1} \sim M/5$.

and its integral over all β, ζ_1, ζ_2 gives unity. Hence,

$$\int F(\alpha)\, d\beta\, d\zeta_1\, d\zeta_2 = 4\pi \int F(\alpha)\alpha^2\, d\alpha = 1. \tag{64}$$

It is convenient to use only a single polar variable ζ, as defined in (62), instead of the Cartesian variables ζ_1 and ζ_2. Then the probability $F^*(\alpha)\, d\beta\, d\zeta$ for finding a given valence quark in the intervals $d\beta$ and $d\zeta$ is

$$F^*(\alpha)\, d\beta\, d\zeta = 2\pi F(\alpha)\, d\beta\, \zeta\, d\zeta. \tag{65}$$

In what follows we neglect the binding forces and consider all bound quarks within the nucleus as a swarm of freely moving quarks with the same[21] momentum distribution (65). Obviously, such an ensemble does not have a definite energy, in contrast to the actual ground state of the nucleon. We simplify our problem further by assuming that the quark masses are so small compared to M that we may consider them massless. This seems to be a good approximation for the u and d valence quarks in the nucleon. Then the fourth component of the quark momentum becomes

$$p_0 = \alpha M. \tag{62'}$$

What does the momentum distribution (65) look like in a rapidly moving frame? We perform a Lorentz transformation from $\mathscr{F}_L$ to a frame in which the nucleon has a momentum $P \gg M$ in the positive z-direction. The velocity v of this frame is

$$v = \frac{P}{\sqrt{P^2 + M^2}} \approx 1, \qquad \gamma \equiv \frac{1}{\sqrt{1 - v^2}} = \frac{\sqrt{P^2 + M^2}}{M} \approx P/M. \tag{66}$$

The quark momenta p given by (62) become p' in this frame:[22]

$$\begin{aligned} p'_z &= \gamma(\beta M + \alpha M v) \approx (\beta + \alpha)P, \\ p'_0 &= \gamma(\alpha M + \beta M v) \approx (\alpha + \beta)P. \end{aligned} \tag{67}$$

We may therefore identify the variable x introduced in (16) with $\alpha + \beta$ for each quark,

$$x = \alpha + \beta, \tag{68}$$

because it determines the fraction of the nucleon's momentum that the quark carries, and satisfies

$$0 \leq x \leq 1. \tag{69}$$

---

[21] From the discussion on p. 438 about $a(x)$ and $b(x)$, we know that this is not quite correct.
[22] At this point our approximation suffers from a significant inadequacy. The Lorentz transformation necessarily involves the energy, and therefore the interaction between quarks. Hence the Lorentz transformation of the nucleon's momentum distribution is, in reality, more complicated than that of a swarm of free particles.

7. THEORETICAL CONSIDERATIONS

The first inequality follows directly from (62); the second because no quark can have a momentum larger than P, since negative x are excluded. Note that α and ζ are positive, but β can be negative: $\beta > -1$.

The expressions (62) and (68) determine the relations between the variables (β, ζ) and (x, α). We can therefore transform (65) into the probability $\hat{F}(x, \alpha)\, dx\, d\alpha$ of finding a quark in the intervals dx and $d\alpha$:

$$\hat{F}(x, \alpha)\, d\alpha\, dx = F^*\, d\beta\, d\zeta = F^* J\, d\alpha\, dx,$$

where J is the Jacobian $\partial(\beta, \zeta)/\partial(\alpha, x)$. It is easiest to compute J^{-1}, which equals ζ/α. Equation (65) therefore yields

$$\hat{F}(x, \alpha) = F^* J = 2\pi \alpha F(\alpha). \tag{70}$$

We arrive at the distribution $G(x)\, dx$ of the quarks by integrating (70) over all values of α compatible with a given x:

$$x = \alpha + \beta = \sqrt{\beta^2 + \zeta^2} + \beta. \tag{71}$$

Consider the case where the quark momentum is parallel to the z-axis in $\mathcal{F}_L$. Then $\zeta = 0$, $\alpha = \beta$, and it follows from (71) that $\alpha = \beta \leq \frac{1}{2}$, since $x \leq 1$. But the state is spherically symmetric; we conclude therefore that for all directions of the quark momentum, $\alpha = \sqrt{\beta^2 + \zeta^2} \leq \frac{1}{2}$, since $\sqrt{\beta^2 + \zeta^2}$ would be β in a coordinate system where the momentum is parallel to the z-axis. Thus we get the x-independent upper limit[23] $\alpha_{\max} = \frac{1}{2}$. From (71), $\beta = (x^2 - \zeta^2)/2x$, from which it follows that

$$\alpha = (\beta^2 + \zeta^2)^{1/2} = \frac{x^2 + \zeta^2}{2x}.$$

Since $\zeta^2 \geq 0$, we find the x-dependent lower limit to be $\alpha_{\min} = \frac{1}{2}x$. Now we can determine the valence quark x-distribution by integrating (70) over α. These distributions $a(x)$ and $b(x)$ were defined by (33) and (34). In our crude treatment there is no difference between them and we call both $a_0(x)$:

$$a_0(x)\, dx = 2\pi\, dx \int_{x/2}^{1/2} F(\alpha)\alpha\, d\alpha. \tag{72}$$

We emphasize that this relation is a very crude approximation based on a Lorentz transformation of the motion of free massless quarks with a momentum distribution $F(\alpha)$ in the rest system. Bound massless quarks would transform somewhat differently because their energies are not equal to $|\mathbf{p}|$.

Since we do not have a theory of the quark wave function, we cannot determine $F(\alpha)$. We expect a decreasing function of α that reaches zero at the maximum value $\alpha = \frac{1}{2}$. The following analytic form is a plausible expression which, as we will see, leads to quark distributions for relativistic nucleons that

---

[23] This relation is true only in our simplified picture of free massless quarks.

resemble those that are observed:

$$F(\alpha) = \begin{cases} B(1 - 4\alpha^2) & \text{for } \alpha < \tfrac{1}{2} \\ 0 & \text{for } \alpha > \tfrac{1}{2}. \end{cases} \quad (73)$$

Normalization according to (64) gives $B = 15/\pi$. Putting this into (72) we arrive at

$$a_0(x) = \tfrac{15}{8}(1 - x^2)^2. \quad (74)$$

This expression for the quark distribution functions cannot be compared directly with the experimental results of Figs. 1 and 7. First, we have not yet considered the presence of gluons, which will change the quark distribution (74). Second, the gluons produce sea quarks, which will contribute importantly to the distributions at low x-values. That is why we now turn to the effects of the gluon field.⟧

(b) The gluon field in the nucleon

⟦In the preceding subsection we considered the nucleon as composed of three "valence" quarks, and we established a connection between the momentum distribution $F(\alpha)$ of a valence quark and the structure functions $a(x)$ and $b(x)$.

The quarks are the sources of the color gauge fields that keep the quarks bound in hadrons. According to this picture, the discussion of the previous section is incomplete: A fast-moving nucleon is not just the three quarks in motion; it also contains the moving color field of these quarks. As we will see, a part of these fields must be considered as "detached," that is, as free gluons moving along with the quarks. These detached gluons are of great importance because the leptons interact only with the quarks, and not with gluons. This has two consequences. First, the detached gluons carry part of the momentum of the quarks. Hence, the quark distribution responsible for the observed scattering of the electrons is not $a_0(x)$ as calculated before, but a different function, in which the gluon part has been removed. (A quark whose total momentum (including the field) was $x'P$ may appear as having a momentum xP where $x < x'$.) Secondly, the gluons give rise to secondary quark-antiquark pairs that also "run along" with the three valence quarks. We have called these pairs the "sea quarks."

This subsection provides a rough estimate of the momentum distribution of the detached gluons that accompany the valence quarks in a scattering process. We use this distribution to determine the change in the quark distribution $a_0(x)$, and the sea distribution function $c(x)$, as defined in (32).

For the purpose of these rather rough estimates we will assume that QCD is quite similar to QED for processes with 4-momentum transfers much larger than the nucleon mass M. But there are two important differences. First, the coupling constant g^2 between quarks and the gluon field is considerably larger[24] than e^2 and decreases with increasing momentum transfer (asymptotic

---

[24] There are eight distinct gluons. Therefore, the effective coupling constant g^2 used here is somewhat larger than the QCD coupling constant that determines the emission of one specific gluon.

7. THEORETICAL CONSIDERATIONS

freedom). Second, the quarks and the gluons in a nucleon are all confined within a volume ("bag") of a radius R which, roughly, is of the order $R \sim \Lambda_0^{-1}$. Therefore, no gluon fields are produced by a quark at distances larger than Λ_0^{-1}.

Because of the close analogy with electromagnetic phenomena, we start by considering the electromagnetic field[25] of a free particle with mass m, charge e, and momentum γm, moving with a velocity $v \approx c (\gamma \gg 1)$ in the z-direction. The electric field is a Lorentz contracted Coulomb field; the field $\mathbf{E}$ in the direction perpendicular to the motion is enhanced by a factor γ, whereas the field in the direction of motion is reduced by a factor γ^{-1}. The magnetic field is $\mathbf{B} = \mathbf{v} \times \mathbf{E}$. Thus, for large γ, the field strengths are essentially perpendicular to the direction of motion, and to each other. Therefore, they can be approximated by light waves moving in the same direction. (This is called the Weizsäcker–Williams approximation.) The energy contained in those light waves can be calculated in the following way. Consider a point P at the distance r from the trajectory. The electric and magnetic field $E_\perp$ and $B_\perp$ perpendicular to the motion (the other components are negligible) are given as a function of time t by

$$E_\perp = B_\perp = \frac{\gamma r e}{4\pi (r^2 + \gamma^2 t^2)^{3/2}} \tag{75}$$

if $t = 0$ is the time at which the particle is nearest to P. The fields (75) represent a light pulse of a duration $\tau \sim r/\gamma$ and, therefore, contain frequencies ω in the interval $0 < \omega < \gamma/r$. Indeed, a Fourier analysis of the pulse (77) gives rise to a frequency spectrum I (energy per unit area per unit frequency interval $d\omega$) at the distance r from the trajectory

$$I(r, \omega)\, d\omega = \frac{e^2}{4\pi^3} \frac{1}{r^2} F\left(\frac{\omega r}{\gamma}\right) d\omega, \tag{76}$$

where $F(x) = [x K_1(x)]^2$ and K_1 is a modified Bessel function.[26] The function $F(x)$ is unity for $x \ll 1$ and goes exponentially to zero when $x \gg 1$.

The light pulse (75) can be thought of as the Lorentz contracted Coulomb field, compressed into a disk perpendicular to the trajectory and moving with the particle. The field energy $\varepsilon_f(r, \omega)\, dr\, d\omega$ contained in that disk within the radial interval dr and the frequency interval $d\omega$ follows from (76):

$$\varepsilon_f(r, \omega)\, dr\, d\omega = 2\pi r I(r, \omega)\, dr\, d\omega. \tag{77}$$

Approximating $F(x)$ by a step function, $F(x) = 1$ for $x < 1$, $F(x) = 0$ for $x > 1$, we see that the energy vanishes for $\omega > \gamma/r$, whereas

$$\varepsilon_f(r, \omega)\, dr\, d\omega \approx \frac{e^2}{2\pi^2} \frac{dr}{r}\, d\omega \qquad \text{for } \omega < \frac{\gamma}{r}. \tag{78}$$

---

[25] See Feynman (1964), §26–2.
[26] As to the definition of K_1 and the details of the Fourier analysis, see Jackson (1975), pp. 107, 554, 625, 721, 722.

We now make a distinction between the field tied to the particle, and the field consisting of "free" photons accompanying the particle. The energy of the former is part of the energy of the particle itself; the energy of the latter is ε_{ph}. This division of the field depends on what happens to the particle during the scattering process. If the particle is not scattered at all, the whole field must, by the principle of relativity, be considered as tied to the particle. If the particle suffers a 4-momentum transfer Q, part of the field becomes detached. In particular, the parts of the field further away from the particle do not "know" what happens inside, and move on as free photons. Since $Q \equiv |Q_0^2 - \mathbf{Q}^2|^{1/2}$ determines a length $|Q|^{-1}$ within which the scattering process takes place, as it were, we assume that the field outside the distance

$$r_{min} \simeq \frac{1}{|Q|} \tag{79}$$

should be regarded as free photons that have detached themselves from the particle; whereas, the field inside r_{min} is a part of what we consider the particle.

The energy (or the momentum, since they all move parallel to the z-direction) of the detached photons at the distance r is again given by (78) as long as $r > r_{min}$. Let us determine the total momentum P_{ph} carried by the detached photons. We first integrate (78) over ω from zero to the maximum frequency Ω of a detached photon. No detached photon can have a momentum larger than $p = \gamma m$, the 3-momentum of the particle; hence, $\Omega \leq p$. But the expression (78) extends over frequencies up to γ/r. Therefore $\Omega = p$ when $(\gamma/r) > p$, while $\Omega = \gamma/r$ when $(\gamma/r) < p$. The transition occurs when $r = r_0$, where $r_0 = \gamma/p = m^{-1}$. Consequently,

$$P_{ph} = \frac{e^2}{2\pi^2} \left(p \int_{r_{min}}^{r_0} \frac{dr}{r} + \gamma \int_{r_0}^{\infty} \frac{dr}{r^2} \right) = \frac{2\alpha}{\pi} p \left(\ln \frac{Q}{m} + 1 \right), \tag{80}$$

where $\alpha = e^2/4\pi$. We see that the momentum carried by the detached photons increases logarithmically with the momentum transfer Q.

Let us now translate this result to the case of quarks and gluons. First, we replace α by the strong-interaction coupling constant α_s. Then we take the maximum extension $r_{max} = \Lambda_0^{-1}$ of the gluon field into account. The momentum of the quark is given by $p = x\gamma M$. Hence the corresponding $r_0 = \gamma/p$ becomes $r_0 = (xM)^{-1}$. Since $\Lambda_0 \sim M$ and $x < 1$, we find that $r_0 \gtrsim \Lambda_0^{-1}$. Therefore, the first integral in (80) with an upper limit Λ_0^{-1}, will dominate. Thus, the momentum P_g carried by the detached gluons is[27] $P_g = \eta p$, where

$$\eta \approx \frac{2\alpha_s}{\pi} \int_{r_{min}}^{\Lambda_0^{-1}} \frac{dr}{r} = \frac{2\alpha_s}{\pi} \ln \frac{Q}{\Lambda_0}. \tag{81}$$

---

[27] Three remarks are in place here. The finite range of the field implies that the lowest frequency is not zero, but Λ_0; thus the integration of (78) over frequency should really go from Λ_0 to p. This is a negligible correction because $p \gg \Lambda_0$. Furthermore, we have slurred over the fact that there are eight distinct gluons, whereas there is only one photon. Therefore, the coupling constant α_s in (81) should be multiplied by a factor. We left this out since we will never use the actual value of α_s in this Chapter. Finally, the gluons accompanying the quark may, in turn, produce quark pairs, so that the momentum carried by pure gluons will be less than (81). We will neglect this higher order effect, in spite of the fact that η is relatively large. The production of quark pairs will be taken into account in §7(c).

7. THEORETICAL CONSIDERATIONS

The value of η depends on Q, but it is a very slowly varying function. It shows an explicit logarithmic increase with $|Q|$, but α_s is a decreasing function of the momentum transfers involved[28] [see Eq. IV.C(71)]. It is likely that η increases slightly with Q and reaches a constant for very high values. The actual value of η can be taken directly from the experiments. In §5 we learned that the fraction of the momentum of the nucleon carried by the quarks is about half of the total [see Eq. (41)]. Thus we can set

$$\eta \approx 0.5. \tag{82}$$

Let us now determine the number of gluons $n_g(\omega)\,d\omega$ in a frequency interval $d\omega$. We obtain this from (78) by dividing by ω, replacing α by α_s, and integrating over r, with $r_{max} = \Lambda_0^{-1}$:

$$n_g(\omega)\,d\omega = \frac{2\alpha_s}{\pi} \ln \frac{Q}{\Lambda_0} \frac{d\omega}{\omega} = \eta \frac{d\omega}{\omega}. \tag{83}$$

This number depends only on the momentum transfer Q, but not on the 3-momentum p of the quark. The total number of detached gluons is obtained by integrating (83) from Λ_0 to $\omega_{max} = p$:

$$n_g = \eta \ln \frac{p}{\Lambda_0}. \tag{84}$$

An important observation must be made at this point. One may find it contradictory that the number of gluons emitted by a quark depends logarithmically upon the 3-momentum p of the quark. Should this number not be a relativistic invariant? It is not, because these gluons are in virtual states; they are not on the mass shell, but very near to it. The field of the quark beyond the radius (79) is not exactly a beam of gluons. The higher the momentum **p**, the more of the total field becomes "gluonlike" (or "photonlike" in the case of a rapidly moving charge), and therefore the number of gluons increases.

Now we must reinterpret the momentum distribution (74) of the quarks that we had calculated before. It is not (74) that will be observed in deep inelastic lepton scattering, because part of the quark momentum, as defined in (74), is carried by the detached gluons. The distribution $a_0(x)$ would be valid if there were no momentum transfer at all ($Q = 0$). It gives the momentum distribution of the valence quarks, including the whole gluon field. We now obtain an approximate expression for the true valence quark distribution, $a(x)$, excluding the detached gluons. In order to simplify our procedure, we assume that η is small, so that we only need to consider first-order terms in η. Evaluating these terms with the actual value of η will give us a first rough orientation as to the character of the function $a(x)$.

We are considering an effect that can be called gluon "bremsstrahlung," as

---

[28] We must distinguish between the momentum transfer Q between the lepton and the quark, and the one from the quark to the gluons. The latter enters into the value of α_s in (81). However, we ignore these details in our qualitative discussion.

indicated by the Feynman diagram (85a):

(a)   (b)

We only keep terms in first order of the coupling constant α_s. Second-order diagrams, such as (85b), lead to quark pair production by the detached gluons; this is treated in the next subsection.

The distribution $a_0(x)$ is reduced because the quark with momentum $p = xP$ is accompanied by gluons of frequencies between Λ_0 and p, whose total number is given by (84). This reduction amounts to

$$a_0(x) \int_0^p n_g(\omega) \, d\omega,$$

where $n_g(\omega)$ is given by (83). Here we have set the lower limit equal to zero[29] instead of Λ_0, since the latter is small compared to p. On the other hand, $a_0(x)$ will be increased by those quarks with $x' > x$ that are accompanied by gluons with frequency $\omega = (x' - x)P$. This increase is

$$\int_p^P a_0(x') n_g(\omega' - xP) \, d\omega', \quad \left(x' = \frac{\omega'}{P}\right).$$

On combining these changes and using (83), we obtain

$$a(x) = a_0 \left[1 - \eta \int_0^x \frac{dx'}{x'}\right] + \eta \int_x^1 \frac{a_0(x')}{x' - x} \, dx'. \tag{86}$$

This expression[30] can be written in the form

$$a(x) = a_0(x) + Y(x),$$
$$\frac{1}{\eta} Y(x) = a_0(x) \int_0^1 \frac{dx'}{x' - x} + \int_x^1 \frac{a_0(x') - a_0(x)}{x' - x} \, dx'. \tag{87}$$

The result for $a(x)$ is shown in Fig. 9, with $\eta = 0.5$. As expected, we find $a(x)$ diverges like $\log(x^{-1})$ for $x \to 0$, so that $xa(x)$ goes to zero at $x = 0$. In our approximation the distribution $b(x)$, as defined by (34), is identical with $a(x)$.

We emphasize again the very approximate nature of procedure in which higher powers of α_s have been neglected. Indeed, the integral $\int_0^1 a(x) \, dx$ should remain unity, but is 0.98 with our choice of constants.]

---

[29] This introduces an apparent singularity that will be resolved in (87).
[30] Note that the singularity of the first integral in (86) is canceled by that of the second. The first integral in (87) is meant to be the principal value: $\int_0^1 dx'/(x' - x) = \log(1 - x)/x$.

7. THEORETICAL CONSIDERATIONS

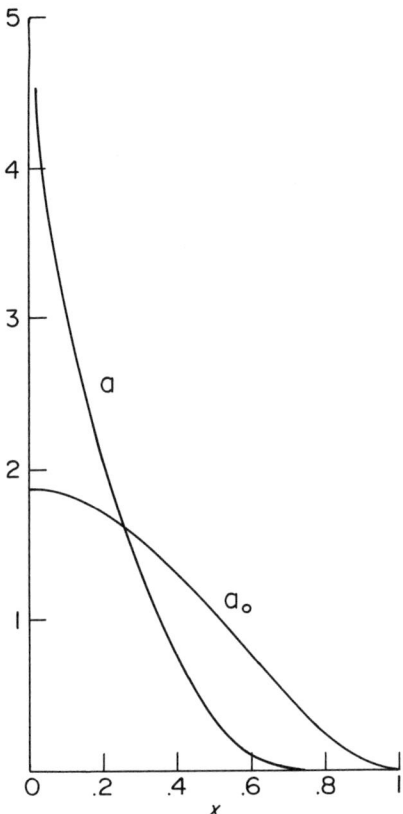

FIG. 9. The distribution function $a_0(x)$ as given by (74), and $a(x)$ as given by (87).

(c) The sea quarks

[We now proceed to a very rough estimate of quark–antiquark pairs contained in a fast-moving nucleon. They are produced by the gluons accompanying the valence quarks, as diagram (85b) shows. This is a second-order process in the coupling constant α_s. We introduce the probability Π_f that a given gluon creates a quark–antiquark pair of flavor f, moving essentially in the same direction. This probability is proportional to α_s, and therefore also to $\eta : \Pi_f = \xi_f \eta$, where ξ_f is somewhat smaller than unity. We assume that Π_f does not depend on the frequency of the gluon. Consider a gluon of momentum ω. The pair into which it may transform must also have the momentum $\mathbf{p}_q + \mathbf{p}_{\bar{q}} = \omega$. We further assume that all sharings between the quark–antiquark momenta are equally probable. Then the probability that the gluon produces a quark of flavor f and momentum p_q in the interval dp_q is $\Pi_f\, dp_q/\omega$. Thus, the probability $w_p^f(p_q)\, dp_q$ to find a sea quark (or antiquark) of flavor f and momentum p_q, originating from a valence quark of total momentum p (including the detached gluons), is

$$w_p^f(p_q)\, dp_q = dp_q \eta \xi_f \int_{p_q}^{p} \frac{n_g(\omega)}{\omega}\, d\omega,$$

where $n_g(\omega)$ is the number of detached gluons as given by (83), which does not depend on p. We then get

$$w_p^f(p_q) = \xi_f \eta^2 \left(\frac{1}{p_q} - \frac{1}{p}\right). \tag{88}$$

The square of η appears here because it is a second-order process.

In order to find the total probability of sea quarks (or antiquarks) of flavor f and momentum ω_q we must multiply (88) with the probability $a_0(x)$ of finding a valence quark with a momentum p larger than p_q before the gluons are detached. Expressing everything in terms of $x_q = p_q/P$, we find that the probability $c_f(x_q)$ that a sea quark of momentum $x_q P$ is produced by the valence quarks is

$$c_f(x_q)\, dx_q = 3\xi_f \eta^2 \int_{x_q}^{1} a_0(x)\left(\frac{1}{x_q} - \frac{1}{x}\right) dx. \tag{89}$$

The factor 3 accounts for the three valence quarks that, in our approximation, have the same probability distribution $a(x)$. We see here the dependence on x_q^{-1}, whereas the valence quark distribution $a(x)$ behaves like $\log(x)^{-1}$ for $x \to 0$.

We expect the quarks with flavors other than u and d to be produced with much lower probability because of their higher mass. We therefore make the approximation

$$\xi_u = \xi_d = \xi, \quad \xi_s = \xi_c = 0,$$

and find from (89)

$$c(x)\, dx = 3\xi \eta^2 \frac{D(x)}{x}\, dx,$$

$$D(x) = \int_x^1 a_0(x')\left(1 - \frac{x}{x'}\right) dx', \tag{90}$$

where $c(x)$ is the probability of finding a sea quark (or antiquark) of flavor u or d in the nucleon. This is the function that was introduced in Eq. (32). A sketch of $xc(x)$, as given by (90), with $a_0(x)$ from (74), and $3\xi\eta^2 = 1.2\eta^2 = 0.3$, is shown in Fig. 10. In our approximation ($\xi_s = 0$) we obviously have $s(x) = 0$. Furthermore, we have neglected all higher approximations, such as the fact that the sea quarks are also accompanied by gluons. That would change $c(x)$ in a fashion similar to the way that $a_0(x)$ was changed into $a(x)$.

7. THEORETICAL CONSIDERATIONS

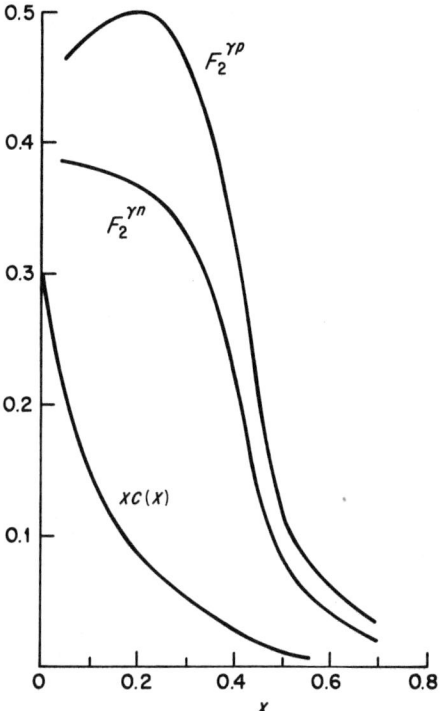

FIG. 10. The calculated structure functions $F_2^\gamma(x)$ for protons and for neutrons, and the calculated sea distribution $xc(x)$ as defined by (90). These should be compared with the observed values in Figs. 1 and 8.

We neglect this effect, since the contribution of the sea quarks is relatively small in itself.[31]]

(d) The structure functions

[We now assemble our theoretical estimates of the structure functions. Let us start with F_2^γ for electron (or muon) scattering by protons and neutrons, as defined by Eqs. (26) and (37):

$$F_2^{\gamma n,p} = xG^{\gamma n,p}(x), \qquad (26')$$

$$G^{\gamma p}(x) = [a(x) + \tfrac{10}{9}c(x)],$$
$$G^{\gamma n}(x) = [\tfrac{2}{3}a(x) + \tfrac{10}{9}c(x)]. \qquad (37')$$

In getting to (37') from (37) we have put $a(x) = b(x)$, $s(x) = 0$, according to the approximations made in our model. The functions $a(x)$ and $xc(x)$, as

---

[31] The approximation (86) used in calculating $a(x)$ from $a_0(x)$ does not work well for $c(x)$, because the latter is concentrated near $x = 0$, where higher order terms in α_s become important.

determined from our model calculations, are plotted in Fig. 9 and Fig. 10, respectively. The resulting structure functions $F_2^{\gamma p}$ and $F_2^{\gamma n}$ are also shown in Fig. 10. They should be compared with the experimental electromagnetic structure functions given in Fig. 1. Note that they reproduce the qualitative aspects fairly well. They are too large at low x and too small at higher x; the difference between the proton and neutron functions is too large for low x and too small for high x. These discrepancies are mainly due to our invalid assumption $a(x) = b(x)$.

We now turn to the neutrino structure functions. Since the measurements are usually done on nuclei containing an equal number of neutrons and protons, we calculate the average of the neutron and proton structure functions, $F_i^{\nu N} = \frac{1}{2}(F_i^{\nu p} + F_i^{\nu n})$. We use (59), and again set $a(x) = b(x)$:

$$F_2^{\nu N}(x) = F_2^{\bar{\nu} N}(x) = x[3a(x) + 4c(x)], \qquad (59')$$
$$F_3^{\nu N}(x) = F_3^{\bar{\nu} N}(x) = 3a(x).$$

Figure 11 shows the structure functions that follow from our estimates of $a(x)$ and $c(x)$. They should be compared with the observed ones in Fig. 7. Again the qualitative features are reproduced, but there are significant quantitative differences.

We repeat that these theoretical predictions are based on extremely crude approximations. The valence quark momentum distributions in a nucleon at rest cannot be determined with our present knowledge of QCD. Expression (74) is only an educated guess. Furthermore, higher powers of the gluon coupling constant were neglected and the detached gluon field was treated in a very cursory way. One should not expect our rough model to provide a quantitative description of the data. The purpose of the model is to show that a plausible momentum distribution in the nucleon at rest can account for the characteristics of the structure functions that have been observed.]

8. Scaling violations

It was shown in §§1–7 that the structure functions F_i depend only on the variable $x = -q^2/2M^2\nu$, whereas, in all generality, they should be functions of q^2 and ν separately. This simplifying property was called *scaling*. We concluded from the approximate validity of scaling, and other features of the structure functions, that an extremely relativistic nucleon can, to a remarkable degree, be described as an assembly of quarks with a distribution of momenta parallel to the motion of the nucleon, and that the quarks are point particles.

Nevertheless, scaling violations have actually been observed, as shown in Figs. 3 and 12. The structure functions for fixed x are not the same for different values of q^2, or of the incident lepton energy ε. Higher ε, of course, leads to higher momentum transfer. Figure 12 shows clearly that, for increasing ε, the structure functions increase at low x and decrease at larger x.

This behavior can be understood in a qualitative way. We must, as we did

8. SCALING VIOLATIONS

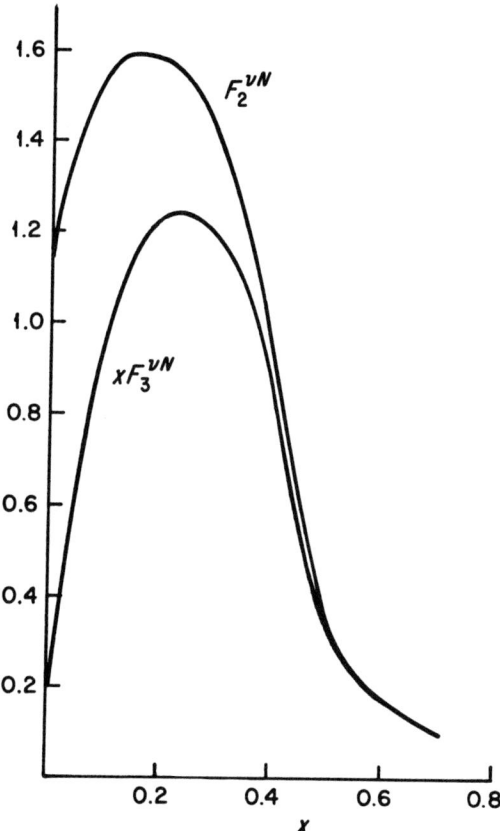

FIG. 11. The calculated structure functions $F_2^{\nu N}(x)$ and $xF_3^{\nu N}$. The plotted curves are the averages of the corresponding functions for protons and neutrons. They should be compared with the observed structure functions shown in Fig. 7.

in §7, consider the color field as well as the quarks. For zero momentum transfer, the total of the two is what we consider to be the quark as such, and the quark momentum is the momentum of that totality. For finite momentum transfer, however, part of the field is detached. The amount of the detached field depends on the momentum transfer; the larger the transfer, the smaller is the part of the momentum to be ascribed to the quark itself. Thus the momentum distribution of the quarks should change towards smaller momenta with increasing momentum transfer. Since the number of valence quarks remains constant, the loss at higher x must be compensated by an increase at lower x. This conservation of quark numbers does not hold for the sea quarks. On the contrary, their number should increase with the growth of the number of detached gluons, since the sea quarks are produced by them. Figure 12 shows that the sea-quark distribution indeed increases with growing ε.

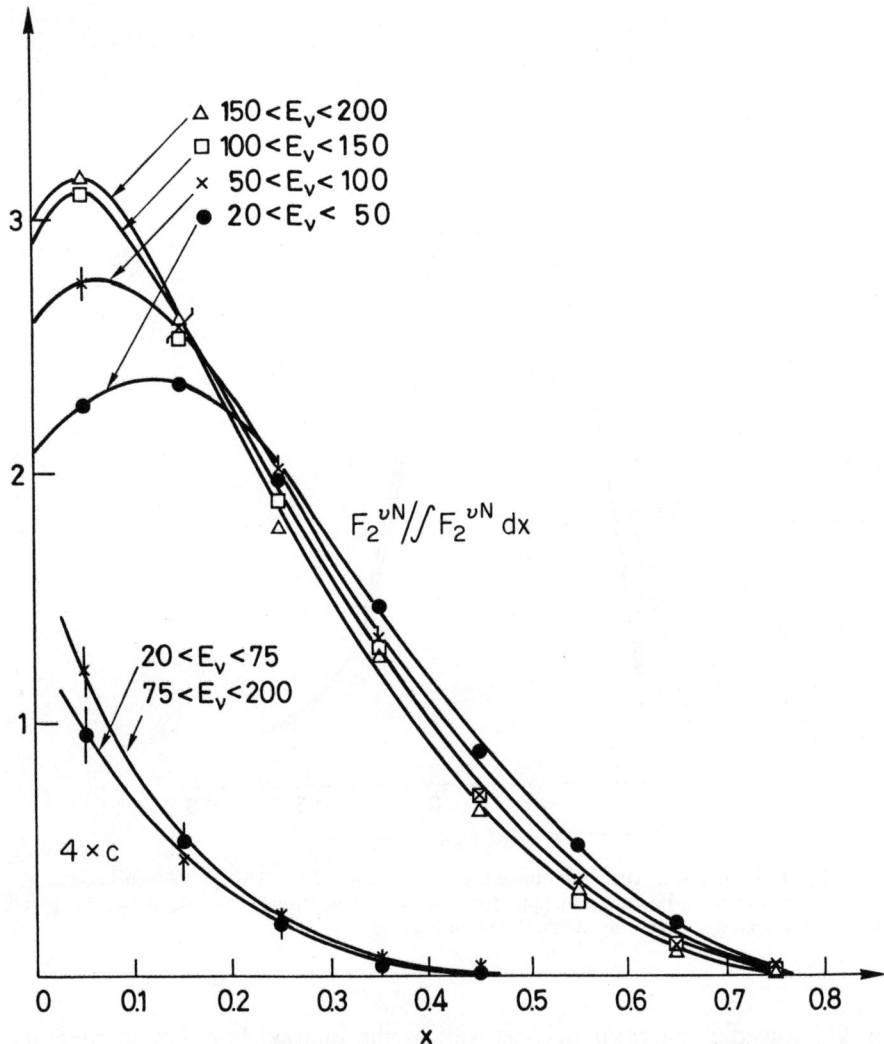

FIG. 12. The structure functions $F_2^{\nu N}(x)$, as measured for different energy intervals, showing the deviation from scaling. The functions F_2 are normalized so that $\int F_2(x)\, dx = 1$. It is seen that with increasing energy F_2 becomes larger for low values of x and smaller for higher values. The figure also shows the change of $xc(x)$, indicating an increase of sea quarks with increasing energy. (J. Steinberger, private communication.)

The violation of scaling is a relatively small effect. We have seen in Eq. (81) that the constant η, which determines the amount of detached field, varies only logarithmically with momentum transfer. The decrease of α_s with increasing momentum transfer (asymptotic freedom) reduces this dependence even more. There are additional scaling violations, because the nucleon mass M is not completely negligible as compared to the momentum transfer, as assumed in the derivation of scaling. These effects will not be treated here.[32]

We have restricted ourselves to these qualitative remarks about scaling violations. One can gain a detailed understanding of them by applying the Feynman graph expansion of QCD to the deep inelastic scattering amplitudes.[33]

---

[32] They are often referred to as "higher twist effects."
[33] See Cheng (1984), p. 312 et seq.

VI

THE ELECTROWEAK INTERACTION

There are few branches of physics that have as tortuous a history as that of the weak interaction. After three-quarters of a century of halting progress, interrupted by occasional breakthroughs, there is good reason to say that we finally have a correct theory of this interaction.[1] According to this theory, electromagnetic and weak processes are almost as intimately related as electric and magnetic phenomena. This electroweak theory therefore incorporates quantum electrodynamics.

This chapter is an introduction to this theory. Our point of departure (Part A) is the coupling of quarks and leptons to the charged bosons W^+ and W^-, as described in §I.E.9. This $W^\pm$ model of the *charge-changing* weak interaction accounts successfully for a vast body of low-energy data. Nevertheless, we will learn that it is not a consistent theory, because the current that carries the weak charge is not conserved.

This defect has serious consequences that cannot be averted without adding further degrees of freedom, as we see in Part B. Given what is known about quarks and leptons, the most economical self-consistent theory introduces a third electrically neutral Bose field that mediates *charge-preserving* weak interactions. The quantum of this field is the neutral boson Z^0 introduced in §I.E.10. Because the electromagnetic field is also a neutral Bose field coupled to quarks and leptons, there are intimate relationships between the charge-changing and charge-preserving weak interactions, on the one hand, and the electromagnetic interaction, on the other. This is a reflection of the electroweak connection, which has some similarity to Maxwell's unification of electric and magnetic phenomena. The latter also followed from the requirement that current conservation be guaranteed by the field equations. Once that was done, the field became a dynamically independent system that could exist without sources—an enlargement of the degrees of freedom.

Part C faces a puzzle that we glossed over in §I.E.10: How can a theory that purports to interrelate weak and electromagnetic phenomena by a symmetry principle account for the dramatic mass difference between the photon, on the one hand, and the W and Z^0, on the other? By borrowing ideas from the theory of superconductivity, we will construct the so-called Higgs mechanism that describes the apparent symmetry breaking of the

---

[1] As in other parts of this book, we seek to maximize clarity, at the expense of a historical perspective. The history can be traced from Kabir (1963), Chang (1977), Brown (1978), Coleman (1979), and Galison (1983).

electroweak interaction, and leads to predictions for the masses of W and Z^0 in terms of the low-energy data.

As shown in Fig. I.37, recent experiments have found numerous events that have all the characteristics expected for the W and Z^0, with masses that are in excellent agreement with these predictions. One can therefore say that the electroweak theory has now made the same transition that electrodynamics underwent when the electromagnetic waves predicted by Maxwell's theory were first discovered a century ago. Whether the symmetry-breaking mechanism of the standard model is fundamentally correct, or only a low-energy phenomenology that glosses over new degrees of freedom at much higher masses, remains an unresolved question [recall §I.E.13(d)].

A. CHARGE-CHANGING WEAK INTERACTIONS

1. The charge-changing weak current

We already exploited the similarity between the weak and electromagnetic interactions in §I.E.9. We continue in this vein here. The electromagnetic interaction Hamiltonian was given in §II.B.1:

$$H_{em} = e \int \mathbf{j}(\mathbf{r}) \cdot \mathbf{A}(\mathbf{r}) \, d^3r, \tag{1}$$

where $\mathbf{j}$ is the electromagnetic current and $\mathbf{A}$ the vector potential. The operator $\mathbf{j}$ creates and destroys $\bar{f}f$ pairs, and produces changes of momentum of f and $\bar{f}$, where f stands for any quark or charged lepton. The operator $\mathbf{A}$ destroys and creates photons.[2] In the language of §I.E.9, the "representative" reaction caused by (1) is $\gamma \leftrightarrow \bar{f}f$, but it also describes the crossed partners thereof: $f \leftrightarrow \gamma f$, $\bar{f} \leftrightarrow \gamma \bar{f}$. As a consequence of (1) a virtual photon whose invariant mass is large compared to fermion masses converts into the state $\Sigma_f Q_f |\bar{f}f\rangle$, where Q_f is the charge.

According to Eq. I.E(45), the charge-changing weak interaction, H_{cc}, must be an operator that can produce a transition from a virtual W^- vector boson to fermion–antifermion pairs whose total lepton and baryon numbers vanish. We therefore assume that it has the same form as H_{em}, i.e., an integral over space of an (Hermitian) interaction energy density that is the product of a Bose field operator that creates and destroys W^+ and W^-, and operators $J_\pm$ that act on quarks and leptons:

$$H_{cc} = \frac{1}{\sqrt{2}} g \int [J_+(\mathbf{r}) W^\dagger(\mathbf{r}) + J_-(\mathbf{r}) W(\mathbf{r})] \, d^3r. \tag{2}$$

Here $W(\mathbf{r})$ is the Bose field: it destroys W^- and creates W^+; in the language of §II.B.5, it is a $\Delta Q = 1$ operator. Its adjoint $W^\dagger(\mathbf{r})$ is therefore a $\Delta Q = -1$ operator; it destroys W^+ and creates W^-. Since H_{cc} must conserve Q, J_+ creates $\bar{f}_\alpha f_\beta$ pairs of $Q = 1$, and causes the transitions

---

[2] Equation (1) only describes the interaction with the transverse radiation field. The longitudinal electrostatic interaction between charges must be added to (1) to obtain the complete Hamiltonian. See Eq. (18) below.

$f_\alpha \to \bar{f}_\beta$ and $\bar{f}_\beta \to f_\alpha$, wherein the fermion charge increases by one unit. The adjoint J_- produces the inverse of these transitions with $\Delta Q = -1$. We shall call $J_\pm(\mathbf{r})$ *the charge-changing weak currents*; the appellation "current" will be justified presently. Finally, g is the unit of "weak charge" introduced in Eq. I.E(45).

The structure of the operators $J_\pm$ must reflect the helicity rule, as enunciated in §I.E.9(f): When extremely relativistic fermions participate in charge-changing weak interactions, their helicities must be $h = -\frac{1}{2}$ and $\frac{1}{2}$ for fermions and antifermions, respectively. As we now see, this rule implies that only a portion of the Dirac field $\psi(\mathbf{r})$ representing any quark or lepton appears in $J_\pm$.

It is best to consider a Fourier component $\psi(\mathbf{p})$ of the Dirac field operator. While $\psi(\mathbf{p})$ is given explicitly in Eq. II.A(43), for our present purpose it is more appropriate to consider the decomposition[3]

$$\psi(\mathbf{p}) = \begin{pmatrix} \psi_R(\mathbf{p}) \\ \psi_L(\mathbf{p}) \end{pmatrix}, \tag{3}$$

where $\psi_R(\mathbf{p})$ and $\psi_L(\mathbf{p})$ are the Fourier components of the upper and lower 2-component spinors defined in Eq. II.A(25). The operators $\psi_{R,L}$ satisfy algebraic equations that are an immediate consequence of the Dirac equation as originally written for the wave functions $\chi^{(\pm)}$, Eq. II.A(22) and (23). The reason for this is that the operator nature of the Dirac field is contained entirely in the creation and destruction operators $d^\dagger$, b, etc., whereas the space-time dependence is carried by plane-wave solutions of the Dirac equation. Let us therefore consider these plane waves:

$$\chi^{(\pm)}(\mathbf{r}, t) = \chi_{R,L}(\mathbf{p}) e^{i(\mathbf{p}\cdot\mathbf{r} - Et)}.$$

From what was just said, the constant spinors $\chi_{R,L}$ and the operators $\psi_{R,L}(\mathbf{p})$ must satisfy the same equations, namely

$$(E - \boldsymbol{\sigma}\cdot\mathbf{p})\psi_R(\mathbf{p}) = m\psi_L(\mathbf{p}), \tag{4_R}$$

$$(E + \boldsymbol{\sigma}\cdot\mathbf{p})\psi_L(\mathbf{p}) = m\psi_R(\mathbf{p}). \tag{4_L}$$

As we know from Eq. II.A(43), we need these solutions for positive and negative E—the former for the destruction of a particle, the latter for the creation of an antiparticle. Equation (4_L) assumes the following forms in

---

[3] Appendix III contains a more detailed treatment of the Dirac field and, in particular, of the significance of this decomposition. The subscripts R and L have precisely the same meaning as the superscripts $(+)$ and $(-)$, respectively, as used in §II.A and Appendix III. The mathematical formalism is more compact with the $(\pm)$ notation, but it has become customary to use the (R, L) notation in the theory of weak interactions, as it emphasizes the lack of reflection invariance.

1. THE CHARGE-CHANGING WEAK CURRENT

these two cases:

$$E > 0: \quad (1 + \boldsymbol{\sigma} \cdot \mathbf{v})\psi_L(\mathbf{p}) = \frac{m}{E}\psi_R(\mathbf{p}), \tag{5}$$

$$E < 0: \quad (1 - \boldsymbol{\sigma} \cdot \mathbf{v})\psi_L(\mathbf{p}) = -\frac{m}{|E|}\psi_R(\mathbf{p}), \tag{6}$$

where $\mathbf{v} = \mathbf{p}/|E|$. When $|E| \gg m$, $|\mathbf{v}| \simeq 1$ and (5) gives a solution with $h = -\frac{1}{2}$, while the solution of (6) has $h = \frac{1}{2}$, except for small contaminations of order $m/|E|$ having the opposite helicity. For large momenta ($p \gg m$), the 2-component operator $\psi_L(\mathbf{p})$ therefore has the property demanded by the helicity rule for both particles and antiparticles: it destroys fermions with $h = -\frac{1}{2}$ and creates antifermions with $h = \frac{1}{2}$. But the inequality $p \gg m$ is not Lorentz invariant; one can make p arbitrarily small by going to a suitable frame. We therefore need a Lorentz-invariant generalization of the helicity rule. This is provided by the following theorem, proved in Appendix III.1: *The separation (3) of the Dirac field into the components ψ_R and ψ_L is invariant under any proper Lorentz transformation, where the qualification "proper" means that spatial reflections are excluded.*[4] Given that the helicity rule selects $\psi_L(\mathbf{p})$ for $p \gg m$, its Lorentz-invariant generalization is now readily formulated: *The portion of the Dirac field that participates in the charge-changing weak interactions is $\psi_L(\mathbf{p})$ for all momenta, whether large or small.* The corresponding coordinate-space portion of the Dirac field is called $\psi_L(\mathbf{r})$.

It has become customary to call $\psi_L(\mathbf{r})$ and $\psi_R(\mathbf{r})$ the "left-handed" and "right-handed" portions of the Dirac field, even though this is a somewhat inaccurate terminology.[5] It is accurate only in the limit $m/E \to 0$. In general, ψ_L destroys not only fermions with $h = -\frac{1}{2}$ but also those with $h = \frac{1}{2}$, although the latter have an amplitude smaller by a factor of m/E. The converse is true of ψ_R. For $p \sim m$, ψ_L and ψ_R create and destroy particles in states that have comparable admixtures of both helicities.

We are now ready to express the operator J in terms of Dirac fields. A convenient abbreviation will be used: The Dirac field will be represented by the symbol of the particle that it destroys. Thus $e(\mathbf{r})$, or more briefly e, is the electron's 4-component Dirac field, while e_L and e_R are its left- and right-handed 2-component parts. Note that $e_{L,R}$ destroys e^- and creates e^+, whereas $e^\dagger_{L,R}$ create e^- and destroy e^+.

Let us focus on the part J_{ud} of J_+ that acts on u and d quarks. It must be

---

[4] In brief, if $\mathbf{p} \to \mathbf{p}'$ under the Lorentz transformation, then $\psi_L(\mathbf{p}') = \Omega_- \psi_L(\mathbf{p})$, where Ω_- is the 2×2 matrix constructed in Appendix III.1; in other words, the transformation does not produce a linear combination of $\psi_L(\mathbf{p}')$ and $\psi_R(\mathbf{p}')$, as Eq. (20) of Appendix III demonstrates.

[5] The operators ψ_L and ψ_R are also called the chiral parts of the Dirac field. This comes from the term *chirality*, which is the signature of the product of helicity with fermion number. Thus a quark or lepton of $h = -\frac{1}{2}$ has negative chirality, while antiparticles of the same helicity have positive chirality, etc. For large p, the particles and antiparticles destroyed and created by $\psi_L(\mathbf{p})$ therefore have negative chirality, while the converse holds for $\psi_R(\mathbf{p})$.

built from the 2-component operators u_L and d_L, and their adjoints. Since J_{ud} raises the charge by one unit and does not change the baryon number, it must have the form $u_L^\dagger K d_L$, where K is some 2×2 matrix. As u_L and d_L are spinors under spatial rotations, $u_L^\dagger K d_L$ can only be a scalar or vector, corresponding to whether K is the unit matrix or a Pauli matrix. Hence J_{ud} is to be formed from the four operators

$$\rho_{ud} = u_L^\dagger d_L, \qquad \mathbf{J}_{ud} = u_L^\dagger \boldsymbol{\sigma} d_L. \tag{7}$$

In Appendix III.1 we show that ρ_{ud} and $\mathbf{J}_{ud}$ together form a Minkowski 4-vector. For that reason it is sometimes convenient to call the 2×2 unit matrix σ_0, and to define $\sigma^\mu \equiv (\sigma_0, \sigma_x, \sigma_y, \sigma_z)$, and $\sigma_\mu \equiv (\sigma_0, -\sigma_x, -\sigma_y, -\sigma_z)$; then for any two fermion fields one can write (7) in the compact form[6]

$$J_{ff'}^\mu = f_L^\dagger \sigma^\mu f_L'. \tag{7'}$$

In §I.E.9(g) we learned that the $Q = -\tfrac{1}{3}$ quarks enter the weak interaction in linear combinations given by Eq. I.E(48). We therefore introduce the corresponding left-handed quark field operators:

$$\begin{aligned} d_L' &= d_L \cos \theta_C + s_L \sin \theta_C, \\ s_L' &= s_L \cos \theta_C - d_L \sin \theta_C, \end{aligned} \tag{8}$$

where θ_C is the Cabibbo angle. In terms of these the lepton pairs (v_e, e) and (v_μ, μ), and the quark pairs (u, d') and (c, s'), enter the weak interaction on a completely symmetric footing.[7] In view of this, Eq. (7) generalizes to

$$\mathbf{J}_+ = v_{eL}^\dagger \boldsymbol{\sigma} e_L + v_{\mu L}^\dagger \boldsymbol{\sigma} \mu_L + u_L^\dagger \boldsymbol{\sigma} d_L' + c_L^\dagger \boldsymbol{\sigma} s_L', \tag{9}$$

$$\rho_+ = v_{eL}^\dagger e_L + v_{\mu L}^\dagger \mu_L + u_L^\dagger d_L' + c_L^\dagger s_L' \tag{10}$$

where the subscript $+$ reminds us that they are $\Delta Q = +1$ operators.

The 4-vector

$$J_+ = (\rho_+, \mathbf{J}_+) \tag{11}$$

is called *the charge-changing weak current*.[8] Since $\boldsymbol{\sigma}$ is Hermitian, the charge-lowering $\Delta Q = -1$ operator is

$$J_- \equiv (J_+)^\dagger; \tag{12}$$

---

[6] In the literature (7') is usually written as $\tfrac{1}{2}\bar{f}\gamma^\mu(1 - \gamma_5)f'$.

[7] We ignore (v_τ, τ) and (b, t).

[8] In the literature $J_\pm$ are called "charged currents." This is a confusing term, since the electromagnetic current carries charge, yet is intimately related to the so-called "neutral weak current" to be introduced in §B.1. As we see, $J_\pm$ are charge-changing operators, whereas the electromagnetic and "neutral weak current" are charge-preserving operators.

1. THE CHARGE-CHANGING WEAK CURRENT

its time and space components are $\rho_- = e_L^\dagger \nu_{eL} + \cdots$, and $\mathbf{J}_- = e_L^\dagger \boldsymbol{\sigma} \nu_{eL} + \cdots$, respectively.

The behavior of the charge-changing weak interaction under space reflection and charge conjugation was discussed in §I.E.9(f). A more complete discussion can be given in terms of the operators just introduced. The motivation for calling ψ_L and ψ_R the "left-" and "right-" handed portions of ψ makes it clear that they are interchanged by a spatial reflection:

$$P: \quad \psi_L \leftrightarrow \psi_R. \tag{13}$$

This can be proved from Eq. (4), since $\mathbf{p} \to -\mathbf{p}$ and $\boldsymbol{\sigma} \to \boldsymbol{\sigma}$ under P. Therefore, P brings in ψ_R, a fermion degree of freedom that has no place in the weak interaction. We may add that ρ_+ is neither a scalar nor a pseudoscalar, and $\mathbf{J}_+$ is neither a polar nor an axial vector, even though they are, respectively, a scalar and a vector under *proper* rotations.[9]

To understand how the charge conjugation operator C acts, we list the action of the ultrarelativistic (UR) portion of the electron field on particles:

	destroys	creates	
$e_L:$	$e^-(h=-\tfrac{1}{2})$	$e^+(h=\tfrac{1}{2})$	(UR)
$e_R^\dagger:$	$e^+(h=-\tfrac{1}{2})$	$e^-(h=\tfrac{1}{2}).$	

Under C, $e^- \leftrightarrow e^+$, but helicities are not changed, and therefore

$$C: \quad \psi_L \leftrightarrow \psi_R^\dagger. \tag{14}$$

Hence C turns a term in ρ_+, such as $u_L^\dagger d_L'$, into $u_R d_R'^\dagger$, which plays no role in the weak interaction. As a consequence, neither P nor C are conserved by the weak interaction, a fact that we already learned in §I.E.9(f).

By combining (13) and (14), one has

$$CP: \quad \psi_L \leftrightarrow \psi_L^\dagger, \quad \psi_R \leftrightarrow \psi_R^\dagger, \tag{15}$$

and therefore the fermion operators that occur in $\mathbf{J}_-$ are transformed by the CP operation into those that appear in $\mathbf{J}_+$. Consequently[10]

$$CP: \quad \rho_+ \leftrightarrow -\rho_-, \quad \mathbf{J}_+ \leftrightarrow \mathbf{J}_-. \tag{16}$$

---

[9] In this connection, see Eq. (31), where the isovector portion of ρ_- is written as a sum of a scalar and a pseudoscalar.

[10] To derive (16), one must use the anticommutation rules obeyed by fermion field operators (see Appendix III). There is also some phase arbitrariness in the definition (14) of C, and therefore of CP. If one uses (14) as it stands, the y-component of $\mathbf{J}_+$ acquires a different sign under CP from the x- and z-components; the transformation (16) uses the "standard" convention: $\psi_L \to \sigma_y \psi_R^\dagger$ under C.

As we see next, J_+ and J_- appear symmetrically in the weak interaction. Consequently, CP is a symmetry of the weak interaction if the weak current is given by[11] Eqs. (9) and (10).

2. The Hamiltonian for charge-changing weak interactions

(a) The Fermi interaction

We return to the Hamiltonian H_{cc} that supposedly describes charge-changing weak interactions. Since $\mathbf{J}$ is now established to be a 3-vector, the field operator $\mathbf{W}(\mathbf{r})$ for the $W^\pm$ particles must be also, and the W's must therefore be spin-1 bosons. As in electrodynamics, H_{cc} has two parts. There is the analogue of the coupling to the radiation field,

$$H_{cc,1} = \frac{1}{\sqrt{2}} g \int [\mathbf{J}_+(\mathbf{r}) \cdot \mathbf{W}^\dagger(\mathbf{r}) + \mathbf{J}_-(\mathbf{r}) \cdot \mathbf{W}(\mathbf{r})] \, d^3r, \qquad (17)$$

which describes the emission and absorption of $W^\pm$'s with helicities 1, 0, and -1. And there is the analogue of the Coulomb interaction:[12]

$$H_{cc,2} = \frac{g^2}{8\pi} \int \rho_+(\mathbf{r}) \frac{e^{-m_W|\mathbf{r}-\mathbf{r}'|}}{|\mathbf{r} - \mathbf{r}'|} \rho_-(\mathbf{r}') \, d^3r \, d^3r', \qquad (18)$$

where m_W is the mass of W. Since $1/m_W$, the W's Compton wavelength, is short compared to all distances of interest in "low-energy" weak interaction phenomena (c.o.m. energies are small compared to m_W), we are interested in the limit $m_W \to \infty$. Due to the exponential in (18), $H_{cc,2}$ reduces to an expression where ρ_+ and ρ_- must be evaluated at the same point, a conclusion we had already reached in §I.E.9(e). In detail, the identity

$$\lim_{m_W \to \infty} \left(\frac{m_W^2 e^{-m_W R}}{4\pi R} \right) = \delta^3(\mathbf{R})$$

implies that (18) can be approximated by

$$H_{cc,2} \simeq \frac{1}{2} \frac{g^2}{m_W^2} \int \rho_+(\mathbf{r}) \rho_-(\mathbf{r}) \, d^3r. \qquad (19)$$

In the low-energy regime W's do not appear in initial or final states, and only have a fleeting existence in intermediate states. Under these circumstances it is appropriate to work with an effective Hamiltonian that does

---

[11] If there is a further quark doublet (b, t), one can construct a J_+ that does not transform into J_-, and then CP is violated; see §I.E.13(b). Whether that is the actual CP-violation mechanism is not yet known.

[12] See Jackson (1975), p. 46.

2. THE HAMILTONIAN FOR WEAK INTERACTIONS

not refer explicitly to the existence of a W-field. Equation (19), which is the analogue of the instantaneous Coulomb interaction, has just this character; it acts only on the sources, and no longer contains the information that there is a field with characteristic range $1/m_W$. The effective Hamiltonian that replaces (17) is the analogue of the Biot–Savart law, i.e., an instantaneous interaction between currents. It is found by the technique sketched in §I.E.9(e): one computes the amplitude for a process with virtual W's, and approximates their propagator $(q^2 + m_W^2)^{-1}$ by m_W^{-2}. Since the Fourier transform of a constant is a δ-function, this leads to an effective interaction where the currents must be in contact:

$$H_{cc,1}^{\text{eff}} = -\frac{1}{2}\frac{g^2}{m_W^2}\int \mathbf{J}_+(\mathbf{r}) \cdot \mathbf{J}_-(\mathbf{r}) \, d^3r. \tag{20}$$

When combined with (19) it gives the so-called *Fermi interaction*:

$$H_F = \frac{1}{2}\frac{g^2}{m_W^2}\int \mathbf{J}_+ \cdot \mathbf{J}_- \, d^3r, \tag{21}$$

where $\mathbf{J}_\pm$ are the 4-currents defined by (9)–(12).

The basic parameter g/m_W that appears in H_F can be measured most accurately in $\mu \to e\bar{\nu}_e\nu_\mu$. One finds[13]

$$\left(\frac{g}{m_W}\right)^2 = 6.5977(2)10^{-5} \, \text{GeV}^{-2}, \tag{22}$$

which can be written in a more suggestive manner as

$$\frac{g^2}{4\pi} \simeq \frac{1}{20}\left(\frac{m_W}{100 \, \text{GeV}}\right)^2. \tag{23}$$

(b) The strength of the semileptonic interaction

In §I.E.9(g) we asserted that the strength of the weak interaction does not depend on whether the source is a lepton pair or a quark pair [provided, of course, that the latter are appropriately defined as in (8)]. The interaction H_F already embodies this assertion because the currents $\mathbf{J}_\pm$ contain all fermion pairs symmetrically. But leptons are emitted and absorbed as free particles, while quarks always occur as constituents of strongly bound hadrons. If this fundamental quark–lepton symmetry is to be put to the test, one must know how to compute the matrix elements that describe the complicated hadronic transition. It is clear that in most instances such matrix elements are sensitive to details of hadronic structure, and ill-suited

---

[13] The so-called Fermi constant G_F is frequently used. It is related to (22) by $G_F = (4\sqrt{2})^{-1}(g/m_W)^2 \simeq 1.17 \times 10^{-5} \, \text{GeV}^{-2} \sim 10^{-5} m_N^{-2}$.

to the determination of parameters in a fundamental Hamiltonian. But there is a special class of transitions where the initial and final states are essentially identical, while the operator causing the transitions is very simple. Under these circumstances, a precise evaluation of the matrix element is possible. Astonishingly enough, certain nuclear decays satisfy these requirements, and also allow high precision measurements. As an example, consider the decay

$$^{14}O \rightarrow {}^{14}N^* e^+ \nu_e, \tag{24}$$

where the ^{14}O ground state and the $^{14}N^*$ excited state are both members of the *same* $J^P = 0^+$ isotriplet. There is an even simpler reaction where the initial and final hadrons have these quantum numbers:

$$\pi^+ \rightarrow \pi^0 e^+ \nu_e. \tag{25}$$

Unfortunately, one cannot measure this rate with the accuracy necessary to compete with (24).

The process that underlies (24) is $u \rightarrow d e^+ \nu_e$. The operator $\boldsymbol{J}_+ \cdot \boldsymbol{J}_-$, when acting on the initial u, must produce the final $d e^+ \nu_e$ state, i.e., it must create a $Q = 1$ lepton pair and decrease the hadron's charge by one. The term in $\boldsymbol{J}_+$ that contributes is therefore of the form $v_{eL}^\dagger e_L$, while that in $\boldsymbol{J}_-$ has the form $d_L^\dagger u_L$. Since $\boldsymbol{J}_-$ only contains u in the combination $d'^\dagger u$, it contributes a term $d_L^\dagger u_L \cos \theta_C$ to the transition operator. In short, for the transition (24) we can replace H_F by

$$\tilde{H}_F = \frac{1}{2}\left(\frac{g}{m_W}\right)^2 \cos \theta_C \int (v_{eL}^\dagger \sigma_\mu e_L)(d_L^\dagger \sigma^\mu u_L)\, d^3r, \tag{26}$$

where we use the notation of Eq. (7').

We must evaluate the matrix element of $\tilde{H}_F$ between the initial and final state of the reaction (24). The latter is a product of the $^{14}N^*$ wave function, and that of the leptons. The matrix element therefore separates into two factors. The leptonic factor is a product of plane waves (to the extent that the positron's interaction with the nuclear Coulomb field is small, an effect that must be taken into account in a detailed analysis of the data). The second factor is the nucleus matrix element

$$M_{AB}^\mu = \int \langle B | K^\mu(\mathbf{r}) | A \rangle e^{i\mathbf{k}\cdot\mathbf{r}}\, d^3r, \tag{27}$$

where $K^\mu \equiv d_L^\dagger \sigma^\mu u_L$, $|A\rangle$ and $|B\rangle$ are the initial and final nuclear states in (24), and $\mathbf{k}$ is the total momentum carried off by $e^+ \nu_e$. The phase factor in (27) stems from the lepton wave functions.

[The states $|A\rangle$ and $|B\rangle$ contain a multitude of strongly interacting quarks. If we are to evaluate (27) with precision, we must take advantage of some special

2. THE HAMILTONIAN FOR WEAK INTERACTIONS

circumstance that allows us to ignore the intricate structure of the nuclear states. This circumstance is that $|A\rangle$ and $|B\rangle$ are members of the same $J = 0^+$ isomultiplet, and that the operator in (27) can be related to the total isospin. One can recognize the possibility of such a relationship by comparing the isospin-lowering operator $I_- = I_1 - iI_2$ with the time component ($\mu = 0$) of (27). For any hadronic system, no matter how complex, the former is given by

$$I_- = \int d^\dagger u \, d^3r, \tag{28}$$

because u and d are isospin "up" and "down" partners, whereas all other quarks carry no isospin. The operator that appears in (27) when $\mu = 0$ is

$$I_-^L(k) = \int d_L^\dagger u_L e^{i\mathbf{k}\cdot\mathbf{r}} \, d^3r. \tag{29}$$

If we could replace (29) by (28), our task would be done, because Eq. I.B(19) tells us that

$$\langle B|I_-|A\rangle = \sqrt{I(I+1) - I_3^A(I_3^A - 1)} \, \delta_{I_3^A, I_3^B + 1} \tag{30}$$

if $|A\rangle$ and $|B\rangle$ are members of the same isomultiplet.
To take advantage of this, we must show that:

1. the phase factor $e^{i\mathbf{k}\cdot\mathbf{r}}$ can be ignored;
2. the spatial components of K^μ do not contribute to (27); and
3. it does not matter that only the left-handed portions of the fields appear in (29), whereas the complete fields contribute to I_-.

The phase $e^{i\mathbf{k}\cdot\mathbf{r}}$ accounts for the nuclear recoil, which is a small effect since the energy released in nuclear β-decay is small compared to nuclear masses.[14] Once recoil is significant, one must know the spatial distribution of the isospin density within the hadron, and that it is sensitive to strong interaction dynamics.

The remaining two points are taken care of by the quantum numbers of $|A\rangle$ and $|B\rangle$. As both have $J^P = 0^+$, the vector $\mathbf{K}$ cannot produce a transition between them. To relate $d_L^\dagger u_L$ to $d^\dagger u$, we note from (3) that $d^\dagger u = d_L^\dagger u_L + d_R^\dagger u_R$, or

$$d_L^\dagger u_L = \tfrac{1}{2} d^\dagger u + \tfrac{1}{2}(d_L^\dagger u_L - d_R^\dagger u_R). \tag{31}$$

Under the spatial reflection (13)

$$P: \; d_L^\dagger u_L \to \tfrac{1}{2} d^\dagger u - \tfrac{1}{2}(d_L^\dagger u_L - d_R^\dagger u_R). \tag{32}$$

Hence, the first term of (31) is a true scalar, the second a pseudoscalar. The latter cannot contribute to a transition between states of the same parity, as is the case here.∎

---

[14] Put another way, in nuclear β-decay $kR \ll 1$, where R is the nuclear radius. In particle decays, on the other hand, recoil is often important, e.g., in $K^+ \to \pi^0 e^+ \nu_e$.

This completes the evaluation of (27):

$$M^\mu_{AB} = \delta_{\mu,0}\frac{1}{2}\langle B|\,I_-\,|A\rangle = \frac{1}{\sqrt{2}}\delta_{\mu,0}, \tag{33}$$

where we used (30), and that $I = 1$ and $I_3 = 1$ for ^{14}O. All factors in the matrix element of (26) except $\cos\theta_C$ are now determined. By comparing the rate for the nuclear decay (24) with that for $\mu \to e\bar{\nu}_e\nu_\mu$, one finds $\cos\theta_C = 0.9737(25)$. As explained in §I.E.9(g), $\sin\theta_C$ is determined independently by the strangeness-changing decays, and agrees with $\sqrt{1 - \cos^2\theta_C}$ to within the experimental error. This shows that nuclear β-decay, and the strangeness-changing decays of hadrons, are manifestations of a universal weak interaction governed by one basic coupling constant g.

3. Defects of the $W^\pm$ model of weak interactions

The effective Hamiltonian H_F successfully describes all low-energy weak-interaction phenomena, with the exception of the elusive "neutral current" processes and CP violation in K^0-decays.[15] As indicated in §I.E.9, and as we will see in detail in this chapter, by "low-energy" one actually means center-of-mass energies small compared to ~ 100 GeV. In this sense even the highest energy neutrino reactions observed thus far are "low-energy" collisions ($E_\nu^{\text{lab}} \lesssim 400$ GeV, or $E_{\text{com}} \lesssim 27$ GeV). Since H_F makes no explicit reference to an underlying field, one can appreciate why the road from the discovery of β-decay to the formulation of the electroweak theory was so arduous.

Despite its phenomenological success, H_F has a dangerous disease: It predicts a high-energy behavior of neutrino cross sections that is in conflict with basic principles. In this Section we acquaint ourselves with this disease. As we will see, it is not cured by returning to the more basic field-theoretic Hamiltonian involving $W^\pm$ bosons. Indeed, the only known cure requires a profound enlargement of the field theory, a topic to which we devote Part B of this Chapter.

The violent high-energy behavior of H_F is a consequence of the fact that its coupling constant $(g/m_W)^2$ has dimension $p = -2$ in the language of §II.B.5. As we learned there, a theory with $p < 0$ has collision amplitudes that grow as some power of the energy, and this power necessarily increases as one increases the order of the perturbation expansion.

Neutrino collisions are treated in Chap. V, and also in Part B of this Chapter. For our present purposes it suffices to know the form of the total neutrino-nucleon cross section $\sigma_{\nu N}$, as derived from H_F. Since $H_F \propto (g/$

---

[15] If a t quark exists, the weak currents $J_\pm$ have a further term involving both b and t, and H_F can then accommodate CP-violation. See §I.E.13(b).

3. DEFECTS OF THE $W^{\pm}$ MODEL

$m_W)^2$, the amplitude is proportional to $(g/m_W)^2$, as $\sigma_{\nu N}$ is to $(g/m_W)^4$. But $\sigma_{\nu N}$ is an area, so $\sigma_{\nu N} = (g/m_W)^4 f$, where f has the dimension (energy)^2. Furthermore, $\sigma_{\nu N}$ is Lorentz invariant, so the energy in question must be frame-independent. At energies that are large compared to all quark and lepton masses the only candidate for f in a theory without W's is therefore the square of the c.o.m. energy, which is denoted by s. Hence

$$\sigma_{\nu N} = \left(\frac{g}{m_W}\right)^4 sC, \qquad (34)$$

where C is a pure number. If E_ν is the neutrino energy in the laboratory frame, $s = 2m_N E_\nu$ when $E_\nu \gg m_N$. Hence, $\sigma_{\nu N}$ has a linear growth with neutrino energy, and the large body of data recorded in Fig. V.6 fully confirms this expectation.

Despite this success, (34) cannot be correct for arbitrarily large E_ν. As we see in (21), H_F describes an interaction between fermion pairs at the same point. This is equivalent to the statement on p. 128 that there is no orbital angular momentum between the interacting pairs. Hence, their total angular momentum J is entirely due to spin, and therefore $J = 0$ or 1. As we saw in §III.A.5, the scattering in a single angular momentum state is bounded by probability conservation:[16] $\sigma_J \leq (2J + 1)\pi/k^2$, where k is the relative momentum, which is related to s by $k = \sqrt{s}$ at high energy. Consequently,

$$\sigma_{\nu q} < \frac{C'}{s}, \qquad (35)$$

where C' is another pure number, of order unity. When $E_\nu \gg m_N$, the binding of quarks can be neglected in computing $\sigma_{\nu N}$, which is then an incoherent sum of cross sections $\sigma_{\nu q}$ for neutrino reactions from the various quarks, and (35) therefore provides a bound on $\sigma_{\nu N}$ itself. In view of (34), this bound is violated when c.o.m. energies $\sqrt{s}$ grow beyond approximately m_W/g, which is ~ 120 GeV in virtue of (22). A more careful calculation that keeps track of C and C' shows that in the first Born approximation probability conservation would be violated for $\sqrt{s}$ larger than

$$\sqrt{s_0} \simeq 300 \text{ GeV}, \qquad (36)$$

which corresponds to laboratory energies of order 50,000 GeV!

The inconsistency we have just uncovered is not disastrous, because it occurs at energies so large that it is not justified to treat m_W as if it were infinite. A finite m_W corresponds to a nonzero force range $R \sim m_W^{-1}$, and this permits collisions with all angular momenta up to $l_{\max} \sim kR \sim \sqrt{s}/m_W$. In short, $l_{\max}$ increases with energy, and the bound (35) is not applicable

---

[16] In scattering theory these restrictions imposed by probability are usually called *unitarity bounds*.

once the W-field is taken into account. As we know from Eq. I.E(42), the field-current interaction H_{cc} leads to scattering amplitudes that contain W-propagators that replace the constant g^2/m_W^2 given by the effective interaction H_F. As a consequence, the W-field theory gives a νN cross section that only grows linearly until $s \simeq m_W^2$, i.e., well below s_0. For $s > m_W^2$, the cross section increases more slowly, and it tends asymptotically to a constant. A detailed calculation[17] shows that H_{cc} predicts the cross section to be

$$\tilde{\sigma}_{\nu N} = \sigma_{\nu N} \frac{m_W^2}{s + m_W^2} \xrightarrow{s \to \infty} \frac{g^4}{m_W^2} C, \tag{37}$$

where $\sigma_{\nu N}$ is the rising cross section predicted by H_F. This cross section does not violate probability conservation, because an interaction of range R can give a cross section as large as that of a black disk of radius R, i.e., $2\pi R^2$. Since $R \sim m_W^{-1}$, we see that $\tilde{\sigma}_{\nu N}$ falls below the bound by a factor of g^4, indicating that we are dealing with an interaction that is far from maximal.

The field-current interaction H_{cc} cures many other afflictions of the Fermi interaction, but it has a disease of its own: The cross sections for certain W-production reactions violate probability conservation. As a consequence, the $W^\pm$-model is, in the last analysis, as ill-behaved as the Fermi interaction.

The simplest example of such a badly behaved process is the weak interaction amplitude for $e^+e^- \to W^+W^-$ arising from the graph

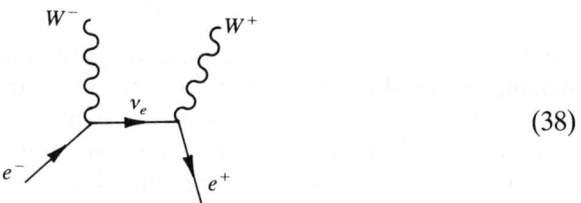

(38)

The offending contributions to the cross section come from the production of helicity zero W's, i.e., from processes having no counterpart in $e^+e^- \to 2\gamma$.

It is easy to understand why $h_W = 0$ has a different energy behavior from $h_W = \pm 1$. As we know from QED [cf. e.g., Eqs. II.B(38) and (39)], the amplitude for any process contains as a factor a polarization vector $\varepsilon_h(\mathbf{k})$ for every emitted spin-1 boson of momentum $\mathbf{k}$ and helicity h. Since W is massive, we can go to its rest frame, where the $\varepsilon_h(0)$ are the familiar spin-1 eigenfunctions. Choosing $\mathbf{k}$ as the z-axis, we have $\boldsymbol{\varepsilon}_{\pm 1}(0) = (\hat{\mathbf{x}} \pm i\hat{\mathbf{y}})/\sqrt{2}$, $\boldsymbol{\varepsilon}_0(0) = \hat{\mathbf{z}}$. Under the Lorentz transformation required to give W the desired momentum $\mathbf{k}$, $\boldsymbol{\varepsilon}_{\pm 1}$ do not change, but the $h = 0$ unit vector becomes $\gamma\hat{\mathbf{z}}$,

---

[17] See Commins (1973), §3.6.

3. DEFECTS OF THE $W^\pm$ MODEL

where $\gamma = (1-v^2)^{-1/2} = \sqrt{(k/m_W)^2 + 1}$. Hence for large k,

$$\boldsymbol{\varepsilon}_0(\mathbf{k}) \simeq \frac{\mathbf{k}}{m_W} + \frac{m_W}{2k}\hat{\mathbf{z}}. \tag{39}$$

Since each emission also carries a factor of g, we conclude that $h = \pm 1$ emission (or absorption) has amplitude $g\boldsymbol{\varepsilon}_{\pm 1}F$, whereas $h = 0$ has amplitude $(g/m_W)\mathbf{k}F$ as $k \to \infty$, where F is a function of the other kinematical variables. The ubiquitous dimensional coupling constant g/m_W of the Fermi interaction therefore reappears, and leads to cross sections that rise unacceptably with energy whenever zero helicity W's are involved.

One may ask why one should be disturbed by a lowest order calculation when it gives results close to the unitarity bound, for that implies that higher order corrections are then important. To show that the high-energy behavior of (38) is of importance, we consider the higher order correction to elastic e^+e^- scattering arising from (38), viz.

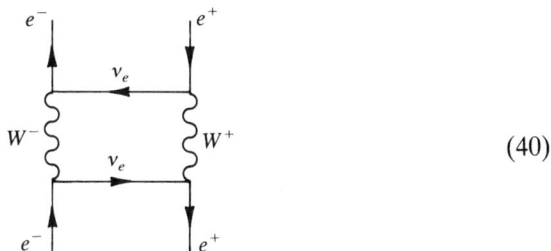

(40)

The integration over the closed-loop momentum brings in W's of arbitrarily high energy, and when these have $h = 0$, the integral diverges more violently than the radiative corrections of QED. As a consequence (40) provides a parity-violating contribution to elastic e^+e^- scattering that depends on the momentum cutoff even after the parameters e, g, and m_W have been renormalized. The same remark applies to *all* electromagnetic phenomena, because closed $W - \nu$ loops can be inserted into any QED graph. In short, the $W^\pm$ model destroys the predictive power of QED (in the standard jargon, it renders QED "unrenormalizable").

One might suppose that this catastrophe can only be averted by removing the $h = 0$ W's from the theory, i.e., by setting $m_W = 0$. Not only would this be in direct conflict with the observed short range of the weak interaction, but to add insult to injury, it would not even cure the problem! As (39) shows, $\boldsymbol{\varepsilon}_0(\mathbf{k})$ is singular as $m_W \to 0$. A successful cure requires the leading term $\mathbf{k}/m_W$ in $\boldsymbol{\varepsilon}_0$ to disappear from the amplitude. *This only happens if the current coupled to the vector boson field is conserved.* A proof of this theorem requires techniques beyond the level of this book. All we can do is to make a remark that illustrates the plausibility of the theorem. Consider the electromagnetic contribution to $e^+e^- \to W^+W^-$:

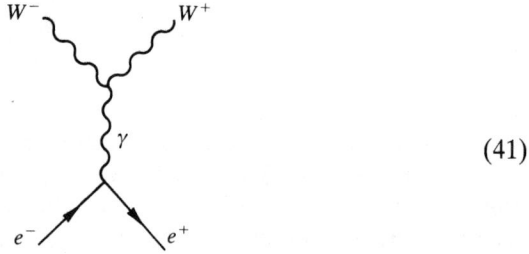

(41)

Like the weak interaction amplitude (38), it is of second order, but in contrast to the latter, it is well behaved at high energies even when the W's have $h = 0$. The reason is that the electromagnetic current, which is active in (41), is conserved, whereas the weak currents $J_\pm$ involved in (38) are not. That the charge-changing currents $J_\pm$ are not conserved will be shown in Part B of this Chapter. There we also learn that a theory with conserved currents requires the existence of a charge-preserving weak current, and an associated electrically neutral vector meson. Such a theory does not contain the divergent terms [see Llewellyn Smith (1973)].

B. NEUTRAL CURRENT WEAK INTERACTIONS AND THE ELECTROWEAK CONNECTION

1. Conservation of the weak current: The neutral current

In this section we demonstrate that the charge-changing currents $J_\pm$ are not conserved, and that their conservation can only be achieved by introducing a weak charge-preserving current, and an associated field whose quanta are electrically neutral partners of W^+ and W^-.

(a) Demonstration that $J_\pm$ are not conserved currents

A current that is not conserved does not satisfy a continuity equation; consequently, the total charge associated with it is not a constant of the motion. Our first task is to show that the weak charges are not conserved within the $W^\pm$-model.

According to Eqs. A(10) and (12), the weak "charges" associated with $J_\pm$ are the operators

$$t_\pm = \int \rho_\pm \, d^3r. \tag{1}$$

Our aim is to show that $\dot{t}_\pm \neq 0$, and we must therefore evaluate

$$i\dot{t}_\pm = [t_\pm, H], \tag{2}$$

where the Hamiltonian is

$$H = H_0 + H_{cc}. \tag{3}$$

The first term is the energy of free fermions and W's while H_{cc} is the charge-changing weak interaction. It turns out that both $[t_\pm, H_0]$ and $[t_\pm, H_{cc}]$ fail to vanish. As the second commutator is an operator that creates and destroys W's, while the first does not, there cannot be any cancellation between these two contributions to $\dot{t}_\pm$. Of these the most significant by far is due to H_{cc}, for it gives rise to the high-energy catastrophe in processes such as $e^+e^- \to W^+W^-$, and in the related radiative corrections. Furthermore, the *entire* structure of the weak interaction, *including* its relationship to electromagnetism, follows from the requirement that H_{cc} be modified in such a manner as to ensure the conservation of the weak currents. For those reasons we postpone our examination of the free Hamiltonian of fermions and W's until §C.2.

In H_{cc} and $t_\pm$ the various lepton and quark pairs appear additively [recall Eq. A(9), etc.]. Since operators pertaining to one fermion species commute with those of any other, all algebraic properties of $t_\pm$ can be found by restricting our attention to any one fermion pair, say (v_e, e). In most of this section we therefore replace the complete operators $t_\pm$ by

$$t_+ = \int v_L^\dagger e_L \, d^3r, \qquad t_- = \int e_L^\dagger v_L \, d^3r, \tag{4}$$

where we have also dropped the subscript on v. In the final results other leptons and quarks can then be incorporated by simple addition.

The computation of $[t_\pm, H_{cc}]$ with the complete field operators that appear in H_{cc} is quite lengthy, yet the final result is remarkably simple. The reason is that all that really matters in determining $t_\pm$ is that a fermion pair transfers its charge to the W-field. The spins and momenta of e, v, and W are of no significance in this context. Therefore, one can obtain the essential results much more easily by simplifying the operator H_{cc}. The simplified calculation is also more transparent, and allows one to grasp the true significance of the final result.

In view of the preceding remarks, we replace the vector fields $\mathbf{W}(\mathbf{r})$ and $\mathbf{W}^\dagger(\mathbf{r})$ by two scalar operators, $\mathcal{W}_+$ and $\mathcal{W}_-$, which do not depend on spatial coordinates. $\mathcal{W}_-$ creates W^- and destroys W^+, but it neither pays heed to where these acts take place nor to what helicity these particles have. The true interaction H_{cc} is then replaced by

$$\hat{H}_{cc} = \frac{1}{\sqrt{2}} g(t_+ \mathcal{W}_- + t_- \mathcal{W}_+), \tag{5}$$

where $\mathcal{W}_+ = (\mathcal{W}_-)^\dagger$. In contrast to H_{cc}, (5) does not conserve linear or angular momentum in the basic reactions $e\bar{v} \leftrightarrow W^-$ and $\bar{e}v \leftrightarrow W^+$, but it does keep track of the particle species in a reaction, and that is enough for our present purpose.

The time derivatives of interest to us are

$$i\dot{t}_+ = [t_+, \hat{H}_{cc}] = \frac{1}{\sqrt{2}} g[t_+, t_-]\mathcal{W}_+, \tag{6}$$

$$i\dot{t}_- = -\frac{1}{\sqrt{2}} g[t_+, t_-]\mathcal{W}_-. \tag{7}$$

We now exploit the resemblance of $t_\pm$ to the isospin raising and lowering

1. CONSERVATION OF THE WEAK CURRENT

operators $I_\pm$ to evaluate the commutator. Recall from Eq. A(28) that

$$I_- = \int d^\dagger u \, d^3r, \tag{8}$$

i.e., it replaces the $I_3 = \frac{1}{2}$ quark by an $I_3 = -\frac{1}{2}$ quark. The raising operator does the opposite

$$I_+ = (I_-)^\dagger = \int u^\dagger d \, d^3r. \tag{9}$$

The third isospin operator is

$$I_3 = \tfrac{1}{2}[I_+, I_-] = \tfrac{1}{2}\int (u^\dagger u - d^\dagger d) \, d^3r \tag{10}$$

which says that u and d have $I_3 = \frac{1}{2}$ and $-\frac{1}{2}$, respectively, since $\int u^\dagger u \, d^3r$ counts the number of u's in any state, etc.

The t's are obtained from the I's by the replacements $u \to v_L, d \to e_L$. Consequently,[1]

$$\tfrac{1}{2}[t_+, t_-] \equiv t_3 = \tfrac{1}{2}\int (v_L^\dagger v_L - e_L^\dagger e_L) \, d^3r, \tag{11}$$

and therefore

$$i\dot{t}_\pm = \pm\sqrt{2}gt_3 W_\pm. \tag{12}$$

Note that *in contrast to t_+ and t_-, the operator t_3 does not change the electrical charge of the fermions.*

Equation (12) shows that the charges $t_\pm$ are not constants of motion, and that *the charge-changing weak current is not conserved if the weak interaction is only mediated by charged fields.*

(b) Weak isospin[2]

Our discussion will be more transparent if we exploit fully the analogy between the lepton doublet (v_e, e) and the isospin doublet (u, d). For this purpose we introduce a three-dimensional *weak isospin space* $\mathscr{E}_3^T$. This space has the same mathematical properties as the hadronic isospin space $\mathscr{E}_3^I$, or, for that matter, as ordinary Euclidean 3-space, but it is distinct from both. The Hermitian generators of rotations in $\mathscr{E}_3^T$ are the operator t_3 as

---

[1] This commutation rule, as well as (10), can also be found by brute force from the commutation rules for Dirac fields given in Appendix III.2(a).
[2] An introduction to this material was provided in §I.E.10(a).

given by (11), and by

$$t_1 = \frac{1}{2}(t_+ + t_-), \qquad t_2 = \frac{1}{2i}(t_+ - t_-). \tag{13}$$

In virtue of (11), these operators satisfy the familiar angular momentum commutation rule

$$[t_1, t_2] = it_3, \tag{14}$$

and cyclic permutations.

In analogy with ordinary spinors and isospinors we combine the left-handed fields into a *weak isospinor* field. In the case of v_e and e, this object is

$$\Psi_e \equiv \begin{pmatrix} v_{eL} \\ e_L \end{pmatrix}. \tag{15}$$

If κ_a, with $a = 1, 2, 3$, are the 2×2 Pauli matrices acting on Ψ_e, we can write the t_a as follows:

$$t_a = \tfrac{1}{2} \int \Psi_e^\dagger \kappa_a \Psi_e \, d^3r + \cdots . \tag{16}$$

Every lepton and quark doublet appearing in $J_\pm$ can also be combined into such weak isospinors:

$$\Psi_\mu = \begin{pmatrix} v_{\mu L} \\ \mu_L \end{pmatrix}, \qquad \Psi_u = \begin{pmatrix} u_L \\ d'_L \end{pmatrix}, \qquad \Psi_c = \begin{pmatrix} c_L \\ s'_L \end{pmatrix}. \tag{17}$$

Each weak isodoublet contributes an identical expression to (16), as indicated by $+ \cdots$ in (16). As explained in connection with Eq. (4), it suffices to keep any one doublet for now, and for that reason we retain only Ψ_e, and drop the subscript e when it is not required for clarity.

The three operators t_a are called the components of the *weak isospin*. They can be combined into the 3-vector

$$\vec{t} = (t_1, t_2, t_3). \tag{18}$$

As with any Euclidean 3-space, the laws of vector algebra hold in $\mathscr{E}_3^T$. If $\vec{A}$ and $\vec{B}$ are vectors of type (18), $\vec{A} \cdot \vec{B} = \Sigma_a A_a B_a$, and $(\vec{A} \times \vec{B})_1 = A_2 B_3 - A_3 B_2$, etc.

There is no (known) relation between weak and hadronic isospin, since leptons carry the former but not the latter. The distinction also holds for quarks, though it is perhaps somewhat confusing at first sight. First, the entire Dirac fields of u and d, and not just their L-portions, contribute to the hadronic isospin. Second, the *L*-portions of all quark fields carry weak isospin, while u and d are the only quarks that carry hadronic isospin. Third, the weak isospinor Ψ_u, defined in (17), has the Cabibbo-mixed field d'_L as its lower component, not just d_L, so that the strange quark belongs, in part, to the same weak isospinor as u; the remainder of s_L then belongs to the weak isospinor Ψ_c containing c_L.

(c) Conservation of weak isospin

We now are ready to examine the physical significance of the equation of motion (12) for the weak isospin components t_+ and t_-. First, we construct Hermitian operators from the operators $\mathcal{W}_\pm$ that act on the *W*'s:

$$W_1 = \frac{1}{\sqrt{2}}(\mathcal{W}_+ + \mathcal{W}_-), \qquad W_2 = \frac{1}{\sqrt{2}i}(\mathcal{W}_+ - \mathcal{W}_-). \tag{19}$$

In terms of these our model Hamiltonian is

$$\hat{H}_{cc} = g(t_1 W_1 + t_2 W_2). \tag{20}$$

The time derivatives (12) then read

$$\dot{t}_1 = gt_3 W_2, \qquad \dot{t}_2 = -gt_3 W_1. \tag{21}$$

The contents of Eqs. (20) and (21) can be most readily understood if we consider an entirely different physical system: a single impurity spin **J** situated in a magnet. The latter can be described by the Heisenberg model, wherein there are spins $\mathbf{J}_i$ at each lattice site i, and the interaction between neighboring spins has the form $-\text{const}\,\mathbf{J}_i \cdot \mathbf{J}_k$. The interaction H_J of the impurity with its host is

$$H_J = -\gamma \mathbf{J} \cdot \mathbf{B} \tag{22}$$

where **B** is the magnetic field due to the spins that compose the magnet, and $\gamma \mathbf{J}$ is the magnetic moment of the impurity.

Consider the conservation of angular momentum in the magnetic system. According to (22),

$$\frac{d\mathbf{J}}{dt} = -\gamma \mathbf{B} \times \mathbf{J}. \tag{23}$$

This describes the precession of the impurity spin due to the torque exerted by the other spins. Obviously, **J** is *not* conserved itself, for the impurity can exchange angular momentum with the lattice spins through the magnetic field. But there is a conserved quantity: the angular momentum of the *whole* system, impurity *plus* host lattice. This conservation law owes its existence to the rotational invariance of the Hamiltonian for the whole system.

We now apply this lesson to the problem of finding a conserved weak current, and "charges" associated therewith that are constant in time. First we note that in contrast to the impurity interaction (22), the weak interaction (20) is not isotropic in the weak isospin space $\mathscr{E}_3^T$, and for that reason $\hat{H}_{cc}$ cannot commute with the generators of rotations in that space. We must, therefore, augment $\hat{H}_{cc}$ so as to make it isotropic. For this purpose we introduce a third field W_3 which, when adjoined to W_1 and W_2, forms the weak isovector

$$\vec{W} = (W_1, W_2, W_3). \tag{24}$$

This allows us to make the correspondence $\vec{t} \leftrightarrow \mathbf{J}$ and $\vec{W} \leftrightarrow \mathbf{B}$ between the weak and magnetic interactions. It is then evident that the desired isotropic generalization of the weak interaction is

$$\hat{H}_{wk} = \hat{H}_{cc} + gt_3 W_3 = g\vec{t} \cdot \vec{W}. \tag{25}$$

Since t_3 does not change the electrical charge of the fermions, as (11) makes clear, W_3 must also be a $\Delta Q = 0$ operator. In contrast to W_1 and W_2, which together create and destroy $W^\pm$, *W_3 must therefore be a field operator whose quanta are electrically neutral bosons W^0*. Because of the symmetry in the weak isospin space, it is clear that W^0 is a spin-1 particle, and that W_3 is a vector field. The three bosons (W^+, W^0, W^-) form a weak isotriplet, analogous to the hadronic isotriplet π^+, π^0, π^-.

The fermionic operator $\vec{t}$ is not conserved itself, because it satisfies a precession equation like that of the impurity spin:

$$\frac{d\vec{t}}{dt} = g\vec{W} \times \vec{t}. \tag{26}$$

T-spin can therefore flow from the fermions into the W-field, and vice versa. Let $\vec{T}_W$ be the weak isospin of the W's; it is the analogue of the host lattice angular momentum. Then the total T-spin of the entire system is

$$\vec{T} = \vec{t} + \vec{T}_W, \tag{27}$$

and it is conserved, just as the angular momentum of the isotropic magnetic lattice, plus that of the impurity, is a constant of motion. An explicit expression for $\vec{T}_W$ is given in Eq. C(29) below.

(d) The symmetric weak interaction

The foregoing considerations were all couched in terms of the simplified model introduced by Eq. (5), which suppresses many of the degrees of freedom of the fermions and W's that appear in the field current interaction $H_{cc,1}$. This simplified model was just generalized to the symmetric form given in (25). We can now write down the corresponding field-current

1. CONSERVATION OF THE WEAK CURRENT

interaction—the symmetrized version of Eq. A(17):

$$H_{\text{wk},1} = g \sum_{a=1}^{3} \int \mathbf{J}_a(\mathbf{r}) \cdot \mathbf{W}_a(\mathbf{r}) \, d^3r. \tag{28}$$

Here a labels the weak isospin components of the W-field and the weak currents. The integrand in (28) is a scalar product of vectors in ordinary 3-space, *and* in the weak isospin space, and is therefore rotationally invariant in both spaces; a totally equivalent way of writing the integrand is $\sum_i \vec{J}_i \cdot \vec{W}_i$, where $i = x, y$, or z. The field $\mathbf{W}_3(\mathbf{r})$ destroys and creates W^0, while $\mathbf{W}_{1,2}$ combine to form the field $\mathbf{W}(\mathbf{r})$ that appears in Eq. A(17):

$$\mathbf{W}(\mathbf{r}) = \frac{1}{\sqrt{2}}[\mathbf{W}_1(\mathbf{r}) + i\mathbf{W}_2(\mathbf{r})], \tag{29}$$

where we used (19), and the fact that $\mathbf{W}(\mathbf{r})$ has the same action as the operator $\mathcal{W}_+$. The charge-changing weak currents $\mathbf{J}_1$ and $\mathbf{J}_2$ are related to $\mathbf{J}_\pm$ by the same relations as (13), e.g., $\mathbf{J}_1 = \frac{1}{2}(\mathbf{J}_+ + \mathbf{J}_-)$. The terms $a = 1$ and 2 in (28) are therefore our old interaction Eq. A(17).

The new weak current appearing in (28) is the $\Delta Q = 0$ operator $\mathbf{J}_3$, coupled to the W^0-field $\mathbf{W}_3$. Its form is easily found, because the corresponding density ρ_3 must integrate to give the weak "charge" t_3, as given by (11):

$$\rho_3 = \tfrac{1}{2}(v_L^\dagger v_L - e_L^\dagger e_L) + \cdots, \tag{30}$$

where $+ \cdots$ indicates contributions from the other weak isodoublets (see Eq. (53) for further details). In view of Eqs. A(9) and (10), we therefore have

$$\mathbf{J}_3 = \tfrac{1}{2}(v_L^\dagger \boldsymbol{\sigma} v_L - e_L^\dagger \boldsymbol{\sigma} e_L) + \cdots. \tag{31}$$

Currents having the charge-preserving ($\Delta Q = 0$) character of $\mathbf{J}_3$ have the name *neutral currents* in the literature, whereas the $\Delta Q = \pm 1$ currents $\mathbf{J}_\pm$ are called *charged currents* (recall p. 472). According to this unfortunate nomenclature the electromagnetic current is a neutral current, even though it carries charge [as does $\mathbf{J}_3$; cf. Eq. (31)].

From the magnetic model, it is clear that (28) cannot be the complete weak Hamiltonian, for it is only the analogue of the interaction between the impurity spin and the host lattice. There must also be a contribution that is the analogue of the host lattice energy, that is, the energy of the W-field. Part of this describes free W^+, W^0, and W^- bosons, but there is also a coupling H_{wk}^W amongst the W's. This weak interaction of the W's themselves is entirely new. It exists because the W-field interacts with all objects that carry weak isospin. We have already concluded that the W's are a weak isotriplet, and therefore carry a unit of weak isospin. Consequently, there is a weak interaction of the field with its own current. Readers of Chap. IV should now recognize an analogy between the strong and weak interactions. In the former, color flows from quarks into the strong interaction gauge field, and since that field is coupled to all carriers of color, it is necessarily self-coupled. It should then come as no surprise that *the W-fields are gauge fields, and the symmetry underlying the conservation of the weak currents is a gauge invariance.* This aspect of the theory will be developed in Part C of this Chapter.

One can proceed considerably further in understanding the implications of the theory without the explicit form for the weak interaction amongst the W's, and we follow this course. But there is one question that we cannot in good conscience postpone, for it impelled us to demand that the weak current be conserved: *How does the inclusion of the neutral boson W^0 cure the high-energy disease of $e^+e^- \to W^+W^-$?* As we shall learn, the operator H_{wk}^W describes, among other things, the transition $W^0 \to W^+W^-$, with a strength that is uniquely specified by the requirement of weak isospin conservation. It is already clear from (11) that t_3W_3 describes, among other things, the transition $e^+e^- \to W^0$. Hence, $e^+e^- \to W^+W^-$ receives a contribution to its amplitude from

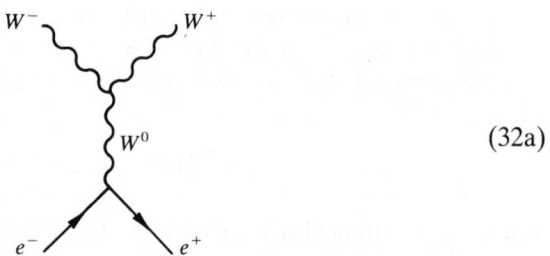

(32a)

This is to be added coherently to the ν_e-exchange graph Eq. A(38), and a detailed calculation then reveals that the offending contributions of order k/m_W from the $h = 0$ polarization vectors precisely cancel.[3] What remains is an amplitude that no longer diverges violently as $k \to \infty$. The divergent higher order corrections to elastic e^+e^- scattering due to Eq. A(40) are also canceled by W^0 diagrams of the same order in g, such as

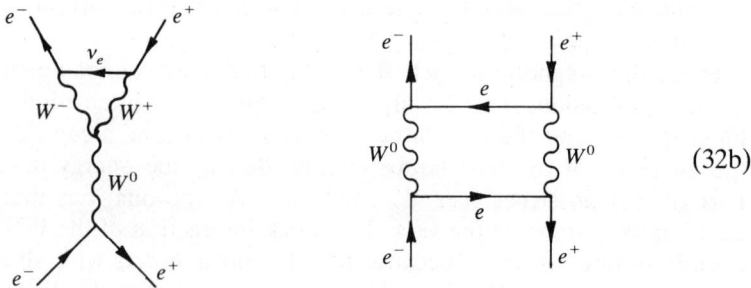

(32b)

---

[3] This conclusion is not changed when one eliminates W^0 in favor of Z^0 and γ, as we shall soon do [see Weinberg (1971)].

2. The electroweak connection

This was to be expected, for we have claimed that a nonconserved current leads to violation of probability conservation, which, in turn, produces divergent radiative corrections. Conversely, the augmented theory with conserved currents has well-behaved higher order corrections.

2. The electroweak connection

In the symmetrized weak interaction that we have just constructed fermions of differing charge belonging to one isodoublet, such as v_e and e, are treated on a symmetric footing. The electromagnetic interaction, on the other hand, distinguishes the neutral v_e from the charged e. The electroweak connection, which we shall now establish, stems from this asymmetry of the electromagnetic interaction in the weak isospin space $\mathscr{E}_3^T$.

(a) Incorporation of electrodynamics

Most of our discussion is phrased in terms of the schematic weak interaction $\hat{H}_{wk} = g\vec{t} \cdot \vec{W}$ defined in (25). The electromagnetic interaction Eq. A(1) is also replaced by an operator having the schematic structure of $\hat{H}_{wk}$, viz.

$$\hat{H}_{em} = eQA.$$

Here A is an operator that creates and destroys photons but, as with $\vec{W}$ in $\hat{H}_{wk}$, it has no space or spin information; for the time being, those are irrelevant complications. Q is the total charge. If we restrict ourselves to (v_e, e) for the moment,

$$Q = -\int e^\dagger e \, d^3r, \tag{33}$$

where e is the complete Dirac field of the electron, and the minus sign conforms with the historic convention that the electron is negatively charged.

It is tempting to assume that the Hamiltonian that describes the weak and electromagnetic interaction of (v_e, e) is

$$H^1_{int} = g\vec{t} \cdot \vec{W} + eQA. \tag{34}$$

Surprisingly enough, Eq. (34) is unacceptable because the total electric charge Q would not be a constant of motion! This is so because Q *is not an invariant in the weak isospin space* $\mathscr{E}_3^T$; Q defines a direction in $\mathscr{E}_3^T$ because it distinguishes v_e from e. Since (v_e, e) are subject to both the weak and electromagnetic interaction, we must formulate both interactions in a manner that guarantees the conservation of Q, while preserving the invariance of the Hamiltonian in $\mathscr{E}_3^T$. The stumbling block that we have encountered in (34) shows that this is a nontrivial task, but its resolution will reveal an unexpected connection between the weak and electromagnetic interaction.

The lack of invariance of Q merits closer examination. For this purpose it is essential to decompose Q, defined by (33), into left- and right-handed portions:

$$Q = -\int (e_L^\dagger e_L + e_R^\dagger e_R)\, d^3r = Q_L + Q_R. \tag{35}$$

Since the rotation generators in $\mathscr{E}_3^T$ act only on the left-handed portions of Dirac fields, it is clear that $[Q_R, \vec{t}\,] = 0$, i.e., that Q_R is invariant in $\mathscr{E}_3^T$, and that the same statement applies to any operator constructed solely from right-handed fields. Any operator involving L-fields must, however, be symmetric in ν_L and e_L. The only expression having the general format of a charge, and having this property, is the operator

$$t_0 = \int (\nu_L^\dagger \nu_L + e_L^\dagger e_L)\, d^3r. \tag{36}$$

That t_0 is an invariant can be seen most clearly by using the notation of Eq. (16):

$$t_0 = \int \Psi_e^\dagger \Psi_e\, d^3r, \tag{37}$$

i.e., it is the scalar product of the weak isospinor Ψ_e with itself.

We can now express the complete charge Q in terms of the invariant operators Q_R and t_0, and the 3-component of the weak isovector $\vec{t}$. Using (11), (35), and (36) we have

$$Q = t_3 - \tfrac{1}{2} t_0 + Q_R. \tag{38}$$

This dissects the electric charge into a piece t_3 that is not invariant in $\mathscr{E}_3^T$, and a remainder that is.

In constructing (38) we only dealt with the (ν_e, e) doublet. Obviously, the doublet (ν_μ, μ) is handled in precisely the same way. But the quark doublets differ somewhat because they have different electric charges. Consider the doublet Ψ_u defined by (17); Ψ_c is treated in the same manner. For this doublet

$$Q_\lambda = \int (\tfrac{2}{3} u_\lambda^\dagger u_\lambda - \tfrac{1}{3} d_\lambda'^\dagger d_\lambda')\, d^3r, \tag{39}$$

where $\lambda = L$ or R, and the electric charge operator is again $Q = Q_L + Q_R$. As before, Q_R is invariant in $\mathscr{E}_3^T$, as is

$$t_0 = \int (u_L^\dagger u_L + d_L'^\dagger d_L')\, d^3r.$$

2. THE ELECTROWEAK CONNECTION

Since $t_3 = \frac{1}{2} \int (u_L^\dagger u_L - d_L'^\dagger d_L') \, d^3r$, we have

$$Q = t_3 + \tfrac{1}{6} t_0 + Q_R \tag{40}$$

for quark doublets, in contrast to (38) for lepton doublets. *For both types of doublets, however, $Q - t_3$ is invariant in the weak isospin space.*[4]

The asymmetric portion $t_3 A$ of QA cannot be added to the weak interaction without spoiling the isotropy of the whole. This would seem to say that we cannot build a consistent theory of the weak interaction unless we ignore electromagnetism! However, on second thought, we note that t_3 already appears in $g\vec{t} \cdot \vec{W}$. Hence *part* of the electric charge operator was included in what we had thought to be the weak "charge" t_3, and *ipso facto*, $g t_3 W_3$ already contains a portion of the electromagnetic interaction! Once this is realized, we can see how to replace (34) by an interaction that contains the $\Delta Q = 0$ Fermi and Bose operators that we need, and that does not spoil invariance in $\mathscr{E}_3^T$: Instead of (34), add to $g\vec{t} \cdot \vec{W}$ an operator that contains only the invariant portion $Q - t_3$ of the charge. Our *Ansatz* for the *electroweak interaction* is therefore

$$H_{\text{int}} = g\vec{t} \cdot \vec{W} + g'(Q - t_3)B. \tag{41}$$

Here B is a $\Delta Q = 0$ Bose field operator having quanta that carry no electric charge; it is *not* the entire electromagnetic field, because it is not coupled to the whole of Q. For that reason, the coupling constant g' is not e. On the other hand, we have also recognized that W_3 is partly involved in electromagnetism. *The electromagnetic field A must therefore be that linear combination of B and W_3 that is coupled to Q.*

(b) The Z^0 and the mixing angle θ_W

To find this linear combination, we need only consider the portion of (41) that involves the $\Delta Q = 0$ Bose fields and source operators:

$$H_{\text{int}}^0 = g'(Q - t_3)B + g t_3 W_3. \tag{42}$$

We wish to write this in the form

$$H_{\text{int}}^0 = eQA + g_Z t_Z Z, \tag{43}$$

where Z is the combination of W_3 and B that is linearly independent of the electromagnetic field A, while t_Z and g_Z are whatever $\Delta Q = 0$ source operator and coupling constant the transformation to (43) manufactures.

The coefficient of t_3 in (42) is $gW_3 - g'B$. This is the combination of fields

---

[4] Indeed, $Q = t_3 + \tfrac{1}{2} \operatorname{tr} Q$ quite generally.

that must be Z, because A is coupled to Q, not t_3:

$$Z = \frac{1}{\sqrt{g^2 + g'^2}} (gW_3 - g'B). \tag{44}$$

The orthogonal, linearly independent combination is the electromagnetic potential:

$$A = \frac{1}{\sqrt{g^2 + g'^2}} (gB + g'W_3). \tag{45}$$

The prefactor $(g^2 + g'^2)^{-1/2}$ is necessary so that the states created and destroyed by Z and A are normalized in the same manner as those of B and W_3. The field Z plays a central role in the sequel; its quanta are neutral and are called Z^0.

For many purposes it is convenient to define the *electroweak mixing angle* θ_W:

$$\tan \theta_W \equiv \frac{g'}{g}. \tag{46}$$

In terms of θ_W

$$A = B \cos \theta_W + W_3 \sin \theta_W, \tag{47}$$
$$Z = W_3 \cos \theta_W - B \sin \theta_W,$$

and

$$B = A \cos \theta_W - Z \sin \theta_W, \tag{48}$$
$$W_3 = Z \cos \theta_W + A \sin \theta_W.$$

On substituting (48) into (42), and comparing with (43), we obtain

$$\frac{e}{g} = \sin \theta_W, \tag{49}$$

and

$$g_Z t_Z = \frac{g}{\cos \theta_W} (t_3 - Q \sin^2 \theta_W). \tag{50}$$

As was already clear in (41), and as (49) emphasizes further, *there are two fundamental coupling constants in the electroweak theory: e and g, or equivalently e and θ_W*.

According to (43) and (50) the complete electroweak Hamiltonian (41)

2. THE ELECTROWEAK CONNECTION

therefore has the following form:

$$H_{\text{int}} = g(t_1 W_1 + t_2 W_2) + \frac{g}{\cos \theta_W}(t_3 - Q \sin^2 \theta_W) Z + eQA. \quad (51)$$

Reading from left to right the three terms are the charge-changing weak interaction, the charge-preserving or neutral current weak interaction, and the electromagnetic interaction.

Let us record the lesson we have now learned: If one demands that the weak charge-changing and electromagnetic currents be conserved (as one must), one is forced to introduce a neutral current weak interaction with an associated neutral vector boson Z^0; further, *the strength and form of the Z^0's interaction are completely determined by the weak charge-changing and electromagnetic interactions, and their coupling constants g and e.*

(c) Universality and unification

At this point some may ask the following penetrating question: Why do leptons have integer electric charges, and quarks fractional charges, whereas the "weak" charge g is the same for all the fundamental fermions? Or, in the language of §I.E.9: Why is the charge-changing weak interaction "universal," while the electromagnetic is not?

This question cannot be answered by examining the schematic model Hamiltonian; H_{int} is a sum of terms pertaining separately to each quark and lepton weak isodoublet, because $\vec{t}$ only appears linearly in it. Consequently, the symmetry of H_{int} in $\mathscr{E}_3^T$ would not be spoiled if each weak doublet appeared in (51) with a different coupling constant g. This is obvious in terms of the magnetic analogy, explored in §1(c); each quark and lepton doublet is analogous to a separate spin-$\frac{1}{2}$ impurity, but these could have different magnetic moments without spoiling rotational invariance. But there is a crucial difference between the magnetic lattice and the weak interaction: in contrast to the electromagnetic field operative in the lattice, the weak interaction fields supposedly form a non-Abelian gauge field.[5] As we learned in §IV.B.3, a non-Abelian gauge field has a coupling to its sources whose structure is uniquely determined by the transformation law of the sources under the symmetry group in question, apart from the fundamental constant g. Since all quarks and leptons belong to weak isodoublets, they transform in the same way in $\mathscr{E}_3^T$, and are therefore coupled to the weak field in a universal manner. The experimental observation of universality therefore provides a strong hint that the weak interaction is propagated by a non-Abelian gauge field.

The electromagnetic field carries no "internal" quantum number like weak isospin. As a consequence, electromagnetic (or Abelian) gauge

---

[5] Recall that the ferromagnetic interaction is $-\mathbf{J}_i \cdot \mathbf{J}_k$. Hence, the spin $\mathbf{J}_i$ at site i cannot be rotated through a different angle than its neighbor $\mathbf{J}_k$, and the Hamiltonian is not locally invariant.

invariance places no restriction on the charges of sources. If one is to understand the various electric charges of quarks and leptons, one must embed the electromagnetic field in a larger non-Abelian gauge field. This embedding must be a true unification of electrodynamics with other interactions, where by "true" we mean that the existence of any one field in the unified theory implies the existence of the others, in the sense that Maxwell's theory requires the simultaneous existence of the electric and magnetic fields. The electroweak theory does not have this character. In (41) we are at liberty to set $g' = 0$; there would then be no electromagnetic interaction, for θ_W would vanish [see Eq. (49)]. In short, the existence of electromagnetism is an empirical observation and not a logical consequence of the existence of the weak interactions. The electroweak model is not a truly unified theory of these interactions, and for that reason we have used the term "the electroweak connection" as the title of this section.

There do exist a class of speculative theories that truly unify the weak, electromagnetic, and strong interactions, in that the existence of any one of the known interactions implies the existence of the others. These are the Grand Unified models briefly described in §I.E.13, and discussed at greater length in §D below.

3. Neutral current phenomena and the determination of the coupling constants

(a) The neutral current weak interaction

The schematic form of the interaction between fermions and Z^0 is given by Eq. (51). The actual interaction has the same format as the charge-changing weak interaction involving $W^\pm$, as given by Eqs. A(17) and (18). That is, the model operator Z in (51) really stands for a vector field $\mathbf{Z}(r)$, while t_3 and Q are abbreviations for the weak and electromagnetic currents $\mathbf{J}_3$ and $\mathbf{J}_{em}$, respectively. The actual neutral current weak interaction is then found from Eq. A(17) by the substitution[6] $g \to g/\cos\theta_W, m_W \to m_Z$, $\mathbf{J}_\pm \to \mathbf{J}_3 - \mathbf{J}_{em}\sin^2\theta_W$:

$$H_{nc,1} = \frac{g}{\cos\theta_W} \int (\mathbf{J}_3 - \mathbf{J}_{em}\sin^2\theta_W) \cdot \mathbf{Z}\, d^3r. \tag{52}$$

There is also a Coulomb-like instantaneous interaction corresponding to Eq. A(18), involving $(\rho_3 - \rho_{em}\sin^2\theta_W)$, which we won't write out, but which will be incorporated in Eq. (56).

The neutral 4-current $\mathbf{J}_3$ has only been given explicitly for leptons [see Eqs. (30) and (31)]. The operator for quarks has an important property that

---

[6] As already mentioned in §I.E.10(c), the weak isospin symmetry is "broken," and $m_W \neq m_Z \neq m_\gamma$. This aspect of the theory will be taken up in Part C of this Chapter. In this Section we treat m_W and m_Z as free parameters.

3. NEUTRAL CURRENT PHENOMENA

we must discuss. For any weak isospinor

$$\Psi_f = \begin{pmatrix} f_{1L} \\ f_{2L} \end{pmatrix},$$

the contribution to ρ_3 is given by the integrand of Eq. (16) with $a = 3$, i.e., by $\frac{1}{2}(f_{1L}^\dagger f_{1L} - f_{2L}^\dagger f_{2L})$. Consequently, the time component of $\mathbf{J}_3$ is

$$\rho_3 = \tfrac{1}{2} \sum_{l=e,\mu} (v_l^\dagger v_l - l^\dagger l)_L + \tfrac{1}{2}(u^\dagger u - d'^\dagger d' + c^\dagger c - s'^\dagger s')_L. \tag{53}$$

In view of the orthogonal nature of the transformation Eq. A(8), $d'^\dagger d' + s'^\dagger s' = d^\dagger d + s^\dagger s$, and therefore

$$\rho_3 = \tfrac{1}{2} \sum_{l=e,\mu} (v_l^\dagger v_l - l^\dagger l)_L + \tfrac{1}{2}(u^\dagger u - d^\dagger d + c^\dagger c - s^\dagger s)_L. \tag{54}$$

The expression for $\mathbf{J}_3$ is obtained by inserting $\boldsymbol{\sigma}$, as in (31).

Equation (54), and the corresponding one for $\mathbf{J}_3$, have the important property of not containing any flavor-changing terms, such as $d^\dagger s$ or $s^\dagger \sigma d$. The electromagnetic current also conserves all flavors. As a consequence *the weak neutral 4-current $\mathbf{J}_3 - \mathbf{J}_{em} \sin^2 \theta_w$ is not flavor changing.* If there were flavor-changing terms in $\mathbf{J}_3$, they would give rise to $d\bar{s} \to Z^0$ transitions. These would combine with $Z^0 \to \mu^+ \mu^-$, which are predicted to exist by the first term of (54), to give the decay $K_L \to \mu^+ \mu^-$:

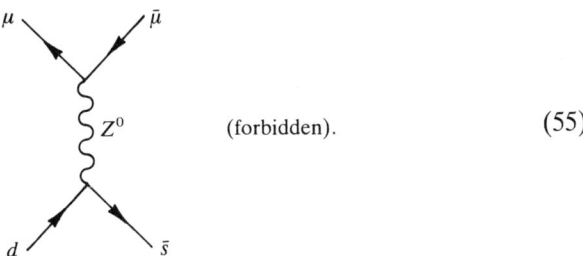

(forbidden). (55)

This amplitude would then be of comparable magnitude to that for the copious decay $K^+ \to \mu^+ \bar{v}_\mu$. But flavor-changing $\Delta Q = 0$ processes like (55) are observed[7] to have rates that are far smaller than flavor-preserving decays, in confirmation of the flavor-preserving structure of $\mathbf{J}_3$.

On the other hand, H_{nc} does predict the existence of many flavor-conserving neutral-current weak processes. At "low" energies they can be handled by an effective interaction of the Fermi type, H_F^{nc}, in which the Z-field does not appear. Its relation to (52) is precisely the same as that of

---

[7] The branching fraction for $K_L \to \mu^+ \mu^-$ is $(9 \pm 2) \times 10^{-9}$ [Particle Data Group (1984)]. This decay is anticipated at roughly this rate as a second-order charge-changing weak process, as explained in Vol. I, p. 136.

H_F, Eq. A(21), to Eq. A(17), and therefore

$$H_F^{nc} = \frac{1}{2} \frac{g^2}{m_Z^2 \cos^2 \theta_W} \int (\mathbf{J}_3 - \mathbf{J}_{em} \sin^2 \theta_W)^2 \, d^3r. \tag{56}$$

It is convenient to introduce the parameter

$$\rho \equiv \left(\frac{m_W}{m_Z \cos \theta_W}\right)^2, \tag{57}$$

because $\rho = 1$ in the "standard" model to be discussed in §C. Then

$$H_F^{nc} = \frac{1}{2} \frac{g^2}{m_W^2} \rho \int (\mathbf{J}_3 - \mathbf{J}_{em} \sin^2 \theta_W)^2 \, d^3r. \tag{58}$$

If ρ is indeed unity, the "strengths" $(g/m_W)^2$ of the neutral and charge-changing Fermi interactions are the same [compare Eq. A(21)].

In contrast to the charge-changing currents $\mathbf{J}_\pm$, the neutral current has a quite intricate structure, because one part ($\mathbf{J}_3$) is parity violating, while the other ($\mathbf{J}_{em} \sin^2 \theta_W$) is not. Furthermore, $\mathbf{J}_{em}$ is quite different for leptons and quarks. Many of these interesting properties of the neutral weak current can be tested by experiments.

The most important neutral-current weak processes involve the collisions of neutrinos or electrons with nuclei or protons. For that reason, it is convenient to break the current that occurs in (58) into a hadronic ($\mathbf{J}_Z$) and a leptonic ($\mathbf{j}_Z$) part:

$$\mathbf{J}_3 - \mathbf{J}_{em} \sin^2 \theta_W = \mathbf{J}_Z + \mathbf{j}_Z. \tag{59}$$

These currents are sums of terms of the type $f_L^\dagger f_L, f_R^\dagger \sigma f_R$, etc. In the sequel, we suppress the Pauli matrix, it being understood that the time and space components have a unit or Pauli matrix, respectively. When we write the scalar product between 4-currents [as in Eqs. (66) and (67) below], we also suppress the Pauli matrices and 4-vector indices, it being understood that the expression really has the form of Eq. A(26).

As the hadron targets in the experiments of interest only contain u and d quarks, J_Z is, in effect, just

$$J_Z = \tfrac{1}{2}(u^\dagger u - d^\dagger d)_L - \omega(\tfrac{2}{3} u^\dagger u - \tfrac{1}{3} d^\dagger d), \tag{60}$$

where we used (54), the fact that u and d have $Q = \tfrac{2}{3}$ and $-\tfrac{1}{3}$, respectively, and the abbreviation

$$\omega \equiv \sin^2 \theta_W. \tag{61}$$

3. NEUTRAL CURRENT PHENOMENA

It will prove convenient to separate (60) into L and R portions, by using[8] $f^\dagger f = f_L^\dagger f_L + f_R^\dagger f_R$, where f is any Dirac field:

$$J_Z = u_L^\dagger u_L(\tfrac{1}{2} - \tfrac{2}{3}\omega) + d_L^\dagger d_L(\tfrac{1}{3}\omega - \tfrac{1}{2}) - \tfrac{2}{3}\omega u_R^\dagger u_R + \tfrac{1}{3}\omega d_R^\dagger d_R, \quad (62)$$

or

$$J_Z = \sum_{\lambda=L,R} \sum_{q=u,d} \varepsilon_\lambda^q q_\lambda^\dagger q_\lambda, \quad (63)$$

where

$$\begin{aligned} \varepsilon_L &= T_3' - Q \sin^2 \theta_W, \\ \varepsilon_R &= -Q \sin^2 \theta_W. \end{aligned} \quad (64)$$

Here T_3' is the eigenvalue of the weak isospin projection T_3, and Q the charge. Because of (59), these formulas also apply to leptons:

$$j_Z = \sum_{l=e,\mu} [\tfrac{1}{2}(v_l^\dagger v_l - l^\dagger l)_L + \omega l^\dagger l]. \quad (65)$$

The effective Hamiltonian (58) has a large number of terms, only some of which contribute to a given process. We collect these to form effective Hamiltonians applicable to the reactions of interest to us. In the case of neutrino collisions with nucleons, this effective interaction is

$$H_{vN}^{nc} = \frac{1}{2} \frac{g^2}{m_W^2} \rho \int J_Z v_L^\dagger v_L \, d^3r, \quad (66)$$

where v is usually v_μ, while for electron-nucleon scattering, it is

$$H_{eN}^{nc} = \frac{g^2}{m_W^2} \rho \sum_{\lambda=L,R} \varepsilon_\lambda^e \int J_Z e_\lambda^\dagger e_\lambda \, d^3r, \quad (67)$$

where ε_λ^e is also given by (64).

The neutral current coupling parameters ε_λ^f, where f is u, d, or e, depend in a quite intricate manner on the electroweak mixing angle θ_W (see Table 1).

We now discuss some experiments that allow one to measure ρ, θ_W, and the ε_λ^f.

(b) Deep inelastic neutrino scattering

Consider the reactions $v_\mu N \to v_\mu X$ and $\bar{v}_\mu N \to \bar{v}_\mu X$, where the incident neutrino has a high energy and scatters with a large transfer of momentum

---

[8] See Appendix III, Eq. (19), and note that $f^\dagger = (f_R^\dagger, f_L^\dagger)$, etc.

TABLE 1
Z^0-fermion coupling constants[a]

	Standard Model	Data
ε_L^u	$\frac{1}{2} - \frac{2}{3}\omega = 0.345$	0.339 ± 0.030
ε_L^d	$-\frac{1}{2} + \frac{1}{3}\omega = -0.423$	-0.425 ± 0.025
ε_R^u	$-\frac{2}{3}\omega = -0.155$	-0.179 ± 0.018
ε_R^d	$\frac{1}{3}\omega = 0.077$	-0.02 ± 0.05
ε_L^e	$-\frac{1}{2} + \omega = -0.267$	-0.25 ± 0.10
ε_R^e	$\omega = 0.233$	0.29 ± 0.10

[a] By "Standard Model" we mean that the ε's are given by Eq. (64), while the numerical values 0.345, etc., are for $\rho \equiv 1$ and $\omega \equiv \sin^2\theta_W = 0.233$ [see Eq. (89) below]. The column marked "Data" is from the analysis of Kim et al. (1981), where references to the original experimental literature can be found. The processes that have been measured for this purpose are listed in Subsection (d) below. This determination of ω via ε_R^e does not compete in precision with those used in Eq. (89).

and energy. As we learned in Chap. V, under these circumstances the target nucleon can be viewed as a set of free quarks that scatter the neutrino independently. The observed *inelastic* cross sections $\sigma(v_\mu N \to v_\mu X)$ and $\sigma(\bar{v}_\mu N \to \bar{v}_\mu X)$ are, therefore, incoherent sums of cross sections for *elastic* $v_\mu q$ and $\bar{v}_\mu q$ scattering, where[9] q is u or d. The evaluation of these elastic cross sections is quite simple at very high energies because the quarks are then ultrarelativistic in the neutrino-quark c.o.m. frame $\mathscr{F}_0$. Under these kinematical conditions, *the neutral current weak interaction conserves the helicity of the struck quark*. This is so because, first, the hadronic neutral current, as given by (63), only contains the operators $q_L^\dagger q_L$ and $q_R^\dagger q_R$; and, secondly, as we learned in §A.1, for ultrarelativistic particles, the Dirac field q_L (or q_R) destroys q's with $h = -\frac{1}{2}$ (or $h = \frac{1}{2}$), while $q_L^\dagger$ (or $q_R^\dagger$) creates q's with $h = -\frac{1}{2}$ (or $h = \frac{1}{2}$). The final result of these calculations is expressions for the ratios of observed cross sections that depend only on the mixing angle θ_W, and the parameter ρ defined in Eq. (57). In a first reading, the derivation of Eqs. (78) and (79) can be skipped.

⟦Let $\sigma_{\text{nc}}(vq_\lambda; \phi)$ and $\sigma_{\text{nc}}(\bar{v}q_\lambda; \phi)$ be the differential cross sections for scattering between the indicated particles through the angle ϕ in $\mathscr{F}_0$. The neutrino helicity need not be specified, since $h = -\frac{1}{2}$ and $\frac{1}{2}$ for v and $\bar{v}$, respectively. From (66) and (63), we then have

$$\sigma_{\text{nc}}(vq_\lambda; \phi) = C\left(\frac{g^2}{m_W^2}\right)^2 \rho^2 |\varepsilon_\lambda^q|^2 |A_{L\lambda}(\phi)|^2. \tag{68}$$

[9] One can gain a sound understanding of the experiments by ignoring everything but u and d quarks in the nucleon. The numerical results finally quoted come from an analysis of the data that incorporates the rather small corrections due to sea quarks [see §V.7(c)].

3. NEUTRAL CURRENT PHENOMENA

Here C depends only on the c.o.m. energy, and $A_{L\lambda}(\phi)$ is the amplitude for scattering of a left-handed (or $h = -\tfrac{1}{2}$) fermion (v) by another fermion (q) of helicity λ (designated by L and R for $h = -\tfrac{1}{2}$ and $\tfrac{1}{2}$, respectively). In Appendix III.3, Eqs. (61) and (62), we show that in the high-energy limit an interaction of the type $v_L^\dagger v_L q_\lambda^\dagger q_\lambda$ produces the amplitudes[10]

$$A_{LL} = A_{RR} = 1, \tag{69}$$

$$A_{LR} = A_{RL} = \cos^2\tfrac{1}{2}\phi. \tag{70}$$

Since the probabilities of finding $h = \tfrac{1}{2}$ and $h = -\tfrac{1}{2}$ quarks in the target are equal, the vq cross section is actually

$$\sigma_{\rm nc}(vq;\phi) = \tfrac{1}{2} C\left(\frac{g}{m_W}\right)^4 \rho^2 \left[|\varepsilon_L^q|^2 + |\varepsilon_R^q|^2 \cos^4\tfrac{1}{2}\phi\right]. \tag{71}$$

The $\bar{v}q$ cross section follows immediately; $\bar{v}$ has $h = \tfrac{1}{2}$, so $A_{L\lambda}$ must be replaced by $A_{R\lambda}$ in (68), which then leads to

$$\sigma_{\rm nc}(\bar{v}q;\phi) = \tfrac{1}{2} C\left(\frac{g}{m_W}\right)^4 \rho^2 \left[|\varepsilon_L^q|^2 \cos^4\tfrac{1}{2}\phi + |\varepsilon_R^q|^2\right]. \tag{72}$$

On integrating over the scattering angle one finds

$$\sigma_{\rm nc}(vq) = C\left(\frac{g}{m_W}\right)^4 \rho^2 \left(|\varepsilon_L^q|^2 + \tfrac{1}{3}|\varepsilon_R^q|^2\right), \tag{73}$$

$$\sigma_{\rm nc}(\bar{v}q) = C\left(\frac{g}{m_W}\right)^4 \rho^2 \left(\tfrac{1}{3}|\varepsilon_L^q|^2 + |\varepsilon_R^q|^2\right). \tag{74}$$

The theory developed in Chap. V allows us to evaluate the nucleon cross sections from these quark cross sections. To do so we multiply (73) and (74) by the probability that the quark has a given fraction x of the nucleon's momentum, and we must also replace C by $C(x)$, since the $v - q$ c.o.m. energy depends on x. In the heavy targets that have been used in the most accurate experiments, there are approximately equal numbers of neutrons and protons, and therefore of u's and d's, so we need only specify the probability $f(x)$ of finding some quark with the momentum fraction x. As a result, the neutral current cross sections for a heavy target are

$$\frac{d\sigma_{\rm nc}^v}{dx} = C(x)\left(\frac{g}{m_W}\right)^4 \rho^2 f(x)\left[|\varepsilon_L^u|^2 + |\varepsilon_L^d|^2 + \tfrac{1}{3}(|\varepsilon_R^u|^2 + |\varepsilon_R^d|^2)\right],$$

$$\frac{d\sigma_{\rm nc}^{\bar{v}}}{dx} = C(x)\left(\frac{g}{m_W}\right)^4 \rho^2 f(x)\left[\tfrac{1}{3}(|\varepsilon_L^u|^2 + |\varepsilon_L^d|^2) + |\varepsilon_R^u|^2 + |\varepsilon_R^d|^2\right]. \tag{75}$$

Many systematic errors are eliminated if one measures the ratio of the neutral-current to charged-current cross sections. The latter were already

---

[10] If E_v and E_v' are the initial and final neutrino energies in the lab frame, $\cos^2\tfrac{1}{2}\phi = E_v'/E_v$.

evaluated in §V.6(c); in the notation of Eq. (75), those results were

$$\frac{d\sigma_{cc}^\nu}{dx} = C(x)\left(\frac{g}{m_W}\right)^4 f(x),$$
$$\frac{d\sigma_{cc}^{\bar\nu}}{dx} = \frac{1}{3}C(x)\left(\frac{g}{m_W}\right)^4 f(x),$$
(76)

where the Cabibbo angle has been approximated by $\theta_C = 0$. Note that the neutral-to-charged current cross-section ratios do not depend on $f(x)$. This is only true for targets with equal numbers of u and d quarks.]

Let us now define the following ratios between neutral- and charged-current neutrino cross sections:

$$R^\nu = \frac{\sigma(\nu_\mu N \to \nu_\mu X)}{\sigma(\nu_\mu N \to \mu X)}, \qquad R^{\bar\nu} = \frac{\sigma(\bar\nu_\mu N \to \bar\nu_\mu X)}{\sigma(\bar\nu_\mu N \to \bar\mu X)}. \qquad (77)$$

The final result of the preceding calculations is then

$$R^\nu = \rho^2[|\varepsilon_L^u|^2 + |\varepsilon_L^d|^2 + \tfrac{1}{3}(|\varepsilon_R^u|^2 + |\varepsilon_R^d|^2)]$$
$$= \rho^2(\tfrac{1}{2} - \omega + \tfrac{20}{27}\omega^2), \qquad (78)$$

$$R^{\bar\nu} = \rho^2[|\varepsilon_L^u|^2 + |\varepsilon_L^d|^2 + 3(|\varepsilon_R^u|^2 + |\varepsilon_R^d|^2)]$$
$$= \rho^2(\tfrac{1}{2} - \omega + \tfrac{20}{9}\omega^2), \qquad (79)$$

where we used (64), and $\omega \equiv \sin^2\theta_W$. The data on these cross-section ratios yield the following results:[11]

$$\sin^2\theta_W = 0.232 \pm 0.027, \qquad (80)$$

$$\rho = 0.999 \pm 0.025. \qquad (81)$$

It is remarkable that these data agree so well with the "standard" model prediction, $\rho \equiv 1$ [see §C.2(c)].

(c) Parity violation in inelastic electron scattering

The cross section for deep inelastic electron-nucleon scattering is determined from the amplitude for electron-quark scattering (cf. §V.4). There are two distinct mechanisms at work in electron-quark collisions, photon

---

[11] These results are taken from the comprehensive analysis of Kim et al. (1981). The quoted numbers include corrections for effects ignored in (78) and (79), such as the influence of sea quarks, possible scaling violations, etc.

3. NEUTRAL CURRENT PHENOMENA

exchange and Z^0-exchange, and therefore the electron-quark scattering amplitude has the form $T = T_\gamma + T_Z$. At c.o.m. energies well below m_Z, $|T_Z| \ll |T_\gamma|$, and therefore $|T|^2 \simeq |T_\gamma|^2 + 2\,\text{Re}\,T_\gamma T_Z^*$. The first term gives the familiar Rutherford-type cross section, and is parity conserving. The interference term is parity violating; it is very small, and it can usually be neglected. But it can be isolated if one measures the difference in cross sections for scattering of electrons with $h = \frac{1}{2}$ and $h = -\frac{1}{2}$ from an unpolarized target, for that difference vanishes if the interaction is reflection invariant.[12]

⟦We designate the electron-quark scattering amplitudes by $T_B(e_\mu q_\lambda)$, where $B = \gamma$ or Z, and μ and λ are the electron and quark helicities, with L and R standing for $h = -\frac{1}{2}$ and $\frac{1}{2}$, as before. Both helicities are conserved at high energies, for the reasons already explained in connection with Eq. (68). The electromagnetic amplitude T_γ was discussed in §II.C.1 and §V.4; in the notation of Eqs. (69) and (70), it is given by

$$T_\gamma(e_\mu q_\lambda) = -\frac{e^2 Q_q}{t} A_{\mu\lambda}, \tag{82}$$

where Q_q is the charge of the quark that we designate by q in units of e, $-t$ is the square of the 4-momentum transfer, and $A_{\mu\lambda}$ is given by Eqs. (69) and (70), while the overall minus sign reflects the attraction between an electron and a positively charged quark. The Z^0-exchange amplitude is a matrix element of the effective Hamiltonian (67), and is therefore given by

$$T_Z(e_\mu q_\lambda) = \left(\frac{g}{m_W}\right)^2 \rho \varepsilon_\mu^e \varepsilon_\lambda^q A_{\mu\lambda}. \tag{83}$$

Note that T_γ and T_Z are real.

Let $\sigma(e_\mu q)$ be the c.o.m. cross section for scattering of an electron of the indicated helicity by quarks q that have a random spin orientation, i.e.,

$$\sigma(e_\mu q) = C \sum_\lambda |T(e_\mu q_\lambda)|^2,$$

where C is a kinematical factor that need not concern us, as it will disappear from the quantity that is measured. As we have explained, $\sigma(e_R q) - \sigma(e_L q)$ is solely due to the $Z^0 - \gamma$ interference term:

$$\sigma(e_R q) - \sigma(e_L q) = -\frac{2e^2 Q_q}{t}\left(\frac{g}{m_W}\right)^2 \rho C \sum_\lambda \varepsilon_\lambda^q (\varepsilon_R^e |A_{R\lambda}|^2 - \varepsilon_L^e |A_{L\lambda}|^2). \tag{84}$$

---

[12] The reaction $e\bar{e} \to \mu\bar{\mu}$ can also be used to isolate a $\gamma - Z^0$ interference term, because the reaction can go via two coherent amplitudes: $e\bar{e} \to (\gamma \text{ or } Z^0) \to \mu\bar{\mu}$. Once again, one uses a parity-violating effect, in this instance the asymmetry (if any) between μ's produced in the forward vs. backward hemispheres. Such an asymmetry has been detected (see Fig. II.3), and yields a rough determination of $\sin^2 \theta_W$, which is consistent with the values quoted in Eqs. (80) and (88).

The interference term is negligible in the sum of the cross sections, whence

$$\sigma(e_R q) + \sigma(e_L q) = \frac{e^4 Q_q^2}{t^2} C \sum_{\mu\lambda} |A_{\mu\lambda}|^2. \tag{85}$$

The cross sections for deep inelastic scattering from a nuclear target are found by the procedure of the last subsection: one multiplies (84) and (85) by the probability $f_q(x)$ for finding the quark q with momentum fraction x, and sums over the quark species ($q = u$ or d). The experiment in question uses a deuterium target, in which u and d have the same probability $f(x)$. As a consequence, the cross-section asymmetry does not depend on x:

$$\mathcal{A} \equiv \frac{\sigma(e_R D) - \sigma(e_L D)}{\sigma(e_R D) + \sigma(e_L D)}$$

$$= -2t\left(\frac{g}{m_W}\right)^2 \frac{\rho}{e^2} \frac{\sum_q Q_q[(\varepsilon_R^e \varepsilon_R^q - \varepsilon_L^e \varepsilon_L^q) + (\varepsilon_R^e \varepsilon_L^q - \varepsilon_L^e \varepsilon_R^q) A_{LR}^2]}{2(1 + A_{LR}^2) \sum_q Q_q^2}. \tag{86}$$

The final result for the asymmetry is found by substituting the Z^0-couplings of Table 1 into Eq. (86). This yields

$$\mathcal{A} = -\frac{9}{5} t \left(\frac{g}{m_W}\right)^2 \frac{\rho}{e^2} \frac{\frac{1}{4} - \frac{7}{9}\omega + \frac{2}{9}\omega A_{LR}^2}{1 + A_{LR}^2}, \tag{87}$$

where $A_{LR} = \cos^2 \frac{1}{2}\phi$, ϕ being the scattering angle in the c.o.m. frame, and $-t$ the square of the 4-momentum transfer.

The experimentally measured asymmetry is shown in Fig. 1 [Prescott et al. (1978) and (1979)]. Note the magnitude of the effect: $\mathcal{A} \sim -10^{-4} t$, where t is in GeV^2. If one sets $\rho \equiv 1$ on the basis of the deep inelastic neutrino results one obtains [Kim et al. (1981)]

$$\sin^2 \theta_W = 0.223 \pm 0.015. \tag{88}$$

This is in accord with the result (80) for the electroweak mixing angle extracted from neutrino scattering. The excellent agreement between these two independent measurements of $\sin^2 \theta_W$ is the strongest evidence in favor of the electroweak theory, apart, of course, from the data related to $W^\pm$ and Z^0 themselves.

(d) The basic parameters of the electroweak theory

There are also data on quite a number of other processes due to Z^0 exchange: deep inelastic ν_μ and $\bar{\nu}_\mu$ scattering from hydrogen and deuterium; elastic $\nu_\mu p$ and $\bar{\nu}_\mu p$ scattering; neutral current neutrino reactions with a pion detected in the final state; and elastic $\nu_\mu e$, $\bar{\nu}_\mu e$, and $\bar{\nu}_e e$ scattering (the latter process involves Z^0 and $W^\pm$ exchange). An analysis that takes all these data into account [Kim et al. (1981)] leads to $\rho = 1.002 \pm 0.015$, and $\sin^2 \theta_W =$

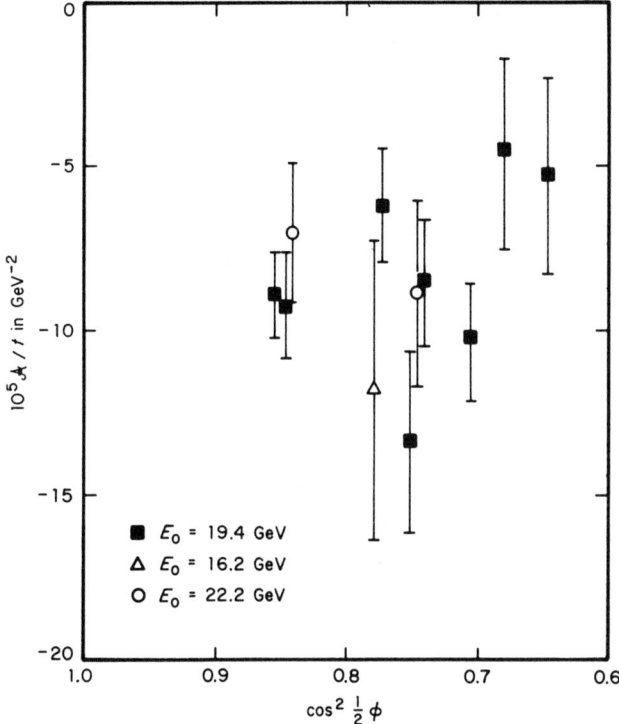

FIG. 1. Parity violation as observed at SLAC in inelastic electron scattering by deuterium [Prescott (1979)]. The plotted quantity is $10^5 \mathcal{A}/t$ as a function of incident energy and the c.o.m. scattering angle ϕ. [Note that $\cos^2 \tfrac{1}{2}\phi = (\boldsymbol{p} \cdot \boldsymbol{k}')/(\boldsymbol{p} \cdot \boldsymbol{k}) = E'/E_0$, where the 4-momenta are defined in Eq. (52) of Appendix III, and E_0 and E' are the incident and final electron energies in the laboratory.] For a description of this elegant experiment, see Perkins (1982), §8.8.3. In a nutshell, an intricate optical system consisting of a laser and a birefringent crystal is used to provide circularly polarized light which then produces electrons of definite helicity. These are then accelerated to high energy. The spin direction of scattered electrons is observed by means of the $(g-2)$ spin precession.

0.234 ± 0.013. These results are close to those of deep inelastic ν-scattering alone, because those measurements have the smallest errors. Setting $\rho \equiv 1$ gives the value

$$\sin^2 \theta_W = 0.233 \pm 0.009, \tag{89}$$

which we will use from here on.

This whole body of data can be used to extract experimental values for the Z^0-coupling constants for the electron, and for the u and d quarks. These values were given in Table 1, and show that at this level there is also impressive agreement with the electroweak theory.

From the measured values of $\sin^2 \theta_W$ and $(g/m_W)^2$, where the latter is given by Eq. A(22), we can compute the other basic parameters of the theory: $g^2/4\pi$, which we might call the fine structure constant of weak interactions, and the masses of $W^\pm$ and Z^0. From (49) we have

$$\frac{g^2}{4\pi} = \frac{\alpha}{\sin^2 \theta_W} \simeq \frac{1}{32}, \qquad (90)$$

and therefore $m_W = 77 \text{ GeV}$. Since $\rho = 1$, Eq. (57) gives $m_Z = 88 \text{ GeV}$. Radiative corrections to these masses, due to processes such as,[13]

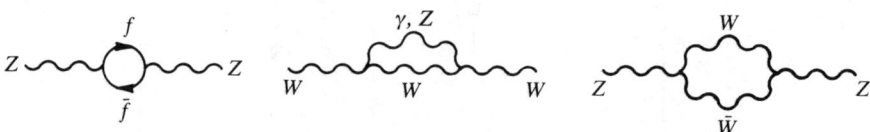

have been computed [Marciano (1979); Wetzel (1983)], and are quite significant. They give the following predictions for the masses:

$$\begin{aligned} m_W &= 82 \text{ GeV}, \\ m_Z &= 93 \text{ GeV}, \end{aligned} \qquad (91)$$

with errors of order 2 GeV due to both theoretical and experimental uncertainties that enter into these calculations. The observed masses are [Particle Data Group (1984)]

$$\begin{aligned} m_W &= 80.8 \pm 2.7 \text{ GeV}, \\ m_Z &= 92.9 \pm 1.6 \text{ GeV}. \end{aligned} \qquad (91')$$

(e) Z^0 production in $e\bar{e}$ annihilation

The reaction $e\bar{e} \to Z^0$ is particularly amenable to a detailed confrontation between theory and experiment, and we therefore give a résumé of what one expects from the "Standard Model."

The diagram of interest is

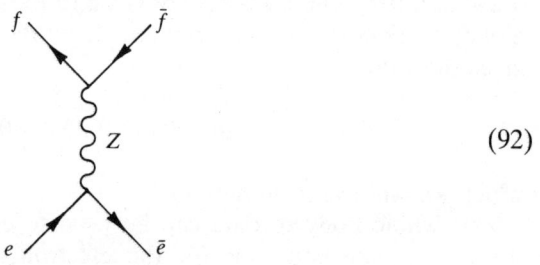

(92)

---

[13] The contribution to Z^0 from fermion vacuum polarization is familiar from QED, as is the contribution of γ to the W self-energy. The other diagrams are analogous to those of QCD (see §VI.C).

where f is any quark or lepton whose mass m_f satisfies $m_f < \frac{1}{2}m_Z$. In the Standard Model there are no flavor-changing neutral currents, and $\bar{f}$ must therefore be the antiparticle of f. If the c.o.m. energy W is in the neighborhood of m_Z, the cross section is dominated by the Z-resonance, which can then be described by the Breit–Wigner formula, Eq. III.A(20), with $s_1 = s_2 = \frac{1}{2}, J_B = 1$, and $k = \frac{1}{2}W$:

$$\sigma(e\bar{e} \to f\bar{f}) = \frac{3\pi}{W^2} \frac{\Gamma(Z \to e\bar{e})\Gamma(Z \to f\bar{f})}{(W - m_Z)^2 + \frac{1}{4}\Gamma_Z^2}. \tag{93}$$

From (52) and (65) we see that the amplitude for $Z \to f_\lambda \bar{f}_\lambda$ is proportional to $(g/\cos\theta_W)\varepsilon_\lambda^f$, where $\lambda = L$ means that f and $\bar{f}$ have helicities $-\frac{1}{2}$ and $\frac{1}{2}$, respectively, and conversely for $\lambda = R$. From these amplitudes one finds that the partial widths are

$$\Gamma(Z \to f\bar{f}) = \frac{1}{24\pi}\left(\frac{g}{\cos\theta_W}\right)^2 m_Z(|\varepsilon_L^f|^2 + |\varepsilon_R^f|^2). \tag{94}$$

The values for the various fermion pairs are then

$$\begin{array}{c|cccc} & \nu\bar{\nu} & l\bar{l} & q_+\bar{q}_+ & q_-\bar{q}_- \\ \hline \Gamma(Z \to f\bar{f}) & 0.18 & 0.09 & 0.31 & 0.40 \end{array}. \tag{95}$$

Here q_+ and q_- designate quarks with $Q = \frac{2}{3}$ and $-\frac{1}{3}$, respectively, all widths are in GeV, and we have used Eq. (64); furthermore, $Z \to q\bar{q}$ contains a factor of 3 because of color. If one assumes that there are three generations of leptons and quarks, and that the t quark has a mass well below $\frac{1}{2}m_Z$, this assumption predicts a total width of $\Gamma_Z = 2.9$ GeV, and a branching fraction for $Z^0 \to$ hadrons of $\sim 70\%$.

Knowing m_Z, Γ_Z and the branching fraction into hadrons permits us to evaluate the cross-section ratio $R(W) \equiv \sigma(e\bar{e} \to \text{hadrons})/\sigma_{\mu\mu}$, where the quantity in the denominator is the cross section for $e\bar{e} \to \mu\bar{\mu}$ given by QED without any Z^0-exchange, i.e., by Eq. II.C(16): $\sigma_{\mu\mu} = (4\pi/3)(\alpha^2/W^2)$. In the vicinity of m_W we therefore obtain[14]

$$R(W) = \frac{9}{4\alpha^2} \frac{\Gamma(Z \to e\bar{e})\Gamma(Z \to \text{hadrons})}{(W - m_Z)^2 + \frac{1}{4}\Gamma_Z^2}. \tag{96}$$

At resonance, (95) and (96) then give $R(m_Z) \simeq 3{,}800$, which is to be compared with the value of $R \sim 4$ observed at currently accessible energies, $W \leq 40$ GeV [see Fig. III.18(a)].

As we see, the Z^0 is expected to appear as an enormous resonance in

---

[14] Nonresonant terms, in particular the contribution due to annihilation into virtual photons, are ignored here, as they are negligible for $W \sim m_Z$.

$e\bar{e} \to$ hadrons. In constrast to the ψ- and Υ-resonances, the Z^0 is broad compared to the intrinsic energy spread of the beams, and the width should be observed directly. With an e^+e^- collider having a typical luminosity of 10^{31} cm^{-2} sec^{-1}, the event rate at the peak is about 1.5×10^3/hr, and many precision tests of the theory would become accessible. It would also be possible to see whether there are yet heavier generations of leptons and quarks having a light neutrino as a member because, as we have seen, every neutrino species contributes 0.18 Gev to Γ_Z if $2m_\nu$ is small compared to m_Z.

C. THE HIDDEN GAUGE INVARIANCE OF THE ELECTROWEAK INTERACTION

In the preceding pages we mentioned that the vector mesons $W^\pm$ and Z^0 are quanta of gauge fields, but we carefully avoided any detailed consideration of this aspect of the theory. Our reason for doing so is clear from QED and QCD: gauge invariance would seem to require that the field quanta be massless (p. 371), whereas we have concluded that the W's and Z^0 have masses that are comparable to those of heavy nuclei! We have, furthermore, claimed that $W^\pm$, Z^0, and the massless photon, are related by a symmetry. The masses of these vector bosons (and also, as we shall see, of the quarks and leptons) pose a paradox for any field theory built on the principle of gauge invariance.

We shall now learn how this paradox is resolved. The essential clues come from phenomena such as ferromagnetism and superconductivity, in which the existence of long-range order hides an underlying symmetry.

1. Hidden symmetry in ferromagnetism and superconductivity

(a) Ferromagnetism

The Hamiltonian of a ferromagnet is rotationally invariant, yet the system manifests a preferred direction when the temperature is below the Curie point. A superficial examination of the "ordered" low-temperature phase could lead one to conclude that the interactions between the constituents of the magnet single out a preferred direction. Therefore, there is a certain analogy between the situation we face and the ferromagnet: The electroweak interaction is supposedly invariant under rotations in the weak isospin space $\mathscr{E}_3^T$, yet the mass spectra of leptons, quarks, and field quanta differentiate the 3-direction of $\mathscr{E}_3^T$ from the 1- and 2-directions. Admittedly this is a very imperfect analogy, because the ferromagnet has no local symmetry like gauge invariance. Nevertheless, ferromagnetism offers a readily visualizable introduction to the concepts that we must come to understand.

For the sake of definiteness, consider once more the Heisenberg model of a ferromagnetic crystal. Each lattice site i has a spin vector $\mathbf{s}_i$, coupled to spins at other sites j by an interaction having the energy

$$H = -\tfrac{1}{2} \sum_{i,j=1}^{N} J_{ij} \mathbf{s}_i \cdot \mathbf{s}_j. \tag{1}$$

J_{ij} tends to zero as the distance between i and j grows, and is positive, so that ordered states, with spins aligned, are energetically favored.

The Hamiltonian (1) is rotationally invariant, but the ground state is not: the lowest eigenvalue of H is achieved by having *all* spins aligned along some direction $\hat{\mathbf{n}}$. The projection of the total angular momentum $\mathbf{J}$ along $\hat{\mathbf{n}}$ is Ns, where N is the number of lattice sites, and s the magnitude of each spin $[(\mathbf{s}_i)^2 = s(s+1)]$. Rotational invariance implies that the ground state is an eigenstate of $(\mathbf{J})^2$, which is $(2Ns + 1)$-fold degenerate. If we consider two degenerate states having the projection Ns of $\mathbf{J}$ along the directions $\hat{\mathbf{n}}$ and $\hat{\mathbf{n}}'$, the number of individual spin flips required to go from one state to the other is of order N. When N is finite, there is nothing remarkable about this situation—in essence it is identical to the ground state of an atom or nucleus that happens to have a nonzero spin. But if the ferromagnet is a macroscopic body, we must draw a distinction between two essentially different types of observations: those that are confined to a fixed finite volume as the size of the body grows, and those that encompass the body as a whole. We call an observer who can only carry out finite volume observations a finite observer, whereas an observer who can examine the entire body is called global.

To a global observer there is no mystery: he can turn the ferromagnet and confirm that this requires no work, as required by the symmetry. But a finite observer lives inside the ferromagnet, so to speak. To him the alignment direction $\hat{\mathbf{n}}$ is an immutable property of nature, because no operation within his powers can alter that direction. For that reason the spherical symmetry of the underlying dynamics is hidden from him. Nevertheless, the finite observer can carry out measurements that uncover this symmetry, although a sophisticated understanding of ferromagnetism is a prerequisite to this undertaking. We do not wish to dwell on ferromagnetism here, so we restrict ourselves to only a few remarks. The low-lying excited states of a ferromagnet are spin waves. Such a wave is a coherent superposition of states wherein one spin is flipped at each lattice site $\mathbf{r}_j$,

$$|\mathbf{k}\rangle \propto \sum_j e^{i\mathbf{k}\cdot\mathbf{r}_j}(s_{jx} - is_{jy})|G\rangle,$$

where $\mathbf{k}$ is the wave vector, and $|G\rangle$ is the ground state wherein all spins point along the z-direction. In the long wavelength limit ($k \to 0$), this state approaches $(J_x - iJ_y)|G\rangle$, which, by symmetry, must have the same energy as $|G\rangle$ itself. Hence the excitation energy $E(\mathbf{k}) \to 0$ as $k \to 0$, and this can be verified to arbitrary precision by our finite observer, for he can detect values of k as small as $V^{-1/3}$, where V is his finite, but arbitrarily large, observation volume.[1] This is but one observable consequence of the underlying rotational symmetry. Others concern the manner in which two spin waves scatter each other.

The relevance of the ferromagnet to our problem emerges if we were to suppose that the apparent lack of weak isospin invariance does not stem from an asymmetry of the electroweak Hamiltonian, but is hidden by a ground state that singles out a particular direction in the space $\mathscr{E}_3^T$. This ground state is the vacuum, which pervades all of space. As finite observers we cannot alter this direction, and could take this to mean that the laws of nature are not rotationally invariant in $\mathscr{E}_3^T$. But a more profound understanding might lead us to conclude that the excitation spectrum (as determined by the masses of the vector bosons), and the amplitudes for scattering and decay of these excitations, obey relations that are imposed by the underlying symmetry. That appears to be the case, as we shall see.

(b) Superconductivity

We have already pointed out that the ferromagnet offers a poor analogue to the problem of interest to us, because it only has a global symmetry, whereas gauge theories have a local symmetry. On the other hand, superconductivity is a phenomenon that hides electromagnetic gauge invariance, and therefore is a far more instructive prototype.

One can glimpse the relevance of superconductivity to our problem from the following two facts: (1) A superconductor is a system of charged particles in interaction with the electromagnetic field, and therefore its Hamiltonian is gauge invariant; (2) low-frequency electromagnetic fields cannot exist within a superconductor—a phenomenon called the Meissner effect. (If such fields are applied externally, they only penetrate a microscopic distance into the surface. When cooling causes a body to become superconducting, any magnetic field that may have been within it is expelled.)

The existence of the Meissner effect implies that the equations of motion for the electromagnetic fields within a superconductor must differ profoundly from those *in vacuo*. The vacuum equations imply that the field quanta have mass zero and helicity $h = \pm 1$, so that all propagating fields are transverse. But within a superconductor the frequency of any disturbance that propagates must exceed a certain threshold value ω_0, and therefore corresponds to a quantum having a finite mass $\hbar\omega_0$. In a superconductor, furthermore, not only transverse, but also longitudinal fields can propagate. The latter correspond to quanta with $h = 0$. The existence of a threshold ω_0, and of longitudinal fields, are intimately connected, because we know that a particle of spin-1 and nonzero mass has the helicity states, $h = 1, 0, -1$, whereas a massless spin-1 particle only has $h = \pm 1$.

Therefore, there is an intriguing similarity between the superconductor, and what we are looking for. On the one hand, it must be possible to

---

[1] This result is a special case of Goldstone's theorem.

describe the superconductor in a gauge invariant manner; on the other, the electromagnetic disturbances that propagate within superconductors have the characteristics of fields whose quanta are massive. By studying superconductivity we can therefore learn how to construct a gauge invariant electroweak theory in which the W's and Z's are massive—an electroweak analogue of the Meissner effect.

For our purpose it is not necessary to have a microscopic description of a superconductor that deals explicitly with the conduction electrons and their mutual interactions.[2] All that the reader must accept is that in certain metals the interactions are such as to allow pairs of conduction electrons to form something like bound states of zero spin when the temperature falls below the critical point T_c. To a large extent one can treat these so-called Cooper pairs as if they were elementary bosons. In this picture superconductivity can be viewed as the Bose-Einstein condensation of these Cooper pairs. When $T < T_c$, a small fraction of the conduction electrons near the Fermi surface condense into the lowest eigenstate. The other electrons are effectively inert; they cannot participate in small departures from equilibrium. The whole many body state can be constructed from the one-pair wave function $\Psi(\mathbf{r})$, where $\mathbf{r}$ is the c.o.m. coordinate of the Cooper pair, and $|\Psi(\mathbf{r})|^2$ is the pair density.

As with any condensed Bose gas, the wave function $\Psi(\mathbf{r})$ extends over the whole body. If there are no applied electromagnetic fields, $|\Psi(\mathbf{r})|^2$ is a constant n_0, except in the immediate vicinity of the surface. When there are fields, $\Psi(\mathbf{r})$ must respond to them in accordance with a Schrödinger equation derived from an effective Hamiltonian H_{eff}, which is a functional of the vector potential and of $\Psi(\mathbf{r})$. Since we are concerned with small departures from the ground state due to weak and slowly varying applied fields, we can expand H_{eff} in powers of $\nabla\Psi$ and $\mathbf{A}$. The leading terms in this expansion are tightly constrained by spherical symmetry and gauge invariance:

$$H_{\text{eff}} = H_{\text{eff}}^0[|\Psi(\mathbf{r})|^2] + \frac{1}{2\mu}\int \left|\left(\frac{1}{i}\nabla - Q\mathbf{A}\right)\Psi\right|^2 d^3r + \frac{1}{2}\int (E^2 + B^2)\, d^3r. \quad (2)$$

The first term is the energy[3] in the absence of fields; it is a functional that depends only on $|\Psi(\mathbf{r})|^2$. The second term represents the kinetic energy of the Cooper pairs, whose effective mass and charge are μ and Q. The last term is the electromagnetic energy. H_{eff} is called the Ginzburg–Landau Hamiltonian.

H_{eff}^0 must be such as to have a minimum at $|\Psi|^2 = n_0$ when the system is

---

[2] For a more detailed treatment of superconductivity along the lines used here, see Schrieffer (1964), pp. 19–23; de Gennes (1966), Chap. 6; Tinkham (1975), Chap. 4; and Weisskopf (1981).

[3] When $T \neq 0$, this is actually the free energy, but that distinction is irrelevant here. Properly speaking, Eq. (2) is an expansion of the free energy for slowly varying Ψ's; the form given has the smallest number of derivatives compatible with rotational and gauge invariance.

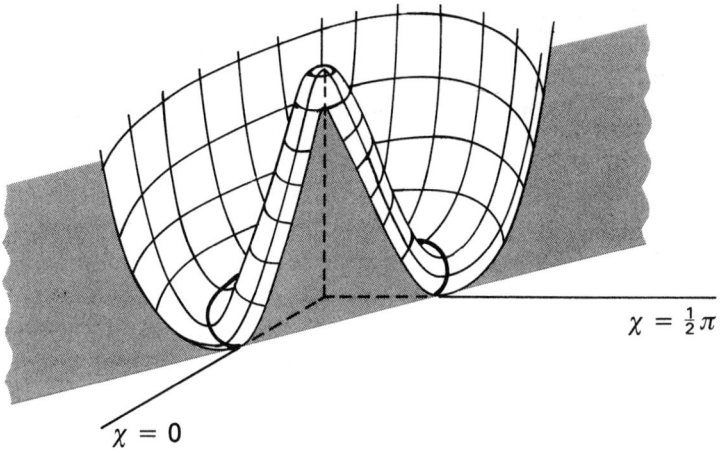

FIG. 2. The energy of the superconductor in the absence of external fields, H^0_{eff}, as a function of the equilibrium wave function $\sqrt{n_0} e^{-i\chi}$.

undisturbed by any electromagnetic field:

$$\left.\frac{\partial H^0_{\text{eff}}}{\partial n}\right|_{n_0} = 0, \quad \left.\frac{\partial^2 H^0_{\text{eff}}}{\partial n^2}\right|_{n_0} > 0, \qquad (3)$$

where $n \equiv |\Psi|^2$. In the superconducting phase there must be a nonvanishing density n_0 of Cooper pairs; a finite energy must be paid to break such a pair, and this yields an observed gap in the superconductor's excitation spectrum. The qualitative shape of H^0_{eff} is shown in Fig. 2; as the temperature rises, the minimum moves inward until it reaches $n_0 = 0$ at T_c, and the condensate disappears. The shape of this curve, its behavior with temperature, and its dependence on the substance, all require a microscopic theory of superconductivity.

Let us examine the field-free spatially uniform equilibrium wave function:

$$\Psi_0 = \sqrt{n_0} e^{-i\chi}. \qquad (4)$$

As we see, there is not one equilibrium point, but a whole circle of equilibria in the complex Ψ-plane! Different values of χ designate distinct quantum states of the superconductor, just as different directions of the spin alignment give distinct, but equivalent, states of the ferromagnet.

In the ferromagnet the degeneracy is due to rotational invariance. In the superconductor, the symmetry principle that underlies the degeneracy is gauge invariance, as one can surmise from (4), for that equation is a special case of the general gauge transformation Eq. IV.B(13). We explore this observation in some detail shortly.

Our goal is to study the departures from the equilibrium condensate (4), and the electromagnetic fields associated with them. As we see from Fig. 2, there is a restoring force for fluctuations in the modulus of Ψ, but not in its phase. We therefore express the departure from equilibrium as

$$\Psi(\mathbf{r}) = [\sqrt{n_0} + \rho(\mathbf{r})]e^{-i\chi(\mathbf{r})} \tag{5}$$

where the real "field" $\rho(\mathbf{r})$ is a small quantity, but $\chi(\mathbf{r})$ is arbitrary. We can now exploit gauge invariance to remove this arbitrary phase field by introducing a special vector potential $\mathscr{A}$ defined by

$$\mathscr{A} = \mathbf{A} - \frac{1}{Q}\nabla\chi. \tag{6}$$

Then

$$(\nabla - iQ\mathbf{A})\Psi = e^{-i\chi}(\nabla - iQ\mathscr{A})(\sqrt{n_0} + \rho) \tag{7}$$

because the $\nabla\chi$ terms cancel.

Since $\Psi = \sqrt{n_0}$ is an equilibrium point, the departure of the energy density from its equilibrium value is

$$\mathscr{H}(\mathbf{r}) = \frac{1}{2}b^2\rho^2 + \frac{1}{2\mu}|\nabla - iQ\mathscr{A}(\sqrt{n_0} + \rho)|^2 + \frac{1}{2}(E^2 + B^2), \tag{8}$$

where b^2 is proportional to the second derivative of H_{eff}^0 at n_0. By hypothesis, the departure from equilibrium ($\rho = \mathbf{E} = \mathbf{B} = 0$) is small, and we can discard terms of the type $\rho^2\mathscr{A}^2$, etc., in (8):

$$\mathscr{H} = \frac{1}{2}b^2\rho^2 + \frac{1}{2\mu}(\nabla\rho)^2 + \frac{n_0 Q^2}{2\mu}\mathscr{A}^2 + \frac{1}{2}(E^2 + B^2). \tag{9}$$

The phase χ was a degree of freedom in (2), yet it has disappeared from (9). At first sight this seems to be unacceptable. But as we will see in a moment, this degree of freedom is not "lost"; it appears in another guise—as a longitudinal mode of vibration of the electromagnetic field.

The electromagnetic current associated with the departure (5) from equilibrium is

$$\mathbf{j} = \frac{Q}{2\mu i}[\Psi^*(\nabla - iQ\mathbf{A})\Psi - \text{c.c.}] \tag{10}$$

$$= -\frac{Q^2}{\mu}(\sqrt{n_0} + \rho)^2 \mathcal{A}. \tag{11}$$

A current proportional to the vector potential is a hallmark of superconductivity. The equation for the magnetic field is

$$\nabla \times \mathbf{B} = -\frac{Q^2}{\mu}(\sqrt{n_0} + \rho)^2 \mathcal{A}, \tag{12}$$

there being no term $\partial \mathbf{E}/\partial t$ as we are dealing with a static departure from equilibrium. If that departure is small, ρ, $\mathcal{A}$, and $\mathbf{B}$ are all small, and we can ignore ρ in the right-hand side of (12); this gives us London's equation:

$$\nabla \times \mathbf{B} = -\frac{n_0 Q^2}{\mu} \mathcal{A}. \tag{13}$$

On taking the curl of (13), and using $\nabla \times \mathcal{A} = \mathbf{B}$, we obtain

$$(\nabla^2 - \kappa^2)\mathbf{B} = 0, \tag{14}$$

where

$$\kappa^2 = \frac{n_0 Q^2}{\mu}. \tag{15}$$

Equation (14) has exponentially decaying solutions, proportional to $e^{-\kappa r}$. This is the Meissner effect: Static magnetic fields cannot penetrate into a superconductor beyond a layer of thickness $\sim \kappa^{-1}$. The persistent current (11) is also confined to this surface layer.

If we could now exploit Lorentz invariance,[4] we could use (14) to show

---

[4] This argument is too simplistic for a superconductor, because the ions in the metal select a preferred frame. But the electroweak field theory must be Lorentz invariant, and we shall be able to use this chain of reasoning to good advantage shortly. In the superconductor, the derivation of the time-dependent equations is a sophisticated undertaking; see Tinkham (1975), §8.3. One must examine the response of the system to both electric and magnetic disturbances, as they are not trivially related, and concern oneself with whether the resulting fluctuations are adiabatic or not. But when all is said and done, electric and magnetic fields can only propagate for frequencies above certain thresholds—loosely speaking, as if the photon had acquired a mass.

that **B** must, quite generally, satisfy

$$\left(\nabla^2 - \frac{\partial^2}{\partial t^2} - \kappa^2\right)\mathbf{B} = 0; \tag{16}$$

further, since **E** and **B** are related by Lorentz transformations, the electric field would also have to satisfy (16). The differential equation (16) gives the connection

$$\omega = \sqrt{k^2 + \kappa^2} \tag{17}$$

between the frequency ω and wavenumber k of a plane wave, and therefore quanta of mass κ. As the fields are vectorial, these quanta are conventional spin-1 bosons with helicities $h = \pm 1$ *and* 0. The "lost" degree of freedom χ has reappeared as the longitudinal mode with $h = 0$.

As we have seen, this remarkable result—and, in particular, London's equation (13)—was derived by starting from the gauge-invariant Ginzburg-Landau Hamiltonian (2). Indeed, the crucial step of eliminating the phase χ to arrive at the expression (11) for the current exploited the gauge invariance. However, once that is done, the explicit, manifest gauge invariance is lost, for we have chosen that *particular* gauge that makes the **r**-dependent Cooper pair wave function *real everywhere*. As a consequence, the current loses its manifestly gauge invariant form $\Psi^*(\nabla - iQ\mathbf{A})\Psi$, and instead acquires the gauge-dependent appearance

$$\mathbf{j} = -\kappa^2 \mathcal{A}. \tag{18}$$

As we learned in the argument that led us to Eq. (9), this elimination of the phase of Ψ is not a formal artifact—an option that we need not exercise. The phase is not an independent degree of freedom of the electrons; in the superconducting state it is the longitudinal degree of freedom of the electromagnetic field. The crucial role of the condensate is also apparent from London's equation (13): If there is no condensate, $n_0 = 0$ and $\nabla \times \mathbf{B} = 0$; the substance will then admit a uniform magnetic field, and there is no Meissner effect.

2. The generation of mass in the electroweak theory

(a) Analogies between superconductivity and electroweak phenomena

We now apply our newly-acquired knowledge to generate masses in the gauge-invariant theory of the electroweak interaction. To do so, we must abstract the following generalizations from the superconductor:[5]

1. The role of the Abelian gauge group $U(1)$ of the superconductor, with

---

[5] In this section we assume an acquaintance with the material in §IV.B concerning non-Abelian gauge theory.

elements $e^{i\chi}$, is to be played by the non-Abelian gauge group $\mathfrak{G}$ of the electroweak theory, to be defined below.
2. The role of the electromagnetic field in the superconductor is to be played by the electroweak fields, whose quanta are $W^\pm$, Z^0 and γ.
3. The role of the pair condensate wave function Ψ is to be played by a so-called Higgs field $\Phi(\mathbf{r}, t)$, to be introduced presently.
4. The role of the superconducting state is to be played by a vacuum state $|\Omega\rangle$, which possesses a nonvanishing vacuum expectation value $\langle\Phi(\mathbf{r}, t)\rangle_\Omega$ of the Higgs field, in analogy to the nonvanishing equilibrium value $\sqrt{n_0}$ of Ψ in the superconducting state.

As we have seen, condensation of Cooper pairs ($n_0 \neq 0$) implies that fields whose frequency is below some threshold cannot exist in superconductors, and as we shall now see, when $\langle\Phi\rangle_\Omega \neq 0$, some of the electroweak gauge fields acquire a mass, and cannot propagate unless their frequency exceeds that mass. That is to say, $\langle\Phi\rangle_\Omega \neq 0$ leads to an electroweak analogue of the Meissner effect, and an apparent breaking of the gauge invariance. Despite this apparent symmetry breakdown, the theory will, at high energy, have all the virtues of a theory where gauge invariance is manifest, because at high energy the symmetry-breaking effects are irrelevant, in analogy to the disappearance of the Meissner effect at high temperature.

(b) The electroweak gauge group

Let us now enact this scenario. Our first task is to identify the electroweak gauge group $\mathfrak{G}$. For this purpose we recall that in QCD the gauge group was found by examining the transformations of the basic quark color-triplet. This group was found to be $SU(3)$, and the requirement of *local* invariance forced us to introduce one gauge field for each of the eight generators of $SU(3)$. In the problem at hand, the basic multiplet is any one of the fermion doublets within one generation; for the sake of concreteness we often use (ν_e, e). As in §B.1, we form a left-handed doublet of fields Ψ_e from the corresponding Dirac fields, and the right-handed singlet field e_R. Under rotations in the weak isospin space $\mathscr{E}_3^T$, the former is a spinor, the latter a scalar, since it does not contribute to the weak isospin. Their local transformation laws are

$$\Psi_e(x) \to \exp[-\tfrac{1}{2}i\vec{\kappa} \cdot \vec{\theta}(x)]\Psi_e(x), \tag{19}$$

$$e_R(x) \to e_R(x), \tag{20}$$

where $\vec{\kappa}$ is the trio of 2×2 Pauli matrices introduced in Eq. B(16), and $\vec{\theta} = (\theta_1, \theta_2, \theta_3)$ are the real parameters, that specify the x-dependent transformation. Obviously, (19) defines an $SU(2)$ group, which we call $SU_L(2)$ to emphasize that it only acts on the left-handed fermions. Insofar as $SU_L(2)$ is concerned, e_R is a singlet.

From our discussion of the electroweak connection in §B.2(a), we also know that the weak isospin does not suffice to identify a fermion, and that a further observable, $Q - t_3$, is needed for that purpose:[6]

$$Y = Q - t_3. \tag{21}$$

We designate its eigenvalues by y. They are given by

state	l_L	v_{lL}	l_R	v_{lR}	q_{+L}	q_{-L}	q_{+R}	q_{-R}	
y	$-\frac{1}{2}$	$-\frac{1}{2}$	-1	0	$\frac{1}{6}$	$\frac{1}{6}$	$\frac{2}{3}$	$-\frac{1}{3}$	(22)

where l stands for any negatively charged lepton, q_{+L} for any quark with weak isospin "up," q_{-L} for its "down" partner [recall Eq. B(17)], and q_{+R}, q_{-R} for the corresponding right-handed singlets.

Since Y commutes with $\vec{t}$, all members of any weak isospin multiplet share a common eigenvalue of Y. As the average of t_3 over any multiplet vanishes (the trace of $\vec{t}$ is zero), this common eigenvalue y is just the average charge in the multiplet, a fact that was already used in constructing Eq. I.E(61). Note that y distinguishes left-handed doublets from right-handed singlets, and that the specification of the eigenvalues of both t_3 and Y identifies a fermion uniquely—apart, of course, from the still-mysterious distinction between generations!

Since Y commutes with all components of the weak isospin, it is the generator of an Abelian group, which we call $U_Y(1)$. This completes the determination of the electroweak gauge group; it is seen to be

$$\mathfrak{G} = SU_L(2) \otimes U_Y(1), \tag{23}$$

in the notation of §IV.A.2. Under a transformation of $\mathfrak{G}$, a fermion state undergoes a linear transformation composed of two factors, corresponding to the two factors in (23). The $SU_L(2)$ factor was already spelled out in (19) and (20). The U_Y factor is familiar from electrodynamics: it is given by $e^{-iy\alpha(x)}$, where y is the Y-eigenvalue of the object in question, and $\alpha(x)$ is the $U_Y(1)$ gauge function. For example, (22) tells us that for a quark doublet, the $\mathfrak{G}$-transformation is

$$\Psi_q \to \exp\left[-\tfrac{1}{6}i\alpha(x)\right] \exp\left[-\tfrac{1}{2}i\vec{\kappa}\cdot\vec{\theta}(x)\right]\Psi_q. \tag{24}$$

The electroweak gauge fields are introduced as in §IV.B.2. There is a triplet of vector fields, $\vec{W}_i$, for the $SU_L(2)$ factor of $\mathfrak{G}$, and a singlet B_i for

---

[6] The definition $Y = 2(Q - t_3)$ is often used. This freedom underscores that the generator of an Abelian gauge symmetry has an arbitrary normalization, in contrast to the generators of a non-Abelian symmetry (such as $\vec{t}$), which are constrained by their nonlinear commutation rules.

2. THE GENERATION OF MASS

the U_Y factor, where $i = x, y, z$. The covariant derivatives[7] are determined by the argument that led to Eq. IV.B(19), viz.

$$D_i \Psi_L(x) = (\nabla_i - \tfrac{1}{2} ig\vec{\kappa} \cdot \vec{W}_i - ig'y_L B_i)\Psi_L(x), \qquad (25)$$

$$D_i f_R = (\nabla_i - ig'y_R B_i)f_R(x), \qquad (26)$$

where Ψ_L is any left-handed doublet, and f_R any right-handed singlet, with Y-eigenvalues y_L and y_R. The coupling constants g and g' are already familiar to us from §B.2. The transformation laws for the fields are then given by Eqs. IV.B(25) and (12):

$$\vec{W}_i \to \vec{W}_i - \frac{1}{g}\nabla_i \vec{\theta} + \vec{\theta} \times \vec{W}_i, \qquad (27)$$

$$B_i \to B_i - \frac{1}{g'}\nabla_i \alpha. \qquad (28)$$

The contribution of W's to the weak isospin [see Eq. B(27) and Eq. IV.C(25)] is given by

$$\vec{T}_W = \int (\vec{W}_i \times \vec{E}_i)\, d^3x, \qquad (29)$$

where $\vec{E}_i$ is the electroweak analogue of the electric field, as given by Eq. IV.C(7) (with $\vec{V}$ replaced by $\vec{W}$). The B-field, being Abelian, carries no attribute, so it does not contribute to anything beyond energy and momentum.

(c) The Standard Model; the masses of Z and W

We now come to the Higgs field Φ—the analogue of the Cooper pair state Ψ in superconductivity. In the latter system, gauge invariance is hidden because the pair carries charge; hence, Ψ undergoes a phase change, and the symmetry is hidden. By the same token, if Φ is to hide the electroweak symmetry, it cannot be invariant under $\mathfrak{G}$. Any quantity that is not an invariant can be expressed as a sum of a spinor, vector, etc., insofar as the $SU(2)$ factor of $\mathfrak{G}$ is concerned. As we shall see, the simplest of all possibilities—where Φ is a spinor—is compatible with the experimental data

---

[7] As we learned in §IV.B.2, by covariant derivative one means that

$$D_i' U(\alpha, \vec{\theta})\Psi_L = U(\alpha, \vec{\theta}) D_i \Psi_L$$
$$U(\alpha, \vec{\theta}) \equiv e^{-iy\alpha} \exp(-\tfrac{1}{2} i \vec{\kappa} \cdot \vec{\theta})$$

when α and $\vec{\theta}$ are arbitrary functions of x, and where D_i' is the differential operator in (25) with $\vec{W}_i$ and B_i replaced by their gauge transforms, (27) and (28). Equation (27) is actually only valid to first order in $\vec{\theta}$ as it stands, but that is all one needs; Eq. (28) holds for any α.

and all theoretical requirements. The theory that is based on a spinor Φ is called the *Standard Electroweak Model,* or the Weinberg-Salam Model.

We require two other properties of Φ: its Y-eigenvalue y_H, and its behavior under Lorentz transformation. Both are determined by our desire to ascribe a nonvanishing, spatially uniform value to Φ in the vacuum state $|\Omega\rangle$. Let Φ_V be this vacuum expectation value. Since the vacuum must be invariant under (proper) Lorentz transformation, Φ must be a scalar field, for anything of lower symmetry would introduce a preferred direction. Furthermore, the vacuum is electrically neutral, and Φ_V must therefore be the expectation value of a $\Delta Q = 0$ operator. Hence, one member of the isodoublet Φ must describe electrically neutral spin-0 bosons. Since $Q = t_3 + y_H$, whence $|y_H| = \frac{1}{2}$. The other member of Φ must describe spin-0 bosons of unit charge; it cannot have a vacuum expectation value since that would destroy charge conservation.

The energy density of a scalar field in the absence of all interactions is given in Appendix II. Let us first examine the "kinetic" term involving the field gradient. This term—like all terms in the energy density—must be invariant under the transformations of $\mathcal{G}$. The ordinary derivative $\nabla_i \Phi$ has no simple transformation law for local transformations, but by its very definition, the covariant derivative $D_i \Phi$ transforms like Φ itself. Once Φ is specified to be an isospinor, the covariant derivative is known to be given by (25), i.e., by

$$D_i \Phi = (\nabla_i - \tfrac{1}{2} i g \vec{\kappa} \cdot \vec{W}_i - i g' y_H B_i) \Phi \equiv \begin{pmatrix} \varphi_i \\ \chi_i \end{pmatrix}, \tag{30}$$

where φ_i and χ_i are the components of the spinor $D_i \Phi$. The gauge invariant contribution of field gradients to the energy density is therefore

$$\mathcal{H}_{\text{kin}}[\Phi] = \varphi_i^\dagger \varphi_i + \chi_i^\dagger \chi_i, \tag{31}$$

where a sum on i is understood.

The energy density will also contain a term $\mathcal{H}_{\text{pot}}$ that plays the role of a potential, i.e., tends to restore Φ to its equilibrium value. $\mathcal{H}_{\text{pot}}$ does not depend on derivatives of Φ, and is akin to the expression H_{eff}^0 in the Ginzburg-Landau Hamiltonian, Eq. (2). It must also be invariant under $\mathcal{G}$. If

$$\Phi \equiv \begin{pmatrix} \varphi \\ \chi \end{pmatrix}, \tag{32}$$

$\sigma = \varphi^\dagger \varphi + \chi^\dagger \chi$ is an invariant, and therefore

$$\mathcal{H}_{\text{pot}} = V(\sigma). \tag{33}$$

As in superconductivity, it is crucial whether V has a minimum at $\sigma = 0$ or

2. THE GENERATION OF MASS

$\sigma \neq 0$. In the former case, the symmetry is manifest, and all gauge fields are massless. If the equilibrium σ_V occurs for $\sigma_V \neq 0$, i.e., for $\Phi \neq 0$, the symmetry is hidden, there is an analogue to the Meissner effect, and some (or all) of the field quanta acquire a mass. We assume that this nontrivial situation obtains, and consider what consequences that has for the mass spectrum. Before doing so we point out that there is an important difference between superconductivity and the problem at hand. In the superconductor we know that the nonvanishing of Ψ is due to the formation of 2-electron Cooper pairs, a process that can be derived from the underlying Hamiltonian of the metal. Here we treat Φ as a fundamental field.[8] There is, therefore, no theory for the construction of $\mathcal{H}_{\text{pot}}$; we simply *assume* that it has a minimum for $\sigma \neq 0$.

In the superconductor, H_{eff} had a minimum on a circle in the complex Ψ-plane (see Fig. 3), and when we chose $\Psi = \sqrt{n_0}$, some (spatially uniform) real number to describe the condensate, we "broke" this circular symmetry that arose from the $U(1)$ gauge invariance (recall Fig. 2 and p. 516). Here we face a similar, but more complicated, situation: Because of the invariance under $\mathfrak{G}$, the value of σ does not specify a definite spinor Φ. Hence, a manifold of spinors gives $\mathcal{H}_{\text{pot}}$ its minimum value. As in the superconductor, we choose just one member Φ_V, and thereby "break" the symmetry (also see footnote 12 on p. 524). This leads us directly to the vector boson mass spectrum, which is our primary concern. The implications of the hidden $\mathfrak{G}$-symmetry (i.e., the existence of a manifold of equilibrium Φ's) will then be examined by an argument patterned after the discussion of the superconductor following Eq. (10).

As our equilibrium (or vacuum) spinor we choose

$$\Phi_V = \frac{1}{\sqrt{2}} \begin{pmatrix} 0 \\ v \end{pmatrix}, \tag{34}$$

where v is real and spatially uniform. The factor $2^{-1/2}$, the reality of v, and the vanishing of the upper component, are conventions, as we shall see presently; we could have interchanged 0 and v. However, since the vacuum cannot carry charge, the choice (34) requires the assignment $y_H = \frac{1}{2}$, so that $Q = t_3 + y_H = 0$ for $t_3 = -\frac{1}{2}$.

Now we return to the "kinetic" energy of the Φ field when the latter is given by (34). Since v is a constant, all derivatives vanish, and (30) and (31)

---

[8] A field theory with elementary (i.e., pointlike) spin-0 particles also has the undesirable feature that it contains quadratically divergent self-energies, in contrast to the logarithmic self-energies of a theory with only spin $\frac{1}{2}$ particles coupled to gauge fields. In consequence, fundamental scalar fields produce theories that are exceedingly sensitive to the parameters in the Hamiltonian. There is, therefore, a widespread belief that the ultimate theory will not contain elementary Higgs particles, but will, instead, have spin-0 bound states composed of fundamental spin-$\frac{1}{2}$ particles, as in superconductivity. No satisfactory theory of this type has been constructed thus far [see §I.E.13(d)].

give[9]

$$\mathcal{H}_{\text{kin}}[\Phi_V] = \tfrac{1}{4}\sum_i (\Phi_V^\dagger, (g\vec{\kappa}\cdot\vec{W}_i + g'B_i)^2\Phi_V)$$
$$= \tfrac{1}{4}(\Phi_V^\dagger, [g^2\vec{W}_i\cdot\vec{W}_i + g'^2 B_i B_i + 2gg'B_i\vec{\kappa}\cdot\vec{W}_i]\Phi_V)$$
$$= \tfrac{1}{8}v^2(g^2\vec{W}_i\cdot\vec{W}_i + g'^2 B_i B_i - 2gg'B_i W_{3i}), \tag{35}$$

where W_{3i} is the ith spatial component of the projection of the isovector W-field along the 3-direction of the weak isospin space $\mathscr{E}_3^T$. This direction is singled out by our arbitrary choice of the vacuum state as (34).

As (35) shows, the existence of a nontrivial vacuum ($v \neq 0$) produces an additional contribution to the gauge field energy density that is a quadratic form in the potentials $\vec{W}_i$ and B_i. This is the direct analogue of the expression $(Q^2 n_0/2\mu)\mathcal{A}^2$ that appears in the superconductor [cf. Eq. (9)]. Any term in the energy density that is quadratic in the fields leads to a linear modification of the field equations and alters the manner in which the field propagates. In particular, the relation between frequency and wavelength (the dispersion law) changes, and that is why (35) gives some of the field quanta a mass.

Because of the term $B_i W_{3i}$ in (35), B_i and W_{3i} are not normal modes. We therefore write

$$\mathcal{H}_{\text{kin}}[\Phi_V] = \tfrac{1}{8}v^2 g^2 \sum_i [(W_{1i})^2 + (W_{2i})^2] + \tfrac{1}{8}v^2 \sum_i (gW_{3i} - g'B_i)^2. \tag{36}$$

Hence $gW_{3i} - g'B_i$ is a normal mode, and its orthogonal partner, $gB_i + g'W_{3i}$, does not appear in this additional term in the energy. These normal modes are just the Z- and electromagnetic fields,[10] as introduced in Eqs. B(44) and (45):

$$Z_i = (gW_{3i} - g'B_i)(g^2 + g'^2)^{-1/2}, \tag{37}$$

$$A_i = (gB_i + g'W_{3i})(g^2 + g'^2)^{-1/2}. \tag{38}$$

Then

$$\mathcal{H}_{\text{kin}}[\Phi_V] = \tfrac{1}{4}v^2 g^2 \mathbf{W}^\dagger \cdot \mathbf{W} + \tfrac{1}{8}v^2(g^2 + g'^2)\mathbf{Z}\cdot\mathbf{Z}, \tag{39}$$

where we again use the field $\mathbf{W}$, defined in Eq. B(29), which destroys W^- and creates W^+. In Appendix II we give the expression for the mass term of a neutral Bose field [Eq. AII(25)]; for a charged Bose field this term is $\mu^2\Phi^\dagger\Phi$. Since $\mathbf{Z}$ and $\mathbf{W}$ are, respectively, neutral and charged Bose fields,

---

[9] The notation $(\Phi_2^\dagger, A\Phi_1)$ means that one is to multiply the spinor Φ_1 by the operator A, and then take the scalar product with the spinor Φ_2.

[10] In the schematic discussion of §B.2 we ignored space and time indices on the fields.

we have

$$m_W = \tfrac{1}{2}vg, \qquad m_Z = \tfrac{1}{2}v\sqrt{g^2 + g'^2}, \tag{40}$$

and therefore

$$\frac{m_W}{m_Z} = \cos\theta_W = \sqrt{1 - \left(\frac{e}{g}\right)^2}, \tag{41}$$

where $\theta_W = \arctan(g'/g)$ is the electroweak mixing angle. The photon does not acquire a mass, since $\mathbf{A}$ does not appear in (39).

The existence of the relationship (40) between the gauge meson masses, on the one hand, and the coupling constants, on the other, comes about because the Higgs field, whose vacuum expectation value hides the symmetry, was introduced into the theory in a gauge invariant manner. This circumstance is reminiscent of the constraints imposed by rotational invariance on the properties of spin waves (see p. 510).

Equation (41) was already used in §B.3(d) to predict the masses of W and Z from the measured values of θ_W and g/m_W. As we learned in §B.3, the ν-scattering experiments show that the quantity $\rho = (m_W/m_Z \cos\theta_W)^2$ is unity to within the 3% accuracy currently available [see Eq. B(81)].

〚One might wonder whether one can, without calculation, tell which gauge bosons will not acquire a mass when a Higgs field Φ acquires a nonzero vacuum expectation value Φ_V. The answer is simply stated. Let $\mathfrak{G}_V$ be the subgroup of $\mathfrak{G}$ that leaves Φ_V invariant. Then any gauge field associated with an infinitesimal generator of $\mathfrak{G}_V$ remains massless. In our case, (34), one can readily determine $\mathfrak{G}_V$. We require

$$e^{-i\alpha/2}\exp(-\tfrac{1}{2}i\vec{\kappa}\cdot\vec{\theta})\Phi_V = \Phi_V.$$

Any rotation involving κ_1 or κ_2 will produce an "up" component in (34), so we must put $\theta_1 = \theta_2 = 0$. The phase factor $e^{-i\alpha/2}$ will compensate the phase resulting from a 3-rotation if we set $\alpha = \theta_3$. Hence, the subgroup of $\mathfrak{G}$ that leaves Φ_V invariant are the matrices $\exp[-\tfrac{1}{2}i(1 + \kappa_3)\theta_3]$, which form an Abelian group. The associated gauge field is the electromagnetic potential.〛

(d) The Higgs boson and its couplings

We must still examine the small oscillations of the weak isospinor Bose field $\Phi(x)$ about the spatially uniform equilibrium vacuum configuration. One can readily see that these oscillations correspond to just one electrically neutral Bose field quantum. In general, an isospinor field has four field quanta per linear momentum state. To understand this, it suffices to think of the kaon doublet: the "upper" component destroys K^+ and creates K^-, the "lower" destroys K^0 and creates $\bar{K}^0$—four linearly independent states. Next, consider the gauge fields. When $\langle\Phi\rangle_\Omega \neq 0$, three field quanta, $W^+, W^-,$

and Z, acquire mass, and each must attain a longitudinal (helicity zero) state that does not exist in the massless case, when the gauge fields are purely transverse. As the total number of degrees of freedom cannot change, these $h = 0$ states must be bought at the expense of three degrees of freedom of Φ, leaving just one. As the latter is alone, it is its own antiparticle, so it cannot carry charge. This particle is called the *Higgs boson*; it is designated by H.

This conclusion can be verified by a generalization of the technique that we used to discuss small departures from equilibrium in the superconductor [cf. Eq. (5) et seq.]. First we note that an arbitrary x-dependent isospinor field can be written as

$$\Phi(x) = e^{-i\alpha(x)/2} \exp[-\tfrac{1}{2}i\vec{\kappa} \cdot \vec{\theta}(x)]\mathring{\Phi}(x), \tag{42}$$

with

$$\mathring{\Phi}(x) = \frac{1}{\sqrt{2}} \begin{pmatrix} 0 \\ v + \rho(x) \end{pmatrix}, \tag{43}$$

where α, $\vec{\theta}$, and ρ are all real.[11] The factor multiplying $\mathring{\Phi}$ is just a local $SU_L(2) \otimes U_Y(1)$ gauge transformation. The potential $\mathcal{H}_{\text{pot}}$ depends on the "stretching" parameter $\rho(x)$, so there is a restoring force that restrains the fluctuations $\rho(x)$. On the other hand, $\mathcal{H}_{\text{pot}}$ does not depend on $\alpha(x)$ and $\vec{\theta}(x)$; they only specify a point on the equipotential surfaces of $\mathcal{H}_{\text{pot}}$, and are unrestrained. As in the superconductor [see Eq. (5) et seq.], this circumstance permits one to remove $\alpha(x)$ and $\theta(x)$ from the Hamiltonian by the corresponding gauge transformation[12] on the fields $\vec{W}_i$ and B_i, as given by (27) and (28). Hence the only portion of Φ that cannot be "gauged away" is $\mathring{\Phi}$, containing the real (or Hermitian) field $\rho(x)$, and as we know from Appendix II, such a field has spin-0 quanta that are their own antiparticles.

Unless one has a theory that allows one to construct $\mathcal{H}[\Phi]$ from some underlying principles, one cannot predict the mass m_H of H. There are loose empirical bounds on m_H, and more sophisticated theoretical arguments entailing radiative corrections also constrain m_H. But this is a topic that is in considerable flux, so we prefer not to dwell further on it.

At this time, there is no experimental evidence for the existence of H. How could one hope to see it? The theory we have described makes quite definite predictions for the interaction of H with the other particles. This can be seen most readily in the case of the coupling of H to the gauge bosons. For this purpose we need only return to (31). We focus on couplings

---

[11] If one writes this out explicitly, one can check that Φ does not depend on the five real functions α, $\vec{\theta}$, and ρ, but only on four independent combinations, in conformity with the foregoing discussion.

[12] By the same token, a global (x-independent) gauge transformation can be used to replace our standard choice (34) for the vacuum expectation value of Φ by any member of the manifold of spinors that give $\mathcal{H}_{\text{pot}}$ its minimum value.

2. THE GENERATION OF MASS

involving Z^0; they are more promising, since Z^0 can be produced singly in e^+e^- colliders. If we substitute (42) into (30), remove the unitary factor in (42) by a gauge transformation, discard terms involving W_{1i} and W_{2i}, and use (37), we obtain

$$\mathcal{H}_{\text{kin}}[\rho, Z] = \tfrac{1}{2}\sum_i |(\nabla_i + \tfrac{1}{2}i\sqrt{g^2 + g'^2}Z_i)(v + \rho)|^2$$
$$= \tfrac{1}{2}(\nabla\rho)^2 + \tfrac{1}{8}(g^2 + g'^2)\mathbf{Z}\cdot\mathbf{Z}(v + \rho)^2.$$

The first term is the kinetic energy of the Higgs field, and the term Z^2v^2 is just the Z^0-mass term already discussed in (39). The H-Z^0 couplings are therefore

$$\mathcal{H}_{\text{int}}(H, Z) = \frac{1}{8}\frac{g^2}{\cos^2\theta_W}\mathbf{Z}\cdot\mathbf{Z}(2v\rho + \rho^2).$$

The first term allows a Z to "radiate" a Higgs particle, the other requires two H's to be radiated. The former is the more important kinematically, so we retain it only. In view of (40), it can be written as

$$\mathcal{H}_{\text{int}}(H, Z) = \frac{1}{2}\frac{gm_W}{\cos^2\theta_W}\mathbf{Z}\cdot\mathbf{Z}\rho. \tag{44}$$

Note that this interaction has no free parameters; everything is determined by the hidden $\mathcal{G}$-invariance. By combining (44) with the $\bar{e}eZ$ vertex, one obtains the process

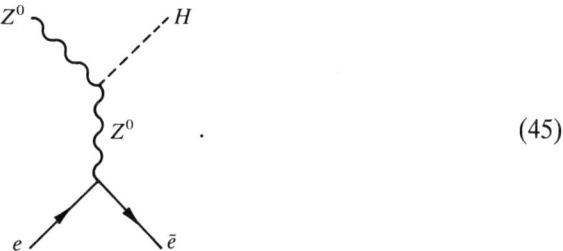

(45)

Hence, an e^+e^- collider, operating at an energy above $m_Z + m_H$, can produce a virtual Z^0, which then emits H by a mechanism akin to bremsstrahlung. If m_H is rather smaller than m_Z, this will be a promising way to establish the existence of the Higgs boson.

[The coupling of H to quarks and leptons also provides a way of finding H, and is of interest in its own right because it reveals another aspect of the requirements imposed by invariance under the gauge group. As we know from everything we have learned about the weak interaction, and as (19) and (20) make explicit, left- and right-handed Dirac fields transform differently under $\mathfrak{G}$. But we also know that the distinction between left- and right-handedness is only Lorentz invariant for massless particles. Hence *explicit* invariance under $\mathfrak{G}$ requires all fermions to be massless! Note the emphasis on explicit; as we shall now see, when we are in the phase where the Higgs field has a nonvanishing vacuum expectation value, and the gauge invariance is *hidden*, it becomes possible for fermions to have a mass, just as it was possible to give some of the gauge bosons a mass. *All* these masses must be introduced in this gauge-invariant manner if the theory is to be renormalizable.

To see this explicitly, consider any Dirac field $f(x)$. The contribution of its mass to the energy density is given by $m\bar{f}f = mf^\dagger\gamma_0 f$, i.e., by

$$\mathcal{H}_m = m_f[f_L^\dagger(x)f_R(x) + f_R^\dagger(x)f_L(x)]. \tag{46}$$

This expression mixes left- and right-handed fields, as we would expect from our construction of the Dirac equation for nonzero mass from the zero-mass Weyl equations (see, in particular, §II.A.3). In contrast to this, the kinetic energy, and all couplings to gauge fields, do not mix f_L and f_R. Under the group $\mathfrak{G}$, f_L and f_R transform differently [cf. Eqs. (19) and (20)], and invariance requires $m_f = 0$.

By using the Higgs field Φ, we can save the situation by introducing an interaction between Φ and the fermions that reduces to a quadratic mass term like (46) when Φ is replaced by its vacuum expectation value Φ_V. This interaction must be invariant under $\mathfrak{G}$, and be bilinear in Fermi fields. Since the left-handed Fermi fields Ψ_L and Φ are weak isospinors, the expression $(\Psi_L^\dagger, \Phi)$ is an invariant, where we use the notation of Eq. (35). Since the right-handed field f_R is a weak isosinglet, $(\Psi_L^\dagger, \Phi)f_R$ is an invariant that assumes the desired form (46) when Φ is replaced by Φ_V. Thus a fermion mass can be introduced in this somewhat underhanded way without spoiling the gauge invariance.

For the sake of concreteness, let Ψ_L be a quark doublet:

$$\Psi_q = \begin{pmatrix} q_{+L} \\ q_{-L} \end{pmatrix}.$$

The gauge invariant trilinear energy density coupling L- and R-fields is then

$$\mathcal{H}_{qH} = \lambda_q[(\Psi_q^\dagger, \Phi)q_{-R} + q_{-R}^\dagger(\Phi^\dagger, \Psi_q)]. \tag{47}$$

As we shall soon see, the coupling constant λ_q is related to the quark mass. The only right-handed quark field that can appear in (47), must have $Q = -\frac{1}{3}$, and is designated by q_{-R}. This is so because $(\Psi_q^\dagger, \Phi)$, though invariant under $SU_L(2)$, is not $U_Y(1)$ invariant: $\Psi_q^\dagger$ raises Y by $\frac{1}{6}$ [see (22)], and Φ destroys particles with $y_H = \frac{1}{2}$, so $(\Psi_q^\dagger, \Phi)$ is a $\Delta Y = -\frac{1}{3}$ operator. This change must be compensated by a $\Delta Y = \frac{1}{3}$ operator; according to (22), this can only be done by a field that destroys $y = Q = -\frac{1}{3}$ right-handed quarks.

Next we substitute (42), and remove the gauge factor by a gauge transformation that now involves Ψ_q as well; in view of (43) this eliminates q_+

2. THE GENERATION OF MASS

and gives us

$$\mathcal{H}_{qH} = \frac{\lambda_q}{\sqrt{2}}[q^\dagger_{-L}q_{-R} + q^\dagger_{-R}q_{-L}](v + \rho). \tag{48}$$

The term proportional to v is the mass term (46) for quarks q_- with $Q = -\frac{1}{3}$; hence, we set

$$\frac{1}{\sqrt{2}}\lambda_q v = m_{q_-}. \tag{49}$$

The coupling of fluctuation quanta (the Higgs particle H) to these quarks is then given by the term linear in ρ:

$$\mathcal{H}_{\bar{q}qH} = \frac{m_{q_-}}{v}(q^\dagger_{-L}q_{-R} + q^\dagger_{-R}q_{-L})\rho. \tag{50}$$

Here v is known from (40) and A(22):

$$v = 2m_W/g = 246 \text{ GeV}. \tag{51}$$

Once again, there are no free parameters beyond m_{q_-} in this interaction.

The interaction (47) only gives mass to the "down" member of a weak isodoublet. But if Φ_1 and Φ_2 are two (weak) isospinors, with components

$$\Phi_i = \begin{pmatrix} \alpha_i \\ \beta_i \end{pmatrix},$$

there are two $SU_L(2)$ invariant bilinear expressions, $(\Phi_1^\dagger, \Phi_2) = \alpha_1^\dagger \alpha_2 + \beta_1^\dagger \beta_2$, and $[\Phi_1, \Phi_2] \equiv \alpha_1\beta_2 - \alpha_2\beta_1$, where the latter is already familiar to us from the $I = 0$ state of Eq. I.D(8). In virtue of this, and the y-eigenvalues, one finds that $\tilde{\lambda}_q\{q^\dagger_{+R}[\Psi_q, \Phi] + \text{h.c.}\}$ is also a $\mathcal{G}$-invariant energy density, where $\tilde{\lambda}_q$ is another coupling constant. When $\Phi \to \Phi_V$, it gives the mass $v\tilde{\lambda}_q/\sqrt{2}$ to q_+, the $Q = \frac{2}{3}$ member of the doublet. By the same token, we can give arbitrary masses to both members of a lepton doublet. Hence, *the theory provides no explanation as to why neutrinos have masses that are either zero, or very small compared to their charged partners, nor does it explain the quark mass spectrum*.

The essential feature of (50) is that it provides an amplitude for $\bar{q}q \to H$ that is proportional to the quark mass. For this mechanism the most propitious source of H is therefore provided by the heaviest mesons available. At the moment these are the $\bar{b}b$ bound states Y (see §III.B.3). If $m_H < m_Y$, they should decay by the mechanism depicted below:

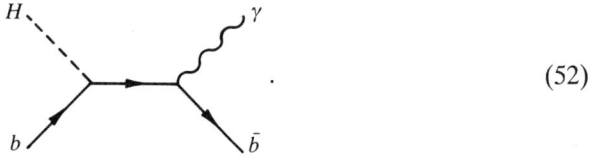

(52)

This has the virtue of having the rather unusual signature of a monochromatic γ-line. The estimated branching fraction for $\Upsilon \to H\gamma$ is of order 10^{-4} if $m_H \sim$ GeV; at this time no evidence for this decay of Υ has been reported. If there is a heavier t quark (the weak isospin partner of b), with $2m_t > m_H$ and $2m_t < m_Z$, the process $e\bar{e} \to t\bar{t} \to H\gamma$ at a $t\bar{t}$ bound state would become a most promising technique for searching for the Higgs boson.[13]

---

[13] The assumption that H is an elementary, structureless particle leads to certain inconsistencies that have stimulated speculations that H is a fermion–antifermion pair bound by a new and very strong force. In that case, the Higgs field would bear a close kinship to the Cooper pair state (recall footnote on p. 521), and its properties, such as the parameters v, λ_q, and m_H, would, at least in part, be determined theoretically. One would then expect H to be the ground state of a whole spectrum, and this could modify our discussion of phenomena involving H.

D. GRAND UNIFICATION[1]

Can some grand principle—some wondrously elegant set of equations—encompass all of physics? This question was first posed and examined in the years following the discovery of the General Theory of Relativity. At that time it was still possible to hope that gravity and electromagnetism were ultimately responsible for all interactions, but before long developments in atomic and nuclear physics convinced virtually everyone that this was not a tenable conjecture. From its inception to the mid-1970s particle physics was in a state of intellectual chaos, and attempts to construct unified theories were bound to be quixotic. The advent of the Standard Model has transformed that situation. It has brought a level of simplicity to the depiction of the fundamental constituents of matter, and to their mutual interactions, that has stimulated a vigorous attack on the unification problem. In our view, this effort has not yet produced a likely candidate for "the key to the Universe." Nevertheless, it has, for the first time, provided concepts that permit a serious examination of questions that no one previously knew how to address in an intellectually responsible manner.

This body of thought goes under the rubric of Grand Unified Theory. Some of the salient features of such theories were sketched in §I.E.13, and are briefly recalled here. In contrast to the Standard Model, which describes the electroweak and strong interactions with unrelated gauge fields (see Fig. I.40), in Grand Unified Theory all these interactions are combined into one inextricably interwoven set of non-Abelian gauge fields. In consequence the three coupling constants of the Standard Model (the two coupling constants of the electroweak interactions and the strong coupling constant) are all related to the sole coupling constant that appears in the unified field theory. In particular, the electroweak mixing angle θ_W is not a free parameter in the Grand Unified Theory. This marriage of the interactions is accompanied by a mingling of quarks and leptons: all the fundamental fermions in one generation are placed into a multiplet (see Fig. I.39). As a result, baryon number is not conserved in the Grand Unified Theory, which yields the remarkable prediction that the proton should be unstable.

The scheme just sketched, though intriguing, suffers from a number of severe flaws. It is intriguing because it provides a calculation "from first

---

[1] Our presentation closely follows Georgi (1974a and b). For a comprehensive treatment, see Cheng (1984), Chap. 14.

principles" of the electroweak mixing angle that is in remarkable agreement with the data. This calculation is rather impressive since it depends not only on symmetry arguments, but also on the variation of the "running" coupling constants that is characteristic of quantum field theory. While the predicted proton lifetime is probably ruled out by the currently available data, one may perhaps take some comfort from the fact that the theory manages to produce a lifetime that is vastly longer than all times known to cosmology. On the other side of the ledger, there are a host of flaws, many of which were already mentioned in §I.E.13: the theory does not determine the number of generations; its symmetry-breaking mechanism introduces a large number of free parameters; and it offers no understanding of why the strong interaction has the discrete symmetries P, C, and T, while the weak interaction does not. As we have said before, a theory that addresses itself to such fundamental questions should not be so undisciplined. Nevertheless, in view of its successes, it is quite possible that some facets of the existing theory will survive in better theories yet to be born, and for that reason some readers may wish to acquaint themselves with the fragment that now exists.

1. Basic assumptions and immediate consequences

We shall take it for granted that QCD and the electroweak gauge theory are correct. This implies that at energies large compared to m_W the symmetries of the electroweak gauge theory will be manifest, and not hidden, as they are at lower energies. The symmetry group of QCD, assumed to be manifest at all energies, will be designated by $SU_{\text{col}}(3)$; that of the electroweak theory by $\mathfrak{G}_{EW}$. From Eq. C(23), we recall that the latter is a direct product of the non-Abelian weak isospin group $SU_L(2)$ and the Abelian weak hypercharge group $U_Y(1)$. The assertion in the opening sentence of this paragraph can therefore be put as follows: *At energies well above m_W the symmetries that should be observed in nature will be described by the gauge group of the Standard Model*,

$$\mathfrak{F} = \mathfrak{G}_{EW} \otimes SU_{\text{col}}(3). \tag{1}$$

Henceforth, we refer to the regime where $\mathfrak{F}$ is the symmetry as "low energies," even though such energies are well in excess of 100 GeV. Note that this is a very different usage of the term than in the electroweak theory, where "low" refers to energies well below m_W.

The direct product in (1) implies that no symmetry operation relates the color and weak isospin degrees of freedom in the Standard Model, as already emphasized in the discussion pertaining to Fig. I.40. To impose a relationship between the strong and electroweak interactions, we replace $\mathfrak{F}$ by some other non-Abelian group $\mathfrak{G}$ that cannot be written as a direct product. Such a nonfactorizable group is said to be *simple*.

1. BASIC ASSUMPTIONS

We have already constructed a prototype of the kind of theory that we need: quantum chromodynamics, which embodies the local symmetries of the simple group $SU(3)$. Many properties of QCD generalize straightforwardly to any simple group $\mathfrak{G}$. In particular, if the theory is to be locally invariant under $\mathfrak{G}$, it must possess a gauge field that is a multiplet under the transformations of $\mathfrak{G}$ having one component per generator of $\mathfrak{G}$. (Each of these generators of $\mathfrak{G}$ will be a constant of the motion in the symmetry limit.) Furthermore, just one coupling constant specifies the amplitudes for emission and absorption of the gauge bosons by all objects that are coupled to the gauge field (including the gauge bosons themselves). Since the gauge field associated with $\mathfrak{G}$ is to contain the electroweak and color fields, which already stand in one-to-one correspondence with the generators of $\mathfrak{F}$, it follows that $\mathfrak{F}$ must be a subgroup of $\mathfrak{G}$.

One might well ask why the Grand Unified Theory must be phrased in the language of gauge theory. No watertight case for this approach can be given. Nonetheless, two considerations point in that direction. First, the "low" energy theory is the gauge theory built on $\mathfrak{F}$, and it is therefore natural to explore whether our present theory can be unified by incorporation into a larger gauge symmetry $\mathfrak{G}$. The second argument is more technical. In §I.E.13(a) we already argued that the known limit on the proton lifetime requires the full symmetry of $\mathfrak{G}$ to be hidden at energies below about 10^{15} GeV. Hence, we need a theory whose scattering amplitudes behave reasonably over an enormous range of energies. Only renormalizable theories have this property; nonrenormalizable theories have amplitudes that violate probability conservation, as the energy tends to infinity. It is a theorem (see Cheng (1984), §10.1) that the only renormalizable field theories that contain fermions and at least some charged vector bosons are gauge theories. For these reasons virtually all attempts to go beyond the Standard Model have exploited gauge theory.

The conjecture that the Grand Unified Theory is to be a gauge theory built on a simple group $\mathfrak{G}$, which contains $\mathfrak{F}$ as a subgroup, has an immediate consequence: *The sum of the electrical charges of the members of any multiplet must vanish*. Before establishing this result, let us examine its implications. The quarks within each generation have charges $\frac{2}{3}$ and $-\frac{1}{3}$, so taking color into account their total charge is $3(\frac{2}{3}) - 3(\frac{1}{3}) = 1$. But the leptons within a generation have charges 0 and -1, or a total charge of -1. Thus, neither the quarks nor the leptons can by themselves belong to a multiplet of $\mathfrak{G}$. On the other hand, *the total charge of the quarks and leptons belonging to one generation vanishes, and the set of all these fermions could therefore be assigned to one multiplet*!

We are already familiar with two analogues of this theorem. The first is that the sum of the magnetic quantum numbers $(m = j, m = j - 1, \ldots, m = -j)$ belonging to any angular momentum multiplet vanishes. The second is that all the eight 3×3 matrices λ_a of $SU(3)$ have vanishing traces [recall §IV.A.2(b)]. The proof of the first example is instructive: one merely computes the trace of the angular momentum commutation rule within a multiplet:

$$\sum_m \langle jm |[J_x, J_y]| jm \rangle = i \sum_m \langle jm |J_z| jm \rangle. \tag{2}$$

In any finite-dimensional space $\mathscr{S}$, $\text{Tr}_{\mathscr{S}}[A, B]$ vanishes, so $\text{Tr} J_z = 0$, as desired. Furthermore, the other angular momentum commutation rules show that $\text{Tr} J_x = \text{Tr} J_y = 0$, a fact that also follows from the invariance of the trace under the unitary transformations that rotate the coordinate axes into each other. This $SU(2)$ argument generalizes to any simple group. The generators $\{T_a\}$ of $\mathfrak{G}$ satisfy

$$[T_a, T_b] = iC_{abc} T_c. \tag{3}$$

Since $\mathfrak{G}$ is simple, there is no generator that commutes with all the T_a. Hence, for any choice of a and b

$$C_{abc} \text{Tr}_{\mathscr{M}} T_c = 0,$$

where $\mathscr{M}$ is any multiplet of $\mathfrak{G}$. One can then show[2] that this set of homogeneous algebraic equations requires that

$$\text{Tr}_{\mathscr{M}} T_a = 0 \tag{4}$$

for every generator T_a. But the electrical charge, like any nonkinematical conserved quantity, must be a linear combination of the generators of $\mathfrak{G}$, so this establishes the result on which we relied in the preceding paragraph.[3]

Since the fermions within a generation have vanishing total charge, we conjecture that *all the fermions within a generation belong to a multiplet $\mathscr{M}_f$ of $\mathfrak{G}$*. This hypothesis has the virtue of economy, since it asserts that there are no unobserved fermions, aside from further complete generations.[4] By the same token, the discovery of fermions with eigenvalues of weak isospin, charge, or color other than those of the known quarks and leptons would be a fatal blow to this simplest form of the Grand Unified Theory.

Once we know that quarks and leptons must be put into a multiplet $\mathscr{M}_f$ of $\mathfrak{G}$, we can show that baryon number is not conserved in the Grand Unified

---

[2] For the groups $SU(N)$ the proof follows directly from the requirement that the unitary matrices representing infinitesimal transformations have unit determinant, as shown for $N = 3$ in §IV.A.2(b).

[3] This theorem impedes another unification scenario which, at first sight, seems more plausible. That is, one might suppose that as one descends in energy, the "Grand" group $\mathfrak{G}$ decomposes into $SU_{\text{col}}(3) \times \mathfrak{G}'$, where $\mathfrak{G}'$ is a *simple* group that provides a true unification of the electromagnetic and weak interactions, in contrast to $\mathfrak{G}_{\text{EW}}$, which does not. Such a $\mathfrak{G}'$ would impose a relationship between e and g, that is, it would determine θ_W. But our theorem would require the sum of the charges within any multiplet of $\mathfrak{G}'$ to vanish, so the multiplets of $\mathfrak{G}'$ cannot coincide with those of $\mathfrak{G}_{\text{EW}}$. In short, this route could only be followed by introducing unobserved leptons and quarks.

[4] Here we assume that the t quark will be found.

1. BASIC ASSUMPTIONS

Theory. To motivate the argument we again turn to angular momentum. We know that if we are given any *one* eigenstate, $|jm\rangle$, an infinitesimal rotation will produce an admixture of neighboring states with $m \pm 1$ and the *same j*. Hence a finite sequence of infinitesimal rotations can be used to reach *all* members of the $(2j + 1)$-dimensional multiplet from any *one* member. By the same token,[5] if we are given any member of the fermion multiplet $\mathcal{M}_f$, say a lepton, a sequence of infinitesimal transformations of $\mathfrak{G}$ will produce a state in which all quarks in $\mathcal{M}_f$ have a nonzero amplitude. Every generator of an infinitesimal transformation of $\mathfrak{G}$ appears in the interaction, and causes the emission or absorption of a gauge boson as represented by

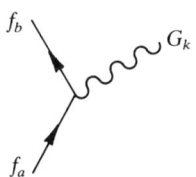

where f_a and f_b are two fermion states in $\mathcal{M}_f$, and G_k is a gauge boson. As a consequence, there are amplitudes that connect all members of $\mathcal{M}_f$ in some finite order of perturbation theory.[6] In particular, a baryon composed of three quarks will have an amplitude for turning into a state containing a lepton, and there is therefore a nonzero probability for proton decay.

In the high-energy limit the symmetry of $\mathfrak{G}$ is not hidden and group theory alone provides relationships between the familiar coupling constants. In particular, it determines a value $\bar{\theta}$ of the electroweak angle. This value does *not* coincide with what we observe at laboratory energies [i.e., with Eq. B(89)], because all the coupling constants "run" in the sense of §IV.C.4. Before we turn to the evaluation of this "running" electroweak angle, we shall calculate its symmetry limit.

The relationship between the electric charge Q and the generators of $\mathfrak{F}$ was given in Eq.C(21): $Q = T_3 + Y$. But Y is the generator of the Abelian factor $U_Y(1)$, and its appearance with coefficient 1 in Q is therefore arbitrary (recall the footnote on p. 518). Non-Abelian generators do not have such freedom since they satisfy nonlinear commutation rules that fix the scale of all eigenvalues, a fact that is familiar to us from angular

---

[5] Here we have slid over a small complication. The argument as given only holds for multiplets $\mathcal{M}_f$ that belong to irreducible representations of $\mathfrak{G}$, that is, to multiplets that span a d-dimensional vector space within which it is impossible to find a basis that reduces the matrices representing the $\{T_a\}$ to a block-diagonal form with blocks having a smaller dimension than d (see p. 358). If $\mathcal{M}_f$ spans a reducible representation [as is the case for the simplest Grand Unified model based on the group $SU(5)$], the argument goes through nonetheless, since then $\mathcal{M}_f = \mathcal{M}_{f1} \oplus \mathcal{M}_{f2} \oplus \cdots$ in the notation of Eq. IV.A(39), where each $\mathcal{M}_{fi}$ is irreducible, and therefore contains both quarks and leptons.

[6] To illustrate this, consider the gauge theory built on $SU(2)$, and the $I = \frac{3}{2}$ multiplet. An amplitude that mixes $I_3 = \frac{3}{2}$ with $I_3 = -\frac{3}{2}$ is obtained by emitting (or absorbing) three gauge bosons with $I = 1$ and $I_3 = 1$ (or $I_3 = -1$).

momentum. We must therefore rewrite Q in terms of the generators $\{T_a\}$ of $\mathcal{G}$. It is convenient to choose a basis in the space spanned by the multiplet $\mathcal{M}_f$, so that Y is proportional to just one generator, which we call T_y:

$$Q = T_3 + cT_y. \tag{5}$$

We first show that c is related to the electroweak angle in the symmetry limit, $\bar{\theta}$. For this purpose, we consider the covariant derivative

$$\boldsymbol{\nabla} - ig T_a \mathbf{V}^a = \boldsymbol{\nabla} - ig(T_3 \mathbf{W}^3 + T_y \mathbf{B} + \cdots), \tag{6}$$

where the $\mathbf{V}^a$ are the vector potentials associated with the gauge group $\mathcal{G}$, T_a is the matrix representing that generator when acting on the multiplet $\mathcal{M}_f$, g is the universal coupling constant of the Grand Unified Theory, $\mathbf{W}^3$ and $\mathbf{B}$ are the two neutral electroweak gauge fields, and $+\cdots$ alludes to the other gauge fields that do not concern us at this moment. In this symmetry limit the two neutral electroweak fields are related to the electromagnetic and Z^0 fields by Eq. B(48), *but* with θ_W *replaced* by $\bar{\theta}$. Hence the term in (6) that contains the electromagnetic interaction is

$$\boldsymbol{\nabla} - ig \sin \bar{\theta} (T_3 + T_y \cot \bar{\theta}) \mathbf{A}.$$

But here the coefficient of $\mathbf{A}$ is, by definition, $-ieQ$. Since the charge operator is given by (5), we have that

$$c = \cot \bar{\theta}. \tag{7}$$

Hence, the electromagnetic fine structure constant α_e is given by

$$\alpha_e = (g^2/4\pi) \sin^2 \bar{\theta}. \tag{8}$$

The evaluation of c relies on

$$\text{Tr}_{\mathcal{M}}(T_a T_b) = \delta_{ab} N_{\mathcal{M}}, \tag{9}$$

where $\mathcal{M}$ is any multiplet of $\mathcal{G}$, and $N_{\mathcal{M}}$ is a number that only depends on the multiplet. This identity expresses the invariance of the trace under all transformations of $\mathcal{G}$ in that δ_{ab} is the only invariant second-rank tensor. Applied to (5), it yields

$$\text{Tr}\, Q^2 = \text{Tr}\, T_3^2 + c^2 \text{Tr}\, T_y^2,$$

$$= (1 + c^2) \text{Tr}\, T_3^2. \tag{10}$$

As $\mathcal{M}_f$ is supposed to contain all the fermions in a generation, these traces are

	u_L	u_R	d_L	d_R	e_L	e_R	v_L	v_R	Trace
Q^2	$3(\frac{2}{3})^2$	$3(\frac{2}{3})^2$	$3(\frac{1}{3})^2$	$3(\frac{1}{3})^2$	1	1	0	0	$\frac{16}{3}$
T_3^2	$3(\frac{1}{2})^2$	0	$3(\frac{1}{2})^2$	0	$(\frac{1}{2})^2$	0	$(\frac{1}{2})^2$	0	2

Hence, $(1 + c^2) = \frac{8}{3}$, or

$$c = \sqrt{\frac{5}{3}} \tag{12}$$

and (7) determines the following value for the electroweak angle in the symmetry limit:

$$\sin^2 \bar{\theta} = \tfrac{3}{8}. \tag{13}$$

This is in gross disagreement with the observed value of 0.23. As we shall see, this difference is accounted for by the "running" of the coupling constants.

2. Evaluation of the electroweak angle at low energies

As we learned in §IV.C.4(e), the coupling constants are not really constant. Due to radiative corrections the amplitude for the emission or absorption of a virtual gauge boson whose propagator carries a 4-momentum q is obtained from the naive amplitude by replacement of the naive constant by a "running" constant that varies logarithmically with $s = q^2$. Furthermore, we saw that the coupling constant of a non-Abelian gauge field decreases as s increases (i.e., is asymptotically free), whereas the coupling constant of an Abelian field has the opposite variation with s. The subgroup $\mathfrak{F}$ that describes the symmetry at "low" energies has two non-Abelian factors describing the coupling of weak isospin and color, and one Abelian factor for the weak hypercharge interaction. As we go to higher energy regimes, the typical values of s that appear in scattering and decay amplitudes will increase accordingly. The strong- and weak-isospin interactions will weaken at different rates, since they are due to different gauge groups, while the weak hypercharge interaction will grow stronger. The disparity in strength of the familiar interactions at familiar energies, which could lead one to conclude that unification is not possible, will therefore diminish as the energy is raised. The requirement that the three running coupling constants of $\mathfrak{F}$ are to merge at one value of $s = M^2$ determines the characteristic energy M above which the symmetries of the unified theory become manifest.[7] Furthermore, this requirement yields a relationship amongst the coupling constants at low energies, and thereby determines the electroweak angle at currently accessible energies.

In this Section it will be convenient to designate the three running coupling constants of $\mathfrak{F}$ that merge at the unification point by $g_N(s)$, where $N = 1, 2$, and 3 refer to the weak hypercharge, weak isospin, and strong interactions, respectively. N also specifies the dimension of the corresponding unitary subgroups of $\mathfrak{G}$. Furthermore, we define the "fine structure" constants

$$\alpha_N = g_N^2/4\pi.$$

The connection with our previous notation is $g_3 = g_s$ and $g_2 = g$, but it is *not* true that $g_1 = g'$, the conventional weak hypercharge coupling. This is clear from the discussion following (5). The coupling constants that merge are the coefficients of the generators of $\mathfrak{F}$ that are also generators of the unifying group $\mathfrak{G}$, and therefore the strength of the Abelian factor in $\mathfrak{F}$ must be measured by the coefficient of the generator T_y of $\mathfrak{G}$. This determines the relationship between g' and g_1: $g'Y = g_1 T_y$, or in view of (5) and (12), $g' = \sqrt{\tfrac{3}{5}} g_1$. Hence, the running electroweak angle is defined as

$$\cos^2 \theta_W(s) = \left(\frac{e(s)}{g'(s)}\right)^2 = \frac{5}{3}\frac{\alpha_e(s)}{\alpha_1(s)}. \tag{14}$$

The variation of all three coupling constants is given by Eq. IV.C(69):

$$\frac{1}{\alpha_N(s)} = \frac{1}{\alpha_N(s_0)} + b_N \ln\left(\frac{s}{s_0}\right). \tag{15}$$

This shows that the inverse of the "fine structure" constants run along straight lines when plotted as functions of the variable $\ln s$ (see Fig. 3); for that reason the b_N are called slope parameters. In the case of $N = 3$, the slope is given by the QCD calculations that culminated in Eq. IV.C(63), which we rewrite as

$$12\pi b_3 = 33 - 4N_g = 21. \tag{16}$$

Here the contribution 33 is due to the gauge bosons, while $-4N_g$ arises from the vacuum polarization of N_g complete generations of fermions; the value

---

[7] One might well ask why a simultaneous merging of all three constants is required. It is not, but it is the only consistent scenario that does not require the introduction of new fermions. For example, if one makes the plausible conjecture that the two electroweak couplings merge at a lower energy than M, above which $\mathfrak{G}_{EW}$ is replaced by a simple group with coupling constant f, and that f merges with the color coupling g_3 at M, one encounters the hurdle explained in the footnote on p. 532

2. EVALUATION OF THE ELECTROWEAK ANGLE

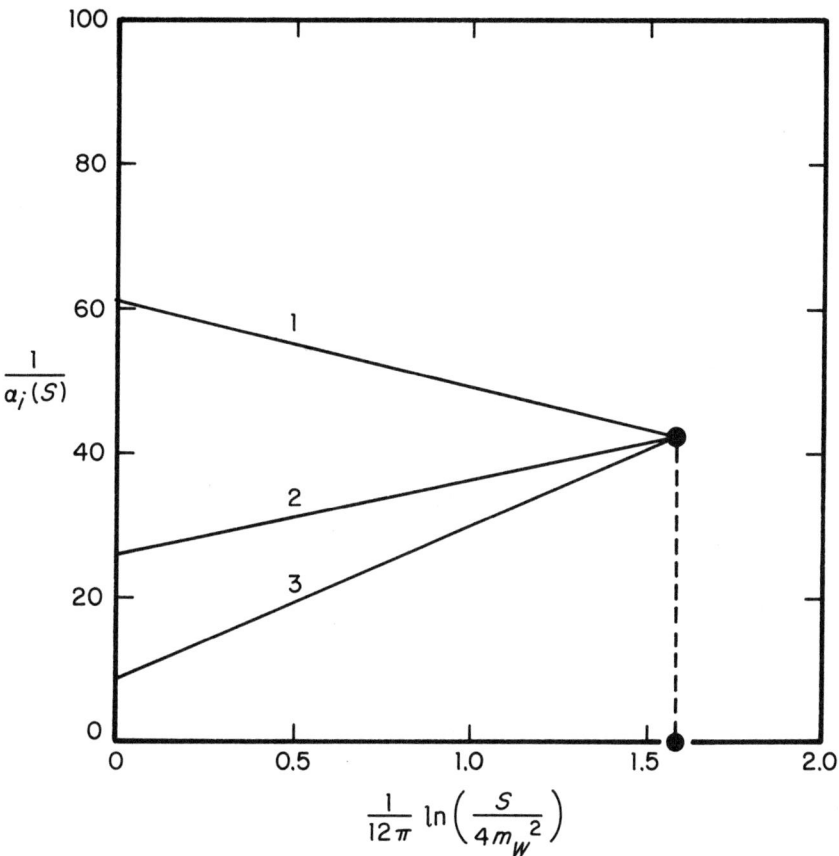

FIG. 3. The variation with mass scale of the three "fine structure constants" for the strong and electroweak interactions. [See Eqs. (20), (22), and (23).] The requirement that these lines meet determines the electroweak mixing angle and the unification mass, as tabulated on p. 539.

21 is the assumption that there are no generations beyond the three now observed. A similar calculation for $N = 2$ yields

$$12\pi b_2 = 22 - 4N_g - \tfrac{1}{2} = 9\tfrac{1}{2}, \qquad (17)$$

where the extra $\tfrac{1}{2}$ comes from the Higgs boson that produces the spontaneous symmetry breakdown in the electroweak theory.

To find b_1, we exploit the fact that the s-variation of the renormalized charge can be cast into the form of (15):

$$\frac{1}{\alpha_e(s)} = \frac{1}{\alpha_e(s_0)} + b_e \ln\left(\frac{s}{s_0}\right).$$

The slope parameter b_e that appears here can be extracted from the vacuum polarization in QED:

$$12\pi b_e = -2N_g \, \mathrm{Tr}\, Q^2. \tag{18}$$

The negative sign is the statement that QED is not asymptotically free. For convenience the trace in (18) runs separately over the right- and left-handed states in $\mathcal{M}_f$, so that we may use the result listed in (11). Naturally, one can only set $N_g = 3$ in (18) if s is large compared to all the fermion masses, so that they contribute fully to vacuum polarization. Equation (18) holds for any $U(1)$ gauge theory; in particular, it gives the slope of the fine structure constant corresponding to g_1 *provided* Q is replaced by the properly normalized generator T_y:

$$12\pi b_1 = -2N_g \, \mathrm{Tr}\, T_y^2 = -2N_g \, \mathrm{Tr}\, T_3^2 = -4N_g. \tag{18'}$$

To exploit (15) we must insert our knowledge of the coupling constants at some convenient point s_0 in the "low" energy regime. For that purpose we choose the point where the electroweak angle is determined empirically. As we recall from §§A.2 and B.3, that determination is based on an effective Hamiltonian from which the W and Z^0 fields have been eliminated, and only fermion fields survive. As one sees quite explicitly in Eq. A(18), the characteristic 4-momenta flowing through the boson propagator have squares that are of order m_W^2, so that is the typical value of s where the weak coupling is known. A more detailed analysis leads to the conclusion that $s_0 = 4m_W^2$ best characterizes the point where the cited value of θ_W is determined. Consequently, we take whatever knowledge we have about the couplings at currently accessible energies and use (15) to shift them to this common value of s_0.

At this time the best determination of α_3 comes from Υ-decay [see Eq. III.B(14)], which yields $\alpha_3(M_\Upsilon^2) = 0.16 \pm 0.02$. At the desired value of s_0 this coupling constant is therefore given by

$$\frac{1}{\alpha_3(4m_W^2)} = \frac{1}{\alpha_3(M_\Upsilon^2)} + \frac{1}{12\pi}(33 - 4N_g) \ln\left(\frac{4m_W^2}{M^2}\right). \tag{19}$$

The s-variation of α_3 at all energies of concern to us is therefore given by

$$\frac{1}{\alpha_3(s)} = \frac{1}{\alpha_3(M_\Upsilon^2)} + (33 - 4N_g)(L + \delta) \tag{20}$$

2. EVALUATION OF THE ELECTROWEAK ANGLE

where

$$L \equiv \frac{1}{12\pi} \ln\left(\frac{s}{4m_W^2}\right), \qquad \delta \equiv \frac{1}{12\pi} \ln\left(\frac{4m_W^2}{M^2}\right). \tag{21}$$

The familiar value $\frac{1}{137}$ of α_e is obtained from electromagnetic processes at threshold, i.e., at $s = 0$; at $s = 4m_W^2$ it has grown to $\frac{1}{129}$ [(Ross & Goldman, (1980); Marciano & Sirlin, (1983)]. Equations (14), (15) and (18'), therefore, tell us that

$$\frac{1}{\alpha_1(s)} = \frac{3}{5} \cdot 129 \cos^2 \theta_W - 4N_g L \tag{22}$$

where θ_W is now the value that can be compared to the one determined empirically. By the same argument

$$\frac{1}{\alpha_2(s)} = 129 \sin^2 \theta_W + (22 - 4N_g - \tfrac{1}{2})L. \tag{23}$$

We now have the desired equations for the three running coupling constants. The requirement that the three couplings merge at some mass M, $\alpha_N(M^2) = \bar{\alpha}$, gives us two equations in the unknowns $\sin^2 \theta_W$ and $\ln(M^2/4m_W^2)$. The results from this simple calculation are tabulated below, and compared there to those of a "professional" calculation. The agreement between them is seen to be quite satisfactory.

	"Naive"	Ross (1980)
$\sin^2 \theta_W$	0.21	0.21 ± 0.01
M	6×10^{14} GeV	$(2 - 8) \times 10^{14}$ GeV
$\bar{\alpha}$	0.024	0.024

This calculation of the electroweak angle from "first principles" is astonishingly successful and, as we see, also quite insensitive to the finer details of the computation. On the other hand, the calculation only provides a rather rough value of the unification mass M, as one would expect, since the coupling constants vary logarithmically with M.

3. Other aspects of the Grand Unified Theory

(a) The gauge group

The foregoing analysis made as little reference to the detailed properties of the unifying group $\mathfrak{G}$ as possible. This was a purposeful evasion, since we suspect that this aspect of the theory has poor prospects for survival. Nevertheless, a cursory glance at some of the group-theoretic details is in order. In particular, one must respond to the question of whether there exists any group with the properties that we have blithely assumed.

To answer this question we examine the behavior of the fermion multiplet $\mathcal{M}_f$ under the subgroup $SU_L(2) \otimes SU_{\text{col}}(3)$ of $\mathfrak{G}$ [cf. Eq. (1)]. Since the first factor distinguishes between the left- and right-handed portions of the fermion fields, it is convenient to describe all states belonging to $\mathcal{M}_f$ by left-handed 2-component fields. This can be done since we know from Eq. A(14) that the charge conjugate of a right-handed 2-component spinor field is the Hermitian conjugate of a left-handed field. In consequence, right-handed quark fields in $\mathcal{M}_f$ belong to the complex-conjugate color triplet 3* of $SU_{\text{col}}(3)$, while their left-handed partners belong to 3. Insofar as transformations under $SU_L(2)$ are concerned, there is no such distinction, as we learned on p. 356. The subgroup content of $\mathcal{M}_f$ is therefore

	u_L	d_L	u_R	d_R	ν_L	e_L	e_R	ν_R
$SU_{\text{col}}(3)$	3	3	3*	3*	1	1	1	1
$SU_L(2)$	2		1	1	2		1	1

(24)

Here the right-handed neutrino field has been included in case $m_\nu \neq 0$. Hence the dimension of $\mathcal{M}_f$ is 15 if m_ν vanishes and 16 if it does not. By hypothesis, each generation has an identical multiplet. Georgi and Glashow (1974) have shown that the smallest groups that have multiplets having such a composition are the group $SU(5)$, and the group $SO(10)$, which describes rotations in a 10-dimensional Euclidean space.[8]

The gauge boson multiplet depends on the unifying group $\mathfrak{G}$. As we saw in §IV.A, the groups $SU(N)$ have $N^2 - 1$ generators; hence, the vector field that mediates all interactions in the $SU(5)$ model has 24 components. Of these, 12 are the gauge bosons of the Standard Model: the octet of gluons,

---

[8] In the $SU(5)$ model, $\mathcal{M}_f$ transforms with a 15-dimensional reducible representation composed of 5- and 10-dimensional irreducible representations. To draw an analogy with a familiar situation, it is as if the multiplet of the four Δ-resonances had to be assigned to two $I = \frac{1}{2}$ isospin multiplets, instead of the correct assignment, $I = \frac{3}{2}$. This is an ugly feature of the $SU(5)$ model, and undermines the hoped for austerity of a unifying group. In concrete terms, it means that this model has more free parameters than one could have, at least in principle. In this regard the $SO(10)$ model is more attractive, because it has a 16-dimensional irreducible

3. THE GRAND UNIFIED THEORY

the two W's and Z^0, and the photon. The other 12 mediate interactions that are fundamentally different from anything yet seen in the laboratory, for their emission or absorption produces transitions between quarks and leptons, and for that reason these bosons must have fractional charge! In the sequel, we refer to all gauge bosons that appear in the Grand Unified Theory, and which have no counterpart in the Standard Model, as X bosons.[9] The $SO(10)$ model has a larger set of gauge fields than the 24 of the $SU(5)$ model.[10]

(b) Symmetry breaking

The large degree of symmetry embodied in the unifying group $\mathfrak{G}$ can be reduced to the symmetry below the unification mass scale M by the Higgs mechanism [recall §C.2(c)]. As in the electroweak theory, one must introduce a scalar field Ξ that transforms nontrivially under $\mathfrak{G}$ (i.e., is not a $\mathfrak{G}$-singlet), and assume that the Hamiltonian has a form that forces it[11] to have a nonvanishing vacuum expectation value Ξ_V. This Higgs field Ξ cannot be the field Φ that was used to hide the symmetry of the electroweak gauge group $\mathfrak{G}_{EW}$. The masses of the W and Z^0 bosons were found to be of

---

(*Footnote 8 continued*)
representation that neatly accommodates all the states listed in (24). At first sight, this would lead one to fear that the $SO(10)$ model would predict comparable neutrino and charged lepton masses, but Witten (1980) has shown that this need not be the case.

Note that the complex-conjugate multiplet, $\mathcal{M}_f^*$, does not transform in the same manner as $\mathcal{M}_f$, since the latter has the quark content $(3, 2) \oplus (3^*, 1)$, whereas the former has $(3^*, 2) \oplus (3, 1)$. This means that $\mathfrak{G}$ must have complex representations, which imposes considerable restrictions. As we already know from $SU(3)$, the unitary groups larger than $SU(2)$ have such representations. In the case of $SO(10)$, there is an analogy with the ordinary rotation group $SO(3)$. The latter has odd-dimensional, single-valued tensor representations, and double-valued, even-dimensional spinor representations. A similar division occurs for $SO(10)$: There are single-valued tensors, and double-valued spinors. The spinor of lowest dimension has 16 components. In contrast to the familiar spinors, it does not transform in the same manner as its complex conjugate. (A further difference with 3-dimensional rotations is that the spinors associated with rotations in ten dimensions are not multiplets of any unitary group; the intimate relationship between $SO(3)$ and $SU(2)$ is an "accident" that does not recur in higher-dimensional spaces.)

[9] These bosons have charges $\pm\frac{4}{3}$ and $\pm\frac{1}{3}$; in the literature the former are called X, the latter Y, but we have no need to distinguish between them. Since they convert quarks into leptons, it is also clear that they must be color triplets. In consequence, the confinement of color implies that one would never see isolated X and Y bosons, even if one could achieve energies of order their mass, 10^{15} GeV; the only isolated objects that could exist would have the compositions $X\bar{X}$, $X\bar{Y}$, Xq, etc.

[10] In an N-dimensional Euclidean space the number of independent angles needed to specify an arbitrary rotation is $N(N-1)/2$. Hence, the $SO(10)$ gauge field has 45 components, and therefore 33 field quanta that have no counterparts in the Standard Model.

[11] Here "it" is a purposeful simplification. In the electroweak case, only the uncharged member of the Higgs doublet has a vacuum expectation value, since electromagnetic gauge invariance must survive that symmetry breaking. By the same token, the member of the multiplet Ξ that is arranged to have a nonvanishing vacuum expectation value must be invariant under the subgroup $\mathfrak{F}$, which is to survive the "grand" symmetry breaking. It must, in other words, carry no color, weak isospin, or charge. As very few multiplets of $\mathfrak{G}$ have such a member, this requirement serves as a useful constraint on the model.

order $g_2\Phi_V$ [see Eq. C(40)], which sets the energy scale above which the electroweak symmetry becomes manifest. For the same reason, the masses of the X bosons that mediate the novel interactions that arise in the Grand Unified Theory must be of order the unification mass M, which is vastly larger than m_W or m_Z. If the Higgs mechanism is to produce such masses for the new field quanta X, it is necessary that

$$\Xi_V \sim M/\sqrt{4\pi\bar{\alpha}} \tag{25}$$

where $\bar{\alpha}$ is the common fine structure constant at the unification point. Therefore,

$$\left(\frac{\Xi_V}{\Phi_V}\right)^2 \sim \frac{\alpha_2}{\bar{\alpha}}\frac{M^2}{m_W^2} \sim 10^{26}. \tag{26}$$

This astronomic ratio leads to the so-called gauge hierarchy disease. The Hamiltonian that forces the two Higgs fields, Ξ and Φ, to have nonzero vacuum expectation values is a quartic polynomial in these fields. Even if it is assumed to contain no term $\Xi^2\Phi^2$ that couples the two fields, radiative corrections will inevitably produce such couplings with appreciable magnitudes, and their influence on low-energy processes can be represented by adding such a term to the Hamiltonian. When Ξ is replaced by its vacuum expectation value, this term becomes $\Xi_V^2\Phi^2$. In view of (26), this will produce a prodigious shift in Φ_V, and a correspondingly huge mass for W and Z^0, unless the parameters of the Grand Unified Theory are adjusted with incredible care. Put more generally, if the Higgs mechanism is asked to account for symmetry breaking at two vastly different mass scales, one obtains a theory that is intolerably sensitive to fine-tuning of its basic parameters. This gauge hierarchy disease might, at least in principle, be cured by replacing the Higgs fields by bound states of fermion–antifermion pairs, in analogy to superconductivity where the role of the Higgs field is played by the Cooper electron pair. A new strong force, and new fermions, must be introduced to flesh out such a picture, but no satisfactory scheme that embodies these ideas has been proposed thus far.

(c) Proton decay

The fractionally charged X bosons with masses of order 10^{15} GeV that mediate the "new" interactions predicted by the Grand Unified Theory can

produce remarkable transitions such as

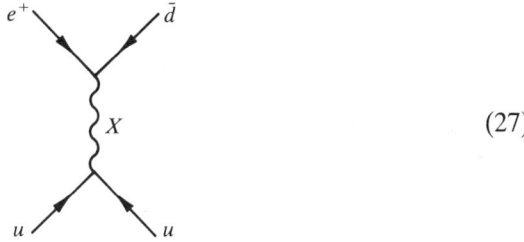 (27)

Hence the process $p \to e^+ d\bar{d}$ can occur, and would manifest itself in proton decays such as

$$p \to e^+\pi^0, \ e^+\rho^0. \qquad (28)$$

The decay rate is proportional to the square of the amplitude (27), and therefore to $(g/m_X)^4$. In other words, the proton lifetime is proportional to the fourth power of the unification mass M, and therefore very sensitive to the errors inherent to the evaluation of that quantity. Smaller (but nevertheless appreciable) uncertainties arise in the computation of the hadronic transition amplitudes for specific processes such as (28). As already stated in §I.E.13(a), the best theoretical estimates for the proton lifetime are *shorter* than the empirical *lower* limit, and indicate trouble for the simple grand unification scenarios.[12]

Note Added in Proof (May 1986)

[At the outset of our description of Grand Unification Theory, on p. 529, we posed the question "Can some grand principle—some wondrously elegant set of equations—encompass all of physics?," and this immediately brought the General Theory of Relativity into the discussion. Einstein's theory has always been viewed as the intellectual structure that best epitomizes the goal that physics seeks: an axiomatic basis of austere but compelling force which, when elegantly formulated, leads inexorably to phenomena that are complex and that were previously inexplicable, or not even anticipated. Everyone has long agreed that Grand Unified Theory, though very instructive, fails that acid test, for as we said on p. 530, "a theory that addresses itself to such fundamental questions should not be so undisciplined."

Recently a far more ambitious attempt at unification that has a conceptual kinship with General Relativity has received a crescendo of attention [see Green (1985); Schwarz (1985)]. Called Superstring Theory, it proposes to encompass the interactions of particle physics *and* gravity in what is essentially

---

[12] For a recent survey of Grand Unification, see Langacker (1985).

a geometric theory. In contrast to conventional quantum field theory, in which the elementary constituents are points in space-time, Superstring Theory postulates that the basic elements are curves (or strings) in a certain D-dimensional curved space, where D is considerably larger than 4, with $D = 10$ being the value that current theory requires. The extra dimensions are not directly observable in macroscopic phenomena, however, because the D-dimensional space is supposed to "curl up" into D-4 compact dimensions, just as an ordinary infinite cylinder defines a 2-space with one compact dimension running around the cylinder and the other noncompact dimension being along the cylinder.

The classical action of a free point particle is simply the proper length of its world-line (see Jackson (1975), p. 573). The action for a free string is the obvious generalization, i.e., the area of the world-sheet (in D dimensions) swept out by the moving string. This area is invariant under a wide class of transformations that merely reparametrize points on the world-sheet, but leave its intrinsic curvature and topology unchanged. As we see, this defines a theory that can be said to be purely geometrical in a sense analogous to General Relativity. It is hoped that the theory will produce the reduction from D to 4 dimensions "by itself," so to speak, in a fashion similar to that by which Einstein's theory leads to models of the universe with certain nontrivial geometries and topologies. The general coordinate transformations in the compact dimensions then correspond to the gauge transformations of particle physics, whereas the others yield the gravitational field.

At the classical level, there are an infinity of such theories. But the requirement that the theory be renormalizable at the quantum level is a draconian constraint. Only a handful of possibilities survive. They are not just renormalizable, but even *finite,* and possess the gauge group of the Standard Model as a subgroup!

The low-energy features of these candidates (where "low" means small compared to the Planck mass!) are being explored with great vigor, but at this time it is still far from clear whether contact with the real world will be made. In particular, how the reduction to 4 space-time dimensions occurs is still a mystery. Nevertheless, Superstring Theory is remarkable for, at the very least, it demonstrates that models exist in which gravity and matter can be consistently quantized, and wherein the fermion multiplets, their various interactions, and even the dimensionality of space-time, may have a common origin.]

Appendix II

BOSE FIELDS

The purpose of this Appendix is to explain the origin and significance of the expressions for the electromagnetic field, and of the field that describes particles with spin-0, Eqs. II.A(40) and (43).

We begin by analyzing a well-known problem in which the fields do not yet appear: the harmonic oscillator and its quantization. As we shall see, the simple oscillator is the basic building block of quantum field theory. It leads naturally to fields that describe bosons. That integer spin particles must obey Bose–Einstein statistics, while half-integer spin particles obey Fermi–Dirac statistics, is proved in Appendix IV.

1. The harmonic oscillator

(a) The classical oscillator

The Hamiltonian is

$$H = \frac{1}{2\mu}p^2 + \frac{\mu\omega^2}{2}q^2. \tag{1}$$

Here q is the coordinate of the oscillator, p is the corresponding momentum, μ is the mass, and $\mu\omega^2$ is the restoring force constant. From the fundamental equations

$$\dot{q} = \frac{\partial H}{\partial p}, \qquad \dot{p} = -\frac{\partial H}{\partial q}, \tag{2}$$

one readily finds the equation of motion

$$\ddot{q} + \omega^2 q = 0. \tag{3}$$

The solutions are

$$q = A\cos(\omega t + \delta), \qquad p = -\mu\omega A\sin(\omega t + \delta),$$

where the amplitude A and the phase δ are arbitrary real constants.

It is useful for our later considerations to introduce a new complex variable α:

$$\alpha = \sqrt{\frac{\omega\mu}{2\hbar}}q + i\sqrt{\frac{1}{2\hbar\omega\mu}}p. \tag{4}$$

At this stage, $\hbar$ is some numerical constant of the same dimension as Planck's constant; i.e., [energy/frequency]. The reason for the particular choice (4) will become clear when we reach Eq. (15). We then have

$$q = \sqrt{\frac{\hbar}{2\mu\omega}}(\alpha + \alpha^*), \qquad p = \frac{1}{i}\sqrt{\frac{\hbar\omega\mu}{2}}(\alpha - \alpha^*), \tag{5}$$

and

$$\alpha = \sqrt{\frac{\omega\mu}{2\hbar}}Ae^{-i(\omega t+\delta)}. \tag{6}$$

The variable α has the time-dependence $\exp(-i\omega t)$, and α^* has $\exp(+i\omega t)$. The purpose of introducing α is to separate q and p into terms having these distinct variations in time. It often is useful to exhibit that time-dependence explicitly:

$$\alpha = ae^{-i\omega t}, \qquad \alpha^* = a^*e^{+i\omega t}. \tag{7}$$

Then

$$q = \sqrt{\frac{\hbar}{2\mu\omega}}(ae^{-i\omega t} + a^*e^{+i\omega t}) \tag{8a}$$

$$p = \sqrt{\frac{\hbar\omega\mu}{2}}\frac{1}{i}(ae^{-i\omega t} - a^*e^{+i\omega t}). \tag{8b}$$

Inserting (8) into (1) gives

$$H = \tfrac{1}{2}\hbar\omega(a^*a + aa^*). \tag{9}$$

Here we have distinguished the two ways of ordering a and a^*, although in a classical treatment $a^*a = aa^*$.

(b) The quantized oscillator

In quantum mechanics we assume that p and q are canonically conjugate operators that fulfill the commutation relations

$$\tilde{q}\tilde{p} - \tilde{p}\tilde{q} \equiv [\tilde{q},\tilde{p}] = i\hbar. \tag{10}$$

The quantum treatment of the oscillator problem is well known. We only quote the results: The Hamiltonian has the eigenvalues

$$E_n = (n + \tfrac{1}{2})\hbar\omega, \tag{11}$$

where $n = 0, 1, 2, \ldots$, and $\hbar$ now is the true constant of Planck. The corresponding stationary states are written as $|n\rangle$.

The operators $\bar{p}$ and $\bar{q}$ possess matrix elements differing from zero only between adjacent states:

$$\langle n |\bar{q}| n - 1 \rangle = -\frac{i}{\mu\omega} \langle n |\bar{p}| n - 1 \rangle = \left(\frac{\hbar n}{2\mu\omega}\right)^{1/2},$$

$$\langle n - 1 |\bar{q}| n \rangle = \frac{i}{\mu\omega} \langle n - 1 |\bar{p}| n \rangle = \left(\frac{\hbar n}{2\mu\omega}\right)^{1/2}. \tag{12}$$

Again we introduce operators in analogy to (4) and (8):

$$\bar{a} = \sqrt{\frac{\omega\mu}{2\hbar}}\, \bar{q} + i\sqrt{\frac{1}{2\hbar\omega\mu}}\, \bar{p}, \tag{13}$$

while $\bar{a}^\dagger$ is the Hermitian conjugate operator. (The operators $\bar{p}$ and $\bar{q}$ represent observables, and are Hermitian.) According to (12)

$$\langle n |\bar{a}| n - 1 \rangle = 0, \quad \langle n - 1 |\bar{a}| n \rangle = \sqrt{n + 1},$$

$$\langle n |\bar{a}^\dagger| n - 1 \rangle = n^{1/2}, \quad \langle n - 1 |\bar{a}^\dagger| n \rangle = 0. \tag{14}$$

Here we see the utility of the $\bar{a}$'s: The matrix elements of $\bar{a}$ differ from zero only for the transition $n + 1 \to n$, and those of $\bar{a}^\dagger$ only for $n \to n + 1$. We therefore call $\bar{a}^\dagger$ the "raising" operator and $\bar{a}$ the "lowering" operator. Furthermore, the commutation relation (10) is equivalent to

$$[\bar{a}, \bar{a}^\dagger] = 1. \tag{15}$$

Now the reason for the peculiar factors in (4) and (13) is clear: with this choice the commutator for $\bar{a}$ and $\bar{a}^\dagger$ is unity, and the matrix elements are square roots of the quantum number n.

The Hamiltonian takes the form

$$\bar{H} = \tfrac{1}{2}\hbar\omega(\bar{a}^\dagger\bar{a} + \bar{a}\bar{a}^\dagger) = \hbar\omega\bar{a}^\dagger\bar{a} + \tfrac{1}{2}\hbar\omega \tag{16}$$

because of (15). Here we have been careful about the order of $\bar{a}$ and $\bar{a}^\dagger$, because they do not commute. The operator $\bar{a}^\dagger\bar{a}$ is diagonal in the $|n\rangle$-representation,

$$\langle n'| \bar{a}^\dagger\bar{a} |n\rangle = n\delta_{nn'}, \tag{17}$$

as is easily seen, for example, from (14), or by comparison with (11).

In a stationary state $|n\rangle$ of the oscillator, the expectation value of the amplitude $\bar{q}$ is zero. It is possible, of course, to build wave packets $|t\rangle$, which are linear combinations of different $|n\rangle$, for which the expectation value of $\tilde{q}$ is nonzero. Then there will be a time-dependence of $\langle t|\,\tilde{q}\,|t\rangle$ corresponding to a periodic vibration with the frequency ω.

2. The scalar field

Nature provides us with particles with zero spin, whose motion can be described by a scalar field, in analogy to the description of the motion of photons by vector fields, and of electrons by spinor fields. The most familiar examples in particle physics are the pion and kaon. As we know, these are not structureless "elementary" particles like the photon and electron; they are bound states of a quark and an antiquark coupled to total angular momentum zero. Nevertheless, a system of pions and/or kaons can be treated as if it is composed of elementary particles provided all relative momenta are low enough so that excited states (such as ρ or K^*) are inaccessible. An analogous situation occurs in atomic physics, when a gas of helium atoms can be treated as a system of elementary particles provided the temperature is so low that thermal excitations of the atoms are unlikely. In either case, such a system of unexcitable spin-0 particles can be described by a scalar field[1] $\Phi(\mathbf{x}, t)$.

The Hamiltonian for a scalar field must lead to a wave equation whose plane wave solutions $\Phi = \exp[i(\mathbf{k}\cdot\mathbf{x} - \omega t)]$ satisfy the energy-momentum relation[2]

$$\omega_k = \sqrt{k^2 + \mu^2}, \tag{18}$$

where μ is the mass of the field quantum—the particles that are to be described by this theory. On squaring (18), and making the familiar correspondence $\omega \to i\,\partial/\partial t$, $\mathbf{k} \to -i\,\partial/\partial\mathbf{x}$, we arrive at the Klein–Gordon equation,

$$\left(\frac{\partial^2}{\partial t^2} - \nabla^2 + \mu^2\right)\Phi = 0. \tag{19}$$

One can derive this equation from Hamilton's principle, $\delta \int L\,dt = 0$. For a noninteracting field, the Lagrangian L must be an integral over all space of a Lorentz-invariant quadratic form in Φ and its derivatives:[3]

$$L = \int d^3x (|\dot{\Phi}|^2 - |\boldsymbol{\nabla}\Phi|^2 - \mu^2 |\Phi|^2).$$

---

[1] At this time elementary spin-0 particles only occur in the theory, not in the laboratory. These are the Higgs bosons that occur in the theory of the electroweak interaction: see §VI.C.2.

[2] Henceforth we set $c = \hbar = 1$.

[3] See Bjorken (1965), Chap. 12. Here we assume that Φ is complex, i.e., that Φ and Φ^* are independent variables.

The total energy H, and total momentum $\mathbf{P}$, are then found in the usual manner by carrying out an infinitesimal displacement in time and space, respectively, and exploiting the invariance of L. This leads to

$$H = \int d^3x [|\dot{\Phi}|^2 + |\nabla \Phi|^2 + \mu^2 |\Phi|^2], \tag{20}$$

$$\mathbf{P} = -\int d^3x [\dot{\Phi}^* \nabla \Phi + (\nabla \Phi^*)\Phi] \tag{21}$$

where H is the Hamiltonian of the scalar field.

We can also construct current and charge densities, given by

$$\mathbf{j} = \frac{1}{i}[\Phi^* \nabla \Phi - (\nabla \Phi^*)\Phi],$$

$$\rho = i\left[\Phi^* \frac{\partial \Phi}{\partial t} - \frac{\partial \Phi^*}{\partial t} \Phi\right], \tag{22}$$

which satisfy the continuity equation

$$\dot{\rho} + \nabla \cdot \mathbf{j} = 0 \tag{23}$$

in virtue of (19).

As we see from (22), $\mathbf{j} = \rho = 0$ if Φ is real. We designate such a real scalar field by ϕ, to distinguish it from the complex field Φ. A real field can describe spin 0 particles that carry no "internal" quantum number such as charge and strangeness, e.g., π^0 and η. Electrically neutral particles, such as K^0 and $\bar{K}^0$, which carry equal and opposite values of an attribute (in this case strangeness), are described by a complex field Φ. (By "describe" we must remember that objects with internal structure can only be described by fields in low momentum-transfer processes.)

(a) Real fields

We can expand the real field ϕ in plane waves. Our considerations become somewhat simpler if we enclose the whole system in a macroscopic cube of volume V. We assume periodic boundary conditions; i.e., the values of the fields should be the same at corresponding points at opposite sides of the cube. The wave numbers $\mathbf{k}$ allowed by these conditions are such that

$$\frac{1}{V} \int_V d^3x e^{i(\mathbf{k}-\mathbf{k}')\cdot \mathbf{x}} = \delta_{kk'}.$$

We do not need to know the exact values of $\mathbf{k}$; for $|\mathbf{k}| \gg V^{-1/3}$ they lie close to each other and practically form a continuum.

We now write the field ϕ in the form[4]

$$\phi = \sum_k \frac{1}{\sqrt{2\omega_k V}} [c_k e^{i(\mathbf{k}\cdot\mathbf{x}-\omega_k t)} + c_k^* e^{-i(\mathbf{k}\cdot\mathbf{x}-\omega_k t)}]. \tag{24}$$

This expression is the field analogue to Eq. (8a) for the oscillator. The corresponding expression for $\dot{\phi}$ would be the analogue of (8b). Since ϕ is real, the second term in (24) must be the complex conjugate of the first.

The Hamiltonian (and momentum) for the real field are obtained from that for the complex field by writing $\Phi = 2^{-1/2}(\phi + i\chi)$, and setting the imaginary part χ to zero. This cuts the number of degrees of freedom by half. The factor $2^{-1/2}$ is needed because the transformation from the variables (Φ, Φ^*) to (ϕ, χ) must be canonical. As a result one finds

$$H = \frac{1}{2} \int d^3x (\dot{\phi}^2 + \mu^2\phi^2 + |\nabla\phi|^2) \tag{25a}$$

$$\mathbf{P} = -\int d^3x \dot{\phi}\nabla\phi. \tag{25b}$$

The energy and momentum of the field can be written as a quadratic form in the c_k analogue of (9) for the oscillator. The expressions are obtained[5] by substituting (24) into (25):

$$H = \tfrac{1}{2}\sum_k \omega_k [c_k c_k^* + c_k^* c_k], \tag{26a}$$

$$\mathbf{P} = \tfrac{1}{2}\sum_k \mathbf{k}[c_k c_k^* + c_k^* c_k]. \tag{26b}$$

We keep the terms cc^* and c^*c separated because the order will be important after quantization. Comparing (26a) with (9) we see that the scalar field is equivalent to a system of independent oscillators with frequencies ω_k. The amplitude c_k is analogous to the a of the oscillator.

We are now ready for quantization. As in the oscillator, the conjugate variables c_k and c_k^* become operators $\bar{c}_k$ and $\bar{c}_k^\dagger$. We postulate that these operators satisfy the commutation relations

$$[\bar{c}_k, \bar{c}_{k'}^\dagger] = \delta_{kk'}. \tag{27}$$

---

[4] The factor $(2V\omega_k)^{-1/2}$ is introduced in order to get a simple form for the Hamiltonian. It is analogous to the factors introduced in the definition (4) of α.

[5] In performing the integration over space in (25a) we must be aware that the integrals over the three terms contain two types of nonzero terms. One comes from products of $\exp(i\mathbf{k}\cdot\mathbf{x})$ and $\exp(-i\mathbf{k}'\cdot\mathbf{x})$, and the other from products of $\exp(i\mathbf{k}\cdot\mathbf{x})$ and $\exp(i\mathbf{k}'\cdot\mathbf{x})$. The second type of products cancel since the integral over $|\dot{\phi}|^2$ gives a term proportional to $\omega^2(c_{-k}c_k + c_k^* c_{-k}^*)$, while the integral over $|\nabla\phi|^2 + \mu^2\phi^2$ gives the same term, but with ω^2 replaced by $-(k^2 + \mu^2)$. The first type of products have no such cancellation.

All other combinations, such as $[\tilde{c}_k, \tilde{c}_{k'}]$, vanish for *all* values of **k** and **k'**. This generalization of (15) follows if every plane wave mode is considered to be an independent oscillator. The field operator therefore is

$$\tilde{\phi} = \sum_k \frac{1}{\sqrt{2\omega_k V}} [\tilde{c}_k e^{i(\mathbf{k}\cdot\mathbf{x}-\omega_k t)} + \tilde{c}_k^\dagger e^{-i(\mathbf{k}\cdot\mathbf{x}-\omega_k t)}], \tag{28}$$

while the energy-momentum operator is

$$\tilde{H} = \tfrac{1}{2}\sum_k \omega_k[\tilde{c}_k \tilde{c}_k^\dagger + \tilde{c}_k^\dagger \tilde{c}_k], \quad \tilde{\mathbf{P}} = \tfrac{1}{2}\sum_i \mathbf{k}[\tilde{c}_k \tilde{c}_k^\dagger + \tilde{c}_k^\dagger \tilde{c}_k]. \tag{29}$$

Relation (27) allows us to write these as

$$\tilde{H} = \sum_k \omega_k \tilde{c}_k^\dagger \tilde{c}_k + E_0, \tag{30a}$$

$$\tilde{\mathbf{P}} = \sum_k \mathbf{k} \tilde{c}_k^\dagger \tilde{c}_k, \tag{30b}$$

where

$$E_0 = \tfrac{1}{2}\sum_k \omega_k. \tag{30c}$$

There is no such constant in $\tilde{\mathbf{P}}$ because $\sum \mathbf{k} = 0$ by symmetry, whereas the ω_k are all positive. The divergent sum E_0 stems from the zero-point energy $\tfrac{1}{2}\omega_k$ of each oscillator. As the zero of the energy scale is irrelevant, it would seem that we are free to drop E_0 from (30a). After Eq. (37) we shall provide a more compelling reason for doing this.

The stationary states of an oscillator are defined by the integer excitation number n. In the case of a field the stationary states are defined by a set of integers $\{n_k\}$ giving the "excitations" of each of the plane waves. The numbers n_k are appropriately called "occupation numbers," since they must be interpreted as the number of particles in the mode **k**. In states describing a finite number of particles, only a finite number of oscillators will be excited, while $n_k = 0$ for all others.

Let us now look in greater detail at the operators $\tilde{c}_k$ and $\tilde{c}_k^\dagger$. In full analogy with the oscillator, they are "lowering" and "raising" operators. In the case of a field $\tilde{\phi}$ it is more appropriate to call $\tilde{c}_k^\dagger$ the "creation" operator of a particle of momentum **k**, and $\tilde{c}_k$ the "destruction" operator of such a particle. We illustrate this in the following example. Let the vacuum state be $|\Omega\rangle$. It contains no particles whatsoever. Then the operation of $\tilde{c}_k^\dagger$ on $|\Omega\rangle$ should produce a state of one particle with a momentum **k**. We call such a state $|1_k\rangle$, and the state with n such particles is called $|n_k\rangle$.

Obviously, we have

$$\tilde{c}_k^\dagger |\Omega\rangle = |1_k\rangle, \qquad \tilde{c}_k |1_k\rangle = |\Omega\rangle,$$
$$\tilde{c}_k |\Omega\rangle = 0. \tag{31}$$

The last equation states that one cannot remove a particle from the vacuum. If there is more than one particle in a given state, we find the matrix elements from the corresponding oscillator matrix elements, Eq. (14):

$$\tilde{c}_k^\dagger |n_k\rangle = (n_k + 1)^{1/2} |n_k + 1\rangle,$$
$$\tilde{c}_k |n_k\rangle = n_k^{1/2} |n_k - 1\rangle. \tag{32}$$

These relations are characteristic of Bose statistics: The creation of a particle into an occupied state is more probable than into an empty one. The Bose character of these particles is even more evident from a 2-particle state, which is constructed as follows:

$$\tilde{c}_k^\dagger \tilde{c}_{k'}^\dagger |\Omega\rangle = |1_k 1_{k'}\rangle. \tag{33}$$

Since $\tilde{c}_k^\dagger$ and $\tilde{c}_{k'}^\dagger$ commute, this equation tells us that the 2-particle state is symmetric:

$$|1_k 1_{k'}\rangle = |1_{k'} 1_k\rangle. \tag{34}$$

This is readily generalized: all multiparticle states are symmetric under the exchange of any particle pair. Thus spin-0 particles are bosons if they are described by field oscillators.

An arbitrary state of the scalar field is specified by the set $\{n_k\}$ of integers n_k for each and every mode. In the state (33), for example, all but two of the $\{n_k\}$ are zero, and are not indicated explicitly. We write this arbitrary state as $|\{n_k\}\rangle$. From (32) we see that

$$\tilde{c}_q^\dagger \tilde{c}_q |\{n_k\}\rangle = n_q |\{n_k\}\rangle. \tag{35}$$

According to (30), the eigenvalues E and $\mathbf{P}$ of the Hamiltonian and momentum operators in $|\{n_k\}\rangle$ are therefore

$$E = \sum_k \omega_k n_k, \tag{36}$$

$$\mathbf{P} = \sum_k \mathbf{k} n_k. \tag{37}$$

This is precisely what we would expect from the interpretation of the n_k as occupation numbers.

In writing the energy eigenvalue E in (36) we have dropped the constant E_0 of Eq. (30a). We had remarked that such an additive constant in the energy scale is arbitrary. That is true in nonrelativistic physics, but is no longer so in relativity. If we retain E_0, the vacuum state $|\Omega\rangle$ would be at rest ($\mathbf{P}|\Omega\rangle = 0$), and have mass E_0. Thus, the vacuum would not be a Lorentz-invariant object, and the whole theory would lose its Lorentz invariance. So we must set $E_0 = 0$.

One might be alarmed by this cavalier mutilation of the theory, but one must remember that the transition from the classical to the quantum theory is only constrained by two general requirements:[6] (1) the latter must reduce to the former when the quantum numbers are large; (2) the quantum theory must be internally consistent. In any one mode the zero-point energy $\frac{1}{2}\omega_k$ is negligible compared to the total energy $(\frac{1}{2} + n_k)\omega_k$ for large n_k, and therefore its existence cannot be inferred from the classical theory. Furthermore, in the classical theory, we could, in (26a), have written $c_k^* c_k$ instead of $\frac{1}{2}(c_k^* c_k + c_k c_k^*)$, and we would never have encountered the zero-point energy.[7]

In short, we can use (36) and (37), safe in the knowledge that they lead to the correct classical limit, and give a properly Lorentz-invariant quantum theory.

(b) Complex fields

As the expressions (22) for the current and charge density indicate, the field Φ must be complex in order to carry charge.[8] Hence we rewrite (24) so that the second term is no longer the complex conjugate of the first:

$$\Phi = \sum_k \frac{1}{\sqrt{2\omega_k V}}[c_k e^{i(\mathbf{k}\cdot\mathbf{x}-\omega_k t)} + f_k^* e^{-i(\mathbf{k}\cdot\mathbf{x}-\omega_k t)}]. \tag{38}$$

Here $f_k^* \neq c_k^*$. Inserting this into (20) gives

$$H = \tfrac{1}{2}\sum_k \omega_k[c_k c_k^* + f_k^* f_k] \tag{39}$$

To carry out the quantization, we turn Φ into an operator by replacing c_k by the operator $\tilde{c}_k$, and f_k^* by $\tilde{f}_k^\dagger$. As our objective is a theory of charged particles, we must assume that the coefficient $\tilde{c}_k$ in (38) destroys a particle of momentum $\mathbf{k}$, mass μ, and charge e. Thus neither $\tilde{\Phi}$ nor $\tilde{\Phi}^\dagger$ are $\Delta Q = 0$

---

[6] Aside from the confrontation with experiment.

[7] Although we can drop the zero-point energy, we cannot drop the zero-point motion. In the ground state of the simple oscillator $\bar{p}$ and $\bar{q}$ fluctuate, no matter what energy scale one uses. In the same way, $\tilde{\phi}(\mathbf{x}, t)$ fluctuates in the vacuum state $|\Omega\rangle$.

[8] By "charge" we here mean any Abelian attribute, in the language of §IV.A.2; i.e., electrical charge, strangeness, charm, etc. Hence Φ can describe the spin 0 particle pairs $\pi^\pm$, $K^\pm$, D^0 and $\bar{D}^0$, etc. Non-Abelian attributes, such as color or weak isospin, cannot be described by a single complex field.

operators, in the language of §II.A.5, since the total charge Q is changed when $\tilde{\Phi}$ or $\tilde{\Phi}^\dagger$ operates on a state. On the other hand, we recall that $\tilde{\mathbf{j}}$ and $\tilde{\rho}$ must be $\Delta Q = 0$ operators. We must, therefore, arrange the charge-changing operators $\tilde{\Phi}$ and $\tilde{\Phi}^\dagger$ so that there is no overall charge-change produced by expressions like $\tilde{\Phi}^\dagger \boldsymbol{\nabla} \tilde{\Phi}$ or $\tilde{\Phi}^\dagger \tilde{\Phi}$, which occur in $\tilde{\mathbf{j}}$ and $\tilde{\rho}$. This clearly requires that any amount of charge created by $\tilde{\Phi}^\dagger$ must be destroyed by $\tilde{\Phi}$. Since the terms with $\tilde{c}_k^\dagger$ in $\tilde{\Phi}^\dagger$ increase Q by e, while those with $\tilde{c}_k$ in $\tilde{\Phi}$ decrease Q by e, $\tilde{\Phi}$ and $\tilde{\Phi}^\dagger$ will have the desired property if the remaining terms in $\tilde{\Phi}$ and $\tilde{\Phi}^\dagger$, which involve the operators $\tilde{f}_k^\dagger$ and $\tilde{f}_k$, have precisely the same charge-changing properties. That is,

$$\tilde{f}_k^\dagger \text{ destroys charge } e,$$
$$\tilde{f}_k \text{ creates charge } e. \tag{40}$$

But we also know from the real field ϕ that the coefficient of the wave $\exp[-i(\mathbf{k}\cdot\mathbf{r} - \omega t)]$ in a field operator is an operator that creates an object of momentum $\mathbf{k}$ and energy ω. Thus

$$\tilde{f}_k^\dagger \text{ creates a particle of momentum } \mathbf{k},$$
$$\tilde{f}_k \text{ destroys a particle of momentum } \mathbf{k}. \tag{41}$$

Upon comparing (40) and (41) we realize that the operators $\tilde{f}_k^\dagger$ create a new species of particles—particles that also have mass μ, but a charge *opposite* to that carried by the particles created by $\tilde{c}_k^\dagger$. These are the antiparticles—their existence is an inevitable consequence of the basic principles of relativistic quantum mechanics.

We must still specify the commutation rules for $\tilde{f}_k$ and $\tilde{f}_k^\dagger$. As the antiparticles are distinct from the particles created by $\tilde{c}_k^\dagger$, operators referring to the two types of particles commute:

$$[\tilde{c}_k, \tilde{f}_k^\dagger] = 0. \tag{42}$$

As for the operators referring to one species alone, we adopt the commutation rules (27):

$$[\tilde{f}_k, \tilde{f}_{k'}^\dagger] = \delta_{kk'},$$
$$[\tilde{f}_k, \tilde{f}_{k'}] = [\tilde{f}_k^\dagger, \tilde{f}_{k'}^\dagger] = 0. \tag{43}$$

The Hamiltonian (20) now becomes

$$H = \sum_k \omega_k (\tilde{c}_k \tilde{c}_k^\dagger + \tilde{f}_k^\dagger \tilde{f}_k)$$

which, because of (43), can be written as

$$\tilde{H} = \sum_k \omega_k (\tilde{c}_k^\dagger \tilde{c}_k + \tilde{f}_k^\dagger \tilde{f}_k), \qquad (44)$$

where the zero-point energy is dropped for the same reason as in the case of neutral particles.

Let us call n_k the number of particles of the "c" kind, and $\bar{n}_k$ the number of the "f" kind. In a state given by the occupation numbers $\{n_k\}$, $\{\bar{n}_k\}$, as in (35), the eigenvalue of the Hamiltonian will be

$$E = \sum_k \omega_k (n_k + \bar{n}_k). \qquad (45)$$

In the discussion following (40), we already saw that $\tilde{c}_k^\dagger$ and $\tilde{f}_k^\dagger$ must create objects of opposite charge. We wish to verify this more explicitly by constructing the total charge operator $\tilde{Q}$, which is obtained by integrating the total charge density $\tilde{\rho}$ over all space. (We always express charges in units of e; the true charge density is $e\tilde{\rho}$, etc.) Clearly $\tilde{\rho}$ (and therefore $\tilde{Q}$) must change sign if particles and antiparticles are interchanged. This operation is called charge conjugation, and is carried out by the substitutions $\tilde{c}_k \leftrightarrow \tilde{f}_k$, $\tilde{c}_k^\dagger \leftrightarrow \tilde{f}_k^\dagger$. Under this substitution, $\tilde{\Phi} \leftrightarrow \tilde{\Phi}^\dagger$, as one sees from (38). Recall the expression (22) for $\tilde{\rho}$: $i[\tilde{\Phi}^\dagger \dot{\tilde{\Phi}} - \dot{\tilde{\Phi}}^\dagger \tilde{\Phi}]$. It is not odd under $\tilde{\Phi} \leftrightarrow \tilde{\Phi}^\dagger$, because $\tilde{\Phi}$ and $\dot{\tilde{\Phi}}^\dagger$ do not commute. This is rectified by changing the definition to read

$$\tilde{\rho} = \tfrac{1}{2} i [(\tilde{\Phi}^\dagger \dot{\tilde{\Phi}} - \tilde{\Phi} \dot{\tilde{\Phi}}^\dagger) - (\dot{\tilde{\Phi}}^\dagger \tilde{\Phi} - \dot{\tilde{\Phi}} \tilde{\Phi}^\dagger)], \qquad (46)$$

with a corresponding modification for the current density $\tilde{\mathbf{j}}$. These new expressions for $\tilde{\rho}$ and $\tilde{\mathbf{j}}$ also satisfy the continuity equation, and when all quantities commute, (46) reduces to the "classical" expressions (22).

To evaluate $\tilde{Q}$, we integrate (46) over space and readily find

$$\tilde{Q} = \sum_k (\tilde{c}_k^\dagger \tilde{c}_k - \tilde{f}_k^\dagger \tilde{f}_k). \qquad (47)$$

In a state with definite occupation numbers this has the eigenvalue

$$Q = \sum_k (n_k - \bar{n}_k),$$

which makes it clear that the "c" and "f" particles have opposite charge.

If the interaction between these particles and the electromagnetic field is introduced, we obtain the processes of pair creation and annihilation. This will not be discussed for the case of charged scalar particles, since it was treated in detail for spinor particles in Chap. II. We can, in a very cursory way, see how these phenomena appear. The interaction with the electromagnetic field is given by the integral $\int \tilde{\mathbf{j}} \cdot \tilde{\mathbf{A}} \, d^3x$, where $\tilde{\mathbf{A}}$ is the vector potential. Equation (22) shows that $\tilde{\mathbf{j}}$ is bilinear in $\tilde{\Phi}$ and $\tilde{\Phi}^\dagger$. When (38) is inserted into $\tilde{\mathbf{j}}$, one obtains terms of the type $\tilde{c}_k^\dagger \tilde{c}_{k'}$, $\tilde{f}_k^\dagger \tilde{f}_{k'}$, $\tilde{c}_k \tilde{f}_{k'}$, $\tilde{c}_k^\dagger \tilde{f}_{k'}^\dagger$. The first two give rise to changes of state (i.e., to scattering), the third to pair annihilation, and the fourth to pair creation.

3. The electromagnetic field

The quantization of the electromagnetic field uses the same concepts as those used in the case of the neutral scalar field. Indeed, when no charges or currents are present, Maxwell's classical theory is equivalent to a system of uncoupled harmonic oscillators. The electric field $\mathbf{E}(\mathbf{x}, t)$ and the magnetic field $\mathbf{B}(\mathbf{x}, t)$ are just linear combinations of oscillator amplitudes.

The Hamiltonian (total energy) of the free electromagnetic field is

$$H = \frac{1}{2} \int (\mathbf{E}^2 + \mathbf{B}^2) \, d^3x. \tag{48}$$

The field strengths are conveniently expressed by introducing an auxiliary field, the vector potential $\mathbf{A}$:

$$\mathbf{E} = -\frac{\partial \mathbf{A}}{\partial t}, \tag{49}$$

$$\mathbf{B} = \nabla \times \mathbf{A}.$$

Then

$$H = \frac{1}{2} \int [\dot{\mathbf{A}}^2 + (\nabla \times \mathbf{A})^2] \, d^3x, \tag{50}$$

which is a quadratic form in the derivatives of $\mathbf{A}$. It is very similar to the Hamiltonian (20) of the scalar field, but there is no mass term in (50).

In the absence of sources, the fields are superpositions of plane waves $\exp[\pm i(\mathbf{k} \cdot \mathbf{x} - \omega t)]$, where $\omega = |\mathbf{k}|$. There are two independent modes for each $\mathbf{k}$, which we label by the helicity, $h = \pm 1$.

A plane circularly polarized electromagnetic wave has the vector potential[9]

$$\mathbf{A} = a_h \boldsymbol{\varepsilon}_h e^{i(\mathbf{k} \cdot \mathbf{x} - \omega t)} + \text{c.c.}, \tag{51}$$

where a_h is a complex amplitude and $\boldsymbol{\varepsilon}_h$ is a circular polarization vector; i.e., if $\hat{\boldsymbol{\varepsilon}}_\alpha$, $\hat{\boldsymbol{\varepsilon}}_\beta$, and the direction of propagation $\hat{\mathbf{k}}$ are a right-handed set three real

---

[9] c.c. (h.c.) means the complex (Hermitian) conjugate of the preceding expression.

mutually orthogonal unit vectors[10]

$$\left.\begin{array}{l} \boldsymbol{\varepsilon}_{\pm 1} = \mp \dfrac{1}{\sqrt{2}}(\hat{\boldsymbol{\varepsilon}}_\alpha \pm i\hat{\boldsymbol{\varepsilon}}_\beta), \\ \boldsymbol{\varepsilon}_h^* \cdot \boldsymbol{\varepsilon}_{h'} = \delta_{hh'}. \end{array}\right\} \quad (52)$$

The latter expresses the linear independence of the two senses of circular polarization, or equivalently, of the two helicity states.

The most general electromagnetic field is a superposition of all possible plane waves (51), each with an arbitrary amplitude. We again enclose the whole system in a volume V and impose periodic boundary conditions. The general field then has the form

$$\mathbf{A}(\mathbf{x}, t) = \sum_{kh} (2\omega V)^{-1/2} [a_h(k)\boldsymbol{\varepsilon}_h(k) e^{i(\mathbf{k}\cdot\mathbf{x} - \omega t)} + \text{c.c.}]. \quad (53)$$

Here $\boldsymbol{\varepsilon}_h(k)$ and $\omega = |\mathbf{k}|$ are the polarization vectors and frequencies belonging to the wave characterized by h and $\mathbf{k}$. We calculate the corresponding expressions for $\mathbf{E}$ and $\mathbf{B}$ from (49), and insert them into (48). Using the orthogonality relations (52), we find the simple result

$$H = \sum_{kh} \omega a_h^*(k) a_h(k). \quad (54)$$

In (54) we have replaced $\frac{1}{2}(aa^* + a^*a)$ by a^*a; as we learned from the example of the scalar field, the difference between these expressions is the zero-point energy, which we discard for the reason already given.

A similar calculation can be performed for the total momentum $\mathbf{P}$ of the electromagnetic field, $\mathbf{P} = \int d^3x (\mathbf{E} \times \mathbf{B})$; it leads to

$$\mathbf{P} = \sum_{kh} \mathbf{k} a_h^*(k) a_h(k), \quad (55)$$

which says that each wave has a momentum equal to $\mathbf{k}$.

We can now quantize the electromagnetic field. The complex-conjugate variables $a_h(k)$ and $a_h^*(k)$ become operators $\tilde{a}_h(k)$ and $\tilde{a}_h^\dagger(k)$ that are postulated to fulfill the commutation relations

$$[\tilde{a}_h(k), \tilde{a}_h^\dagger(k')] = \delta_{hh'} \delta_{kk'}. \quad (56)$$

All other combinations commute for all values of the indices. This is an obvious generalization of (27). The operators for the energy and momentum are obtained from (54) and (55) by the replacements $a \to \tilde{a}$, $a^* \to \tilde{a}^\dagger$.

---

[10] The overall sign $\mp$ (52) is a convention chosen to agree with the standard phase convention for spherical harmonics for $l = 1$.

As in the example of the scalar field, the electromagnetic field can be viewed as an ensemble of quantized oscillators. Indeed, the expressions for the energy and momentum of the electromagnetic field are identical to those for the neutral scalar field, if one substitutes $\bar{c}_k \to \bar{a}_h(k)$, etc., and sums over helicities as well as momenta. The states of the electromagnetic field are likewise specified by a set $\{n_h(k)\}$ of occupation numbers, where $n_h(k)$ is the eigenvalue of $\bar{a}_h^\dagger(k)\bar{a}_h(k)$. Thus,

$$E = \sum_{kh} \omega_k n_h(k), \qquad \mathbf{P} = \sum_{kh} \mathbf{k} n_h(k). \tag{57}$$

The one-photon state $|\mathbf{k}h\rangle$ was already given in Eq. II.A(39), and the n-photon state is

$$\bar{a}_{h_1}^\dagger(k_1)\bar{a}_{h_2}^\dagger(k_2)\cdots\bar{a}_{h_n}^\dagger(k_n)|\Omega\rangle. \tag{58}$$

Since the $\bar{a}^\dagger$ commute among themselves, it follows that a multiphoton state is symmetric under the interchange of any pair, and therefore photons obey Bose statistics.

The field operator $\mathbf{A}$ is given by (53) except that a and a^* are replaced by the operators $\bar{a}$ and $\bar{a}^\dagger$. We then get the expression [II.A(40)] introduced in Chap. II as the operator $\tilde{\mathbf{A}}$. The operators of $\tilde{\mathbf{E}}$ and $\tilde{\mathbf{B}}$ follow from $\tilde{\mathbf{A}}$ via (49).

The field strengths are linear combinations of creation and destruction operators. Acting with $\tilde{\mathbf{E}}$ or $\tilde{\mathbf{B}}$ on a state with a definite number n of photons produces a superposition of states with $n+1$ and $n-1$ photons. Thus $\tilde{\mathbf{E}}$ and $\tilde{\mathbf{B}}$ have vanishing expectation values in all states with a definite number of photons—i.e., $\langle\tilde{\mathbf{E}}\rangle$ and $\langle\tilde{\mathbf{B}}\rangle$ vanish in all stationary states of the free electromagnetic field. From this the converse follows: A state in which $\tilde{\mathbf{E}}$ (and/or $\tilde{\mathbf{B}}$) has a nonzero expectation value is a coherent superposition of states with various numbers of photons. It is not stationary. In the limit of large photon numbers it passes smoothly to the classical limit of a free electromagnetic wave.[11]

What we have just seen is a most striking illustration of Bohr's principle of complementarity: The quantum theory contains within it elements of two complementary classical conceptions of electromagnetism, the corpuscular and the wave theories. The escape from the classical paradox is, as always, provided by the uncertainty principle, which, in this case, states that field strengths and photon occupation number cannot be measured simultaneously to arbitrary precision.

---

[11] Note the analogy to the simple oscillator: In any stationary state $|n\rangle$, both $\bar{p}$ and $\bar{q}$ have vanishing expectation values. States where $\langle\bar{q}\rangle$ and $\langle\bar{p}\rangle$ oscillate harmonically are superpositions of energy eigenstates. It is these superpositions, or wave packets, that pass smoothly to the classical limit.

Appendix III

THE DIRAC FIELD

In this Appendix we consider the significance of the Dirac equation as an expression of the covariance of a spinor under Lorentz transformations; the modification of the commutation rules required by Fermi–Dirac statistics; the construction of the Dirac field operator; and the evaluation of certain matrix elements that arise in weak and electromagnetic processes.

1. The Lorentz transformation of spinors and the Dirac equation

(a) Three-vectors as 2×2 matrices

In §I.B.3(d) we discussed the transformations of an arbitrary spinor in the two-dimensional complex vector space $\mathscr{C}_2$, and explained why the transformation of a spinor can be parametrized by the three angles that specify a rotation in a Euclidean 3-space, $\mathscr{E}_3$. There is another formulation of the $\mathscr{C}_2$-$\mathscr{E}_2$ relationship that is very illuminating in its own right, and that has the further advantage of providing a natural framework for an analysis of the behavior of spinors under Lorentz transformations.

We are accustomed to designating a point P in $\mathscr{E}_3$ by a position vector $\mathbf{r} = (x, y, z)$. But P can be specified equally well by the matrix

$$R = \begin{pmatrix} z & x - iy \\ x + iy & -z \end{pmatrix} = \mathbf{r} \cdot \boldsymbol{\sigma}. \tag{1}$$

Note that R is Hermitian and traceless, and that $\det R = -r^2$. Under a rotation P is turned into P', where the latter point is specified by a matrix R' having the form (1), which is again Hermitian and traceless, and with the same determinant. From a theorem of algebra we know that if U is a unitary matrix, the matrix

$$R' = URU^\dagger \tag{2}$$

has the same determinant and trace as R, and is still Hermitian. Hence, there must exist a U that carries out the rotation. Indeed, U is precisely the matrix [I.B.(40)] that transforms a two-component spinor χ when a 3-vector $\mathbf{r}$ is rotated into $\mathbf{r}'$: $\chi \to U\chi$.

This establishes the relation between the rotation of vectors in $\mathscr{E}_3$, and the unitary transformation of spinors in $\mathscr{C}_2$.

(b) Four-vectors as 2×2 matrices

We can generalize the foregoing to 4-vectors and Lorentz transformations. Insofar as rotations are concerned, the essential property of (1) was that its determinant, $-r^2$, is the quadratic form that is left invariant. We therefore ask: Given a 4-vector (t, r), can one construct a 2×2 matrix whose determinant is the Lorentz-invariant quadratic form $t^2 - r^2$? It is not difficult to see that this can be done. In constructing R, we only used the three Pauli matrices, so that R contained the three parameters $\mathbf{r}$. But there is a fourth independent 2×2 matrix—the unit matrix, and it can be used to incorporate a fourth parameter, t, into our 2×2 matrix. There are actually two generalizations of (1) that allow us to construct two distinct 2×2 matrices given one 4-vector $(t, \mathbf{r})$; they are[1]

$$X_\pm = t \mp \mathbf{r} \cdot \mathbf{\sigma} = \begin{pmatrix} t \mp z & \mp x \pm iy \\ \mp x \mp iy & t \pm z \end{pmatrix}. \tag{3}$$

Observe that

$$\det X_\pm = t^2 - (x^2 + y^2 + z^2) \tag{4}$$

is the desired Lorentz-invariant quadratic form. Note, furthermore, that

$$\operatorname{tr} X_\pm = 2t, \tag{5}$$

and that $X_\pm$ are Hermitian if $(t, \mathbf{r})$ is a real 4-vector. Therefore a spatial rotation leaves both $\det X_\pm$ and $\operatorname{tr} X_\pm$ invariant, while a Lorentz transformation (abbreviated by LT henceforth) leaves only $\det X_\pm$ invariant.[2]

One might wonder whether one needs both X_+ and X_-. We will examine this point in detail shortly. For the moment, we point out that X_+ and X_- are interchanged under a spatial reflection, $\mathbf{r} \to -\mathbf{r}$. Therefore, one needs both matrices if one intends to incorporate reflections into the theory.

Under the LT that takes $(t, \mathbf{r})$ into $(t', \mathbf{r}')$,

$$X_\pm \to X'_\pm = t' \mp \mathbf{r}' \cdot \mathbf{\sigma}, \tag{6}$$

where $X'_\pm$ is again Hermitian, and has the same determinant as $X_\pm$. We now ask whether there are 2×2 matrices $\Omega_\pm$ that can transform $X_\pm$ into $X'_\pm$ in a manner similar to (2). To this end we note that if $X_\pm$ and $X'_\pm$ are related by

$$X'_\pm = \Omega_\pm X_\pm \Omega_\pm^\dagger, \tag{7}$$

---

[1] Note that the following discussion applies to any 4-vector. That is, if $V = (V_0, \mathbf{V})$, one can define the matrices $V_\pm = V_0 \mp \mathbf{V} \cdot \mathbf{\sigma}$, and proceed in the same manner.

[2] By definition, an LT is a linear transformation on $(t, \mathbf{r})$ that leaves (4) invariant, and a rotation is therefore a special case of the general LT. We shall, however, reserve the nomenclature "LT" for transformations involving two frames in relative motion, i.e., for $t \neq t'$.

$X'_\pm$ is Hermitian if $X_\pm$ is. Furthermore,

$$\det X'_\pm = |\det \Omega_\pm|^2 \cdot \det X_\pm, \tag{8}$$

$$\operatorname{tr} X'_\pm = \operatorname{tr} \Omega_\pm^\dagger \Omega_\pm X_\pm. \tag{9}$$

Hence an LT requires $|\det \Omega_\pm| = 1$, while a rotation, which leaves the trace invariant, imposes the stronger condition that $\Omega_\pm$ be unitary, as we already know from our discussion of Eq. (2).

The most general 2×2 matrix $\Omega_\pm$ having $|\det \Omega_\pm| = 1$ can be written as

$$\Omega_\pm = e^{i\phi_\pm}(Q^0_\pm + i\mathbf{Q}_\pm \cdot \boldsymbol{\sigma}), \tag{10}$$

where

$$Q^0_\pm = +(1 - \mathbf{Q}_\pm \cdot \mathbf{Q}_\pm)^{1/2}, \tag{11}$$

where $\mathbf{Q}_\pm$ are arbitrary *complex* 3-vectors, and $\phi_\pm$ is real. Since the phase factor drops out of (7), we set $\phi_\pm = 0$. Hence, six real parameters are required to specify $\Omega_\pm$. This is as it should be, because a general LT is defined by the relative velocity $\mathbf{v}$ of the two frames (three parameters), and their relative orientation (three angles). If the vector $\mathbf{Q}$ is real, Ω is unitary. A real $\mathbf{Q}$, then, gives us Ω's corresponding to spatial rotations. As we wish to focus on those LTs that correspond to moving frames of the same orientation, we confine ourselves to purely imaginary $\mathbf{Q}$'s.

Let us first determine Ω_+. It will suffice to consider the case where $x = y = 0$, and the LT is along the z-axis. As we shall now show, in this case $\mathbf{Q}_+$ points along the z-axis, and is most conveniently expressed as $\mathbf{Q}_+ = -i\hat{\mathbf{z}} \sinh \tfrac{1}{2}\zeta$, so that $Q^0_+ = \cosh \tfrac{1}{2}\zeta$, and

$$\Omega_+ = \Omega_+^\dagger = \cosh \tfrac{1}{2}\zeta + \sigma_z \sinh \tfrac{1}{2}\zeta = e^{\tfrac{1}{2}\zeta \sigma_z}. \tag{12}$$

(One should compare this with the unitary matrix [I.B(38)] that describes a rotation about the z-axis.) After inserting (12) and (3) into (7), we obtain

$$\begin{aligned} X'_+ &= e^{\tfrac{1}{2}\zeta \sigma_z}(t - z\sigma_z)e^{\tfrac{1}{2}\zeta \sigma_z} = e^{\zeta \sigma_z}(t - z\sigma_z) \\ &= (t \cosh \zeta - z \sinh \zeta) - \sigma_z(z \cosh \zeta - t \sinh \zeta). \end{aligned}$$

According to (6), the transformed space-time coordinates are the two expressions in parentheses. When these are compared with the familiar formulas for a Lorentz transformation, we obtain the following relationships between ζ and v:

$$\cosh \zeta = (1 - v^2)^{-1/2}, \qquad \tanh \zeta = v. \tag{13}$$

The parameter ζ is called the *rapidity*. It provides the most convenient characterization of Lorentz transformations because, as we see from (12), two successive LTs along one axis, with rapidities ζ_1 and ζ_2, are equivalent to one LT with rapidity $\zeta_1 + \zeta_2$. The same considerations lead to Ω_-, which differs from Ω_+ by $\mathbf{Q}_- = \mathbf{Q}_+^*$; e.g., if the LT is along $\hat{\mathbf{z}}$, $\Omega_- = \exp(-\tfrac{1}{2}\zeta\sigma_z)$.

For an LT with a velocity $\mathbf{v}$ along an arbitrary direction $\hat{\mathbf{n}}$, the matrices $\Omega_\pm$ are the obvious generalization of (12), viz.,

$$\Omega_\pm = \cosh \tfrac{1}{2}\zeta \pm \hat{\mathbf{n}} \cdot \boldsymbol{\sigma} \sinh \tfrac{1}{2}\zeta = e^{\pm \tfrac{1}{2}\zeta \hat{\mathbf{n}} \cdot \boldsymbol{\sigma}}. \tag{14}$$

Thus far we have only concerned ourselves with *proper* LTs, i.e., those that involve no reflections. As we had already remarked above Eq. (6), under a reflection through the origin

$$X_+ \leftrightarrow X_-. \tag{15}$$

Can this be accomplished by a transformation like (7), i.e., by $\mathcal{R} X_+ \mathcal{R}^{-1} = X_-$, where $\mathcal{R}$ is a 2×2 matrix? As we see from (3), this would require $\mathcal{R}\boldsymbol{\sigma} = -\boldsymbol{\sigma}\mathcal{R}$. But there is no 2×2 matrix that anticommutes with all three Pauli matrices, so there is no matrix that can produce the spatial reflection (15).

(c) Spinors

The existence of 2×2 matrices U that can describe rotations of real Euclidean 3-vectors allows one to introduce spinors—that is, complex 2-vectors—that transform linearly under U. Since we have learned that U can be generalized to matrices that describe the Lorentz transformations of real 4-vectors, we now know how such 2-component spinors transform under LTs.

As there are two distinct matrices, Ω_+ and Ω_-, for each proper LT, there are two distinct types of 2-component spinors, $\chi^{(\pm)}$, insofar as LTs are concerned. Under a proper LT in which the space-time point $(\mathbf{r}, t)$ goes into $(\mathbf{r}', t')$, they transform as follows:

$$\begin{aligned}\chi^{(+)}(\mathbf{r}, t) &\to \Omega_+ \chi^{(+)}(\mathbf{r}', t'), \\ \chi^{(-)}(\mathbf{r}, t) &\to \Omega_- \chi^{(-)}(\mathbf{r}', t').\end{aligned} \tag{16}$$

For an LT with velocity $\mathbf{v} = v\hat{\mathbf{n}}$, Eq. (14) tells us that

$$\chi^{(\pm)} \to e^{\pm \tfrac{1}{2}\zeta \hat{\mathbf{n}} \cdot \boldsymbol{\sigma}} \chi^{(\pm)}. \tag{17}$$

In the case of a rotation, $\Omega_+ = \Omega_-$, so $\chi^{(+)}$ and $\chi^{(-)}$ transform in the same way. This must be so, since we found only one type of spinor when we considered pure rotations.[3]

---

[3] Put another way, once one restricts oneself to rotations, one can choose the origin of time so that $t = 0$. Then both X_+ and X_- reduce to the matrix R, apart from a trivial overall sign.

THE DIRAC FIELD

While the proper LTs of, say, $\chi^{(+)}$, do not produce a linear combination of $\chi^{(+)}$ and $\chi^{(-)}$, the same cannot be true of a reflection because of the interchange $X_+ \leftrightarrow X_-$ under reflection. To see this explicitly, we note that if we first carry out an LT with velocity $-\mathbf{v}$. In the case of $\chi^{(+)}$, for example, what we would need is

$$\chi^{(+)} \xrightarrow[\text{LT}]{} e^{\frac{1}{2}\zeta \hat{\mathbf{n}} \cdot \boldsymbol{\sigma}} \chi^{(+)} \xrightarrow[\text{reflect}]{} e^{-\frac{1}{2}\zeta \hat{\mathbf{n}} \cdot \boldsymbol{\sigma}} \chi^{(+)}. \tag{18}$$

But, as we already saw, there is no 2×2 matrix $\mathcal{R}$ for the second step, because there is no $\mathcal{R}$ that anticommutes with $\boldsymbol{\sigma}$. On the other hand, the last expression in (18) is just the transformation law for $\chi^{(-)}$. In short, just as X_+ and X_-, as defined by (3), are interchanged by a reflection, so are $\chi^{(+)}$ and $\chi^{(-)}$. Hence, if we are to include reflections, as well as proper LTs, in our symmetry transformations, we must introduce 4-component spinors, as in Eq. II.A(25):

$$\psi = \begin{pmatrix} \chi^{(+)} \\ \chi^{(-)} \end{pmatrix}. \tag{19}$$

Under a proper LT, $\psi \to \Omega \psi$, where the 4×4 matrix Ω is

$$\Omega = \begin{pmatrix} \Omega_+ & 0 \\ 0 & \Omega_- \end{pmatrix} \tag{20}$$

in terms of 2×2 elements. Under a reflection, $\chi^{(+)} \leftrightarrow \chi^{(-)}$, and therefore the reflection operation on ψ is[4]

$$\psi \to \gamma_0 \psi, \qquad \gamma_0 \equiv \begin{pmatrix} 0 & 1 \\ 1 & 0 \end{pmatrix}. \tag{21}$$

(d) The Dirac equation

We now show that the Dirac equation for a free particle follows directly from (17). Let ξ_h be the familiar Pauli spinor of a particle of mass m in its rest frame, $\mathcal{F}_0$, with spin projection $h = \pm\frac{1}{2}$ along the z-axis of $\mathcal{F}_0$. In a frame having the velocity $\hat{\mathbf{n}} \tanh \zeta$ with respect to $\mathcal{F}_0$, this particle will have the 4-momentum

$$E = m \cosh \zeta, \qquad \mathbf{p} = m\hat{\mathbf{n}} \sinh \zeta. \tag{22}$$

We want to exploit our newly won knowledge of the LT to "boost" the rest frame spinors ξ_h to momentum $\mathbf{p}$. Some care must be exercised if one wishes to construct spinors for moving particles that have a well-defined

---

[4] The statement, made after Eq. II.A(35), that γ_μ transforms like a 4-vector means that $\Omega \gamma_\mu \Omega^{-1} = a_{\mu\nu} \gamma_\nu$, where the coefficients $a_{\mu\nu}$ constitute the familiar 4×4 matrix for the LT of a 4-vector.

helicity, because an LT only commutes with the rotation about the axis defined by **v**. If we apply (17) straightforwardly to go from rest to momentum **p**, we therefore will construct a state that has no well-defined helicity unless **p** happens to be parallel to the z-axis. We can overcome this difficulty by first boosting ξ_h along z with a velocity v so that we attain the desired magnitude of **p**. This produces a state with momentum $p\hat{z}$, and with angular momentum h along $\hat{z}$, i.e., with helicity h. But h is a pseudoscalar, so it is invariant under a rotation, and we can therefore obtain the desired helicity state by rotating $p\hat{z}$ into $p\hat{n}$. In view of (17), there are two distinct spinors for accomplishing this:

$$\eta_h^{(\pm)}(\mathbf{p}) = U(\hat{n})e^{\pm\frac{1}{2}\zeta\sigma_z}\xi_h, \tag{23}$$

where $U(\hat{n})$ is the unitary matrix that rotates the z-axis into $\hat{n}$. Next we solve (23) for ξ_h,

$$\xi_h = e^{\mp\frac{1}{2}\zeta\sigma_z}U^\dagger(\hat{n})\eta_h^{(\pm)}(\mathbf{p})$$

and eliminate ξ_h by inverting (23), which gives us

$$\eta_h^{(\pm)}(\mathbf{p}) = U(\hat{n})e^{\pm\zeta\sigma_z}U^\dagger(\hat{n})\eta_h^{(\mp)}(\mathbf{p}) = e^{\pm\zeta\hat{n}\cdot\boldsymbol{\sigma}}\eta_h^{(\mp)}(\mathbf{p}),$$

since U represents the rotation of $\hat{z}$ into $\hat{n}$. Then we express $\exp(\pm\zeta\hat{n}\cdot\boldsymbol{\sigma})$ in terms of hyperbolic functions, and use (22) to obtain, for both values of h,

$$\begin{aligned} m\eta^{(+)}(\mathbf{p}) &= (E + \boldsymbol{\sigma}\cdot\mathbf{p})\eta^{(-)}(\mathbf{p}), \\ m\eta^{(-)}(\mathbf{p}) &= (E - \boldsymbol{\sigma}\cdot\mathbf{p})\eta^{(+)}(\mathbf{p}). \end{aligned} \tag{24}$$

These constitute the Dirac equation in momentum space.[5] They are nothing else than a description of how spinors transform when we go from the rest frame to a frame where the state has the momentum **p**. Any linear combination of plane waves $\exp[i(\mathbf{p}\cdot\mathbf{r} - Et)]\eta^{(\pm)}(\mathbf{p})$ satisfies the differential version of Dirac's equation, Eqs. II.A(22) and (23). The derivation given here makes it clear that the equation is covariant under Lorentz transformations. Furthermore, when $m = 0$, we see that the helicity is invariant under proper Lorentz transformations, because Eqs. (24) decouple.

The 4-component spinors $u_h(\mathbf{p})$ that appear in Eq. II.A(42) et seq. are

$$u_h(\mathbf{p}) = \begin{pmatrix} \eta_h^{(+)}(\mathbf{p}) \\ \eta_h^{(-)}(\mathbf{p}) \end{pmatrix}, \tag{25}$$

where it is understood that the $\eta^{(\pm)}$ are solutions of (24) for positive E.

[5] These equations are identical to Eq. VI.A(4), where we used the notation R and L instead of $(+)$ and $(-)$, respectively, for reasons explained there. Note that the superscripts $(\pm)$ do not specify the helicity once $m \neq 0$; both helicities have $(+)$ and $(-)$ components unless $m = 0$.

The 4-spinor v is the coefficient of $\exp[-i(\mathbf{p}\cdot\mathbf{r} - Et)]$ in the solution of Dirac's differential equation [II.A(32)]. Therefore, the antiparticle spinors $v_h(\mathbf{p})$ also have the form (25), but with $\eta^{(\pm)}$ replaced by solutions of (24) with $E \to -E, \mathbf{p} \to -\mathbf{p}$. However, some care must be exercised in constructing these. First, we define the $E < 0$ spinors for motion along the positive z-direction, so as to preserve the significance of helicity:

$$m\bar{\eta}^{(+)}(p\hat{\mathbf{z}}) = (-|E| + \sigma_z p)\bar{\eta}^{(-)}(p\hat{\mathbf{z}}),$$
$$m\bar{\eta}^{(-)}(p\hat{\mathbf{z}}) = -(|E| + \sigma_z p)\bar{\eta}^{(+)}(p\hat{\mathbf{z}}). \tag{26}$$

We must now rotate $p\hat{\mathbf{z}}$ into $-p\hat{\mathbf{n}}$. To construct the appropriate matrix $\bar{U}(\hat{\mathbf{n}})$, we exploit (2) to relate $\bar{U}(\hat{\mathbf{n}})$ to $U(\hat{\mathbf{n}})$. We know that $U\sigma_z U^\dagger = \boldsymbol{\sigma}\cdot\hat{\mathbf{n}}$, but require $\bar{U}\sigma_z\bar{U}^\dagger = -\boldsymbol{\sigma}\cdot\hat{\mathbf{n}}$. This can be achieved by $\bar{U} = Ui\sigma_y$, since $i\sigma_y$ generates a rotation through π about the y-axis. This leads us to define the v's as

$$v_h(\mathbf{p}) = \begin{pmatrix} U(\hat{\mathbf{n}})i\sigma_y\bar{\eta}_h^{(+)}(p\hat{\mathbf{z}}) \\ U(\hat{\mathbf{n}})i\sigma_y\bar{\eta}_h^{(-)}(p\hat{\mathbf{z}}) \end{pmatrix}. \tag{27}$$

Explicit expressions for u and v for arbitrary $\mathbf{p}$ are readily constructed from (25) and (27), but are rarely needed.[6] On the other hand, we need explicit expressions for ultrarelativistic (UR) motions, as they have important consequences for interactions of relativistic fermions with vector fields in the electromagnetic, weak, and strong interactions. For our present purpose it will suffice to consider motions along $+z$. Since $1 - (p/E) = (m^2/2E^2)$ when $E \gg m$, Eq. (24) reduces to

$$(1 \pm \sigma_z)\eta^{(\mp)} = \frac{m}{E}\eta^{(\pm)} \simeq 0. \tag{28}$$

Defining the basis spinors as

$$\xi_{1/2} = \begin{pmatrix} 1 \\ 0 \end{pmatrix}, \quad \xi_{-1/2} = \begin{pmatrix} 0 \\ 1 \end{pmatrix}, \tag{29}$$

---

[6] If one compares the expressions so constructed with standard texts, one must bear footnote 10 on p. 203 in mind. If u^{st} is a spinor in the "standard" representation, it is related to our u by

$$u^{\text{st}} = \frac{1}{\sqrt{2}}\begin{pmatrix} 1 & 1 \\ 1 & -1 \end{pmatrix}u,$$

where each 1 stands for a 2×2 unit matrix. Furthermore, most texts use spinors u_s and v_s obtained from the rest frame spinors by boosting along the direction of $\mathbf{p}$, and then their 2-valued spin label s has no simple relation to helicity, nor to the spin projection in $\mathscr{F}_0$. Our normalization convention is defined in Eq. (44).

we see that (28) implies $\eta_h^{(\pm)} \simeq \xi_{\pm 1/2}$, and therefore the UR limit of (25) is

$$u_{1/2}(p\hat{\mathbf{z}}) = \begin{pmatrix} \xi_{1/2} \\ 0 \end{pmatrix}, \qquad u_{-1/2}(p\hat{\mathbf{z}}) = \begin{pmatrix} 0 \\ \xi_{-1/2} \end{pmatrix}. \tag{30}$$

The normalization is arbitrary, since the Dirac equation is homogeneous.

The $E < 0$ solutions are constructed in the same way; Eq. (26) reduces to

$$(1 \mp \sigma_z)\bar{\eta}^{(\pm)} \simeq 0. \tag{31}$$

The momentum inversion demanded by (27) is $i\sigma_y \bar{\eta}^{(\pm)} = \mp \bar{\eta}^{(\mp)}$, so our final result for the UR antiparticle 4-spinors is

$$v_{1/2}(p\hat{\mathbf{z}}) = \begin{pmatrix} 0 \\ \xi_{-1/2} \end{pmatrix}, \qquad v_{-1/2}(p\hat{\mathbf{z}}) = \begin{pmatrix} \xi_{1/2} \\ 0 \end{pmatrix}. \tag{32}$$

Equations (30) and (32) were already given on p. 215, though there we purposely overlooked the rotation $p\hat{\mathbf{z}} \to -p\hat{\mathbf{z}}$.

(e) Four-currents

In both QED and the electroweak theory, we encounter charge and current densities that are bilinear combinations of Dirac spinors. We now show that these quantities form 4-vectors. Let $\chi^{(\pm)}$ and $\varphi^{(\pm)}$ be two pairs of spinors referring to two distinct fermions, for example, χ may refer to a $Q = \frac{2}{3}$ quark, and φ to a $Q = -\frac{1}{3}$ quark, as in a charge-changing weak process. A special case occurs when $\chi = \varphi$, as in the electromagnetic or neutral weak current. We wish to show that[7]

$$K_0^{(\pm)} = \varphi^{(\pm)*}\chi^{(\pm)}, \qquad \mathbf{K}^{(\pm)} = \pm\varphi^{(\pm)*}\boldsymbol{\sigma}\chi^{(\pm)}, \tag{33}$$

form two separate 4-vectors $\mathbf{K}^{(\pm)} = (K_0^{(\pm)}, \mathbf{K}^{(\pm)})$. Examples of such expressions are the electromagnetic current [II.A.28 and 29], and the $Q = 1$ weak currents [VI.A(9) and (10)]. Consider a LT of $K_0^{(+)}$. Both $\chi^{(+)}$ and $\varphi^{(+)}$ transform as in (17); if M is any 2×2 matrix, $(M\varphi^{(+)})^* = \varphi^{(+)*}M^\dagger$, and since $\boldsymbol{\sigma}$ is Hermitian, the LT produces the result

$$K_0^{(+)} \to \varphi^{(+)*}(e^{\frac{1}{2}\zeta\hat{\mathbf{n}}\cdot\boldsymbol{\sigma}})^2\chi^{(+)}$$
$$= K_0^{(+)} \cosh \zeta + \hat{\mathbf{n}} \cdot \mathbf{K}^{(+)} \sinh \zeta. \tag{34}$$

This is precisely the transformation law for the time component of a 4-vector. A similar calculation shows that $\mathbf{K}^{(+)}$ is the spatial part of the 4-vector $\mathbf{K}^{(+)}$. The same applies to the 4-vector $\mathbf{K}^{(-)}$.

Thus we conclude that the "left"-$(\mathbf{K}^{(-)})$ and "right"-handed $(\mathbf{K}^{(+)})$

---

[7] Our notation is $\varphi M\chi = \sum_{\alpha\beta} \varphi_\alpha M_{\alpha\beta}\chi_\beta$, $(\alpha, \beta = 1, 2)$, for any 2×2 matrix M.

THE DIRAC FIELD

quartets of quantities defined in (33) are, *separately*, 4-vectors under proper LTs. Under a spatial reflection, $\varphi^{(+)} \leftrightarrow \varphi^{(-)}, \chi^{(+)} \leftrightarrow \chi^{(-)}$, and therefore $K_0^{(\pm)} \rightarrow K_0^{(\mp)}, \mathbf{K}^{(\pm)} \rightarrow -\mathbf{K}^{(\mp)}$. This establishes all the transformation laws used in Chaps. II and VI.

2. The Dirac field operator

In Appendix II we constructed the Bose field operators by first expressing the classical field as a Fourier series, and then replacing the amplitude of each term in that series by an operator. From these field operators, we then constructed operators for the total energy (the Hamiltonian), the total momentum, and the charge.

(a) The field operator: Anticommutation rules

We do the same with the Dirac field. A general term of the Fourier series for the field was already given in Eq. II.A(43); we recall that in terms with positive frequency the amplitude operators[8] ($b_h(\mathbf{p})$ and $d_h(\mathbf{p})$) destroy a fermion, whereas in terms with negative frequency, they create a fermion ($b_h^\dagger(\mathbf{p})$ and $d_h^\dagger(\mathbf{p})$). The sum of all terms gives us the desired expression for the Dirac field operator

$$\psi(\mathbf{r}t) = \sum_{h\mathbf{p}} \frac{1}{\sqrt{2EV}} [u_h(\mathbf{p})e^{i(\mathbf{p}\cdot\mathbf{r}-Et)}b_h(\mathbf{p}) + v_h(\mathbf{p})e^{-i(\mathbf{p}\cdot\mathbf{r}-Et)}d_h^\dagger(\mathbf{p})]. \quad (35)$$

As in the case of the charged scalar field [see Eq. AII(38)], $b_h(\mathbf{p})$ destroys a particle, while $d_h^\dagger(\mathbf{p})$ creates an antiparticle.

Despite this similarity to the charged scalar field, there must be an essential difference in the algebra of the b's and d's, since they refer to spin-$\frac{1}{2}$ particles obeying Fermi–Dirac statistics,[9] whereas the spin-zero quanta of the scalar field are bosons. As we recall, the symmetry of multi-boson wave functions is guaranteed by commutation rules such as $[c^\dagger(\mathbf{k}), c^\dagger(\mathbf{k}')] = 0$. The two fermion state $b_h^\dagger(\mathbf{p})b_{h'}^\dagger(\mathbf{p}')|\Omega\rangle$ must, however, be antisymmetric; this is guaranteed if the operators anticommute, i.e., if $b_h^\dagger(\mathbf{p})b_{h'}^\dagger(\mathbf{p}') = -b_{h'}^\dagger(\mathbf{p}')b_h^\dagger(\mathbf{p})$. The other commutation rules have a similar structure, in that they follow from the Bose–Einstein rules if all commutators $[A, B] = AB - BA$ are replaced by anticommutators

$$\{A, B\} \equiv AB + BA. \quad (36)$$

---

[8] Here we do not designate operators, such as ψ or $b^\dagger$, by a tilde. As in Appendix II, we impose periodic boundary conditions on the surface of V.

[9] Here we assume the connection between spin and statistics. In Section 4 of this Appendix, and in Appendix IV, we show that this connection can be derived from other principles.

The complete list of rules is

$$\{b_n(\mathbf{p}), b_{h'}^\dagger(\mathbf{p}')\} = \delta_{hh'}\,\delta_{\mathbf{pp}'}, \tag{37}$$

$$\{b_h(\mathbf{p}), b_{h'}(\mathbf{p}')\} = 0, \tag{38}$$

$$\{d_h(\mathbf{p}), b_{h'}(\mathbf{p}')\} = \{d_h(\mathbf{p}), b_{h'}^\dagger(\mathbf{p}')\} = 0, \tag{39}$$

as well as those obtained therefrom by Hermitian conjugation, and by the replacement of b's by d's.

We now show that these rules lead to the Pauli principle. Consider (37) for $h = h', \mathbf{p} = \mathbf{p}'$:

$$b_h^\dagger(\mathbf{p})b_h(\mathbf{p}) + b_h(\mathbf{p})b_h^\dagger(\mathbf{p}) - 1 = 0.$$

If we multiply by $N_h(\mathbf{p}) \equiv b_h^\dagger(\mathbf{p})b_h(\mathbf{p})$, and use $[b_h(\mathbf{p})]^2 = 0$, which follows from (38), we have

$$N_h(\mathbf{p})[N_h(\mathbf{p}) - 1] = 0.$$

This states that the operators $N_h(\mathbf{p})$ have the eigenvalues 0 or 1, and can be interpreted as number operators. As we would expect from the Pauli principle, the only allowed occupation numbers are 0 or 1. The same applies to the antiparticle operator $\bar{N}_h(\mathbf{p}) = d_h^\dagger(\mathbf{p})d_h(\mathbf{p})$. The Hermitian conjugate of (38) was already discussed; it guarantees the antisymmetry of many particle states, and as a special case it requires $(b^\dagger)^2 = 0$, which simply says that two identical fermions cannot be put into the same state.

The spin-statistics connection, by itself, cannot impose conditions on the commutation rules between b's and d's, because particles are distinguishable from antiparticles. As we will see in Appendix IV, the principle of causality requires these operators to satisfy (39). The rules (39) show that a fermion and its antiparticle are not as "independent" as, say, electrons and muons.

(b) Energy, momentum, and charge

The Hamiltonian of the Dirac field can be inferred from the form [II.A(32)] of the Dirac equation, which can be written as

$$i\frac{\partial}{\partial t}\psi = (-i\boldsymbol{\alpha}\cdot\boldsymbol{\nabla} + \gamma_0 m)\psi, \quad \boldsymbol{\alpha} = \gamma_0\boldsymbol{\gamma}. \tag{40}$$

If this were a one-particle wave equation, the energy operator would be $(-i\boldsymbol{\alpha}\cdot\boldsymbol{\nabla} + \gamma_0 m)$. For a field, i.e., a system with an infinite number of degrees of freedom, the corresponding Hamiltonian is

$$H = \int \psi^\dagger(-i\boldsymbol{\alpha}\cdot\boldsymbol{\nabla} + \gamma_0 m)\psi\, d^3x. \tag{41}$$

We then insert the field operator (35) into H, and recall that u and v are solutions of the Dirac equation:

$$(\boldsymbol{\alpha}\cdot\mathbf{p} + \gamma_0 m)u(\mathbf{p}) = Eu(\mathbf{p}),$$
$$(-\boldsymbol{\alpha}\cdot\mathbf{p} + \gamma_0 m)v(\mathbf{p}) = -Ev(\mathbf{p}). \quad (42)$$

The orthogonality of different spinors, and the anticommutation rules (37), then lead to

$$H = \sum_{h\mathbf{p}} E(p)[b_h^\dagger(\mathbf{p})b_h(\mathbf{p}) + d_h^\dagger(\mathbf{p})d_h(\mathbf{p})]$$

$$= \sum_{h\mathbf{p}} E(p)[N_h(\mathbf{p}) + \tilde{N}(\mathbf{p})] \quad (43)$$

with our normalization[10]

$$\sum_{\alpha=1}^{4} |u_{h\alpha}(\mathbf{p})|^2 = \sum_{\alpha=1}^{4} |v_{h\alpha}(\mathbf{p})|^2 = 2E. \quad (44)$$

A constant (infinite) term has been omitted from (43), as in the Bose cases. The total momentum of the field is

$$\mathbf{P} = \sum_{h\mathbf{p}} \mathbf{p}[N_h(\mathbf{p}) + \tilde{N}_h(\mathbf{p})]. \quad (45)$$

That the particles and antiparticles in Dirac's theory have opposite charge was already demonstrated in §II.A.5. We have nothing to add to that. But we must still evaluate the charge operator for the Dirac field. We already have the expression [II.A(28)] for the charge density in terms of Dirac spinors. This becomes an operator when ψ is replaced by (35) and ψ^* by $\psi^\dagger$. As it stands, ρ does not change sign when C, the charge conjugation operation $b \leftrightarrow d$, $b^\dagger \leftrightarrow d^\dagger$, is carried out. We therefore face the same situation as in the case of a charged spin zero particle, where the most naive classical expression [AII(22)] for ρ also did not have the desired property under C. Once again we require that the charge and current densities satisfy the continuity equation by virtue of Dirac's equation, and be odd under C. For this purpose we define the charge conjugate Dirac field ψ_c, which is obtained from ψ [i.e., Eq. (35)] by the replacements $b \to d$, $d^\dagger \to b^\dagger$. The expression [II.A(34)] for the 4-current is then replaced by

$$j_\mu = \tfrac{1}{2}(\bar{\psi}\gamma_\mu\psi - \bar{\psi}_c\gamma_\mu\psi_c) \quad (46)$$

where $\bar{\psi}_c = \psi_c^\dagger \gamma_0$; this has the same structure as the modified charge density for the scalar field, Eq. AII(46). The charge operator (in units of e) is the

---

[10] If one defines the adjoint spinor $\bar{u} = u^*\gamma_0$ as in Eq. II.A(33), one finds the invariant norm is $\bar{u}u = 2m$ [whereas (44) is $u^*u = 2E$] in the notation of Eq. II.A(35).

integral of j_0 over all space; it is found to be

$$Q = \sum_{h\mathbf{p}} [N_h(\mathbf{p}) - \bar{N}_h(\mathbf{p})]. \tag{47}$$

As expected, it has precisely the same form as the charge operator for the scalar field. In a state with n particles and $\bar{n}$ antiparticles, the eigenvalue of Q is $n - \bar{n}$.

(c) The relative parity of particle and antiparticle

〚In §II.A.5 we already gave a fairly complete explanation of how the Dirac theory predicts that fermions and antifermions have opposite intrinsic parities.[11] All that we add here are a few remarks to clarify precisely what is meant by the arrows in (21) and in Eqs. II.A(53), (57), and (58). Let P be the unitary operator that represents space reflection in the Hilbert space of states. Then the precise form of the transformation law (21) is

$$P\psi(\mathbf{r}, t)P^{-1} = \gamma_0 \psi(-\mathbf{r}, t). \tag{48}$$

Applied to a single term in the Fourier series (35), Eq. (48) implies

$$u_h(\mathbf{p})e^{i(\mathbf{p}\cdot\mathbf{r}-Et)}Pb_h(\mathbf{p})P^{-1} = \gamma_0 u_h(\mathbf{p})e^{-i(\mathbf{p}\cdot\mathbf{r}+Et)}b_h(\mathbf{p}), \tag{49}$$

and a similar relation involving $vd^\dagger$. Now Eqs. II.A(56) and (57) tell us that $\gamma_0 u_h(\mathbf{p}) = u_{-h}(-\mathbf{p})$, and therefore (49) becomes

$$u_h(p)e^{i(\mathbf{p}\cdot\mathbf{r}-Et)}Pb_h(\mathbf{p})P^{-1} = u_{-h}(-\mathbf{p})e^{-i(\mathbf{p}\cdot\mathbf{r}+Et)}b_h(\mathbf{p}).$$

Taking advantage of the sum over h and $\mathbf{p}$ in (35), and comparing terms having the same exponential, we find that

$$Pb_h(\mathbf{p})P^{-1} = b_{-h}(-\mathbf{p}). \tag{50}$$

The same argument applied to the term containing $vd^\dagger$ gives

$$Pd_h^\dagger(\mathbf{p})P^{-1} = -d_{-h}^\dagger(-\mathbf{p}). \tag{51}$$

These equations are the precise form of Eq. II.A(58).〛

---

[11] We might point out that one cannot really define the relative parity of states having different eigenvalues of a strictly conserved quantum number, because a determination of their relative parity requires the measurement of an interference phenomenon. This is called a *superselection rule*. Hence the relative parity of e^+ and e^- is actually arbitrary! But the parity of the electrically neutral e^+e^- state can be compared to, say, that of a $\pi^+\pi^-$ state. What we are really proving, therefore, is that in a state of given orbital angular momentum, e^+e^- and $\pi^+\pi^-$ have opposite parities.

3. Amplitudes for weak and electromagnetic scattering

Many weak and electromagnetic phenomena are described by scattering amplitudes that correspond to the following Feynman diagram:

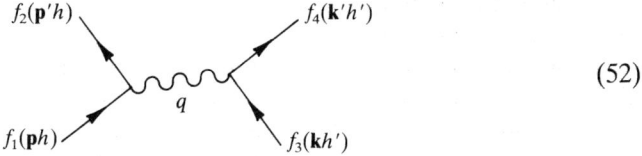

(52)

Here f_i is a quark or lepton with the indicated momenta and helicities, and the propagator represents a vector boson ($W^\pm, Z^0$, or γ). The interaction responsible for the emission and absorption of the boson is of the form $\int \mathbf{J} \cdot \mathbf{V} \, d^3x$, where $\mathbf{J}$ is a current carried by the fermions, and $\mathbf{V}$ the vector field that creates and destroys the bosons. In all cases—whether W, Z^0, or γ is exchanged—the 4-currents involved are of the form (33); in the UR regime the helicity selection rules stated on p. 214 therefore hold for all electroweak processes. We confine ourselves to this regime, and for that reason diagram (52) already incorporates helicity conservation. The amplitude therefore has the form of Eq. II.B(46):

$$T_{hh'} = \frac{\Phi_{hh'}}{q^2 - m^2}. \tag{53}$$

The numerator in this expression is the product of two matrix elements, one for each vertex:[12]

$$\Phi_{hh'} = \tfrac{1}{2} M_h \cdot \tilde{M}_{h'}, \tag{54}$$

where the factor $\tfrac{1}{2}$ is inserted for convenience, since we are not concerned with the absolute magnitude of the amplitude. In an obvious notation, these matrix elements are

$$M_h = \langle f_2(\mathbf{p}'h) | J | f_1(\mathbf{p}h) \rangle, \tag{55}$$

$$\tilde{M}_{h'} = \langle f_4(\mathbf{k}'h') | J^\dagger | f_3(\mathbf{k}h') \rangle. \tag{56}$$

In view of the helicity selection rules, once the interaction, and the helicity and fermion number are specified, one knows which portion of J enters the

---

[12] While Eq. (54) has a plausible form, a number of steps are involved in its derivation. First, the two field operators $\mathbf{V}$ associated with the vertices produce the Feynman propagator $(q^2 - m^2)^{-1}$, and the matrix elements M and $\tilde{M}$ are to be viewed as amplitudes for the transitions fermion $\to$ fermion plus vector boson. Second, a particular choice of gauge allows one to cast the result into the manifestly covariant form (54).

matrix element. For example, in an electromagnetic process, **J** has the form

$$\mathbf{J} = \chi^{(+)\dagger}\boldsymbol{\sigma}\chi^{(+)} - \chi^{(-)\dagger}\boldsymbol{\sigma}\chi^{(-)}.$$

Only the first term enters M_h if $h = \frac{1}{2}$ and f_1 and f_2 are particles, whereas the second term enters if $h = \frac{1}{2}$ and f_1 and f_2 are antiparticles, etc. For a W-mediated process, **J** has the form $\varphi^{(-)\dagger}\boldsymbol{\sigma}\chi^{(-)}$; hence, M_h vanishes unless $h = -\frac{1}{2}$ when f_1 and f_2 are particles, or $h = \frac{1}{2}$ when f_1 and f_2 are antiparticles.

We evaluate (55) and (56) in the c.o.m. frame, where $\mathbf{k} = -\mathbf{p}, \mathbf{k}' = -\mathbf{p}'$, the scattering angle is θ, and the z-axis is along **p**. Consider first the case where all f's are particles, and the states are described by spinors in UR limit. Then if $\lambda = 2h = \pm 1$,

$$\begin{aligned}\chi^{(\lambda)}|f_1(\mathbf{p}h)\rangle &= \xi_h|\Omega\rangle, & \chi^{(\lambda)}|f_2(\mathbf{p}'h)\rangle &= U_y(\theta)\xi_h|\Omega\rangle, \\ \chi^{(\lambda)}|f_3(\mathbf{k}h)\rangle &= U_y(\pi)\xi_h|\Omega\rangle, & \chi^{(\lambda)}|f_4(\mathbf{k}'h)\rangle &= U_y(\pi + \theta)\xi_h|\Omega\rangle.\end{aligned} \quad (57)$$

Here ξ_h are the spinors of Eq. (29), $U_y(\theta) = \exp(-i\sigma_y\theta/2)$ rotates **p** into **p**', and **k** into **k**', while $U_y(\pi)$ takes into account that $\mathbf{k} = -\mathbf{p}$.

The components of M_h are therefore

$$\begin{aligned}M_{h0} &= (\xi_h, U_y^\dagger(\theta)\xi_h) = \cos\tfrac{1}{2}\theta, \\ \mathbf{M}_h &= \lambda(\xi_h, U_y^\dagger(\theta)\boldsymbol{\sigma}\xi_h),\end{aligned} \quad (58)$$

or

$$(M_{hx}, M_{hy}, M_{hz}) = (\sin\tfrac{1}{2}\theta, i\lambda\sin\tfrac{1}{2}\theta, \cos\tfrac{1}{2}\theta), \quad (59)$$

Here it is understood that **J** must contain the term demanded by the helicity rules; otherwise, M_h vanishes. The matrix element $\tilde{M}_h$ is determined by M_h, since $U_y(\pi)$ is just a rotation about the y-axis; hence,

$$\tilde{M}_h = (M_{h0}, -M_{hx}, M_{hy}, -M_{hz}). \quad (60)$$

Using (58)–(60) to evaluate the scalar product (54), one finds

$$\Phi_{hh} = 1 \equiv \Phi_0, \quad (61)$$

$$\Phi_{h,-h} = \cos^2\tfrac{1}{2}\theta \equiv \Phi_1. \quad (62)$$

Observe that $|h - h'|$ is the magnitude of the projection of the total angular momentum along both the incident and final direction (because of helicity conservation). Hence, Φ_0 is isotropic, because two-body states of total momentum zero with $h = h'$ behave as if they had no spin. On the other hand, Φ_1 vanishes like $(\theta - \pi)^2$ for backward scattering, since $|h - h| = 1$ requires a change of two units of angular momentum along the direction of **p**.

One readily convinces oneself that the results (61) and (62) do not depend on whether the particles involved are fermions or antifermions; all that matters is whether $|h - h'|$ is 0 or 1, because the angular distribution depends only on angular momentum considerations.

Equation (53), with the expressions (61) and (62) for $\Phi_{hh'}$, provides a derivation of Eq. II.C(7), since $\cos^2 \frac{1}{2}\theta = (\mathbf{p} \cdot \mathbf{k}')/(\mathbf{p} \cdot \mathbf{k})$, as shown in the derivation of Eq. II.C(9). [Note that $m = 0$ and $t = q^2$ in §II.C].

[The angular distribution of the annihilation amplitude [II.C(12)] follows from simpler considerations. Because of the helicity rule, the initial pair and final pair are both in two-body states of opposite helicity. Hence, the projections μ and μ' of total angular momentum along the initial and final directions are either 1 or -1, but need not be equal. Further, the intermediate state has $J = 1$. The annihilation amplitude $B_{\mu\mu'}(\theta)$ therefore has an angular distribution that is given by the amplitude for finding a state of $J = 1$ and projection μ' along the vector $\hat{\mathbf{n}}$ in a $J = 1$ state with projection μ along $\hat{\mathbf{z}}$, where θ is the angle between $\hat{\mathbf{z}}$ and $\hat{\mathbf{n}}$:

$$B_{\mu\mu'}(\theta) = N d^{(1)}_{\mu\mu'}(\theta), \tag{63}$$

where $d^{(j)}_{mm'}(\theta)$ is an element of the spin j rotation matrix (see, for example, Gottfried (1966), §32), N is a constant, and

$$\begin{aligned} d^{(1)}_{\mu\mu'}(\theta) &= \tfrac{1}{2}(1 + \cos\theta) & \mu = \mu' = \pm 1 \\ &= \tfrac{1}{2}(1 - \cos\theta) & \mu = -\mu' = \pm 1. \end{aligned} \tag{64}$$

For a randomly polarized initial state, and undetected final spins, the cross section is therefore proportional to

$$\tfrac{1}{4}(1 + \cos\theta)^2 + \tfrac{1}{4}(1 - \cos\theta)^2 = \tfrac{1}{2}(1 + \cos^2\theta),$$

as stated in [II.C(16)].]

4. Spin and statistics

The evidence given thus far for the connection between spin and statistics has been purely empirical. Our purpose here is to show why integer spin particles cannot obey Fermi–Dirac statistics, whereas half-integer spin particles cannot obey Bose–Einstein statistics. There are two ways of doing this. The first is based on the necessity of having a sensible energy spectrum. The other proof is based on the fundamental requirement of local causality and is discussed in Appendix VI.

We first consider the neutral scalar field. Since we want to find out what statistics the spin-0 particles should have, we consider the two alternative commutative relations for their creation and destruction operators $c_k^\dagger$ and c_k:

$$[c_k, c_{k'}^\dagger]_\pm = \delta_{kk'}, \qquad [c_k, c_{k'}]_\pm = 0, \tag{65}$$

where we define

$$[a, b]_\pm \equiv ab \pm ba. \tag{66}$$

The two alternatives correspond to the two cases of interest: the commutation rules [ ]_ to Bose–Einstein statistics, the anticommutation rules [ ]_+ to Fermi–Dirac statistics.

We first show that for the neutral scalar field only the Bose rules lead to a sensible result for the energy. In Appendix II.2(a) we already learned that the assumption of Bose statistics leads to the energy eigenvalues

$$E = \sum_k \omega_k n_k + C,$$

where the n_k are 0 or a positive integer, and $C = \frac{1}{2} \sum \omega_k$. Hence, there is a state of lowest energy E, wherein all n_k vanish, and by redefining the energy scale, we can set this vacuum energy to zero. If, however, we choose Fermi anticommutation rules in passing to the quantum theory from the classical Hamiltonian, Eq. AII(20), we would find that H has the eigenvalue C for all states! Obviously, this is a senseless result, and we therefore conclude that we must choose Bose statistics for a neutral scalar (spin-zero) field. The same conclusion follows straightforwardly for the charged scalar field discussed in Appendix II.2(b). Furthermore, the argument can be generalized to the electromagnetic field, and to non-Abelian gauge fields. We therefore draw the conclusion that fields with integer spin have quanta that obey Bose statistics.

Finally, we come to spin-$\frac{1}{2}$ fields. On inserting the field operator (35) into the Hamiltonian (41), and exploiting the Dirac equations (42), but *without* any use of commutation rules, one arrives at

$$H = \sum_{\mathbf{p}h} |E(p)| \, [b_h^\dagger(\mathbf{p}) b_n(\mathbf{p}) - d_h(\mathbf{p}) d_h^\dagger(\mathbf{p})]. \tag{67}$$

The relative minus sign between the first and second terms is important; it stems from the appearance of the eigenvalues $+|E|$ and $-|E|$ on the right-hand side of (42). Then d and $d^\dagger$ are to be interchanged by using the Fermi anticommutation rules. This gives the positive definite expression (43) for the energy, apart from a constant. On the other hand, if we were to assume that the b's and d's obey Bose commutation rules, such as

$$[b_n(\mathbf{p}), b_{h'}^\dagger(\mathbf{p}')]_- = \delta_{hh'} \, \delta_{\mathbf{pp}'}, \tag{68}$$

etc., we would find that the Hamiltonian is

$$H = \sum_{h\mathbf{p}} |E(p)| [N_h(\mathbf{p}) - \tilde{N}_h(\mathbf{p})], \qquad (69)$$

apart from a constant, where the Bose occupation numbers $N_h(\mathbf{p})$ and $\tilde{N}_h(\mathbf{p})$ have the eigenvalues $0, 1, 2, \ldots$. Evidently, the energy spectrum of (69) has no lower bound, and is unacceptable. Consequently, one must assign Fermi-Dirac statistics to spin-$\frac{1}{2}$ particles.

Appendix IV

CAUSALITY AND ITS CONSEQUENCES

In this Appendix we demonstrate the power of the principle of causality with a few relatively simple examples, and describe some important theorems of quantum field theory that follow from causality.

Causality is not a principle that is nearly restrictive enough to define a theory—i.e., to determine the interactions, let alone the spins and other quantum numbers of the particles. It is therefore astonishing that causality leads to profound results that explain some of the most striking regularities found in nature. Among these regularities we might single out the connection between spin and statistics and the existence of antiparticles.

1. The basic axiom of quantum field theory[1]

We begin by considering the uncertainty relations involving the electromagnetic field strengths. Recall that for any two observables $\mathcal{A}$ and $\mathcal{B}$, the uncertainty relation is

$$\Delta\mathcal{A}\,\Delta\mathcal{B} \geq \tfrac{1}{2}|\langle[\mathcal{A},\mathcal{B}]\rangle|. \tag{1}$$

Here

$$(\Delta\mathcal{A})^2 = \langle(\mathcal{A} - \langle\mathcal{A}\rangle)^2\rangle,$$

where $\langle\mathcal{A}\rangle$ is the expectation value of $\mathcal{A}$ in any state of the system and, on the right-hand side of (1), the expectation value of the commutator is taken in this same state. Therefore, an uncertainty product such as $\Delta E_x(\mathbf{r}t)\,\Delta B_y(\mathbf{r}'t')$ can be found by calculating the commutator $[E_x(\mathbf{r}t), B_y(\mathbf{r}'t')]$. This is a relatively straightforward exercise if one uses the expansions for the fields in terms of creation and destruction operators, and the basic commutators [AII(56)] for the latter. The essential result of this calculation is that all such commutators vanish throughout all of space-time except on

---

[1] In this Appendix we do not distinguish operators by a tilde.

the surface[2]

$$(\mathbf{r} - \mathbf{r}')^2 = c^2(t - t')^2. \tag{2}$$

If we pick $\mathbf{r}'$ and t' as the origin of our coordinate system, the points $(\mathbf{r}, t)$ satisfying (2) lie on the light cones having the origin at their apex. On second thought, this should have been expected. It says that if we measure a field at $(\mathbf{r}', t')$, the test charge involved will undergo uncontrollable accelerations because its coordinates and momenta are only known simultaneously to a precision restricted by the uncertainty principle. Such accelerations cause the test charge to radiate, and this radiation will disturb the measurement of another field at $(\mathbf{r}, t)$. As this "uncertainty radiation" travels with the speed of light, two measurements are only in conflict if they can be connected by a light signal. Equation (2) defines just those space-time points that can be connected by light signals.

To exploit this observation, it will prove convenient to introduce a standard terminology from the theory of relativity. Let $x = (\mathbf{r}, t)$ and $x' = (\mathbf{r}', t')$ be two arbitrary space-time points. If x and x' can be connected by a light signal, these two points are said to have a lightlike relation. This was true of the points discussed in the preceding paragraph. If x and x' can be connected by a signal moving slower than c, we say the points are timelike, because there then exists a Lorentz frame where x and x' have the same spatial coordinates, and differ only in time. Conversely, if x and x' can only be connected by signals traveling faster than c, we call them spacelike, because there then exists a frame where x and x' have the same time coordinate, but are spatially separated.

As we just saw, the commutation rules for the fields incorporate the principle of causality. This being understood, it is to be expected that this causal property of the uncertainty relation holds for all observables, not just the fields. That this is indeed so can be seen from the example of the Poynting vector $(\mathbf{E} \times \mathbf{B})$ and the energy density $\frac{1}{2}(E^2 + B^2)$. To determine the uncertainty product for such operators, we need commutators of the type

$$[F_i(x)F_j(x), F_k(x')F_l(x')], \tag{3}$$

where $F_i(x)$ is any one of the six quantities $\mathbf{E}, \mathbf{B}$. But the algebraic identity

$$[AB, CD] = A[B, C]D + AC[B, D] + [A, C]DB + C[A, D]B \tag{4}$$

reduces (3) to commutators of the basic fields, and these vanish unless x and x' are lightlike. Consequently, if $\mathscr{F}(x)$ is any operator built from fields $\mathbf{E}$ and $\mathbf{B}$ at the point x, and $\mathscr{F}'(x')$ is another such operator at x', $\mathscr{F}$ and $\mathscr{F}'$ will satisfy a causal uncertainty relation.

[2] Here we have inserted c for the moment so as to emphasize the physical principles involved. We shall return to our usual units ($c = 1$) shortly.

It is natural to suppose that the causal nature of the uncertainty relations transcends pure electrodynamics. Let us suppose that all physical phenomena, and not just electromagnetism, are describable by operators like $\mathbf{E}$, ρ, etc.; or operators like the energy or current density, which are polynomials in the field operators evaluated at the same point. These are all observables attached to infinitesimal space-time regions. Let $\mathcal{O}(x)$ and $\mathcal{O}'(x')$ be any two such *local observables* associated with space-time points x and x'. Then the principle of relativity requires that measurements of $\mathcal{O}$ and $\mathcal{O}'$ cannot interfere with each other if x and x' are at spacelike separation. There is a small but important difference in wording between this last sentence and the discussion following Eq. (2). There we were considering pure electrodynamics, where all disturbances travel with the speed of light. Consequently, interference between measurements could only occur if x and x' were lightlike. But in a more realistic description there can also be slower disturbances. For example, the act of measuring a field at x can produce an e^+e^- pair whose members travel outward from x with velocities ranging from 0 to c, and thereby disturb measurements of other quantities at all x' that have a timelike separation from x.

In general, therefore, any pair of local observables $\mathcal{O}(x)$ and $\mathcal{O}'(x')$ can be measured without mutual disturbance provided x and x' are at spacelike separation. In terms of commutation rules this requirement becomes

$$[\mathcal{O}(x), \mathcal{O}'(x')] = 0 \quad \text{if} \quad (\mathbf{r} - \mathbf{r}')^2 > c^2(t - t')^2. \tag{5}$$

We shall raise this physical requirement to the status of the basic axiom of all quantum field theory. As we shall see, it is a remarkably powerful axiom; it leads to fundamental results such as the connection between spin and statistics.

2. The connection between spin and statistics

We have already shown in Appendix III.4 that the connection between spin and statistics can be inferred from the requirement that the free Hamiltonian of spin-0 or spin-$\frac{1}{2}$ particles has a physically reasonable spectrum. Here we derive the connection from the causality principle. This has the advantage of not requiring a knowledge of the Hamiltonian.[3]

As in Appendix III.4, we begin with the neutral spin 0 field operator ϕ, as given by Eq. AII(28). Let us evaluate both the commutator and the anticommutator of $\phi(x)$ with $\phi(y)$ when $x - y$ is spacelike. By using the Bose and Fermi commutation rules, Eq. AIII(65), and the notation of Eq.

---

[3] For rigorous proofs of this and other theorems in this Appendix, see Streater and Wightman (1964).

AIII(66), one finds

$$[\phi(x), \phi(y)]_\pm = \sum_\mathbf{k} (2\omega_k V)^{-1}[e^{-ik\cdot(x-y)} \pm e^{ik\cdot(x-y)}] \tag{6}$$

where $k = (\omega_k, \mathbf{k})$, and $k \cdot x = \omega_k t - \mathbf{r}\cdot\mathbf{k}$. Next we define

$$\Delta_\pm(x - y) = \sum_\mathbf{k} (2\omega_k V)^{-1} e^{\mp ik\cdot(x-y)}. \tag{7}$$

These are Lorentz-invariant functions of the 4-vector argument $(x - y)$. This fact is not manifest in (7), since the definition of the states was based upon a fixed reference volume V. That $\Delta_\pm$ are, nevertheless, invariant, can be understood by turning the sum over $\mathbf{k}$ into an integral: $\sum_\mathbf{k} = V(2\pi)^{-3} \int d^3k$. But for any 4-vector $(p_0, \mathbf{p})$ the combination d^3p/p_0 is a Lorentz-invariant volume element in momentum space.[4] Hence

$$\Delta_\pm(x - y) = \int \frac{d^3k}{(2\pi)^3 2\omega_k} e^{\mp ik\cdot(x-y)}, \tag{8}$$

where the Lorentz invariance has now become manifest. Since we are concerned with the case where $(x - y)$ is spacelike, we choose a system where $x^0 = y^0$ for the evaluation of (8), which then becomes

$$\Delta_\pm = \int \frac{d^3k}{(2\pi)^3 2\omega_k} e^{\pm i\mathbf{k}\cdot(\mathbf{x}-\mathbf{y})}.$$

We see immediately that $\Delta_\pm$ is real, and that

$$\Delta_+(x - y) = \Delta_-(x - y). \tag{9}$$

Both statements follow because the integral contains equivalent terms for $+\mathbf{p}$ and $-\mathbf{p}$. The identity (9) holds whenever $x - y$ is spacelike. It plays an important role in what follows.

We now write (6) in terms of Δ_+ and Δ_-:

$$[\phi(x), \phi(y)]_\pm = \Delta_+(x - y) \pm \Delta_-(x - y). \tag{10}$$

When $x - y$ is spacelike,[5] we can use (9), and then we immediately see that (10) only vanishes if one assumes the commutation rule with the minus sign for the basic creation and destruction operators. Hence if ϕ is an

---

[4] This follows from $\int d^4p\, \delta(p^2 - m^2) = \int d^3p/2p_0$, where the obviously invariant left side extends only over 4-momenta satisfying $p_0 > 0$.

[5] For timelike $x - y$, the commutator $[\phi(x), \phi(y)]$ is nonzero, and as we said before, describes disturbances due to measurements that travel with speeds up to c.

observable, *causality requires the spin-0 quanta of the neutral scalar field to be bosons.*

It is easy to apply the same conclusions to photons. We start with the corresponding expression [Eq. AII(53)] for the vector potential. It differs from ϕ only by containing the polarization vectors. Therefore the same procedure can be applied to prove that the commutation relations vanish for the potential **A**, and for the fields **E** and **B**, at two points with spacelike separation, if the commutation relations between the destruction and creation operators are those appropriate to Bose statistics.

We also forego a detailed calculation of the commutator for charged fields Φ:

$$[\Phi(x), \Phi^\dagger(y)]_\pm = \Delta_+(x - y) \pm \Delta_-(x - y). \tag{11}$$

The proof goes along the same lines as before, but one must not only posit the rules [AIII(65)] for the c's and f's, but also

$$[c_\mathbf{k}, f_{\mathbf{k}'}^\dagger]_\pm = 0, \quad [c_\mathbf{k}, f_{\mathbf{k}'}]_\pm = 0, \quad \text{etc.} \tag{12}$$

Again one finds that these charged spin-0 particles must be bosons if ϕ is an observable, and if causality is not to be violated. Note that $c^\dagger$ and $f^\dagger$ create particles that are distinguishable, because they have opposite charge. Nevertheless, causality requires the wave function to be symmetric under the interchange of a particle and its antiparticle.

This argument does not quite prove that particles described by scalar or vector fields must have Bose statistics. It only shows that if we impose the Fermi anticommutation rules on the creation and destruction operators, the fields at different spacelike points would not anticommute. But this in itself does *not* yet establish the connection between spin and statistics, because it could be that ϕ and **A** are not observable! This need not be fatal—as we shall soon see, the Dirac field is not an observable. Indeed, our classical view of ϕ or **A** as a field has gone out the window anyway once we say that the particles are fermions, because it is not possible to build up a macroscopic field strength if the occupation numbers are bounded by one. All that we can now demand is that operators that have a direct physical significance, such as the energy, charge, or current densities, be local observables.

The observables just mentioned have a common characteristic: they are bilinear forms in the basic fields. Now there is a simple identity, similar to (4), that relates the commutator of bilinear operators to the anticommutators of the factors:[6]

$$[AB, CD] = A\{B, C\}D - \{A, C\}BD + CA\{B, D\} - C\{A, D\}B. \tag{13}$$

---

[6] Here we again use { } as the symbol for the anticommutator, not [ ]$_+$.

This identity shows an interesting and important fact: If the fields (and their derivatives) satisfy causal *anti*commutation rules, bilinear expressions would still be local observables. By "causal anticommutation rule" one simply means that

$$[A(x), A(y)]_+ \equiv \{A(x), A(y)\} = 0$$

if the separation of x and y is spacelike. But we know from (9) that this expression is not zero for ϕ or $\mathbf{A}$, but $2\Delta_+(x - y)$. *Therefore it is not possible to impose Fermi statistics on spin-0 particles or photons.* Thus we have established the spin-statistics connection in these cases.

We now come to spin-$\frac{1}{2}$ fields. We use the expression [AIII(35)] to evaluate the commutation relations for ψ and $\psi^\dagger$ at two points x and y with spacelike separation. We find

$$[\psi_\alpha(x), \psi_\beta^\dagger(y)]_\pm = \sum_\mathbf{p} \frac{1}{2EV}[U_{\alpha\beta}(\mathbf{p})e^{-i\mathbf{p}\cdot(x-y)} \pm V_{\alpha\beta}(\mathbf{p})e^{i\mathbf{p}\cdot(x-y)}] \quad (14)$$

where

$$U_{\alpha\beta}(\mathbf{p}) = \sum_h u_{h\alpha}(\mathbf{p})u^*_{h\beta}(\mathbf{p}), \qquad V_{\alpha\beta}(\mathbf{p}) = \sum_h v_{h\alpha}(\mathbf{p})v^*_{h\beta}(\mathbf{p}). \quad (15)$$

In (14) the anticommutation rules [AIII(37)–(39)] are used for the upper (+) sign, and the corresponding Bose commutation rules for the lower (−) sign.

To evaluate (15), we return to [AIII(23)], and compute the 4×4 matrix $U(\mathbf{p})$, when $\mathbf{p} = p\hat{\mathbf{z}}$, bearing in mind the normalization [AIII(44)]. This gives

$$U(p\hat{\mathbf{z}}) = E + \alpha_3 p + m\gamma_0,$$

where it is understood that E is proportional to the unit matrix. For an arbitrary orientation of $\mathbf{p}$, rotational invariance requires that $\alpha_3 p$ generalizes to $\boldsymbol{\alpha}\cdot\mathbf{p}$. A similar calculation for V then gives the final results

$$U = E + \boldsymbol{\alpha}\cdot\mathbf{p} + m\gamma_0, \qquad V = E + \boldsymbol{\alpha}\cdot\mathbf{p} - m\gamma_0. \quad (16)$$

These matrices can be extracted from the sum in (14) by differentiation with respect to the components of $x = (t, \mathbf{x})$:

$$[\psi_\alpha(x), \psi_\beta^\dagger(y)]_\pm = (i\partial_t - i\boldsymbol{\alpha}\cdot\boldsymbol{\nabla} + m\gamma_0)_{\alpha\beta}$$

$$\cdot \sum_\mathbf{p} \frac{1}{2EV}(e^{-i\mathbf{p}\cdot(x-y)} \mp e^{i\mathbf{p}\cdot(x-y)}).$$

Observe the crucial change of sign that occurred here; it comes about because U and V are replaced by first-order differential operators acting on $\exp(\pm ip \cdot x)$. On recalling the definition (7), we therefore obtain the final result

$$[\psi_\alpha(x), \psi_\beta^\dagger(y)]_\pm = (i\partial_t - i\boldsymbol{\alpha} \cdot \boldsymbol{\nabla} + m\gamma_0)_{\alpha\beta}[\Delta_+(x-y) \mp \Delta_-(x-y)]. \quad (17)$$

Since Δ_+ and Δ_- are identical functions of t and $\mathbf{x}$ when $x - y$ is spacelike, expression (17) vanishes for the anticommutator and *not* for the commutator! Therefore the only way to get a vanishing commutator for local observables bilinear in the operators ψ and $\psi^\dagger$ at spacelike separations is to choose the upper sign in (17), that is, to have the creation and destruction operators *anticommute*. Thus, we have shown that causality requires spin-$\frac{1}{2}$ particles to be fermions.[7]

Furthermore, this derivation also imposes the anticommutation rule [AIII(39)] on the fermion and antifermion operators, even though they refer to distinguishable particles! As we showed in §II.C.7 (see, in particular, the footnote on p. 250), the rule

$$\{b_h^\dagger(\mathbf{p}), d_{h'}^\dagger(\mathbf{p}')\} = 0, \quad (18)$$

demanded by causality, plays an essential ingredient in the derivation of the charge conjugation eigenvalue [II.C(37)] for fermion–antifermion states. The 2- and 3-photon decays of positronium, and the copious decays $\psi \to e^+e^-$, etc., all confirm this eigenvalue, and rule out the opposite C-eigenvalue that would be a consequence of the assumption that $[b_h^\dagger(\mathbf{p}), d_{h'}^\dagger(\mathbf{p}')] = 0$.

3. The need for antiparticles: Crossing symmetry

We now substantiate our oftmade claim that a causal theory cannot do without antiparticles. Consider the charged scalar field [AII(38)] as an example, which we write as

$$\Phi(\mathbf{x}, t) = \Phi_+(\mathbf{x}, t) + \Phi_-^\dagger(\mathbf{x}, t). \quad (19)$$

Here Φ_+ destroys particles, while $\Phi_-^\dagger$ creates antiparticles. We demonstrate that: (1) in any process the operator for emission of a particle has an amplitude that equals that of the operator for the absorption of an antiparticle; and (2) a particle and its antiparticle must have the same mass.

---

[7] By similar arguments one can show that causality requires antifermions to have the opposite intrinsic parity from fermions, whereas bosons and antibosons must have the same intrinsic parity, where the physical meaning of these parities is explained in footnote 11. Detailed proofs, valid for any spin, of this and other theorems that follow from causality can be found in Weinberg (1964).

As it stands, (19) has both these properties. To see that one cannot tamper with these, we modify (19) as follows:

$$\Phi(\mathbf{x}, t) = \alpha_+ \Phi_+(\mathbf{x}, t; \mu_+) + \alpha_- \Phi_-^\dagger(\mathbf{x}, t; \mu_-). \tag{20}$$

Here $\alpha_\pm$ are arbitrary constants, and $\mu_\pm$ are the masses of particle and antiparticle, momentarily taken as different. The expressions for $\Phi_\pm$ are now

$$\Phi_+(\mathbf{x}, t) = \sum_\mathbf{k} \frac{1}{\sqrt{2\omega_+ V}} c_\mathbf{k} e^{i(\mathbf{k}\cdot\mathbf{x}-\omega_+ t)}, \quad \Phi_-(\mathbf{x}, t) = \sum_\mathbf{k} \frac{1}{\sqrt{2\omega_- V}} f_\mathbf{k} e^{i(\mathbf{k}\cdot\mathbf{x}-\omega_- t)}, \tag{21}$$

where $\omega_\pm = \sqrt{k^2 + \mu_\pm^2}$. We again calculate the basic commutator, as in (6). Since we are only interested in spacelike separations, we can set the times equal, and exploit space-time translation invariance:

$$[\Phi(\mathbf{x}, 0), \Phi(0, 0)] = |\alpha_+|^2 \Delta_+(\mathbf{x}, \mu_+) - |\alpha_-|^2 \Delta_-(\mathbf{x}, \mu_-), \tag{22}$$

where, according to (8),

$$\Delta_\pm(\mathbf{x}, \mu_\pm) = \int \frac{d^3 k}{2(2\pi)^3 \sqrt{k^2 + \mu_\pm^2}} e^{\pm i \mathbf{k}\cdot\mathbf{x}}.$$

It is clear that $\Delta_\pm(\mathbf{x}, \mu_\pm) = g(\mathbf{x}, \mu_\pm)$, i.e., does not depend on the subscript on $\Delta_\pm$, whence

$$[\Phi(\mathbf{x}, 0), \Phi(0, 0)] = |\alpha_+|^2 g(\mathbf{x}, \mu_+) - |\alpha_-|^2 g(\mathbf{x}, \mu_-). \tag{23}$$

The function $g(\mathbf{x}, \mu)$ depends on the value of the mass μ; for example, for $x\mu \to \infty$, it behaves like $e^{-x\mu}$ times a power of $(x\mu)$. Hence (23) can only vanish for *all* $\mathbf{x}$ if *both* $\mu_+ = \mu_-$, and $|\alpha_+| = |\alpha_-|$.

The significance of $\mu_+ = \mu_-$ is obvious: particle and antiparticle must have the same mass. The requirement $\alpha_+ = \alpha_-$ means that the particle destruction portion of the field (Φ_+), and antiparticle creation portion ($\Phi_-^\dagger$), must appear in the symmetric form[8] (19). Causality requires all observables to be constructed from such symmetric operators. This delivers the *coup de grace* to all attempts to build a theory of charged particles without antiparticles.

While the conclusion $\alpha_+ = \alpha_-, \mu_+ = \mu_-$, was derived for the spin-0 field, it is evident from the derivation of Eq. (17) for the Dirac field that causality

---

[8] Putting $\alpha_+ = \alpha_-$, but different from unity, does not alter the physical content of the theory. It merely amounts to a rescaling of the field, and this can be compensated by a corresponding rescaling of the coupling constant. Thus, in electrodynamics we can let $\mathbf{E} \to f\mathbf{E}, e \to e/f$.

also requires equal masses for fermions and their antiparticles, and equal amplitudes for the creation of a fermion and the destruction of the corresponding antifermion. This equality of the amplitude for emission of a particle, and the destruction of its antiparticle, is a universal requirement of causality, and leads immediately to *crossing symmetry*.

[Further insight into why Φ_+ and Φ_- are not local operators can be gained by deriving an equation of motion for Φ_+ (or Φ_-). By taking the time derivative of Φ_+ in (21), we obtain the equation of motion:

$$i\frac{\partial}{\partial t}\Phi_+(\mathbf{x}, t) = \int K(\mu |\mathbf{x} - \mathbf{x}'|)\Phi_+(\mathbf{x}', t) d^3x' \qquad (24)$$

where

$$K(\mu R) = \int \frac{d^3p}{(2\pi)^3} e^{i\mathbf{p}\cdot\mathbf{R}}\sqrt{p^2 + \mu^2}. \qquad (25)$$

(Φ_- satisfies the same equation.) Thus Φ_+ satisfies the Schrödinger-like equation (24) with a Hamiltonian K that is just the Fourier transform of the energy—hardly an astonishing result. Indeed, the nonrelativistic limit is found by letting μ tend to infinity, which replaces $\sqrt{p^2 + \mu^2}$ by $\mu + (p^2/2\mu)$ in (25), and gives

$$K(\mu R) \to \mu \delta^3(\mathbf{R}) - \frac{1}{2\mu}\nabla^2 \delta^3(\mathbf{R}). \qquad (26)$$

When this is substituted into (24), we retrieve the ordinary[9] Schrödinger equation. This familiar equation has a property that will now prove to be of vital significance; it is a *local* relation between Φ_+ and $\dot{\Phi}_+$, that is, $\Phi_+(\mathbf{x}, t)$ is determined completely by Φ_+ and its spatial derivatives at the *same* space-time point. In contrast to this, the relativistic "theory", Eq. (24), is *nonlocal*. Instead of a δ-function and its derivatives, the "relativistic" kernel K has an exponential tail

$$K(\mu R) \xrightarrow[\mu R \to \infty]{} -\left(\frac{1}{2\pi}\right)^{3/2} \frac{\mu^4}{(\mu R)^{7/2}} e^{-\mu R}. \qquad (27)$$

Now suppose that we insert a field Φ_+ into (24) that is localized within an infinitesimal volume; then because of (27) our equation of motion will instanteneously spread Φ_+ throughout all space.

This now raises the question of how the field $\Phi = \Phi_+ + \Phi_-^\dagger$ manages to be local. The answer is that Φ_+ and Φ_- separately satisfy equation (24), which is nonlocal in space, whereas Φ only obeys the local Klein–Gordon equation.

---

[9] The extra term $\mu\Phi_+$ just corresponds to the fact that here we include the rest mass in the energy, which is usually not done in ordinary wave mechanics.

Hence, there is no expression of the form (24) for the time derivative of Φ. This is just a clumsy way of saying that not only Φ, but also $\partial_t\Phi$, must be supplied as initial data, i.e., that Φ satisfies a second-order equation in ∂_t—the Klein–Gordon equation. That equation computes $\Phi(\mathbf{x}, t)$ an instant later, and it does so with data taken solely from the infinitesimal neighborhood of $\mathbf{x}$. That is why Φ is a local operator.]]

Appendix V

VACUUM POLARIZATION

The phenomenon of vacuum polarization appears most simply if one studies the interaction energy of two fixed charges.

The Feynman approach to field theory, as described in Chap. II, leads most directly to scattering amplitudes, and we must therefore learn how to extract an interaction energy from a scattering amplitude. For this purpose, consider the electromagnetic (Rutherford) scattering of two spin-0 objects of charge Q. The lowest order graph is

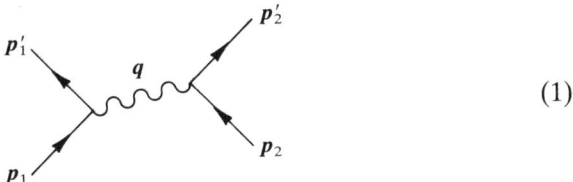
(1)

where $q = p_1 - p'_1$. The scattering amplitude is given by the rules of §II.2.4:

$$T(q) = \text{const.} \frac{Q^2}{q^2} = -\text{const.} \frac{Q^2}{\mathbf{q}^2}, \tag{2}$$

with the latter expression holding in the c.o.m. frame, where $q = (0, \mathbf{q})$. The connection between the scattering amplitude and the interaction potential $\Phi_0(r)$ is given by the Born approximation formula,

$$T(\mathbf{q}) \propto \int e^{-i\mathbf{q}\cdot\mathbf{r}} \Phi_0(r) \, d^3r. \tag{3}$$

By Fourier inversion Φ_0 can therefore be found from the scattering amplitude—i.e., the Feynman propagator when q is purely spatial. Applying this to (2) gives

$$\Phi_0(r) = \int \frac{d^3q}{(2\pi)^3} e^{i\mathbf{q}\cdot\mathbf{r}} T(q) = \int \frac{d^3q}{(2\pi)^3} e^{i\mathbf{q}\cdot\mathbf{r}} \frac{Q^2}{(\mathbf{q})^2} = \frac{Q^2}{4\pi r}, \tag{4}$$

the Coulomb potential between two charges.

The lowest order radiative correction to the Coulomb potential is found by replacing the "bare" photon line in (1) by the graph:

$$\text{(5)}$$

Here the loop represents a virtual e^+e^- pair; pairs of all other charged particles also contribute, but as we shall see, they are not important because their relative contribution decreases like $(m_e/m)^2$, where m is the mass of these other particles. The evaluation of the Feynman graph (5) is rather complicated, and the essential features can be understood from a more heuristic approach. The transient e^+e^- pair in (5) has a total mass M that is somewhere above $2m_e$, depending on the relative e^+e^- momentum. One can therefore replace every particular transient e^+e^- pair by a particle having the appropriate mass M. As we also know, such an object has a propagator $(p^2 - M^2)^{-1}$, whose momentum p is found by 4-momentum conservation at the vertices. From (5) it is clear that $p = q$. Thus, we expect that the e^+e^- loop in (5) will contribute a factor

$$e^2 \int_{2m_e}^{\infty} \frac{\rho(M)\,dM}{q^2 - M^2} \equiv \Pi(q^2). \tag{6}$$

Here $\rho(M)$ is the probability that the transient e^+e^- pair has total mass M; the detailed form of ρ requires the evaluation of the Feynman integral, but we do not need that here.

The contribution of diagram (5) is therefore $(1/q^2)\Pi(q^2)(1/q^2)$. On combining this with (2), we obtain the scattering amplitude corrected for vacuum polarization:

$$T(q) = Q^2\left[\frac{1}{q^2} + \frac{1}{q^2}\Pi(q^2)\frac{1}{q^2}\right]. \tag{7}$$

As before, we get the interaction energy $\Phi(r)$ as a function of separation between the charges by replacing q by $(0, \mathbf{q})$ in this amplitude, and taking the Fourier transform.

Concerning $\Phi(r)$ we know one thing with very high precision:[1] as $r \to \infty$, the force between two charges falls like $1/r^2$. The coefficient of this $1/r^2$ asymptote is written as $Q_{\text{obs}}^2/4\pi$, where Q_{obs} is, by definition, the observed or "physical" charge. As we shall soon see, Q_{obs} differs from the "input"

---

[1] E. R. Williams, J. E. Fuller, and H. A. Hill, *Phys. Rev. Lett.* **26**, 721 (1971), have performed a modern version of the Cavendish experiment, and shown that the force varies like $1/r^{2+\varepsilon}$, where $\varepsilon = (2.7 \pm 3.1) \times 10^{-16}$. For further discussion see Jackson (1976), §I.2.

charge Q with which we began our discussion. The large distance behavior of the potential is determined by grazing collisions, i.e., by the small momentum transfer behavior of T, and if Φ is to vary asymptotically like $1/r$, then T must have the Rutherford behavior[2] $1/q^2$ as $q^2 \to 0$. Hence, it is imperative that the correction term $\Pi(q^2)/q^4$ also varies like $1/q^2$, and this requires $\Pi(q^2)$ to be proportional to q^2 as $q^2 \to 0$. If we examine the expression (6) for Π at $q^2 = 0$, it is clear that it does not vanish because $\rho(M)$, being a probability, is positive definite. Thus, our corrected $T(q)$ does not lead to a Coulomb force at large distance! The reason for this calamity is that the virtual e^+e^- pairs described by Π have given the photon a mass of order e^2. That is, although we set out assuming a massless photon, there is a photon self-energy, just as there is an electron self-energy.

To see why a nonzero value of $\Pi(0)$ corresponds to a photon mass, we first consider the propagator of a hypothetical photon, the square of whose mass is $e^2 m^2 a$, which we then expand in powers of e^2:

$$\frac{1}{q^2 - e^2 m^2 a} \simeq \frac{1}{q^2} + \frac{1}{q^2} e^2 m^2 a \frac{1}{q^2} + 0(e^4). \tag{8}$$

On the other hand, (7) tells us that the photon propagator $D(q^2)$ is the expression in parentheses, i.e.,

$$D(q^2) = \frac{1}{q^2} + \frac{1}{q^2} \Pi(q^2) \frac{1}{q^2} + 0(e^4). \tag{9}$$

The first term is the photon propagator without the loop, and is independent of e^2, while the second term due to the loop is proportional to e^2. To compare (9) with (8), we expand Π about $q^2 = 0$:

$$\Pi(q^2) = \Pi(0) + q^2 \Pi'(0) + \cdots. \tag{10}$$

When the leading term is inserted into (9), we see that it has the physical significance of being the square of a photon mass ($\Pi(0) = e^2 m^2 a$). The previously stated limit on the departure from Coulomb's law implies that the photon mass $\sqrt{\Pi(0)}$ must be smaller than $2 \times 10^{-20} m_e$. It takes but little courage to insist on $\Pi(0) \equiv 0$.

We remove this unwanted photon mass in a manner reminiscent of fermion mass renormalization. That is, we begin once more from scratch, but give the photon a "bare" or input mass μ_0; then we calculate the radiative correction $\Pi(q^2)$ as before; and finally, we require the total "physical" mass $\mu_0^2 + \Pi(0)$ to vanish, i.e., $\Pi(0) = -\mu_0^2$. Instead of (7), we

---

[2] Henceforth, q^2 means $\mathbf{q}^2$.

then have

$$T(q) = Q^2 \left\{ \frac{1}{q^2} + \frac{1}{q^2} [\Pi(q^2) + \mu_0^2] \frac{1}{q^2} \right\}$$

$$= Q^2 \left\{ \frac{1}{q^2} + \frac{1}{q^2} [\Pi(q^2) - \Pi(0)] \frac{1}{q^2} \right\}. \quad (11)$$

If $\Pi(0) + \mu_0^2 = 0$, this leads to a pure Coulomb field for $q^2 \to 0$.

We return to (6), which gives us

$$\Pi(q^2) - \Pi(0) = e^2 q^2 \int_{2m_e}^{\infty} \frac{\rho(M) \, dM}{(M^2 - q^2)M^2} \equiv e^2 q^2 \Lambda(q^2). \quad (12)$$

By definition, $\Lambda(q^2)$ vanishes for $q^2 \to 0$. We exploit the identity

$$\Lambda(q^2) = \int \frac{\rho(M) \, dM}{(M^2 - q^2)M^2} \equiv \int \frac{\rho(M)}{M^4} \, dM + q^2 \int \frac{\rho(M) \, dM}{(M^2 - q^2)M^4}$$

to write the amplitude as follows:

$$T(q) = T_0(q) + T_1(q),$$

with

$$T_0(q) = \frac{Q_{\text{obs}}^2}{q^2}, \qquad T_1(q) = e^2 Q^2 \int_{2m_e}^{\infty} \frac{\rho(M) \, dM}{(M^2 - q^2)M^4}, \quad (13)$$

and

$$Q_{\text{obs}}^2 = Q^2 \left(1 + e^2 \int \frac{\rho(M)}{M^4} \, dM \right). \quad (14)$$

Since Q_{obs}^2 does not depend upon q^2, the first part T_0 of T is the same as our original Coulombic scattering amplitude (2), but with a changed (renormalized) charge Q_{obs}.

Let us see how the difference between (13) and (2) changes the interaction energy $\Phi(r)$, i.e., the Fourier transform (4). In so doing, we use

$$\int \frac{e^{i\mathbf{q}\cdot\mathbf{r}}}{M^2 + \mathbf{q}^2} d^3q = \frac{e^{-Mr}}{r}.$$

This is the so-called Yukawa potential for the mass M; it is a shielded Coulomb potential with a range of M^{-1}. With $T(q)$ as given by (13), we find

$$\Phi(r) = \frac{Q_{\text{obs}}^2}{4\pi r} + \frac{e^2 Q^2}{4\pi r} \int_{2m}^{\infty} e^{-Mr} \frac{\rho(M)}{M^4} \, dM.$$

Since the difference between Q_{obs} and Q is of order e^2, we obtain, to first order in e^2,

$$\Phi(r) = \frac{Q_{\text{obs}}^2}{4\pi r}\left(1 + e^2 \int_{2m}^{\infty} e^{-Mr}\frac{\rho(M)}{M^4}\,dM\right). \tag{15}$$

For distances r much longer than $1/2m$ the second term vanishes, and $\Phi(r)$ becomes the Coulomb potential,

$$\Phi(r) = \frac{Q_{\text{obs}}^2}{4\pi r}, \qquad r \gg \frac{1}{2m},$$

i.e., between the renormalized charges Q_{obs}, as given by (14), but *not* between the input charges Q. Clearly, Q_{obs} is what we would call the charge of the objects under investigation, since we measure charges at macroscopic distances. But when the objects approach each other to distances comparable to or smaller than $(2m)^{-1}$, the second term of (15) becomes important, and increases the interaction energy. The charges become more effective at closer distance. Comparing (15) with Eq. II.D(15), we see that

$$C(r) = e^2 \int_{2m}^{\infty} e^{-Mr}\frac{\rho(M)}{M^4}\,dM.$$

A detailed evaluation of this integral gives the function $C(r)$ as plotted in Fig. II.6.

Appendix VI

THE MAGNETIC SUSCEPTIBILITY OF A MASSLESS VECTOR FIELD

We wish to evaluate the contribution of orbital motion to the magnetic susceptibility of the vacuum state of a massless charged spin-1 Bose field. The contribution of the magnetic moments was found in §IV.C.4(c). We first solve a simpler problem, namely the magnetic susceptibility of the vacuum state of a massless *scalar* Bose field with a charge e_0. The zero-point amplitude of an oscillator corresponds to a zero-point energy of $\frac{1}{2}\omega$. Therefore, the vacuum is equivalent to a state where all modes in a volume V are occupied with 50% probability. But there are two particles, with charges $+e_0$ and $-e_0$, per mode; thus the magnetic properties of the vacuum are those of an ensemble with one particle per mode, since only the square of the charge appears in the susceptibility.

We begin by calculating the dependence of the energy E_{sc} of the vacuum state on a constant applied magnetic field **B**. The magnetic susceptibility of the scalar field is then found from

$$\chi_m^{sc} = -\frac{1}{V}\frac{\partial^2 E_{sc}}{\partial B^2}. \tag{1}$$

In order to calculate E_{sc}, we recall that the energy operator E of a free massless relativistic particle is

$$E^2 = p^2 \tag{2}$$

In the presence of a magnetic field **B**, **p** must be replaced by $\mathbf{p} - e_0\mathbf{A}$, where **A** is the vector potential at the particle's position. Let **B** be parallel to the z-axis. Then

$$E^2 = (p_z^2 + v_x^2 + v_y^2), \tag{3}$$

where $v_x = p_x - e_0 A_x$, etc. Since $\mathbf{B} = \nabla \times \mathbf{A}$, we get $[v_x, v_y] = ie_0 B$. After introducing $u = v_x/\sqrt{e_0 B}$ and $p_u = v_y/\sqrt{e_0 B}$, this commutator becomes $[p_u, u] = -i$. Hence p_u is the canonical momentum conjugate to the variable u. Therefore, we write (2) in the form

$$E^2 = p_z^2 + \frac{1}{2}(u^2 + p_u^2)2e_0 B. \tag{4}$$

The second term is the Hamiltonian of an oscillator whose eigenvalues are $2e_0B(n + \tfrac{1}{2})$, where $n = 0, 1, 2, \ldots$; the first term refers only to the motion in the z-direction which is unaffected by the magnetic field.

Consider now the states i of a particle in a large cubic volume of linear dimension L. We characterize the states by the value of p_z and by the oscillator eigenvalue n that determines the motion in the x-y plane. According to (4), the energy ε_i of such state is given by

$$\varepsilon_i^2 = p_z^2 + 2e_0B(n + \tfrac{1}{2}). \tag{5}$$

The energy E_{sc} is the sum of energies ε_i over all states. In order to calculate this sum, we count the number of these states within an interval dp_z, and for a given value of n. The number of p_z-values in the interval dp_z is $L\,dp_z/(2\pi)$. The number of states with momenta perpendicular to z between p_1 and $p_1 + dp_1$ is $L^2 p_1\,dp_1/(2\pi)$. How many of the latter states correspond to one unit of n in (5)? We observed that $p_1^2 = p_x^2 + p_y^2 = 2(n + \tfrac{1}{2})e_0B$; therefore, $2p_1\,dp_1 = 2e_0B\,dn$. Thus the number of states per unit n is $L^2 e_0 B/2\pi$.

The total energy E_{sc} is then

$$E_{sc} = L\int_{-K}^{+K} \frac{dp_z}{2\pi} \frac{e_0 BL^2}{2\pi} \sum_{n=0}^{n^*} [p_z^2 + 2e_0B(n + \tfrac{1}{2})]^{1/2}. \tag{6}$$

Here we introduce the momentum cutoff $|\mathbf{p}| \leq K$ which, for each p_z, also determines the maximum value n^*:

$$p_z^2 + 2e_0B(n^* + \tfrac{1}{2}) = K^2. \tag{7}$$

In order to evaluate the sum over n we use the approximate Euler–McLaurin formula[1]

$$\sum_{n=0}^{b-1} f(n + \tfrac{1}{2}) = \int_0^b f(x)\,dx - \tfrac{1}{24}[f'(b) - f'(\tfrac{1}{2})], \tag{8}$$

appropriate when $|f''| \ll |f'|$. Using (8) we have

$$\sum_{n=0}^{n^*} \left[p_z^2 + 2e_0B\left(n + \frac{1}{2}\right)\right]^{1/2} = \frac{1}{3e_0 B}(K^3 - p_z^3) + \frac{1}{24(p_z^2 + e_0 B)^{1/2}}. \tag{9}$$

The upper limit $f'(b)$ in (8) can be neglected compared to $f'(\tfrac{1}{2})$. When (9) is inserted into (6), the first term will give rise to an expression independent of B. The magnetic field produced the discreteness of the spectrum, and

---

[1] This differs from the conventional expression in that the summation variable is $n + \tfrac{1}{2}$ [see Nielsen, (1981)].

therefore the classical approximation, as described by the integral, does not depend on B. The second term of (9) gives

$$E_{sc} = \frac{L^3 e_0^2 B^2}{4\pi^2} \frac{1}{12} \ln \frac{K}{(e_0 B)^{1/2}}. \tag{10}$$

This is the field-dependent part of the vacuum energy of a massless charged scalar boson field. According to (1) the magnetic susceptibility of this state is

$$\chi_m^{sc} = -\frac{e_0^2}{24\pi^2} \ln \frac{K}{(e_0 B)^{1/2}}; \tag{11}$$

as $B \to 0$. The system is diamagnetic, as expected.

We use this expression to determine the magnetic susceptibility of the vacuum of a massless charged spin-1 boson field due to orbital motion. For this purpose we assume that the orbital effects of a charged spin-1 boson field are the same as those of a charged scalar field except that there are two spin states for each particle.[2] That would multiply (10) and (11) by a factor of 2. Hence the orbital contribution χ_m^{orb} to the magnetic susceptibility of the vacuum of a charged spin-1 field is

$$\chi_m^{orb} = -\frac{e_0^2}{12\pi^2} \ln \frac{K}{(e_0 B)^{1/2}}, \tag{12}$$

i.e., $-\frac{1}{12}$ of the spin contribution of the spin-1 field, as claimed in Eq. IV.C(52).

---

[2] For a massless particle, the separation of angular momentum into spin and orbit is not unambiguous. A calculation that takes this into account is given by Nielsen (1981).

Appendix VII

SOLUTIONS OF DIRAC'S EQUATION IN A SPHERICAL ENCLOSURE

We solve the Dirac equation for a scalar potential as given by Eq. IV.D(4)

$$V(r) = 0, \quad \text{for } r < R \qquad (A \to \infty). \tag{1}$$
$$V(r) = A, \quad \text{for } r > R$$

In order to do so, we use the "standard" representation of the Dirac matrices, not the one given by Eq. II.A(31). The latter is appropriate for solving problems when the mass is negligible, but the potential (1), is equivalent to a mass going to infinity for $r > R$. The standard matrices γ_μ^s are related to Eq. II.A(31) by $\gamma_\mu^s = A\gamma_\mu A^{-1}$, where A is the matrix given in the footnote on p. 565.

As most textbooks show [e.g., Bjorken and Drell (1964), §4.4], the four components u_i of the wave function ψ in a spherically symmetric potential, in the standard representation, are given by

$$\begin{pmatrix} u_1 \\ u_2 \end{pmatrix} = \frac{g(r)}{r} y^+, \qquad \begin{pmatrix} u_3 \\ u_4 \end{pmatrix} = \frac{f(r)}{r} iy^-, \tag{2}$$

where $y^\pm$ are two-component spinors depending only on the angles, and related by

$$(\hat{\mathbf{r}} \cdot \boldsymbol{\sigma}) y^\pm = y^\mp. \tag{3}$$

For the lowest $j = m = \tfrac{1}{2}$ state, one finds

$$y^+ = \begin{pmatrix} (4\pi)^{-1/2} \\ 0 \end{pmatrix}, \qquad y^- = \begin{pmatrix} (3)^{-1/2} Y_{10} \\ (\tfrac{2}{3})^{1/2} Y_{11} \end{pmatrix}. \tag{4}$$

Here Y_{lm} are the conventional normalized spherical harmonics. The functions $g(r)$ and $f(r)$ satisfy differential equations

$$[E + (m + V(r))]f = \frac{dg}{dr} - \frac{g}{r}$$
$$[E - (m + V(r))]g = -\frac{df}{dr} - \frac{f}{r}. \tag{5}$$

In contrast to the Coulomb potential, the scalar potential appears as an addition to the mass.

Let us solve (5) for the case of massless quarks, using the potential (1):

$$\left.\begin{array}{l} Ef = g' - g/r \\ Eg = -f' - f/r \end{array}\right\} \quad \text{for } r < R. \tag{6}$$

For $r > R$ we neglect E and $1/r$ compared to A:

$$\left.\begin{array}{l} Af = g' \\ Ag = f' \end{array}\right\} \quad \text{for } r > R. \tag{7}$$

The solution of (7) is

$$g = Ce^{-A(r-R)}, \quad f = -Ce^{-A(r-R)}, \quad (r > R),$$

where we exclude a solution with a growing exponential, and C is a normalization constant. Therefore,

$$f = -g, \quad (r > R). \tag{8}$$

The following relation[1] is derived from (2) and (3):

$$(\mathbf{r} \cdot \boldsymbol{\gamma}^s)\psi = \begin{pmatrix} 0 & 1 \\ -1 & 0 \end{pmatrix}\begin{pmatrix} \hat{\mathbf{r}} \cdot \boldsymbol{\sigma} & 0 \\ 0 & \hat{\mathbf{r}} \cdot \boldsymbol{\sigma} \end{pmatrix}\begin{pmatrix} gy^+ \\ ify^- \end{pmatrix} = \begin{pmatrix} ify^+ \\ -gy^- \end{pmatrix}. \tag{9}$$

Because of (8), when $r = R$

$$(\hat{\mathbf{r}} \cdot \boldsymbol{\gamma}^s)\psi(R) = -i\psi(R), \tag{10}$$

which is the boundary condition of Eq. IV.D(5). Since

$$\bar{\psi}\psi = \psi^*\gamma_0^s\psi = (f^2 - g^2)/4\pi r^2 = 0, \quad \text{for } r = R, \tag{11}$$

---

[1] Here γ^s is the product of the two 4×4 Dirac matrices in Eq. (9).

we find that the current density $\bar{\psi}(\hat{\mathbf{r}} \cdot \boldsymbol{\gamma}^s)\psi$ perpendicular to the bag surface vanishes at R as expected.

We now solve Eq. (6). By inserting the first equation into the second, we get $E^2 g = -g''$, or

$$g = \sin x,$$

with $x = Er$. The first equation gives

$$f = \cos x - \frac{\sin x}{x}.$$

The boundary condition (8) leads to

$$\tan x_0 = \frac{x_0}{1 - x_0},$$

where $x_0 = ER$, which gives $x_0 = 2.04$. This is the value of β_0 used in Eq. IV.D(7). The solution of (5), including the mass term, would lead to Eq. IV.D(6), and the boundary condition (10) then gives the function β shown in Fig. IV.11.

BIBLIOGRAPHY

Alvensleben, H. et al. (1972), *Phys. Rev. Lett.* **28,** 66.
Amaldi, U. (1973), *Scientific American* **229,** p. 36.
Arenton, M. W. et al. (1982), *Phys. Rev.* **D25,** 2241.
Armstrong, T. A. et al. (1972), *Phys. Rev.* **D5,** 1640.
Ballam, J. et al. (1970), *Phys. Rev. Lett.* **24,** 960.
Barbaro-Galtieri, A. (1977), Proc. 1977 Intl. Symp. Lepton & Photon Interactions at High Energy, Hamburg; F. Gutbrod, editor, DESY, Hamburg.
Behrends, S. et al. (CLEO Collaboration) (1983), *Phys. Rev. Lett.* **50,** 881.
Berestetskii, V. B., E. M. Lifshitz, and L. P. Pitaevskii, (1971), *Relativistic Quantum Theory,* Pengamon Press, Oxford.
Berkelman, K. (1983), *Phys. Reports* **98,** No. 3.
Biagi, S. F. et al. (1981), *Zeits. f. Phys.* **C9,** 305.
Bjorken, J. D. and S. D. Drell (1964), *Relativistic Quantum Mechanics,* McGraw-Hill, New York.
Bjorken, J. D. and S. D. Drell (1965), *Relativistic Quantum Fields,* McGraw-Hill, New York.
Blatt, J. M. and V. F. Weisskopf (1952), *Theoretical Nuclear Physics,* J Wiley, New York.
Boyarski, A. M. et al. (1975), *Phys. Rev. Lett.* **34,** 1357.
Brown, L. M. (1978), *Physics Today,* September 1978, p. 23.
Cahn, R. N. (1984), *Semi-Simple Lie Algebras and Their Representations,* W. A. Benjamin, New York.
Carruthers, P. (1966), *Introduction to Unitary Symmetry,* Wiley-Interscience, New York.
Chang, N. P. (1977), *Five Decades of Weak Interactions,* New York Academy of Sciences.
Cheng, T.-P. and L.-F. Li (1984), *Gauge Theory of Elementary Particle Physics,* Oxford University Press, Oxford.
Close, F. E. (1979), *Quarks and Partons,* Academic Press, London.
Coleman, S. (1979), *Science* **206,** 1290.
Commins, E. D. (1973), *Weak Interactions,* McGraw-Hill, New York.
Cox, P. T. et al. (1981), *Phys. Rev. Lett.* **46,** 877.
Dally, E. B. et al. (1980), *Phys. Rev. Lett.* **45,** 232.
Dally, E. B. et al. (1982), *Phys. Rev. Lett.* **48,** 375.
Daum, C. et al. (1981), *Phys. Lett.* **100B,** 439.
Davier, M. (1982), Proc. XXI, Intl. Conf. High Energy Physics, Paris, P. Petiau, editor; Les éditions de physique, Paris, 1982.
DeGrand, T., R. L. Jaffe, K. Johnson, and J. Kiskis (1975), *Phys. Rev.* **D12,** 2060.
Delfosse, A. et al. (1981), *Nucl. Phys.* **B183,** 394.
Dittman, P. and V. Hepp (1982), *Zeit. f. Phys.* **C10,** 283.

Edwards, C. et al. (Cal Tech, Harvard, Princeton, SLAC, Stanford Collaboration) (1982), *Phys. Rev. Lett.* **48,** 70.
Eichten, E., K. Gottfried, T. Kinoshita, K. D. Lane and T.-M. Yan (1980), *Phys. Rev.* **D21,** 203.
Feynman, R. P., R. B. Leighton, and M. Sands (1964), *The Feynman Lectures on Physics,* Vol. II, Addison-Wesley, Reqding, Mass.
Flaminio G. et al. (1979a), CERN High Energy Reactions Analysis Group, Report HERA 79-01.
Flaminio G. et al. (1979b), ibid., Report HERA 79-02.
Franzini, P. and J. Lee-Franzini (1983), *Ann. Rev. Nucl. Part. Sci.* **33,** 1.
Galison, P. (1983), *Rev. Mod. Phys.* **55,** 477.
de Gennes P. G. (1966), *Superconductivity of Metal and Alloys,* W. A. Benjamin, New York.
Georgi, H. and S. L. Glashow (1974a) *Phys. Rev. Lett.* **32,** 438.
Georgi, H., H. Quinn, and S. Weinberg (1974b), *Phys. Rev. Lett.* **33,** 451.
Georgi, H. (1982), *Lie Algebras in Particle Physics,* W. A. Benjamin, New York.
Giacomelli, G. et al. (1969), CERN Report, HERA 69-1.
Gottfried, K. (1966), *Quantum Mechanics,* W. A. Benjamin, New York.
Gottfried, K. (1978), *Phys. Rev. Lett.* **40,** 598.
Green, M. B. (1985), Proc. 1985 Intl. Symp. Lepton & Photon Interactions at High Energy, Kyoto; M. Konuma and K. Takahashi, editors, Yukawa Hall, Kyoto; referred to henceforth by Kyoto Symposium.
Greyer, G. et al. (1974), *Nuc. Phys.* **B75,** 189.
Hansson, T. H., K. A. Johnson, and C. Peterson (1982), *Phys. Rev.* **D26,** 2069 and 2070.
Hoehler, G. et al. (1976), *Nuc. Phys.* **B114,** 505.
Holmgren, S. D. et al. (1977), *Nuc. Phys.* **B119,** 261.
Jackson, J.D. (1975), *Classical Electrodynamics,* 2nd ed., Wiley, New York.
Johnson, K. A. and C. B. Thorn (1976), *Phys. Rev.* **D13,** 1934.
Kabir, P. K. (1963), *The Development of Weak Interaction Theory,* Gordon & Breach, New York.
Kim, J. E., P. Langacker, M. Levine, and H. H. Williams (1981), *Rev. Mod. Phys.* **53,** 211.
Kuang, Y.-P. and T.-M. Yan (1981), *Phys. Rev.* **D24,** 2874.
Landau, L. D. and E. M. Lifshitz (1977), *Quantum Mechanics,* 3rd ed., Pergamon Press, London.
Langacker, P. (1985), Kyoto Symposium.
LeFrancois, J. (1971), Proc. Intl. Symp. on Electron and Photon Interactions at High Energy, Cornell University, Ithaca, New York; N. Mistry, editor, p. 51; Laboratory of Nuclear Studies, Cornell University, 1972.
Lifshitz, E. M. and L. P. Pitaevskii (1974), *Relativistic Quantum Theory,* Part 2, Pergamon Press, London.
Llewellyn Smith, C. H. (1973), *Phys. Lett.* **46B,** 233.
Mackenzie, P. B. and G. P. Lepage (1981), *Phys. Rev. Lett* **47,** 244.
Marciano, W. (1979), *Phys. Rev.* **D20,** 274.
Marciano, W. J. and A. Sirlin (1983), *Phys. Rev.* **D27,** 552.
Mark J Collaboration (1982), *Phys. Reports* **63,** No. 7.
McClellan, G., et al. (1971), *Phys. Rev. Lett.* **26,** 1593.
Mess, K. H. and B. H. Wiik (1982), DESY Report No. 82/011.
Messiah, A. (1966), *Quantum Mechanics,* Wiley, New York.

Nagy, E. et al. (1979), *Nuc. Phys.* **B150,** 221.
Nielsen, N. K. (1981), *Am. J. Phys.* **49,** 1171.
Particle Data Group (1984), *Rev. Mod. Phys.* **56,** No. 2, Part II.
Partridge, R. et al. (1980), *Phys. Rev. Lett.* **45,** 1150.
Perkins, D. H. (1982). *Introduction to High Energy Physics,* 2nd ed., Addison-Wesley, Reading, Mass.
Prescott, C. Y. et al. (1978), *Phys. Lett.* **77B,** 347.
Prescott, C. Y. et al. (1979), ibid. **84B,** 524.
Rapidis, P. et al. (1977), *Phys. Rev. Lett.* **39,** 526.
Ross, D. A. and T. J. Goldman (1980), *Nuc. Phys.* **B171,** 273.
Schrieffer, J. R. (1964), *Theory of Superconductivity,* W. A. Benjamin, New York.
Schwarz, J. H. (1985), *Superstrings,* World Scientific, Singapore.
Siegrist, J. L. et al. (1982), *Phys. Rev.* **D26,** 969.
Silverman, A. (1984), Proc. XXII Intl. Conf. High Energy Phys., Leipzig.
Streater, R. F. and A. S. Wightman (1964), *PCT, Spin and Statistics, and All That,* W. A. Benjamin, New York.
TASSO Collaboration (1982), DESY Report No. 82/32.
Tinkham, M. (1975), *Introduction to Superconductivity,* McGraw-Hill, New York.
van der Waerden, B. L. (1974), *Group Theory and Quantum Mechanics,* Springer, New York.
Weber, G. (1967), in Proc. 1967 Intl. Symp. Electron & Photon Inst. of High Energy, Stanford, California.
Weinberg, S. (1964), *Phys. Rev.* **133,** B1318; **134,** B882.
Weinberg, S. (1971), *Phys. Rev. Lett.* **27,** 1688.
Weisskopf, V. F. (1981), *Contemporary Physics,* **22,** 375.
Wetzel, W. (1983), *Nuc. Phys.* **B227,** 1.
Witten, E. (1980), *Phys. Lett.* **91B,** 81.
Wu, S. L. (1984), *Phys. Reports* **107,** Nos. 2–5.
Wybourne, B. G. (1974), *Classical Groups for Physicists,* Wiley, New York.
Yan, T.-M. (1980), *Phys. Rev.* **D22,** 1652.

INDEX

Anticommutation rules, 568, 574, 581
Antiparticles, 208
   and causality, 584
Antiscreening in QCD, 331, 386, 387
   See also Confinement
Asymptotic freedom, 380, 396
   See also Running coupling constant;
      Vacuum polarization in QCD

Bag Model, 404–19
   boundary conditions, 406, 411
   center-of-mass motion in, 413
   charge and current density, 408
   eigenvalues in, 406
   hadron masses, 406, 413, 415–17
   hadron radii, 407, 415
   interaction of quarks in, 409–12
   magnetic moments, 408, 416
   and QCD vacuum structure, 398–401
   Regge slope, 417–19
Barn, 289
Baryon resonances, 291–95, 298, 303, 304
$b\bar{b}$ spectrum, 331–32
   See also $Q\bar{Q}$-systems
Bohr formula, 241
Branching ratio (or fraction), 288
Breit-Wigner formula, 288
   in $e\bar{e}$ scattering, 316, 322
Broken symmetry, *See* Hidden symmetry

$\mathscr{C}_3$, 343
Cabibbo angle, 440, 472, 478
Callan-Gross relation, 428, 432, 440, 446
Causality condition, 579
   crossing symmetry, 584–86
   spin and statistics, 581–83
$c\bar{c}$ spectrum, 328–30, 332
   See also $Q\bar{Q}$-systems
Central collisions, 299
Charge conjugation
   and isospin, 305
   for photons, 249
   for pions, 305
   in positronium, 249
   in $SU(3)$, 356
Charge distribution
   of leptons, 240–43

   of mesons, 281
   of nucleons, 279
Charge operator, 208, 209, 555, 570
Charge quantization, 372
Charmonium, 313
Chirality, 471
Color
   "addition" of, 355, 357–59, 361–63, 380
   charge density of, 379
   induced, 382–86
   conservation of, 372
   coupling constant, 369, 373
   electric field, 376
   of gauge field, 370–80
   magnetic field, 376
   singlet, 345
   total, operator for, 380
   variable, 342
   wavefunctions, 346
   See also Bag Model; Confinement; Gauge
      field
Commutation rules
   for antiparticle operators, 581, 583
   and decay of positronium, 583
   for Dirac field, 583
   for scalar field, 580–82
   See also Anticommutation rules
Complementary Principle, 558
Complex fields, 553
Compton scattering, 221
   Feynman diagrams for, 228
Confinement, 397–401
   and scattering, 425
   See also Bag Model
Cooper pairs, 512, 516
Coulomb's law, 590
Coupling constant
   dimension of, 223, 231, 478
   in gauge theories, 372, 373
   in Grand Unified theory, 534–39
   of QCD, 334, 369, 373, 397
   See also Running coupling constant
Covariant derivative
   electromagnetic, 369
   in electroweak theory, 519, 520
   in Grand Unified theory, 534
   in $SU(3)$, 369, 370, 378

INDEX

CP transformation, 473
CP violation, 474
Creation operators
  for fermion, 207, 208, 568
  in oscillator, 547
  for photons, 205, 208, 557
  for spin-0 particles, 550, 554
Cross section
  defined, 219
  evaluation of, 226–28

Decay
  process, 284
  rate, 286
Deep inelastic scattering, 424
  kinematics, 425–27
Deep inelastic scattering, charge-changing weak
  cross section, 439, 443, 445
  data for, 448
  structure functions, 440, 446
Deep inelastic scattering, charge-preserving weak, 499–502
  determination of θ_W, 502
Deep inelastic scattering, electromagnetic
  cross section, 427, 438
  data for, 433–35
  structure functions, 428, 459, 461
Deep inelastic scattering, parity violation in, 502–4
Destruction operators, See Creation operators
Dielectric constant (coefficient)
  in QED, 260, 261, 263
Diffraction scattering of hadrons, 281–83
Dirac current, 203, 204, 566
Dirac equation, 202–4, 563–65
  in momentum space, 564
  in spherical enclosure, 600
Dirac field
  charge operator, 570
  commutation rules, 583
  current operator, 566, 569
  Hamiltonian of, 568
  Lorentz transformation of, 471
  L-R decomposition of, 470
  plane wave expansion of, 207
  space reflection, 473
    intrinsic parities, 570
Dirac matrices, 203, 563, 565
Dirac spinor, 202, 207, 563
  for antiparticles, 565
  plane wave expansion of, 206, 567
  space reflection, 562, 563
  in spherical enclosure, 601

UR limit of, 215, 471, 565, 567
See also Dirac field

$\mathscr{E}_3$, 559
$\mathscr{E}_3^T$, 485
$e\bar{e} \to e\bar{e}$
  data for, 239
  hadronic resonances in, 320
  $\mu\bar{\mu}$-resonances in, 314–17
  theory for, 237–38
$e\bar{e} \to$ hadrons, 318–24, 507
$e\bar{e} \to$ Higgs, 525
$e\bar{e} \to \mu\bar{\mu}$
  data for, 240
  hadronic resonances in, 319, 320
  perturbation theory of, 235–36, 317
$e\bar{e} \to \tau\bar{\tau}$, 241
$e\bar{e} \to 2\gamma$, 315
$e\bar{e} \to W\bar{W}$, 480–82, 490
$e\bar{e} \to Z^0$, 506–8
e_L, 471, 473, 518
e_R, 471, 473, 517
Electromagnetic field
  angular momentum of, 198
  energy of, 197
  Hamiltonian of, 196
  momentum of, 197
  operators for, 537
Electromagnetic interaction
  operator, 212–14
  See also Electroweak theory
Electromagnetic potentials, 195
Electron scattering
  inelastic, 289
  by nucleon, elastic, 280
  by nucleus, 424
  See also Deep inelastic scattering; Lepton-nucleon scattering
Electroproduction, 289
Electroweak theory
  coupling constants in, 494, 506
  $e\bar{e} \to$ fermion pairs, 507
  $e\bar{e} \to Z^0$, 506
  gauge group, 518
  incorporation of QED, 491–93
  mixing angle θ_W defined, 494
    data, 502, 505
  in Grand Unified theory, 535–40
  W and Z masses, 506
Exotic hadrons, 297

$\mathfrak{F}$, gauge group of standard model, 530
Fermi constant, 475
Fermi interaction
  for charge-changing processes, 475

INDEX

for charge-preserving processes, 498, 499
Fermion-Fermion scattering
  in electroweak theory, 571–73
  in QED, 232–35, 571–73
Fermions, 198–204
  helicity of, 200
Ferromagnetism, 509–11
  and gauge invariance, 495
Feynman diagrams
  for bound states, 244, 313
  relation to perturbation theory, 225
  rules for construction of, 229, 257
Field operators, 195, 579
Fine structure, 246
  in He atom, 273
Flavor threshold, 334
Flux tube, in QCD, 400, 418
Formation reactions, 283, 299
Form factor, 242, 279, 423
  of nucleon, 280, 424

$\mathfrak{G}_{EW}$, gauge group of electroweak theory, 530
G-parity
  defined, 305
  of mesons, 309–11
  and quark model, 311
Gauge field
  in Grand Unified theory, 534
  interaction with sources, 372
  quanta of, 371
  self-interaction, 377
  strengths, 376
  $SU(3)$, 369, 371
  in weak interaction, 490
  See also Electroweak theory; Grand Unification; Yang-Mills equation
Gauge invariance
  and boson masses, 371, 511, 516, 523
  electromagnetic, 368
  in electroweak theory, 496, 517–19
  and fermion masses, 526
  in Grand Unification, 540
  and hidden symmetry, 523
  in superconductor, 514–16
  $SU(3)$, 366, 370
Gauge theory, 366
  see also Gauge fields; Gauge invariance; Yang-Mills equations
General Relativity
  and gauge theory, 365–66, 374
  and superstrings, 543
Generation, of fermions, 531, 532
Ginzburg-Landau Hamiltonian, 512, 516, 520
Global symmetry, 365
Gluon, 371

bremsstrahlung, 455, 456
field in nucleon, 452–55
momentum in nucleon, 437, 455
self-coupling, 375
See also Confinement; Gauge field
Gluon-gluon scattering, 375
Goldstone's theorem, 511
Grand Unification
  coupling constant, 534, 536–39
  fermion multiplets in 532, 533, 535, 540
  gauge group, 540
  hierarchy problem in, 542
  Higgs mechanism in, 541
  proton decay, 533, 543
  unification mass, 539
  vertices in, 533
  θ_W, 533, 535–40
Groups
  Abelian, 349
  definition of, 348
  direct product of, 349, 356–59
  non-Abelian, 349
  Poincaré, 349
  simple, 530
  unimodular, 348
  unitary, 348
  weight diagram, 352
  See also $\mathfrak{F}$; $\mathfrak{G}_{EW}$; $SU(2)$; $SU(3)$; $U(N)$

Harmonic oscillator, 545–47
Helicity
  and hidden gauge invariance, 511, 516, 524
  Lorentz transformation of, 200, 201
Helicity rule in UR phenomena, 214–16, 441, 471, 571–73
Hidden symmetry
  in electroweak theory, 520–28
  in ferromagnetism, 510
  in Grand Unification, 541, 542
  in superconductors, 511–16
  See also Higgs field
Higgs boson, 521
  in electroweak theory, 524–28
Higgs field
  in electroweak theory, 517, 519
Hydrogenic states, 245–46
Hyperfine structure, 247
  in hydrogen, 272

Impulse approximation, 424
Invariant mass, 284
Isospin
  determination of, 291, 295
  in πN scattering, 295–97

INDEX

J/ψ, 319, 321
  See also $c\bar{c}$ system

$K_L \to \mu\bar{\mu}$, 497
Klein-Gordon equation, 198, 548, 585
Klein-Nishina formula, 226
KN Scattering, 297, 298

λ-matrices, 351
Lamb shift, 226–70
Lattice QCD, 403
Lepton-nucleon scattering
  kinematics, 425–27
  See also Deep inelastic scattering;
    Electron scattering
Leptons
  charge radii of, 243
  excited states of, 193
Lifetime, 286
Light-by-light scattering, 226
Local observables, 579
Local symmetry, 365–66
  See also gauge invariance
London's equation, 515
Lorentz transformation (LT)
  of Dirac spinors, 562
  of 4-vectors, 560, 566
  proper, 562

Magnetic moment, anomalous
  of leptons, 270–72
Magnetic monopole, 372
Magnetic susceptibility
  of vacuum, 388–89
  of vector field, 595–97
Mass generation
  of fermions, 526–527
  of vector bosons, 523
Maxwell's equations, 199, 368
  in Weyl form, 201
Meissner effect, 511, 515, 517
Meson resonances, 300–302, 306–8
MHz, 246

Mu-mesic atoms, 245
  and QED, 265
$\mu\bar{\mu}$ bound states, 314–17
Multipole expansion in QCD, 337
Muonium, 245
  hyperfine structure in, 273

Neutrino masses, 199
Neutrino scattering
  high energy behavior of, 479–82
  by nucleons, 444–48
  by quarks, 440
  See also Deep inelastic scattering

Occupation numbers, 197, 568

Parity, intrinsic
  of fermions and antifermions, 210, 570
Parity violation
  in $e\bar{e} \to \mu\bar{\mu}$, 240
  in inelastic eN scattering, 504, 505
Pauli catastrophe, 342
Pauli principle, 568
Peripheral collisions, 299, 300, 303
Photon, 196–98
  helicity of, 196
  mass, 371, 591
  multiphoton states, 197
Photoproduction
  of pions, 289–90
  of vector mesons, 303
Pion-nucleon scattering, 291–95
Planck length, 258, 263
Point structure
  of leptons, 243
  of quarks and leptons, 444
Positronium, 245, 247–51
  and causality, 583
  decay of, 250, 251
Positrons, 209
  in Feynman diagrams, 217
Propagator, 230
Proton decay, 533, 543
ψ, 319, 321
$\psi L_{,R}$, 470
  See also $c\bar{c}$ spectrum

$Q\bar{Q}$-systems, 312
  decay into gluons, 333
  hadronic cascades in, 336–37
  phenomenological potential for, 402
  radiative transitions in, 326–31
  $S-D$ mixing in, 321, 336
  spectrum of, 325–27
Quark-antiquark potential
  in perturbative QCD, 392, 393
  phenomenological, 402
  Richardson potential, 402
  See also $Q\bar{Q}$-systems
Quark-gluon coupling, 372
Quark model of inelastic scattering
  cross section
    electromagnetic, 429, 431
    weak, charge-changing, 440
    weak, charge-preserving, 501
  gluon momentum in, 437, 454, 455
  kinematics, 430, 431
  quark momentum distribution in, 433–36,
    449–52
  and scaling, 432
Quarkonium, 313

INDEX 607

Radiative corrections, 246
    in QED, 254, 481
    in weak interaction, 481, 490, 506
Rapidity, 561
Real fields, 549
Regge trajectory
    and $q\bar{q}$ interaction, 402
    See also Bag Model
Renormalization
    of charge in QED, 261
    of coupling constant in QCD, 396
    and dimension of coupling constant, 231
    in Grand Unification, 536–39
    group, 396
    of mass, 257, 258
    and weak interaction, 481, 489
    See also Running coupling constant
Resonance fluorescence, 285
Resonances
    elastic, 295
    in production, 300–303
ρ-parameter, 498, 502, 523
Rotations of spinors, 559
Running coupling constant
    electroweak angle, 536
    in Grand Unified theory, 536–39
    in QCD, 389, 396, 397
    in QED, 390, 395, 539
Rutherford scattering, 235

Scalar field
    classical theory of, 548–50
    operator, 223
    quantum theory of, 551–56
    scattering of, 222–28
    See also Spin-0 field
Scaling in deep inelastic scattering, 428, 432
    data on, 435
Scaling violations, 460–63
Scattering amplitude
    defined, 219
    fermion-fermion, 571–73
    and Feynman diagrams, 229
    perturbation expansion for, 220
    and probability conservation, 479
Sea quarks, 434, 448, 457–59, 460–63
Self-energy
    of electron, 255–58
$SO(10)$, 541
Spin and statistics
    from causality, 579–83
    for Dirac particles, 574
Spin, determination of, 293, 303, 308
Spin-0 field
    anti-particles, 554
    Bose statistics of, 552

    charge carried by, 555
    Hamiltonian of, 551
    operators for, 551, 553, 554
    states, 552
Spin wave, 510
Standard Model
    electroweak, 520
    electroweak and strong, 530
Structure functions. See Callan-Gross relation; Deep inelastic scattering; Scaling in deep inelastic scattering; Sea quarks; quark model of inelastic scattering.
$SU(2)$, 348
    direct products in, 350–59
    generators, 353
$SU(2)$ and $SU(3)$ compared, 359
$SU_L(2)$, 517
$SU(3)$, 350
    antitriplet, (3^*), 356
    direct products in, 357–63
    generators, 351, 357, 359, 361
    infinitesimal transformations in 354
    octets, 355, 359, 363
    "vector" algebra of, 355
    triplet, 352
$SU(5)$, 540
Superconductivity, 511–16
Superstring theory, 543
Symmetry breaking, See Hidden symmetry

θ_W (electroweak mixing, or Weinberg, angle). See Electroweak theory; Grand Unification
Thomson scattering, 221

$U(N)$, 348
$U_Y(1)$, 518
Uncertainty Principle, and causality, 577–79
Unitarity bound, 479, 481
Y. See $Q\bar{Q}$-spectrum systems, spectrum of
$\Upsilon \to$ Higgs, 527
UR (ultrarelativistic), 214
    spinors, 215

Vacuum polarization in QCD, 381–86
    See also magnetic susceptibility
Vacuum polarization in QED, 253, 587–93
    basic diagram, 259
    compared with QCD, 381
    data on, 265
    modificated Coulomb law, 363, 593
Vacuum state, 205
    in QCD, 397–400
    in QED, 558
Valence quarks, 434

Van der Waals interaction, 347
Vector potential, 195
  as operator, 206
  plane wave expansion of, 205
Vertices
  in Grand Unified theory, 533
  in QCD, 372
  in QED, 216
Virtual states, 220

W-boson
  helicity states, 481
  mass, 506
  generation of 523
W-field, 469, 474, 489
  and renormalization, 481
W^0-field, 489, 522
Weak current, 472–74
  conservation of, 485
  neutral, 489, 497
Weak hypercharge, 518, 520
Weak interaction, charge changing, 467
  Fermi interaction, 475
  Hamiltonian, 469
  and nuclear β-decay, 476, 477
  strength of, 475–78
  See also Deep inelastic scattering
weak interaction, charge preserving
  Hamiltonian, 495, 498
  See also Deep inelastic scattering; Electroweak theory

Weak interaction, symmetric, 489
Weak interaction, universality of, 495
Weak isospin, 485, 519
  conservation of, 487, 488
  and electroweak connection, 491–93
  and hidden symmetry, 509, 511
  see also Electroweak theory
Weinberg angle. *See* Electroweak theory
Weinberg-Salam model, 520
Weizsäcker-Williams approximation, 453
Weyl equations, 201
Width for decay, 227
  of mesons, 335
  partial, 288

X-boson, 541, 543

Yang-Mills equations, 378–80

Z^0-boson
  coupling to fermions, 507
  coupling to Higgs, 527
  decay, 507
  mass, 506
  generation of, 523
Z^0-field, 494
Zero-point fluctuations
  in QCD, 386–390
  in QED, 267, 558
  See also Harmonic oscillator; Vacuum polarization; Vacuum state